THE EDITOR

Roger E. Schirmer, Ph.D., is Vice President for Research & Development, Guided Wave, Inc., El Dorado Hills, California. Dr. Schirmer's previous positions include Group Leader in the Raw Materials Group, Analytical Chemical Development, Eli Lilly & Co., Indianapolis, Indiana until 1977; and Manager, Chemical Methods and Kinetics Section, Battelle Memorial Institute, Northwest Research Laboratories, Richland, Washington until 1985.

Dr. Schirmer graduated in 1965 from Wayne State University, Detroit, Michigan, with a B.S. degree in Chemistry and Mathematics. He obtained his Ph.D. in Physical Chemistry from the University of Wisconsin, Madison in 1970.

Dr. Schirmer is a member of the American Association for the Advancement of Science, American Physical Society, American Chemical Society, Society for Applied Spectroscopy, American Society for Testing and Materials, and Instrument Society of America. He has authored or co-authored more than 25 papers in pharmaceutical and toxicological analysis, applications of nmr to problems in chemical kinetics and structure, and remote chemical analysis through fiber optics.

CONTRIBUTORS

L. J. Lorenz, Ph.D.
Research Scientist
The Lilly Research Laboratories
Eli Lilly and Company
Indianapolis, Indiana

Eugene C. Rickard, Ph.D.
Senior Research Scientist
The Lilly Research Laboratories
Eli Lilly and Company
Indianapolis, Indiana

Rex W. Souter, Ph.D.
Head, Structural and Organic Chemistry Research
The Lilly Research Laboratories
Eli Lilly and Company
Indianapolis, Indiana

TABLE OF CONTENTS

Chapter 1

POLAROGRAPHY

E. C. Rickard

TABLE OF CONTENTS

I. INTRODUCTION

Polarography is a generic term referring to several related voltammetric techniques which use the dropping mercury electrode (DME). The most well known is direct current (DC) or conventional polarography originally developed by Heyrovsky in 1922. However, the most successful developments in theory, instrumentation, and applications have been made within the last 10 to 20 years. This is particularly true for organic electrochemistry.[1-4]

The DME microelectrode consists of a mercury reservoir connected to a very small bore capillary tube upon which a mercury drop grows and drops every few seconds. The applied potential is slowly varied so that each drop is maintained at nearly constant potential, so that a complete polarogram will consist of current-time responses at many drops. Thus the current-potential behavior can be obtained over a wide range of potentials. According to the redox behavior of the species in solution, an electrolysis reaction will occur at some point in this potential scan. The electrolysis current is proportional to the concentration of the species in solution when material is transported to the electrode by diffusion. Thus a polarogram contains both qualitative and quantitative information. That is, the current-potential curve indicates both the nature and the concentration of the electroactive material present in solution.

The change in the area of the mercury drop and the nature of the diffusion process cause the current to alternately increase during the life of the drop and fall towards zero when the drop falls. If the current is measured only at a specific time during the life of the drop, usually at the point just before the mercury drop falls, the technique is called sampled DC polarography or Tast polarography. In addition to the characteristics imposed by the nature of the DME, DC polarography exhibits the following traits:

1. Polarography is essentially a comparative assay, requiring comparison of a sample response to a standard response.
2. Typical concentrations are millimolar. The useful concentration range is about 10 m*M* to 0.01 m*M*.
3. The required sample size is about 10 to 50 mℓ for conventional cells down to about 0.1 mℓ for microcells.
4. Selectivity is determined by the relative ease of oxidation or reduction of the individual components.

DC polarography can be applied to either reduction or oxidation processes. However, the mercury electrode itself is rather easily oxidized and limits the investigation of oxidation processes to those species which are more easily oxidized than the electrode. The very high overvoltage for hydrogen reduction on mercury makes the DME especially suitable for the investigation of reductive processes. Ultimately, the reduction properties of the sample solvent/supporting electrolyte determine the useful negative potential limit. Some of the electroactive species and functionalities which can be reduced are listed in Table 1; those which can be oxidized at the dropping mercury electrode are listed in Table 2.[5-9] It should be noted that polarography of inorganic species typically involves a change in the oxidation state of a central atom, usually a metal. On the other hand, electrochemical oxidation or reduction of organic compounds frequently involves changes in covalent bonds leading to interconversion of functional groups or substitution, addition, elimination, coupling, or cleavage reactions.[6]

Besides DC and sampled DC polarography, techniques which use the DME include pulse polarography, differential pulse polarography, alternating current (AC) polar-

Table 1
SPECIES AND FUNCTIONALITIES WHICH CAN BE REDUCED AT THE DME

Hydrogen ion and water
Higher oxidation states of many metals and their complexes
Inorganic oxidants such as bromate, iodate, chlorite, hypochlorite, hydrogen peroxide, and sulfur dioxide
Carbon-carbon bonds
- Aromatic hydrocarbons
- Alkenes
- Aryl alkynes
- Aromatic nitriles

Carbon-oxygen bonds
- Quinones
- Aldehydes
- Aromatic or other activated ketones
- Dicarboxylic acids
- Keto-acids
- Epoxides

Carbon-heteroatom bonds
- Imines (azomethine, −C=N−)
- Oximes (−CO−C=N−)
- Carbon-halogen (Cl, Br, or I)
- Nitriles (−CN)
- Hydrazones (−C=N−NH_2)
- Carbazones (−N=N−CO−N=N−)

Nitrogen-oxygen bonds
- Nitro (−NO_2)
- Nitroso (−NO)
- Hydroxylamine (−NHOH)
- N-oxide (−N→O)

Sulfur containing compounds
- Disulfides (RSSR′)
- Sulfoxides (−SO−)
- Sulfones (−SO_2−)
- Thioethers (RSR′)

Nitrogen, oxygen, and sulfur heterocyclic compounds
Others
- Azo (−N=N−)
- Azoxy (−NO=N−)
- Quaternary ammonium (R_4N^+)
- Peroxy (ROOR′)
- Organometallic

Table 2
SPECIES AND FUNCTIONALITIES WHICH CAN BE OXIDIZED AT THE DME

Lower oxidation states of some metals

Species which form insoluble mercury salts or strong mercury complexes such as halides, mercaptans (thiols), cyanide, azide, thiocyanide, and thiosulfate

Activated hydrocarbons, particularly condensed aromatics

Others
- Hydroquinones
- Dihydroxy or trihydroxy aromatics
- Hydroazo or hydrazines (−NH−NH−)
- Carbazides (−NH−NH−CO−NH−NH−)
- Ene-diols (such as ascorbic acid)
- Phenols
- Uracils
- Barbituric acids

ography, triangular sweep polarography (cathode-ray polarography), square wave polarography, radio frequency polarography, and various derivative and harmonic polarographic techniques. Several other voltammetric techniques which measure the current flowing through an electrolysis cell in response to an applied potential do not use a dropping mercury electrode. These include chronoamperometry, cyclic voltammetry (chronoamperometry with linear potential sweep), and stripping voltammetry which are performed at a stationary hanging mercury drop or a solid (platinum, graphite, etc.) electrode; hydrodynamic voltammetry which utilizes convective mass transfer to a stationary, solid electrode; and amperometric techniques which utilize convective transfer to a rotating (disk, ring-disk, or wire) or stirred (mercury pool) electrode. Amperometric titrations use the DME or other suitable electrode as the detector for a conventional titration when either the substance being titrated and/or the titrant and/or the product of the titration is electroactive. Differential pulse polarography, stripping voltammetry, and AC polarography are the most frequently used techniques for quantitation other than DC or sampled DC polarography. The characteristics and applicability of some of these related techniques will be discussed.[1,2,4,5,10-19]

This chapter will briefly present the fundamentals of polarographic techniques related to their use for quantitative analysis. Then, selection of variables such as electrolyte, cell, and instrumentation will be discussed. Finally, the application of polarography to analysis of selected compounds and group of compounds will be discussed. This chapter will not attempt to duplicate the many excellent reviews available for polarography. Instead, it will provide an overview and guide to more comprehensive reviews of specific aspects which may be of interest to readers. The frequently used symbols in this chapter are listed in Table 3.

II. ADVANTAGES AND LIMITATIONS

In comparison with other analytical techniques, DC polarography has several advantages and disadvantages.[2,4,9-11,14,20-24] One of the well known advantages of electrochemical techniques which employ the DME is its reproducible, constantly renewed surface. Any contamination of the electrode surface or accumulation of insoluble electrolysis products is removed by the fall of the old drop and the formation of a new drop. Also, only one drop is affected by any temporary disturbance of the diffusion process (such as vibration) whereas the entire voltammetric experiment would be affected at a stationary electrode. Some of the other advantages are listed below.

1. Polarography is applicable to a wide variety of reducible and oxidizable functionalities as described above.
2. The precision is good, typically about 1 to 2% for concentrations greater than about $10^{-4}M$.
3. The sensitivity is high, with a detection limit of about $10^{-6}M$ (0.1 μg/mℓ for a molecular weight of 100). It can be extended to about $10^{-8}M$ (1 ng/mℓ for a molecular weight of 100) using differential pulse polarography or about $10^{-9}M$ for stripping voltammetry (generally limited to metals and a few anions). This compares to typical sensitivity ranges of about 1% (in sample) for infrared spectroscopy; mg/mℓ for nuclear magnetic resonance spectroscopy; mg/mℓ to μg/mℓ for high pressure liquid chromatography; μg/mℓ for UV spectroscopy, gas chromatography-mass spectroscopy (total ion current) or gas chromatography (FID detector); μg/mℓ to nanogram/mℓ for fluorescence spectroscopy, atomic absorption spectroscopy or gas chromatography-mass spectroscopy (single or multiple ion detection), and nanogram/mℓ for gas chromatograpby (EC detector) or flame emission spectroscopy (alkali or alkaline earth elements).[25]

Table 3
DEFINITION OF SYMBOLS

Symbol	Definition
A	Electrode area, cm^2
a	Area of adsorbed molecule, $Å^2$
B	Adsorption coefficient, cm^3/mole
C^*	Bulk solution concentration, m*M*
C_i	Concentration of i'th species, m*M*
C_s	Concentration of standard
C_u	Concentration of unknown
D_i	Diffusion coefficient of i'th species, cm^2/sec
d	Density of mercury, 13.534 g/cm^3 at 25°C
E	Electrode potential, V
E^o	Standard electrode potential, V
$E^i_{1/2}$	DC polarographic half-wave potential of the i'th species, V
E_p	Peak potential, differential pulse polarography and cyclic voltammetry, V
ΔE	Pulse height, differential pulse polarography, or peak-to-peak applied AC voltage, AC polarography, V
E_{DC}	DC potential of applied signal in AC polarography
E_{pzc}	Potential of zero charge on the DME, V
E_λ	Potential at which scan direction is reversed in cyclic voltammetry, V
E_1	Initial potential before application of pulse (pulse polarography or differential pulse polarography) or start of scan (cyclic voltammetry), V
E_2	Final potential after application of pulse, pulse polarography, V
F	Faraday's constant, 96487 c/eq
f_i	Molar activity coefficient of i'th species
h_{corr}	Corrected mercury column height, cm of mercury
I	Diffusion current constant in DC polarography, $\mu a\ sec^{1/2}\ mM^{-1}\ mg^{-2/3}$
$I(\omega t)$	AC current in AC polarography, μa
$I_{p,AC}$	Peak current in AC polarography, μa
$\bar{i}$	Instantaneous current, μa
^{-}i	Average current over the drop life, μa
i_d	Diffusion limited current, DC polarography, μa
i_c	Charging current, μa
i_k	Kinetic current, CE mechanism, μa
$\bar{i}_{cat}$	Average current, catalytic regeneration mechanism, μa
i_{ads}	Limiting current for adsorption pre-wave, μa
i_{pp}	Diffusion limited current, pulse polarography, μA
Δi_{DPP}	Current, differential pulse polarography, μA
Δi_{max}	Diffusion limited maximum current, differential pulse polarography, μA
i_L	Limiting current, voltammetry at RDE, μA
i_p	Peak current, cyclic voltammetry, μA
i_{corr}	Diffusion limited current corrected for dilution, μA
K_d	Dissociation constant for metallic complex
K_{eq}	Equilibrium constant for a chemical reaction
K'_{eq}	Pseudo-first-order equilibrium constant
k_i	Rate constant of i'th chemical reaction
k'_i	Pseudo-first-order rate constant
k_f and k_b	Forward and reverse heterogeneous electron-transfer rate constants, cm/sec
k_s	Standard heterogeneous electron-transfer rate constant, cm/sec
m,	Mercury flow rate, mg/sec
n or n_i	Number of electrons transferred in an electrode reaction
n_a	Number of electrons transferred in the rate controlling step of an electrode reaction
Q	Charge, μC
q	Stoichiometric coefficient of species
R	Gas constant, 8.314 Joule/(°K-mole)
(RT/F)	ln(10) 0.05915 at 25°C
r	Drop radius, cm

Table 3 (continued)
DEFINITION OF SYMBOLS

Symbol	Definition
T	Absolute temperature, °K
t	Time, sec
t_{const}	Drop time of controlled drop time DME, sec
t_a	Length of time required to reach adsorption equilibrium in pulse polarography, sec
t_1	Time interval from start of drop to current measurement, pulse polarography and differential pulse polarography, sec
t_2	Time interval from application of pulse to current measurement, pulse polarography and differential pulse polarography, sec
t_3	Time interval from start of drop to first current measurement, differential pulse polarography, sec
t_+	Transport number of cation
V	Volume of standard, standard addition, m*ℓ*
v	Volume of sample, standard addition, m*ℓ*
x	Distance from electrode, cm
y	Kinetic parameter, rate-controlled electron transfer and CE mechanisms
z_i	Ionic charge of i'th species
A, I, M, Ox, O, O_i, Red, R, R_i, X and Z	Symbols for chemical species involved in electrolysis mechanisms
α	Electron transfer coefficient
Γ	Amount of material adsorbed on electrode (surface excess), moles/cm^2
Γ_m	Maximum amount of material which can be adsorbed on electrode, moles/cm^2
Θ	Ratio of equilibrium concentrations of oxidized and reduced species at the electrode surface
$\varkappa$	Integral capacitance, μF/cm^2
λ_i	Equivalent ionic conductivity of the i'th species
μ	Ionic strength, *M*
ν	Scan rate, cyclic voltammetry, V/sec, or kinematic viscosity, cm^2/sec
ω	Angular frequency, radians/sec

4. The linear dynamic range is large, about 10^3 for high precision work with DC polarography, up to 10^6 including differential pulse polarography.
5. The instrumentation is very simple and relatively inexpensive. Both commercial units and circuit descriptions for construction of instrumentation are readily available.
6. Aqueous supporting electrolytes are commonly used for routine assays. However, many nonaqueous or mixed solvents may be used when greater sample solubility is needed.[26-28]
7. Relatively little sample preparation is usually necessary, even in the case of formulations.[5] Polarography is applicable to control assays of raw material drug substances or formulations, stability or dissolution studies and assays of drugs in biological fluids.[23]
8. The analysis time is relatively fast, typically about 15 to 30 min per sample, and the analysis is easily automated.[29-31]
9. The amount of sample actually electrolyzed is very small, less than 0.05 m*ℓ* for reasonable electrolysis times. Thus, multiple polarograms can be obtained without altering the solution composition (for sample volumes greater than a few milliters) and/or the sample may be recovered after analysis.

Some of the limitations of DC polarography include:

1. Current oscillations and charging current due to the constantly changing electrode area add complexity and limit sensitivity.
2. The technique is not as well known or frequently applied as many other instrumental techniques.
3. Although the relative ease of reduction or oxidation imparts some selectivity (more than many titrimetric techniques, for example), it is much less selective than most chromatographic or spectroscopic techniques.
4. Commercial instrumentation for automated analysis is just now becoming available. Like other instrumental techniques, optimal design of digital and microprocessor controlled instrumentation is still in a developmental state.
5. Electrode processes are frequently complex and difficult to understand. This problem is alleviated somewhat by the numerous investigations of model compounds and by careful characterization (during assay development) of factors which may affect the analytical application.
6. It is difficult to investigate oxidation processes due to the ease of oxidation of the mercury electrode. However, oxidation processes can be investigated with other voltammetric techniques and electrode materials.

III. PRINCIPLE

Electrochemical techniques are concerned with the interrelationships of potential (voltage), current or charge, and time.[19] Potentiometry is a static or zero current measurement of electrode potential in response to solution composition such as the measurement of pH. Measurements of conductance or dielectric constants are other zero current techniques. Dynamic electrochemical techniques (nonzero current) consist of controlled potential (voltammetric), controlled current, and controlled charge techniques. As previously stated, polarography is a voltammetric technique based on the measurement of current flowing through a dropping mercury electrode in response to the applied potential.

A. Description of Experiment

The basic polarographic experiment is illustrated in Figure 1. That is, the DME and a second electrode (required to complete the electric circuit) are placed into the electrolysis cell containing the solution to be analyzed, perhaps a 1 m*M* solution of cadmium ions. The potential difference between the DME and the other electrode is varied while the resulting current flow is measured. If the second electrode has a fixed potential (i.e., it is a reference or nonpolarized electrode such as a saturated calomel electrode), the relative potential of the DME can be measured. When the potential initially applied to the DME is sufficiently positive, the electrode and solution will be in equilibrium and no current will flow. If the potential applied to the DME is then scanned negatively, a value will be reached where the cadmium ions near the electrode will no longer be in thermodynamic equilibrium with the electrode. That is, some of the cadmium ions will be reduced to metallic cadmium. Since the metallic cadmium is soluble in mercury but not in the solution, the reduced cadmium will remain at the electrode in the form of cadmium amalgam. In other cases, the electrolysis product might remain in solution or on the electrode surface. As the DME is made even more negative, all the cadmium ions which arrive at the electrode surface will be reduced. The limiting current which results is proportional to the concentration of the electroactive species when diffusion is the mass transfer process which brings material to the electrode. The applied potential function and the resulting polarogram (current-potential plot) are

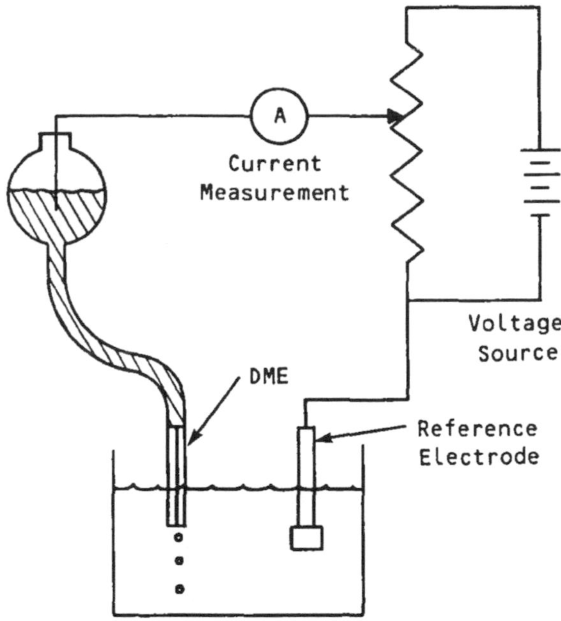

FIGURE 1. Diagram of DME and DC polarographic experiment.

illustrated in Figure 2. The polarogram is illustrated for both a normal and a sampled DC (Tast) mode. In the polarogram, the half-wave potential ($E_{1/2}$) is characteristic of the cadmium ion under these specific conditions, and the diffusion limited current (i_d) is proportional to its concentration. The chemical and mathematical description of the processes which lead to this polarographic behavior are discussed below. In addition, more refined implementations of the experiment will be presented.

In a polarogram, current due to a reduction process at the DME is called a cathodic current and, by convention, is taken as positive.[32] Current due to oxidation processes is called anodic current. The convention for electrode potentials is such that more negative potentials for the DME (vs. the reference electrode) favor cathodic processes while more positive potentials favor anodic processes. Cathodic and anodic refer to electrode processes (direction of current flow), not the electrode potentials.

The DME has two principal modes of operation: naturally dropping and mechanically controlled dropping modes. In the naturally dropping mode, the drop is allowed to grow until its weight causes it to fall. In the mechanically controlled mode, the drop is dislodged before its natural fall time by striking the DME with an object, usually activated by a solenoid. The mechanical dislodgement allows shorter, more reproducible drop times and synchronization of current measurements with the drop time for techniques such as sampled DC polarography, pulse polarography, or differential pulse polarography. In the following discussion, any differences between natural drop times and mechanically controlled drop times will be distinguished.

B. Mass Transport Processes

Mass transport results from the action of electric fields, mechanical forces, chemical potentials, or thermal gradients.[33] At constant temperature, these forces produce migration, diffusion, and convection, respectively, within electrolytes. All three transport mechanisms can bring electroactive material to the electrode.

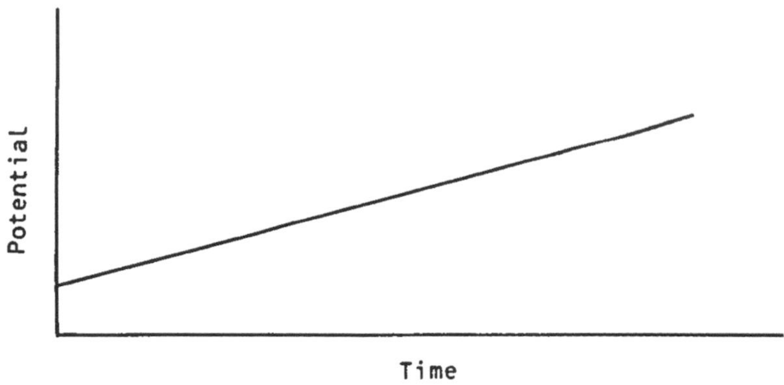

FIGURE 2A

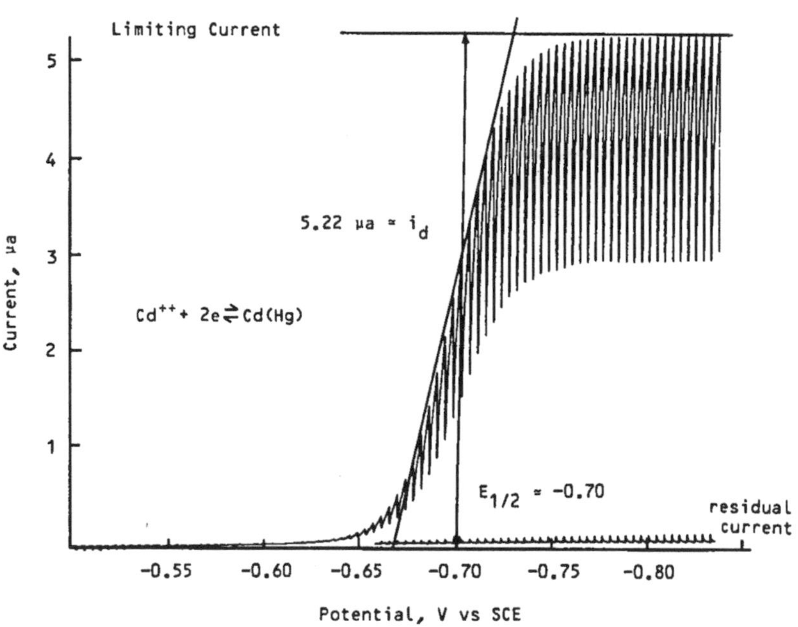

FIGURE 2B

FIGURE 2. Polarogram of 1 mM Cd^{++} in 1 M KCl (reversible, two-electron wave). (A) applied potential. (B) DC polarogram illustrating diffusion current, residual current, and half-wave potential. (C) Sampled DC (Tast) polarogram.

1. Migration

Migration is the transport of charged species under the influence of an electric field. That is, the flow of current through a solution requires that positively charged ions flow towards the negative electrode and negatively charged ions flow towards the positive electrode. The movements of ions due to these electrostatic forces is described by their transport number. The transport number is dependent upon the ions equivalent ionic conductivity (λ) and concentration (C) relative to that of all species. The equivalent conductivities of most ions are approximately the same except for the significantly larger conductivities of the hydronium and hydroxide ions. The transport number for a cation (t_+) is given by the following equation.

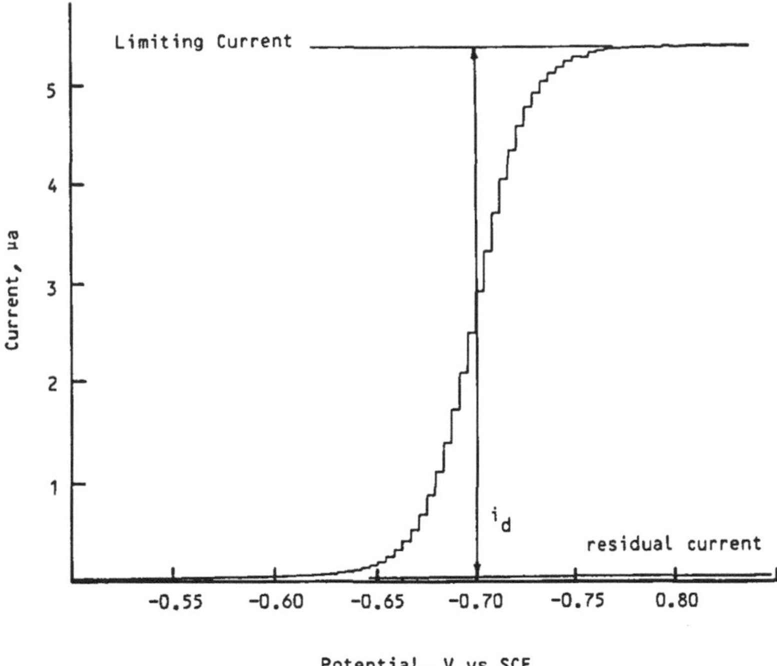

FIGURE 2C

$$t_+ = C_+\lambda_+ / \sum_i (C_i \lambda_i) \qquad (1)$$

It is desirable to eliminate migration effects on the measured current because transport of the electroactive species to the electrode by migration is dependent upon the total solution composition and not just its own concentration. An approximately 100-fold excess of other ionic species is sufficient to suppress the undesired migration of charged electroactive species. Even if the electroactive species is known to be uncharged, it is desirable to maintain a relatively high supporting electrolyte concentration to provide adequate potential control.

2. Convection

Convection is mass transport by physical movement of the solution. Under precisely controlled conditions (hydrodynamic voltammetry), or in the exhaustive electrolysis of a solution, convection is desirable. However, it leads to random or uncontrolled fluctuations in the current at the DME. The only functional purpose of convection in polarography is the replenishment of the depleted solution near the capillary as explained below.

Convection can be caused by density gradients or by unintentional stirring or vibration. Density gradients most frequently arise during the polarography of solutions with high concentrations of the electroactive species where the physical parameters are appreciably changed in the solution near the electrode. Vibration, which can cause solution movement and/or erratic, premature fall of the mercury drop, can be minimized by careful isolation of the polarographic cell from the environment. Vibrational dislodgement of the mercury drop is most often observed with long, natural drop times and least often observed with shorter, mechanically controlled drop times. Movements of the bulk solution also can arise from stirring due to the flow of the gas used to deoxygenate the solution. Several other important convectional processes occur in the solution layers adjacent to the electrode or within the mercury drop itself. These processes, which lead to phenomena called maxima, will be discussed separately.

3. Diffusion

Diffusion is the movement of ions or molecules due to a concentration gradient, i.e., a chemical or electrochemical potential. The concentration gradient arises when the electroactive species is removed from the solution by the electrode reaction. The diffusion coefficient measures the rate at which movement by diffusion occurs. It primarily depends upon the viscosity of the solvent, the size of the species, and its charge.[10] The diffusion coefficient is typically 6×10^{-6} cm^2/sec for metallic ions and 6 to 10×10^{-6} cm^2/sec for many organic molecules. The most important exceptions to these generalizations are the hydronium and hydroxide ions which have significantly larger diffusion coefficients.

C. Current Processes

Current results from the flow of charge into the mercury electrode. This charge may be transferred across the electrode-solution interface to a species in solution and is then called faradaic current. Faradaic current produces an oxidation or reduction of a species in solution. The other type of current is called charging current. Charging current does not involve actual transfer of electrons from the electrode to the solution, but consists of an accumulation of charge at the interface. These current types will be discussed in terms of the processes which occur at the electrode: the diffusion limited current process for the major electroactive species and residual current processes.[7,10,20,21,34]

1. Diffusion Limited Current

The nature of the diffusion process causes the limiting current to be directly proportional to the concentration of the electroactive species in the bulk of the solution (C^*, m*M*), the electrode area (A, cm^2), and the square root of the diffusion coefficient (D, cm^2/sec). It also is proportional to the number of electrons transferred to each molecule of the electroactive species (n) and Faraday's constant (F, 96487.2 C/eq.) which relates the charge on the electron to coulombs. Finally, and most important, the diffusion process to a stationary planar electrode leads to a decrease in the current with time as the solution nearest the electrode is depleted of the electroactive species, i.e., the electroactive species has to travel further to reach the electrode. This decrease in current is inversely related to the square root of time (t, sec) as indicated in Equation 2.

Diffusion to plane:

$$i = n F A C^* \sqrt{D/\pi t} \tag{2}$$

The stationary planar electrode diffusion process, however, is modified by two effects due to the growth and fall of the mercury drop electrode in polarography. First, the surface area of the drop increases with time so that more material can gain access to its surface as the drop grows. The area can be obtained (assuming a constant mercury flow rate) using the linear relationship of volume with time (t) given by

$$\text{volume } (cm^3) = (4/3)\,\pi\, r^3 = 10^{-3}\ m\, t/d \tag{3}$$

where d is the density of mercury (13.534 g/cm^3), r is the radius of the mercury drop (cm), and m is the mercury flow rate (mg/sec). Thus, the area is directly proportional to the two-thirds power of time with mercury flow rate.

$$A = 4\pi r^2 \quad \propto \quad m^{2/3}\, t^{2/3} \tag{4}$$

The second effect which results from the growth of the mercury drop is due to its expansion into the solution. This expansion effectively pushes the drop surface towards fresh, nondepleted solution (compresses the diffusion layer). This phenomenon is usually treated mathematically as an expanding plane diffusional process and results in a multiplication of the predicted current by a constant ($\sqrt{7/3}$) compared to that for a stationary planar electrode (Equation 2). Both effects, the increase in surface area with time and the expansion of the drop into the solution, cause an increase with time in the amount of material which reaches the drop surface and, therefore, an increase in the observed current.

The overall combination of depletion by diffusion and growth of the mercury drop leads to a current (i_d) which increases with the one-sixth power of time and the two-thirds power of the mercury flow rate as given by Equation 5.

Diffusion to
expanding plane:

$$i_d \propto (t^{-1/2})\,(m^{2/3}\, t^{2/3})\,(\sqrt{7/3}) \tag{5}$$

$$\propto m^{2/3}\, t^{1/6} \tag{5a}$$

These relationships (Equations 2 to 5) assume that all transport of the electroactive species to the electrode occurs by diffusion. They also assume that the electrode potential is held in a region where all material reaching the electrode is reduced (or oxidized). If not, the currents are multiplied by the fraction of the electroactive species which undergoes reaction. In all cases, the equations are valid for any time within the life of the mercury drop when time is measured from the fall of the previous drop. When t is taken as the drop time, then the calculated current refers to that current which would be obtained just before the fall of the mercury drop (i.e., the current normally used for most polarographic measurements).

Of course, the current rapidly decreases when the old drop falls off. The extent of the observed decrease depends upon the ability of the electronic measuring circuits to follow the true current and the amount of mercury area left exposed to the solution at the capillary tip after the fall of the drop.

Another effect of the dropping process is a stirring of the solution when the drop falls. This stirring causes the solution for the succeeding drop to appear uniform and very nearly the same concentration as that for the initial drop. That is, all depletion effects in the solution near the electrode are removed. Thus, two successive drops at the same potential should give the same current. It is important to realize that the bulk solution concentration remains unchanged since the overall rate of removal of the electroactive species is very low.

The errors and assumptions with this simplistic description of the diffusion controlled polarographic experiment have been discussed in other texts.[10,20,24] Briefly, these include the following:

1. Diffusion to the mercury drop is spherical, not planar. The mathematical modification for diffusion to an expanding sphere (rather than an expanding plane) results in an additional term proportional to $\sqrt{Dt}/r$ (Lingane-Loveridge approximation) or that term and a third term proportional to Dt/r^2 (Koutecky equation).
2. The drop, however, is not really spherical since it deforms under its own weight. Furthermore, it is not really an expanding sphere because the center of mass moves as the drop grows.
3. Not all areas of the drop are equally exposed to the solution. Those near the top of the drop are shielded by the capillary.
4. The stirring caused by the fall of the old drop does not completely restore the initial solution conditions in the region of the DME capillary. In fact, there is a pronounced difference in the experimentally measured current observed for the first drop and that observed for succeeding drops (which remain nearly constant at a lower value).[35,36]
5. The mercury flow rate (i.e., drop growth) is not constant within the life of the drop.

The combined effects of these assumptions make it essentially impossible to adequately treat this problem mathematically. However, the simplified description results in a predicted current which is very close to that actually observed, especially in terms of its dependence upon the important experimental variables. This relationship, developed for the expanding plane model, is commonly known as the Ilkovic equation and can be obtained by combining Equations 2 to 5 to give:

$$i_d = 708\, n\, D^{1/2}\, C^*\, m^{2/3}\, t^{1/6} \qquad (6)$$

In the Ilkovic equation, i_d is the diffusion limited faradaic current (microamps), n is the number of electrons transferred to each electroactive species, D is the diffusion coefficient of the electroactive species (cm^2/sec), C^* is the bulk solution concentration of the electroactive species (m*M*), m is the mass rate of flow of the mercury (mg/sec), and t is the time (sec) measured from the fall of the previous mercury drop.

Because the current predicted by the Ilkovic equation depends upon the capillary constants, it is difficult to compare absolute magnitudes of current obtained in different laboratories, or even in the same laboratory using different conditions (e.g., capillary, potential, or supporting electrolyte). One method to avoid this situation is to record the capillary constants under the same conditions used to obtain the analytical data. Alternatively, a diffusion current constant (I, $\mu a\ sec^{1/2}\ mM^{-1}\ mg^{-2/3}$) can be defined as the diffusion limited current normalized for concentration and the measured capillary constants:

$$I = i_d / (C^*\, m^{2/3}\, t^{1/6}) = 708\, n\, D^{1/2} \qquad (7)$$

Equations 6 and 7 apply to the instantaneous current at any time in the growth of the drop. When the drop time is used as the value for t, the maximum current obtained just before the drop falls is predicted. Substitution of 607 for the value of the constant in Equation 6 gives the average diffusion limited current over the life of the drop. This form of the Ilkovic equation is used occasionally for instruments which present an averaged current output (e.g., galvanometer or highly damped electronic circuit). However, almost all modern instruments display instantaneous rather than average currents.

Analytically, it is important to characterize the electrode process since either form of the Ilkovic equation relates current to concentration only when the current is controlled by the rate of diffusion of material to the DME. Thus, tests for diffusion controlled limiting current are required during the development of an analytical method. Electrode processes may be complicated by adsorption of material onto the electrode surface, by the rate of electron transfer or by homogeneous chemical reaction kinetics involving electroactive species. However, even with these complicating situations (to be discussed later), diffusion controlled currents are frequently obtained in some electrode potential region.

a. *Tests for Diffusion Limited Current*

The form of the Ilkovic equation allows several direct tests for diffusion control. In addition, linearity of the limiting current with concentration over a sufficiently wide range of concentration can indirectly confirm diffusion control. Both direct tests and linearity tests should be part of the analytical method development.

For direct tests of diffusion control it is necessary to characterize the dependence of the limiting current on experimental parameters. First, the mercury flow rate (m) is directly proportional to the net pressure producing the mercury drop. The net pressure consists of three terms:[10,20] the hydrostatic mercury pressure itself (proportional to the height of the mercury column, usually 30 to 100 cm), the back pressure due to the mercury-solution interfacial surface tension of the drop, and the back pressure due to the hydrostatic pressure of the solution. The last term is negligible (equivalent to about 0.1 cm of mercury) while the second term contributes to a minor effect (equivalent to 1 to 2 cm of mercury). The height of the mercury column corrected for the interfacial tension back pressure is termed the corrected height (h_{corr}, cm of mercury). This net pressure which determines the mercury flow rate is given by:

$$h_{corr} = h - 3.1 / (m\,t)^{1/3} \tag{8}$$

where h (cm of mercury) is the observed height of the mercury column (capillary tip to top of column). Because the interfacial tension back pressure term in Equation 8 is dependent upon time, the mercury flow rate varies slightly during the life of the drop.

The second experimental parameter which affects the limiting current is the drop size. The drop size or drop weight (product of mercury flow rate times drop time) is constant at any given potential in a specific electrolyte.[10,20] Thus, drop time must be inversely proportional to the net pressure or corrected height (for a naturally dropping DME). The combination of height dependencies of flow rate and drop time produce a diffusion current proportional to the square root of the corrected mercury height according to:

Natural drop time DME:

$$\text{Natural drop time DME:}\quad i_d \propto m^{2/3}\,t^{1/6} \propto (h_{corr})^{2/3}\,(1/h_{corr})^{1/6} \propto (h_{corr})^{1/2} \tag{9}$$

Note that when a mechanically controlled constant drop time is used, t is independent of the mercury height. Therefore, observed diffusion currents become proportional to the two-thirds power of the corrected height as given by:

Controlled, constant drop time DME:

Controlled, constant drop time DME:
$$i_d \propto m^{2/3}\, t^{1/6} \propto (h_{corr})^{2/3}\, (t_{const})^{1/6} \qquad (10)$$

With the controlled drop time DME, the variation of the diffusion limited current with drop time may also be determined. However, the dependence on the one-sixth power of time produces relatively small changes in the diffusion limited current with realistic variations of the drop time (e.g., only a 31% change in limiting current would be observed for a fivefold change in drop time).

For a series of polarograms recorded with different mercury column heights, a plot of the limiting current vs. the appropriate power (one-half or two-thirds) of the corrected mercury column height will be linear when the current is controlled by diffusion. In contrast to the power of this test for diffusion controlled currents at a natural drop time DME, it will not discriminate between diffusion limited currents and most kinetically controlled currents at a constant drop time DME (as discussed in the next section). However, the variation of the limiting current with changes in drop time (at the controlled drop time DME) is sufficient to distinguish between several electron transfer mechanisms.

The plot of limiting current vs. the square root of the corrected mercury column height for a natural drop time DME or the value of the ratio $i_d/(h_{corr})^{1/2}$ is the most frequently used diagnostic test for diffusion controlled currents. A range of heights selected to give drop times between about 1 and 10 sec is appropriate. Shorter drop times may give stirring due to the fall of the previous drop. At longer times, convection due to density gradients may cause deviation from the expected behavior. Note that the spherical correction term for the naturally dropping DME actually predicts a small positive intercept for the plot (or a few percent decrease in the ratio $i_d/(h_{corr})^{1/2}$ with increasing height) since it results in an equation of the following form:[10]

Natural drop time DME with spherical correction:

$$i_d \propto (h_{corr})^{1/2} + \text{constant} \qquad (11)$$

For a controlled drop time DME, the spherical correction leads to an equation containing an additional term related to the corrected height.

Controlled, constant drop time DME with spherical correction:

$$i_d \propto (h_{corr})^{2/3}\, (t_{const})^{1/6}\, [1 + k(t_{const})^{1/6}\, (h_{corr})^{-1/3}] \qquad (12)$$

The magnitude of the proportionality constant (k) leads to small, predicted deviations from linearity at low mercury heights.

Another characteristic of diffusion controlled currents is a small temperature dependence. This arises from small temperature dependencies for D, m, and t.[10,20] Typical values for the overall temperature effect range from about 0.75 to 1.6% increase in the diffusion current per degree Celsius increase in temperature. Observed temperature effects of this magnitude do not prove diffusion controlled behavior; however, extremely divergent values may indicate the occurrence of other processes.

According to the Ilkovic equation, the current profile during the life of a single drop should increase with the one-sixth power of time. However, this relationship is highly dependent upon the assumptions made in the derivation of the Ilkovic equation. In fact, it is experimentally found to be invalid, even under conditions where the current is known to be diffusion controlled.[10,20] However, a semiquantitative examination of the current-time profile for a single drop may serve to distinguish diffusion controlled currents from some other types of currents as discussed later.

b. Surface Tension-Drop Time Relationship

The natural drop size is found to be proportional to the mercury-solution interfacial surface tension. Thus, the drop time must be nearly proportional to the interfacial tension since the mercury flow rate is essentially constant (except for minor changes in the back pressure correction term, Equation 8, as the surface tension changes).

The growing mercury drop is subject to two internal forces which affect surface tension: the cohesive van der Waals forces and the repulsive electrostatic forces determined by the electrode potential. The surface tension is also affected by the external solution composition and by adsorption, and is somewhat temperature dependent.

The surface tension may be directly measured with a capillary electrometer or indirectly obtained from measurements of the differential capacitance.[37] The resulting electrocapillary curve indicates a maximum value for the surface tension at the potential where the mercury drop has no charge (corresponding to the minimum electrostatic force and the maximum drop time) and decreases for more negative and more positive potentials.[20,37] Because the diffusion limited current depends upon drop time, the change in surface tension with potential leads to diffusion controlled current plateaus which change with potential, e.g., a somewhat dramatic "droop" or decrease in the diffusion current is observed at very negative potentials where the drop time rapidly decreases.

The interfacial surface tension is very sensitive to the solution composition. Species which are adsorbed onto the mercury surface (termed specific adsorption) change the electrocapillary curve and, therefore, the drop time behavior. In the absence of specific adsorption the maximum value of the electrocapillary curve occurs at about −0.45 V vs. SCE. In the presence of adsorption, it may shift in either direction and/or change in magnitude depending upon the amount of adsorbed material and its charge.

2. Residual Current

Residual current is actually composed of two distinct components: charging current and faradaic currents due to electroactive impurities. Instrumental offset currents and electronic noise also may be present. At low sample concentrations, the residual current (in combination with electronic noise) becomes comparable to the diffusion current, and the sensitivity limit of the polarographic method is reached.

a. Charging Current

Charging current reflects the charge necessary to bring the mercury drop to the applied potential. The charged mercury drop attracts an atmosphere of ions with the opposite charge type within the solution layer adjacent to the electrode. The charge resides on the electrode surface (for metallic electrodes) and a layer of oriented, adsorbed ions or dipoles combined with unequal distribution of ions in a diffuse solution layer in the electrolyte. This electrical double layer, composed of the two charged phases, acts as a capacitor. It is similar to a capacitor in an electric circuit in that a flow of current is necessary to change its potential but different in that its capacitance is dependent upon external conditions. The number of ions present within the double layer is dependent upon the charge on the electrode (i.e., dependent upon potential), upon the constantly changing electrode area and upon the solution concentration. In addition, specific adsorption may occur between ions and the electrode. For instance, many ions which form insoluble mercury salts (e.g., chloride, bromide, iodide, and thiocyanate), some relatively nonpolar cations (e.g., tetraalkylammonium and cesium) and many large, nonpolar organic species are adsorbed within definite electrode potential regions. These specific adsorption processes increase the total double layer capacitance and increase the charging current within the potential regions where adsorption occurs. Because the electrical double layer and specific adsorption are both dependent upon the electrode potential, there will be some potential where the charging current is zero. This potential is termed the potential of zero charge (E_{pzc}). It is sometimes called the electrocapillary maximum since, as noted earlier, it occurs at the potential which has the largest mercury-solution interfacial surface tension. The differential capacitance (derivative of charge with potential plotted versus potential) also exhibits a minimum at this potential.[37] Experimentally, this potential is approximately −0.461 V for 0.1 *M* KCl, −0.535 V for 0.1 *M* KBr, −0.589 V for 0.1 *M* KSCN or −0.693 V for 0.1 *M* KI, all vs. the SCE reference electrode.[10] These values reflect the increasingly strong adsorption of the anion going from chloride to iodide.

If the E_{pzc} and integral capacitance (average capacitance between two potentials) (κ, μf/cm^2) are known, the charging current (i_c, μA) contribution can be calculated from the surface area of the drop (A, cm^2) and the electrode potential (E, volts) using

$$\begin{aligned} i_c &= dQ/dt \\ &= \kappa\,(E_{pzc} - E)\,(dA/dt) - (\kappa A)\,(dE/dt) \end{aligned} \tag{13}$$

where Q is the charge on the electrode (microcoulombs). Under normal conditions, the electrode potential is changing very slowly and the second term may be neglected. Since the drop area changes with time, this equation may be rewritten in terms of the capillary parameters m and t using equations (3) and (4) and substituting for the constants.

$$i_c = 0.005677\,\kappa(E_{pzc} - E)\,m^{2/3}\,t^{-1/3} \tag{14}$$

The integral capacitance is usually about 20 μF/cm^2 for potentials negative of the E_{pzc} and about 40 μF/cm^2 for potentials positive of the E_{pzc}. For typical values of κ, m, and t, the charging current increases about 0.1 μA/V negative of E_{pzc} and about 0.2 μA/V positive of the E_{pzc}. Thus, charging current flows throughout the life of the mercury drop with its sign and magnitude dependent upon the electrode potential and the solution composition.

Note that the charging current decreases with longer times (within the same drop) whereas the faradaic current Equation 6 increases with longer times. Thus, an increased signal-to-noise (faradaic-to-charging current) is obtained near the end of the drop. Instrumental methods of minimizing the effects due to the remaining charging current will be discussed later.

b. Faradaic Impurity Current

Residual currents due to electroactive impurities also are frequently observed. Oxygen is a common electroactive contaminant which is difficult to eliminate entirely. Other impurities may be introduced as minor contaminants in the substances employed in preparation of the supporting electrolyte or from the sample itself (especially with complex sample matrices such as blood or urine). The nature and magnitude of currents from impurities cannot be generalized. However, the removal of oxygen is discussed in a later section and the supporting electrolyte can be preelectrolyzed to remove impurities (see Chapter III, Coulometry).

D. Current-Potential Relationship

In contrast to the mass transfer process which determines the magnitude of the diffusion limited current, the shape of the polarographic wave is determined by the nature of the electron transfer process. The electron transfer process includes the actual electron transfer step and any chemical reactions involving electrode reactants or products. The number of electrons transferred as well as the mechanism may vary with changes in the electrode potential or electrolyte composition. The associated chemical reactions include the dissociation of a metallic ion-ligand complex, proton transfer, or followup chemical reactions such as addition of coupling reactions. Therefore, the chemical reaction is usually quite sensitive to the solution environment, e.g., availability of proton donors, pH, or presence of electrophiles or nucleophiles.

Various electrochemical processes and coupled chemical reactions modify the shape of the polarographic wave in specific manners. Therefore, polarography can be used to obtain information about these events. Additionally, knowledge of electron transfer mechanisms can lead to modifications in experimental procedures to obtain better analytical methods. Thus, a brief description of some of the electron transfer mechanisms are included. Additional information can be obtained from other sources, particularly Guidelli.[34] Much of the early work and current descriptive reports are discussed in terms of phenomena (pre- or postwaves, catalytic waves, kinetic waves, etc.) rather than the underlying mechanisms (adsorption and coupled chemical reactions). This chapter will attempt to provide insight into the mechanistic processes.

1. Reversible Electrochemical Reactions

An electrochemical reaction is termed reversible if the electrode is in thermodynamic equilibrium with the solution at all electrode potentials.[20,24,33,34] That is, a change in the electrode potential must be instantaneously reflected by a corresponding change in the composition of the adjacent solution layer and vice-versa. In practice, this change must be faster than the time scale of the observation (essentially the drop time for DC polarography) to appear reversible. The Nernst equation is the idealized statement of this relationship which can be written (ignoring activity effects) as:

$$\begin{aligned} E &= E^{o} - (RT/nF)\ \ln[C_R]\ /\ [C_O] \\ &= E^{o} - (0.05915/n)\ \log\ [C_R]\ /\ [C_O] \qquad (15) \end{aligned}$$

where it is related to the electrochemical reaction:

$$Ox + n\,e \rightleftharpoons Red \qquad (16)$$

and Ox is the oxidized species which undergoes a n electron reduction to species Red. In the Nernst equation, C_R and C_O are the concentrations of species Red and Ox at the electrode surface, respectively, R is the ideal gas constant, T is the absolute temperature, F is Faraday's constant, E° is the standard thermodynamic potential of the half-reaction, E is the actual electrode potential (measured vs. the same reference electrode used to obtain the value for E°), and [(RT / F) ln (10)] has a value of 0.05915 at 25°C. When activity coefficients are included, the ratio C_R/C_O is replaced by $[(C_R)\,(f_R)\ /\ (C_O)\,(f_O)]$ where f_R and f_O are the molar activity coefficients of species Red and Ox, respectively.

Conditions producing reversible electrochemical reactions are distinctly different than the conditions which lead to diffusion limited current at the dropping mercury electrode. A reversible electrochemical reaction will give a diffusion controlled current when the electrode potential is held in the appropriate region (a potential where essentially all of the electroactive species that reach the electrode are reduced or oxidized). However, a diffusion limited current at the DME which has all of the characteristics defined above (i_d proportional to the square root of h_{corr}, etc.) does not imply a reversible electrochemical reaction at the electrode. A diffusion controlled current only means that the forward electrochemical reaction proceeds at a rate faster than the diffusion process within a particular potential range.

It can be demonstrated that the current during any portion of the polarographic wave for a reversible electron transfer reaction is given by the Ilkovic equation modified by a potential dependent term

$$i = i_d / [1 + (C_O/C_R)_{x=0}\sqrt{D_O/D_R}\,] \qquad (17)$$

In this equation, D_O and D_R are the diffusion coefficients for the oxidized and reduced species, respectively, and the concentrations of oxidized and reduced species are measured at the electrode surface. This ratio is obtained by restating Equation 15 as:

$$(C_O/C_R)_{x=0} = \exp\,[(nF/RT)\,(E - E^O)] \qquad (18)$$

When Equations 17 and 18 are combined and the natural logarithm taken, the following relationship is obtained (ignoring activity coefficients):

$$E = E^O - (0.05915/n)\log(\sqrt{D_O/D_R}) - (0.05915/n)\log[i/(i_d - i)] \qquad (19)$$

The right hand side of this equation has been divided into two constant terms and a potential dependent term involving the current. When the current is one-half of its maximum diffusion limited value, the last term is zero. This potential is called the polarographic half-wave potential and is equal to the standard potential for the half-reaction modified by the relative rates of diffusion.

$$E_{1/2} = E^O - (0.05915/n)\log(\sqrt{D_O/D_R}) \qquad (20)$$

Actually, a rigorous derivation includes the effects of activity coefficients and variation of the diffusion coefficient with concentration.[10,38] The (molar) activity coefficient correction consists of a replacement of $(\sqrt{D_O/D_R})$ by the product $[(f_R/f_O)(\sqrt{D_O}/\sqrt{D_R})]$ in Equations 19 and 20. Since the magnitude of this correction is difficult to determine experimentally and, thus, is usually not known, it will be disregarded in subsequent discussion. However, E^o in Equations 19 and 20 should be considered a formal rather than a standard potential.

These general relationships are valid for almost all reversible polarographic waves (one exception occurs when the product of the electrode reaction is insoluble in both the solution and the electrode). When the electrode reaction contains more than the two species Ox and Red, additional terms must be added to the Nernst equation. There are two specific situations involving reversible electrochemical reactions in which modified forms of Equations 19 and 20 are useful. The first occurs when the electrode reaction involves the hydrogen ion, a frequent event for organic polarographic waves of the type:

$$Ox + qH + ne \rightleftharpoons Hq\cdot Red \qquad (21)$$

In this situation, the Nernst equation contains a term involving the hydrogen ion concentration and Equations 19 and 20 are modified to:

$$E = E^o - (0.05915/n)\log(\sqrt{D_O D_R}) - (0.05915\ q/n)\ pH - (0.05915/n)\log[i/(i_d - 1)] \quad (22)$$

$$E_{1/2} = E^o - (0.05915/n)\log(\sqrt{D_O/D_R}) - (0.05915\ q/n)\ pH \quad (23)$$

Thus, the half-wave potential would shift (59 q/n) mV negatively for each unit increase in pH. If n is known from other information, then the number of hydrogen ions involved in the electrode reaction can be calculated from the slope of a half-wave potential vs. pH plot. This derivation assumes a well buffered solution such that the pH near the electrode remains constant despite the consumption of hydrogen ions in the electrode reaction.

The second specific reversible electrochemical reaction is classically found in the reduction of a metallic ion (M) in a medium containing a complexing agent (X):

$$MX_q^{+n} + ne \rightleftharpoons M + qX \quad (24)$$

Again, the Nernst equation contains additional terms corresponding to the other chemical species. In the presence of a large excess of the complexing agent (concentration C_x) and when the electrode reaction predominantly involves the free metallic ion (M) and the chemical dissociation step is rapid, then the relationships are

$$E = E^o - (0.05915/n)\log(\sqrt{D_O/D_R}) + (0.05915/n)\log(K_d)$$
$$- (0.05915\ q/n)\log(C_X) - (0.05915/n)\log[i/(i_d - i)] \quad (25)$$

$$E_{1/2} = E^o - (0.05915/n)\log(\sqrt{D_O/D_R}) + (0.05915/n)\log(K_d) - (0.05915\ q/n)\log(C_X) \quad (26)$$

K_d is the dissociation constant for the complex. In this case, the half-wave potential shifts negatively with stronger complexes or with the addition of complexing agent.

2. Tests for Reversible Electrochemical Reactions

A polarographically reversible redox couple will give the same half-wave potential regardless of the relative amounts of oxidized and reduced species initially present in the solution (e.g., for the cathodic wave obtained with the oxidized species, for the anodic wave obtained with the reduced species or for the wave obtained on a mixture of the two forms). Of course, other factors which could influence the half-wave potential (pH, ionic strength, temperature, etc.) must be constant for all measurements.

This fundamental criterion for reversibility is not absolutely conclusive since a product of the electrode reaction may undergo electrolysis at nearly the potential expected for the reverse reaction without actually being connected to the reactant through a reversible electrochemical couple. In some cases, this test cannot be performed because the product of the electrode reaction is not stable chemically or is unavailable.

A second test for reversibility may be applied to the shape of the polarographic wave itself. The form of Equation 19, indeed of similar equations for all reversible polarographic waves, suggests that a plot of the log term vs. applied potential must be linear. The slope will be 59/n mV so that n may be evaluated simultaneously. This plot also gives a better estimate of the half-wave potential than can be obtained from the wave itself (the half-wave potential is the point where the line crosses zero on the ordinate or logarithmic axis). This test is valid when either the reduced species or the oxidized species or any combination is present originally. In analyzing the wave shape, points should be taken throughout the entire polarogram (current between 5 and 95% of the limiting current). When the redox couple is not polarographically reversible, the plot usually will be nonlinear or linear with a slope which is not a submultiple of 59 mV.

However, a linear plot with the appropriate slope cannot be taken as conclusive proof of reversibility without supporting evidence.

The measured half-wave potential should be independent of the electroactive species concentration (assuming all other species which participate in the electrode reaction, such as hydrogen ion, are well buffered) except in a few instances with unusual stoichiometry in the electrode reaction. Similarily, the half-wave potential should be independent of the capillary constants, m and t, or the mercury height, h. As before, failure to give the expected behavior probably indicates nonreversible behavior while reversible behavior is not conclusively proven by these limited tests.

Electrochemical reversibility can also be investigated using other electrochemical techniques. With this approach, one must recall that reversibility is a relative concept subject to the conditions used for its evaluation. Also, seemingly small changes in the experimental parameters may result in major changes in behavior. Within these limitations, techniques such as differential pulse polarography, AC polarography, cyclic voltammetry, or voltammetry at the rotating disk electrode can be employed to evaluate reversibility.

3. Other Electron Transfer Mechanisms

Electrochemical processes involving organic molecules, such as pharmaceuticals, tend to be more complicated than the simple reversible model just described. The effect of various kinetic and equilibrium processes on the polarographic wave has historically been termed overpotential or overvoltage. The contributions to overpotential can be attributed to diffusion, reaction, adsorption and charge-transfer processes including homogeneous chemical reactions in the bulk of the solution, heterogeneous chemical reactions on the surface of the electrode, adsorption processes and the actual charge transfer process itself. The relative importance of the individual processes contributing to overpotential may be dependent upon the electrolyte composition and/or the electrode potential as well as the actual equilibrium constants and rates of reaction.

Specifically, the current can be controlled by kinetic effects such as the rate of diffusion of material to the electrode in the reversible case, the rate of a chemical reaction which involves the actual electroactive species, the rate of adsorption of material onto the electrode surface or the rate of the actual electron transfer step. Thermodynamic equilibria (such as coupled chemical reactions or adsorption) may alter the current-potential relationship or cause shifts in the half-wave potential. Thus, both kinetic and equilibrium factors tend to affect the height, shape, and half-wave potential of the polarographic wave.

The classification of these processes will be discussed briefly with references to more extensive treatments. The description will present some limiting cases for theoretical models which describe the system response with variations in concentration of the electroactive species, dependence of the limiting current on the mercury column height, the current-potential relationship and the current-time profile within individual drops. Although the analytical usefulness of a polarographic assay does not depend upon a complete elucidation of the electron transfer process, recognition of specific behaviors should lead to development of more reliable analytical methods in a shorter time.

The reversible electrochemical reaction will be used as a starting point to describe changes which occur with other electron transfer mechanisms. The symbols O and R will be used to denote the initial oxidized and final reduced species, respectively, while other letters will denote other chemical species involved in coupled chemical reactions. All descriptions will be based on reduction processes, but analogous situations also occur for oxidation processes.

a. Rate-Controlled Electron Transfer (Irreversible)

When the electron transfer process is so slow that the electrode is not in equilibrium with the solution, it is termed a rate-controlled or irreversible electron transfer mechanism. Thus, the Nernst equation (Equation 15) no longer gives the relationship between electrode potential and concentrations within the solution layer surrounding the electrode. Instead, the polarogram exhibits a current-potential relationship determined by the heterogeneous kinetics for the electron transfer between the electrode phase and the species in solution. The electrode reaction can be stated as:

$$O + ne \underset{k_b,\ 1-\alpha}{\overset{k_f,\ \alpha}{\rightleftharpoons}} R \tag{27}$$

where k_f and k_b are potential dependent forward and reverse rate constants (cm/sec) for electron transfer, and alpha (α) is called the electron transfer coefficient. The Eyring absolute reaction rate formulation for this mechanism leads to the relationships

$$k_f = k_s \exp\left[(-\alpha n_a F/RT)(E - E^o)\right] \tag{28}$$

$$k_b = k_s \exp\left[(1-\alpha)(n_a F/RT)(E - E^o)\right] \tag{29}$$

$$i = n_a F A k_s (C_O)_{x=0} \exp\{(-\alpha n_a F/RT)(E - E^o)\} - (C_R)_{x=0} \exp\left[(1-\alpha)(n_a F/RT)(E - E^o)\right] \tag{30}$$

where k_s is the standard heterogeneous rate constant (the value of k_f and k_b at the standard potential, E^o) and the concentrations are those at the electrode surface. The surface concentration of O is equal to the bulk concentration for very small values of k_f (more positive potentials where O does not react), essentially zero in the diffusion controlled region (k_f large, E very negative, rapid electrode reaction), and intermediate at moderate values of k_f (potentials on the rising portion of the wave). The surface concentrations of R varies in an opposite manner. In these equations, n_a is the actual number of electrons involved in the rate determining step. Thus, n_a will be equal to or less than the overall number of electrons transferred, n, which appears in Equation 27.

The solution for the current-potential relationship in terms of known parameters is very complex, but several characteristics can be noted. First, the forward rate constant increases as the electrode potential is made more negative and the reverse rate constant increases as the potential is made more positive. Thus, a normal diffusion controlled limiting current will be observed at sufficiently negative (reduction) or positive (oxidation) potentials.

The second general characteristic is the effect of k_s on the current-potential relationship. The value of k_s determines the current-potential relationship of a polarogram obtained with a solution containing both species O and R. A k_s value greater than about 0.05 cm/sec yields a normal, reversible polarographic wave while values less than about 10^{-5} cm/sec yield a complete separation of the reduction and oxidation waves (termed a totally irreversible system). An equivalent definition of a totally irreversible system is one where the half-wave potentials for the reduction and oxidation processes differ by more than $120/n_a$ mV. The half-wave potential for the cathodic process will be negative of the value expected for a reversible electron transfer (Equation 20) and negative of the half-wave potential for the anodic process. An intermediate value of k_s will give a partially overlapped wave with the cathodic branch shifted to slightly more negative potentials and the anodic branch shifted to slightly more positive potentials compared to the reversible case. However, most electron transfer processes tend to be either reversible or totally irreversible because of the relatively small transition range of k_s. Finally, the parameter alpha influences the symmetry of the polarogram obtained on a mixture of O and R. Some representative polarograms are illustrated in Figure 3.[39]

When a polarogram is obtained on a solution initially containing only species O, the current will be given by

$$i = 821\, n\, C^{*}\, m^{2/3}\, t^{2/3}\, (k_f/y) \sum_i (\gamma_i z_i) \tag{31}$$

where

$$y = \sqrt{12\, t/7\, D}\; k_f \tag{32}$$

and t is the drop time, the gamma terms are coefficients which include the gamma function, the z terms are functions of y, the diffusion coefficients for both species are assumed to be equal to D, and k_f is a function of potential (Equation 28). The series in Equation 31 exhibits limiting values when y is small (i.e., when k_s is small and/or the potential is very positive of the half-wave potential) and when y is large (i.e., when k_s is large and/or the potential is very negative of the half-wave potential). These limiting values and the corresponding currents are given by:

$$\text{y small:} \quad \sum_i (\delta_i z_i) \to y \qquad i \to 821\, n\, C^{*}\, m^{2/3}\, t^{2/3}\, k_f$$

$$\text{y large:} \quad \sum_i (\delta_i z_i) \to \sqrt{7D/3\pi t}\; (y/k_f) \qquad i \to i_d = 708\, n\, D^{1/2}\, C^{*}\, m^{2/3}\, t^{1/6} \tag{33}$$

Note that at low values of k_f, the current is controlled by the heterogeneous kinetics while diffusion control is observed at sufficiently negative potentials such that k_f is large.

The polarographic wave for a reduction process of a totally irreversible system is nearly identical to that of a reversible wave except that it is stretched along the potential axis by an amount determined by alpha and it is displaced to more negative potentials by an amount dependent upon k_s and the drop time. The approximate relationship between the potential and the current is given by[10]

$$E = E^{o} + (0.05915/\alpha n_a) \log\, [1.349\, k_s \sqrt{t/D_o}\,] - (0.0542/\alpha n_a) \log\, [i\,/\,(i_d - i)]$$

$$= E_{1/2} - (0.0542/\alpha n_a) \log\, [i\,/\,(i_d - i)] \tag{34}$$

Equation 34 is valid for instantaneous currents (i.e., the current observed at the end of the drop if the drop time is used for t); when average currents are used, and the coefficients of k_s and the current ratio are 0.886 and about $(0.0569/\alpha n_a)$, respectively. Therefore, the cathodic half-wave potential will shift $59/\alpha n_a$ mV negative for each decade decrease in the value of k_s and $28.5/\alpha n_a$ mV negative for each decade decrease in drop time. Similarily, the half-wave potential measured from average currents for electrode reactions involving hydrogen ions in the limiting step will be shifted $(59q/\alpha n_a)$ mV negative per pH unit (see Equation 23); reductions of metallic complexes which involve a rapid preceding chemical equilibrium will be shifted $(59q/\alpha n_a)$ mV negative per decade increase in concentration of the complexing agent (see Equation 26).

The log plot for a totally irreversible system will be linear with a slope of about $54/\alpha n_a$ mV when maximum currents are used or $57/\alpha n_a$ mV when average currents are used compared to a slope of $59/n$ mV observed for the reversible system. A more exact statement of the current-potential relationship reveals that the coefficient of the current ratio term is actually slightly potential dependent.[40] Thus, the log plot is slightly nonlinear (sigmoid), and the constants given above are for the slope of the best fit straight line. A linear or slightly sigmoid log plot which appears to correspond to a noninteger n value is characteristic of the heterogeneous kinetic mechanism. However, a heterogeneous mechanism with a n value of 2 and an alpha of 0.5 could be confused with a one electron reversible mechanism until the n value is estimated independently (see Section III.D.6.).

Finally, the current at the foot of the wave (kinetically controlled current) is proportional to the area of the electrode, or the product $m^{2/3}$ $t^{2/3}$. Thus, the current is independent of the corrected mercury height at the foot of the wave but proportional to the square root of the mercury height in the limiting current (diffusion controlled, $m^{2/3}$ $t^{1/6}$) region for a naturally dropping DME. For a controlled drop time DME, it will always be dependent upon the two-thirds power of the corrected mercury height but will change from a two-thirds power to a one-sixth power of time. In both cases, the current-time profiles obtained on individual drops will exhibit a $t^{2/3}$ dependence at potentials corresponding to the foot of the polarographic wave changing to a $t^{1/6}$ dependence in the limiting current region. Further relationships can be obtained from more extensive treatments of this mechanism.[10,11,13,20,24,41]

b. Consecutive Electron Transfer (EE)

In many cases, the number of electrons transferred to a species is dependent upon the electrode potential. For example, n_1 electrons may be transferred at one potential; but a second wave with an additional n_2 electron transfer may occur at more negative potentials. A homogeneous chemical equilibrium which involves the three species also occurs. These reactions can be written as:

$$O + n_1 e \rightleftharpoons I \qquad E_1^o \qquad E_{1/2}^1 \tag{35}$$

$$I + n_2 e \rightleftharpoons R \qquad E_2^o \qquad E_{1/2}^2 \tag{35a}$$

$$n_1 = n_2 = n: \; O + R \underset{k_2}{\overset{k_1}{\rightleftharpoons}} 2I \tag{35b}$$

$$k_1/k_2 = \exp\left[(nF/RT)(E_1^o - E_2^o)\right] \tag{35c}$$

where I is the intermediate chemical species. This process is termed a consecutive electron transfer or an EE (electrochemical-electrochemical) mechanism. Metallic ions frequently exhibit consecutive electron transfer mechanisms. For example, the one electron reduction of cupric to cuprous occurs at more positive potentials than the two electron reduction to metallic copper in many basic electrolytes containing ammonium ion.

The EE mechanism exhibits very simple behavior when the homogeneous chemical reaction does not occur or is very slow. When $E_{1/2}^1 \gg E_{1/2}^2$, then the consecutive electron transfer mechanism exhibits a limiting current corresponding to n_1 electrons for the first wave, n_2 for the second and $(n_1 + n_2)$ for the combined or total current. This behavior is illustrated in Figure 4; the currents are given in units corresponding to the apparent n value, i.e., the diffusion limited current normalized for other parameters.

In the absence of the homogeneous chemical reaction, the current-potential relationship for each wave, the slope of the log plot and the current-time profile for individual drops will be determined by the degree of reversibility of each electron transfer and the separation of the half-wave potentials. The limiting current for each wave will be proportional to the bulk concentration of species O and exhibit normal dependencies on mercury column height. The polarogram obtained with a solution initially contain-

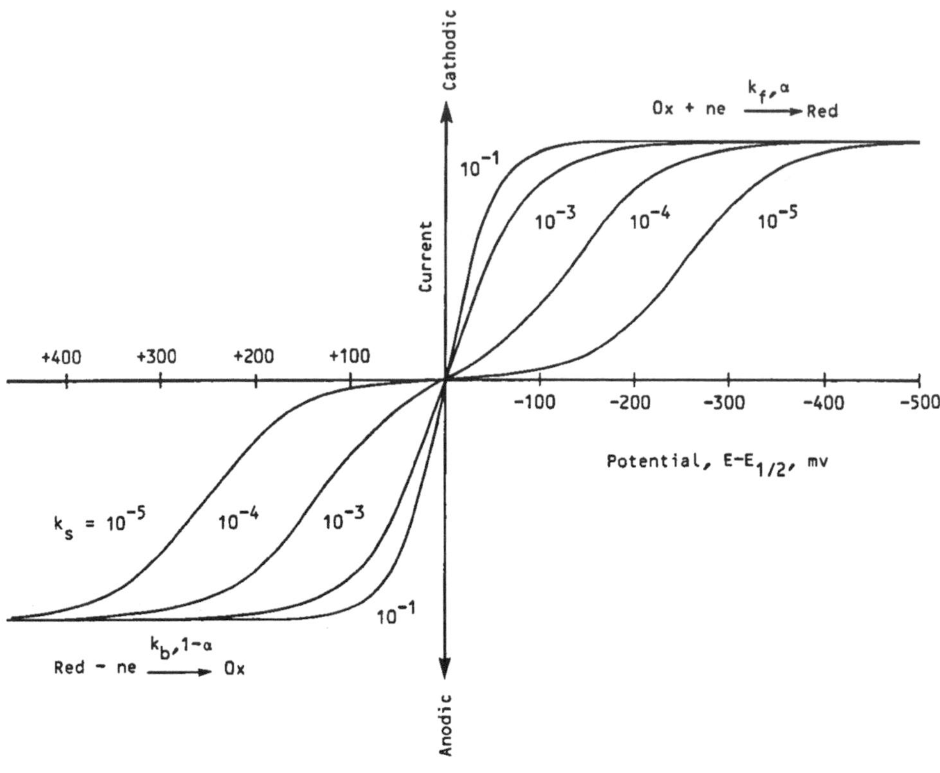

FIGURE 3A

FIGURE 3. DC polarogram, rate controlled electron transfer (irreversible) mechanism.[39] Parameter values: n_a = 1; α = 0.5; D = 10^{-5} cm^2/sec; t = 3 sec; k_s (cm/sec), as indicated. (A) Polarogram for equal concentrations of Ox and Red. (B) Polarogram for solution containing Red only. (Data calculated from Nishihara, C. and Matsuda, H., *J. Electroanal. Chem.*, 73, 261, 1976.)

ing species I may be identical to the second wave of the polarogram of O. However, the conditions are not identical since, in the first case, the species I is present in a homogeneous solution and, in the second case, it is generated at the electrode surface where it can immediately react. That is, in the consecutive electron transfer case, the intermediate chemical species (I) may be in a nonequilibrium chemical state (electronic, vibrational, rotational, polarization, or spatial configuration) which can affect the subsequent electrochemical reaction.

As $E^2_{1/2}$ becomes more positive, the two waves will begin to overlap. The overall height in the diffusion limited region will correspond to (n_1 + n_2) electrons. When $E^2_{1/2} > E^1_{1/2}$ by about 180/n mV, a single nondistorted wave will be observed (assuming that both electron-transfer processes are reversible). Its half-wave potential will be equal to that for O ($E^1_{1/2}$) and the height will correspond to (n_1 + n_2) electrons.

The EE mechanism becomes more complicated when the homogeneous chemical reaction is important, corresponding to large values for the quantity (k_1 + k_2) t. The transition range between almost no effect and almost maximum effect for typical conditions (concentration, time, and diffusion coefficient) corresponds to a second order rate constant (k_1) value of about 10^{+3} to 10^{+5}/(M-sec) when the electrode reaction is reversible.[42] The net direction of the homogeneous chemical reaction will depend upon the relative values of the formal potentials while the observed polarographic half-wave potentials depend upon the formal potentials and the degree of reversibility of the

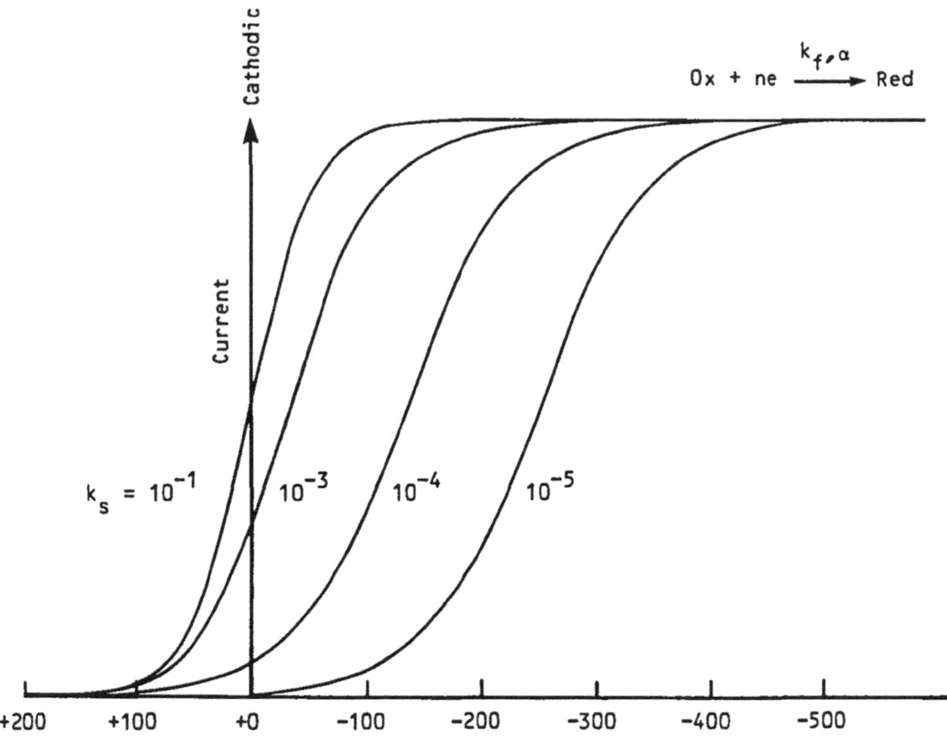

FIGURE 3B

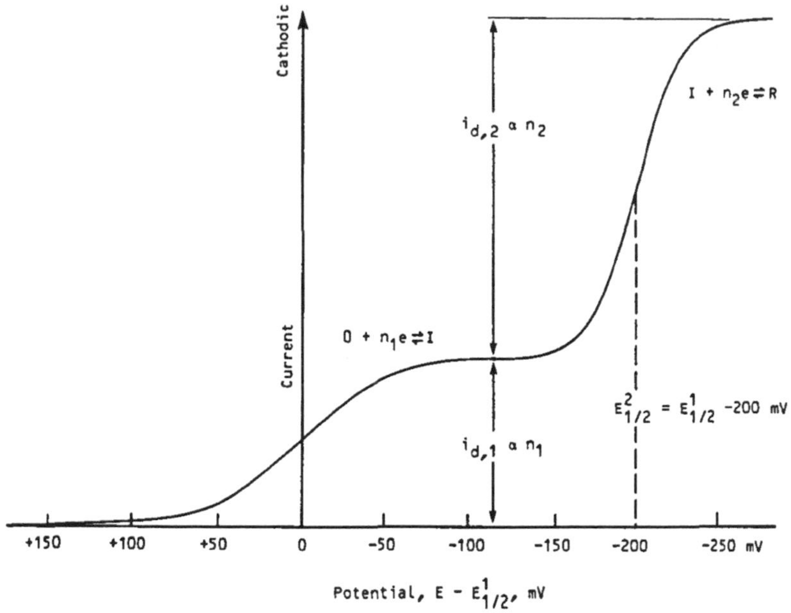

FIGURE 4. DC polarogram, EE mechanism, reversible electron transfers. Parameter values: $n_1 = 1$; $n_2 = 2$; $E^1_{1/2} - E^2_{1/2} = 200$ mV.

electrode reaction. When species I is more difficult to reduce than O and the half-wave potentials are well-separated but in the same order as the formal potentials (i.e., $E_1^o \gg E_2^o$, $E_{1/2}^1 \gg E_{1/2}^2$ and $k_1 \gg k_2$), the recombination (reproportionation) reaction of product with reactant can occur if the electrode potential is sufficiently negative to produce R (on the second wave). The recombination reaction produces no change in the limiting currents if the diffusion coefficients of all species are assumed to be equal. The chemical reaction does not affect the current-potential relationship of the second wave if it is a reversible electron transfer or causes minor changes if the second electron transfer process is rate controlled. Thus, the normal relationships for the limiting currents will be valid, while the shape and half-wave potential of the second wave will be somewhat changed when the second electrode reaction is rate controlled. If $E_{1/2}^1 \ll E_{1/2}^2$ (half-wave potentials in opposite order as the formal potentials; one wave), then species I will be removed by the second electrolysis reaction as soon as it is formed. The limiting current will be unaffected by the homogeneous reaction in this case, the wave shape will be slightly changed for a rate controlled electron transfer reaction. The behavior of the system which exhibits an EE mechanism with a homogeneous chemical reaction can be compared to the electrochemical-chemical-electrochemical (ECE) case, presented later, where the intermediate chemical species has a finite lifetime even when the electrode potential is negative of both half-wave potentials.

When the formal potentials are in the reverse order, then the homogeneous chemical reaction will proceed in the direction of disproportionation. Disproportionation can occur when I is produced at potentials where it is not removed by the second electron transfer, that is, on the first wave for the conditions $E_1^o \ll E_2^0$, $E_{1/2}^1 \gg E_{1/2}^2$ and $k_1 \ll k_2$.

Because the disproportionation produces additional amounts of O which can diffuse to the electrode and react, the height of the first wave will be increased by an amount dependent upon k_2 while the height of the second wave will be decreased. Thus, the height of the first wave will be greater than that expected for n_1 electrons and less than that for $(n_1 + n_2)$ electrons, but the total limiting current for both waves will remain constant. Therefore, only the total limiting current will be proportional to concentration and related to the mercury column height independent of other variables. These solution equilibria are unlikely to be important in cases such as the nitro/hydroxylamine/amine system but can become important in systems such as the Cu(II) reduction or the quinone/hydroquinone system depending upon the relative relationships of the formal potentials for the homogeneous reaction and the half-wave potentials for the electrode reactions.

In summary, the EE mechanism (uncomplicated by a homogeneous redox reaction) produces two waves of heights n_1 and n_2 when the second electron transfer is more difficult than the first; but generates only one wave of height $(n_1 + n_2)$ when the second electron transfer is easier. When two waves are present, the homogeneous (redox) recombination will affect the shape but not the height of the waves. The homogeneous disproportionation will enhance the height of the first wave while decreasing the second. Additional discussions on the EE mechanism[10,20] and the effect of the homogeneous chemical reactions[8,13,42] are available.

c. Preceding Chemical Reaction (CE)

When the electroactive species is in equilibrium with another species in solution, the electrode mechanism is termed a preceding chemical reaction or CE (chemical-electrochemical) mechanism. The current from this process traditionally has been referred to as a kinetic current, but it is not the only case of kinetically controlled current. The CE reactions can be written as

$$A \underset{k_2}{\overset{k_1}{\rightleftharpoons}} O \qquad K_{eq} = k_1/k_2 = C^*_O/C^*_A$$

$$O + ne \rightleftharpoons R \qquad (36)$$

where K_{eq} is the equilibrium constant for the chemical reaction with first order rate constants k_1 and k_2 for the production and disappearance of O, respectively. It will be assumed that K_{eq} is small, i.e., that species A is the predominant species present in the solution. The electrochemical reaction can be either reversible, totally irreversible, or partially irreversible. In the following discussion, a reversible electron-transfer process will be assumed unless otherwise stated. The CE mechanism was originally described for the reduction of pyruvic acid, a partially dissociated weak acid which has a dynamic equilibrium between the free acid and the anion. The CE mechanism occurs also for complexed metallic ions, for dehydration reactions preceding the charge transfer, for enol-keto equilibria, etc.

Two cases arise depending upon whether species A is nonelectroactive or electroactive. The first case occurs when species A is electroinactive (nonelectroactive) and species O undergoes a reversible electrochemical reaction. When k_1 is very small on the polarographic time scale (the drop time), no appreciable amounts of A will be converted to O. Thus, the wave height will correspond to the very small equilibrium concentration of species O; the wave shape and half-wave potential will be identical to the reversible reduction of species O. On the other hand, when k_1 is sufficiently large to maintain equilibrium, the wave height will correspond to the total solution concentration of species A and O. With large k_1, the wave will be shifted negatively (for a reduction) by an amount dependent upon K_{eq}, i.e., the reactant is in a lower potential energy configuration and, therefore, more difficult to reduce. For the reaction as written in Equation 36, the shift in half-wave potential will be 59 $\log(1 + 1/K_{eq})$ mV negative. The large k_1 case was previously described for the reduction of a complexed metallic ion in which the complex is the predominant species present in solution, the uncomplexed metallic ion is the electroactive species and the chemical reactions for the dissociation of the complex are rapid (Equations 24 to 26). In Equations 25 and 26, it was assumed that K_d (or K_{eq}) was much less than unity. If the electroactive species is a partially complexed ion (MX_p), then the parameter q is replaced by the difference $(q - p)$ in Equations 25 and 26.

When k_1 is moderately large, the wave height will correspond to a concentration between the equilibrium concentration of O and the total solution concentration of A and O due to the dynamic conversion between A and O. Mathematically, this case is described by equations equivalent to Equations 31 and 32 for the heterogeneous kinetic case in which k_f has been replaced by $k_1K_{eq}D_A$ and C* refers to the bulk concentration of species A. Although the mathematical formulation is similar, note that k_1 refers to a homogeneous chemical rate constant for the CE mechanism while k_f referred to a heterogeneous, potential-dependent rate constant for the rate controlled electron transfer case. Therefore, k_1 in the CE case is potential independent and remains constant throughout the entire polarogram and for a set of specific chemical conditions.

When y or $(k_1K_{eq}t)$ is moderately small (k_1 and/or K_{eq} small), the kinetically controlled limiting current (i_k) for the CE mechanism is given by the sum of the kinetic contribution (first term) and the diffusion controlled current contribution for the amount of O initially present in solution (second term):

$$\text{y small:}\ i_k = 821\, n\, D_A^{1/2}\, C_A^*\, m^{2/3}\, t^{2/3} \sqrt{k_1 K_{eq}} + 708\, n\, D_O^{1/2}\, C_O^*\, m^{2/3}\, t^{1/6}$$

$$= 708\, n\, m^{2/3}\, t^{1/6}\, (C_A^* + C_O^*) \quad [\sqrt{3\pi D_A/7}\, \sqrt{k_1 K_{eq}\, t} + K_{eq} \sqrt{D_O}]\ /\ [K_{eq} + 1] \qquad (37)$$

where D_A and C_A* refer to species A and the total concentration of the electroactive species is (C_A* + C_O*). The kinetically controlled current is smaller than the corresponding value calculated for the diffusion controlled case by the ratio

$$\text{y small: } (i_k/i_d) = [\sqrt{3\pi/7}\,\sqrt{k_1 K_{eq}\, t} + K_{eq}] \,/\, [(K_{eq} + 1)] \qquad (38)$$

when the diffusion coefficients are assumed to be equal. Note that some authors consider the kinetic current to be only the first term in Equations 37 and 38. However, Equation 37 gives the actual current which will be observed for the wave. Equations 37 and 38 are valid to within about 0.2, 0.7, and 2% of the true value for (k_1 K_{eq}) less than 10^{-6}, 10^{-5}, and 10^{-4}, respectively, (t equal to 3 sec). Comparison of the first and second terms of Equation 38 indicates that when (k_1t/K_{eq}) is greater than 10^{-4}, a kinetic effect will be observed.

The wave height is independent of the mercury column height (directly proportional to area) for a naturally dropping DME and proportional to the two-thirds power of height and two-thirds power of time for a controlled drop time DME (assuming that the first term of Equation 37 predominates). The current profile within each drop will be proportional to the two-thirds power of time at all potentials. The half-wave potential will be shifted negatively compared to the value expected for a simple reversible wave (given by Equation 20). The exact amount can be calculated using several relationships.[39,43,44] Finally, Equation 37 can be used to evaluate the forward rate constant of the chemical reaction if the homogeneous equilibrium constant, n value, and the diffusion coefficient are known from other data. The n value and D frequently can be obtained by altering the solution conditions so that only species O is initially present or (when A is electroactive) by using the height of the combined wave.

When the electrochemical reaction is totally or partially irreversible, the polarographic limiting current will be unaffected. However, the current-potential characteristics of the wave and the current-time profiles within individual drops will be affected by the degree of irreversibility of the electron transfer process. The log plot will be linear with a slope of 59/n mV for a reversible electron transfer and $59/\alpha n_a$ mV for a completely irreversible electron transfer (average current). Because the kinetic current is a function of k_1, the temperature dependence of the current may be larger than that observed for a diffusion controlled process. However, this is not a conclusive test for the CE mechanism.

The second case for the preceding chemical reaction mechanism occurs when species A is electroactive, but has a half-wave potential more negative than that for species O (A is more difficult to reduce than O). When k_1 and k_2 are very small, two waves will be observed; the height and shape of each wave will correspond to the equilibrium concentration of the corresponding species and the nature of its electrochemical reaction. When k_1 is very large, only one wave for the reduction of species O will be observed. Its height will correspond to the total solution concentration of A plus O and its half-wave potential will be shifted negative dependent upon the value of the equilibrium constant.

When k_1 is moderately large on the polarographic time scale, the limiting current of the first wave will be enhanced by the conversion of species A to O. However, the second wave will be diminished in height so that the total wave height remains constant at the value expected for a diffusion controlled wave. The behavior of the first wave is exactly as described for the CE mechanism, case 1 (A electroinactive, Equations 37 and 38). This behavior is illustrated in Figure 5.[39,43,44]

The enhancement of the height of the first wave was initially observed with pyruvic acid.[20] In that investigation, the ratio of the wave heights for the more easily reduced

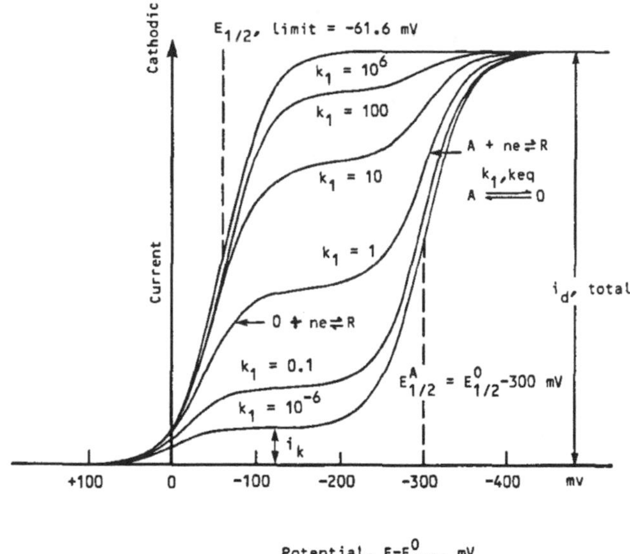

FIGURE 5. DC polarogram, CE mechanism, reversible electron transfer.[39,43,44] Parameter values: n = 1; t = 3 sec; K_{eq} = 0.1; $E^0_{1/2} - E^A_{1/2}$ = 300 mV; k_1 (sec^{-1}), as indicated.

free acid and the less easily reduced anion did not correspond to their equilibrium solution concentrations as calculated from the acid dissociation constant. Further investigations led to the first realization of the CE mechanism in polarography.

When K_{eq} is near unity (both A and O initially present), the behavior will be analogous to the cases described above except that the reference polarogram (k_1 small) will contain a wave for the equilibrium concentration of species O. A minor mathematical complication for the predicted wave heights results unless the diffusion coefficients of A and O are assumed to be equal (this is not a valid assumption when one of the species is the hydronium or hydroxide ion). When K_{eq} is large, there is no preceding chemical reaction. When species A is more easily reduced (oxidized) than species O, the roles of A and O are merely interchanged.

In summary, the height of the wave of the electroactive species O is related to its equilibrium concentration in the original solution and the extent of conversion from its nonelectroactive (at that potential) precursor. As the rate constant for the production of O increases, the wave height increases until it reaches the diffusion controlled limit for the total concentration of O plus A. The negative shift of the half-wave potential of a reduction wave reflects the increased stability of the precursor. More extensive references are available for the CE mechanism.[10,13,20,24,34,41]

d. Following Chemical Reaction (EC)

When the product of the electrode reaction is converted to another species in solution by a homogeneous chemical reaction, the electrochemical mechanism is termed a following chemical reaction or EC mechanism. The electrochemical and chemical reactions can be written as

$$O + ne \rightleftharpoons R \qquad (39)$$

$$R \underset{k_2}{\overset{k_1}{\rightleftharpoons}} Z \qquad K_{eq} = k_1 / k_2 \qquad (39a)$$

where k_1 and k_2 are the first order chemical rate constants and K_{eq} is the equilibrium constant. The product of the chemical reaction, Z, is assumed to be electroinactive. The EC mechanism is frequently observed in the polarography of organic species in protic solvents. In many instances, the initial product of the electrode reaction (e.g., a one-electron reduction to a radical anion or a two-electron reduction to a carbanion) will react with the solvent or other hydrogen ion donor to produce a protonated species.

The limiting current for the EC wave is equal to that expected in the absence of the following chemical reaction since that reaction does not affect the rate of diffusion of the electroactive material to the electrode. The limiting current is given by the Ilkovic equation (Equation 6) and exhibits the normal dependencies on mercury column height and drop time. However, the chemical removal of the electrolysis product makes it easier to reduce the oxidized species. Thus, the half-wave potential (for a reduction) will be shifted to more positive potentials by an amount dependent upon the characteristics of the chemical reaction. For a rapid, reversible chemical reaction and a reversible electron transfer process, the half-wave potential will be shifted $(59/n) * \log(1 + K_{eq})$ mV positive of the half-wave potential which would be observed in the absence of a chemical reaction while the wave shape will remain identical to the reversible case.

For a kinetically controlled chemical reaction coupled to a reversible electron transfer process, the half-wave potential will be shifted positive by an amount dependent upon k_1 and the drop time. The shift in half-wave potential will become significant when the quantity (k_1t) is greater than about unity. The shape of the polarographic wave will be somewhat distorted for a slow chemical reaction. When the electron transfer process and the chemical reaction both occur on the polarographic time scale, a combination of heterogeneous and homogeneous kinetic effects will be observed. There has been little theoretical interest in this mechanism for polarography since the observed currents display a relatively small dependence on the chemical reaction. However, some information is available.[24,34,44]

e. *Electrochemical-Chemical-Electrochemical Reaction (ECE)*

When the product of the chemical reaction in the EC mechanism is itself electroactive, the process is termed an ECE mechanism. That is, a homogeneous chemical reaction involving the product of the first electron transfer (R_1) gives a substance (O_2) which is electroactive also. The sequence of electrochemical and chemical reactions can be stated as

$$O_1 + n_1e \rightleftharpoons R_1 \qquad E_1^0 \quad E_{1/2}^1 \tag{40}$$

$$R_1 \underset{k_2}{\overset{k_1}{\rightleftharpoons}} O_2 \qquad K_{eq} = k_1/k_2 \tag{40a}$$

$$O_2 + n_2e \rightleftharpoons R_2 \qquad E_2^0 \quad E_{1/2}^2 \tag{40b}$$

where k_1 and k_2 are first order or pseudo-first order rate constants and K_{eq} has its usual definition. The ECE mechanism, like the EC mechanism, is frequently observed in the electrochemistry of organic molecules. For example, the product of the first reduction can undergo protonation, elimination, substitution, etc. to give a second electroactive species.

The two limiting electrochemical cases exhibit behavior which depend upon whether species O_2 is more or less easily reduced (oxidized) than species O_1. In either situation, the effect of the chemical reaction is observed if an appreciable conversion occurs on the time scale of the experiment, i.e., the drop time for polarography. The electrode and homogeneous reactions may be any combination of slow (kinetically controlled) or fast (diffusion controlled electrochemical or rapid chemical equilibria) processes.

When the second electron transfer is more difficult than the first and the polarographic waves are well separated ($E_1^\circ \gg E_2^\circ$ and $E_{1/2}^1 \gg E_{1/2}^2$), two waves are observed. The limiting current of the first wave is diffusion controlled and retains the normal height and concentration relationships. The current-time behavior will be the same as the EC case, a $t^{1/6}$ dependence. The half-wave potential and shape of the first wave will be determined by the same factors which control these parameters in the EC mechanism (heterogeneous k_f, homogeneous k_1, K_{eq}, and t). The height of the second wave (reduction of O_2) will be dependent upon the chemical reaction, ranging from no wave (k_1 small) to the normal diffusion controlled wave height (k_1 large). When k_2 is zero, the height of the second wave varies from 10 to 90% of its maximum value as the quantity (k_1t) varies from 0.2 to 30, respectively. However, the height of the second wave is also dependent upon K_{eq} (or k_2), becoming smaller for smaller values of K_{eq} (larger k_2) at a given value of k_1. Essentially, K_{eq} determines the equilibrium distribution of the intermediate product between the R_1 and O_2 forms, while k_1 determines the rate of conversion of R_1 to O_2. As K_{eq} increases, the first electron transfer becomes easier (shifts positive of its normal position). The second electron transfer approaches its reversible value when K_{eq} is large. (Its half-wave potential is negative of this value for $K_{eq} \ll 1$.) The behavior of the half-wave potential and shape of the second wave will qualitatively resemble those of a CE mechanism; however, they are quantitatively different since the starting material is produced at the electrode and diffuses into the solution rather than starting with a homogeneous solution. Some limiting behaviors are illustrated in Figure 6 for the indicated values of (k_1t) when k_2 is equal to zero (a typical experimental situation) and the electron transfers are reversible.

When the second electron transfer is easier than the first ($E_{1/2}^1 \ll E_{1/2}^2$), the product of the chemical reaction (O_2) will be reduced at the potential where it is produced. Thus, only one wave will be observed. The current for this wave will be increased over that expected for n_1 electrons by an amount dependent upon the extent of formation of O_2, a function of k_1, K_{eq}, and the drop time. The wave height, therefore, will correspond to a n value between n_1 and ($n_1 + n_2$) electrons, the limiting values for a slow and a fast chemical reaction, respectively. The limiting current for this case is exactly the same as the total limiting current for waves one and two that is observed for the $E_{1/2}^1 \gg E_{1/2}^2$ case. The limiting current is proportional to concentration when factors affecting k_1 and K_{eq} are maintained constant. However, no simple relationship exists between the limiting current and the mercury column height or for the dependence of current on time. When the polarographic waves are only partially overlapped, the behavior is more complex but the total limiting current retains the same characteristics. When the electrode reactions are rate controlled, the waves tend to elongate causing a higher probability of overlapping waves and introducing additional parameters (k_s and alpha for each electron transfer process) which affect the overall current-potential and current-time behavior.

The ECE case may be complicated by the homogeneous redox reaction given by

$$n_1 = n_2 = n: \qquad O_1 + R_2 \underset{k_4}{\overset{k_3}{\rightleftharpoons}} R_1 + O_2 \tag{41}$$

$$k_3/k_4 = \exp\left[(nF/RT)(E_1^\circ - E_2^\circ)\right] \tag{41a}$$

where k_3 and k_4 are second order chemical rate constants. The equilibrium constant for this reaction may be calculated from the observed half-wave potentials when the

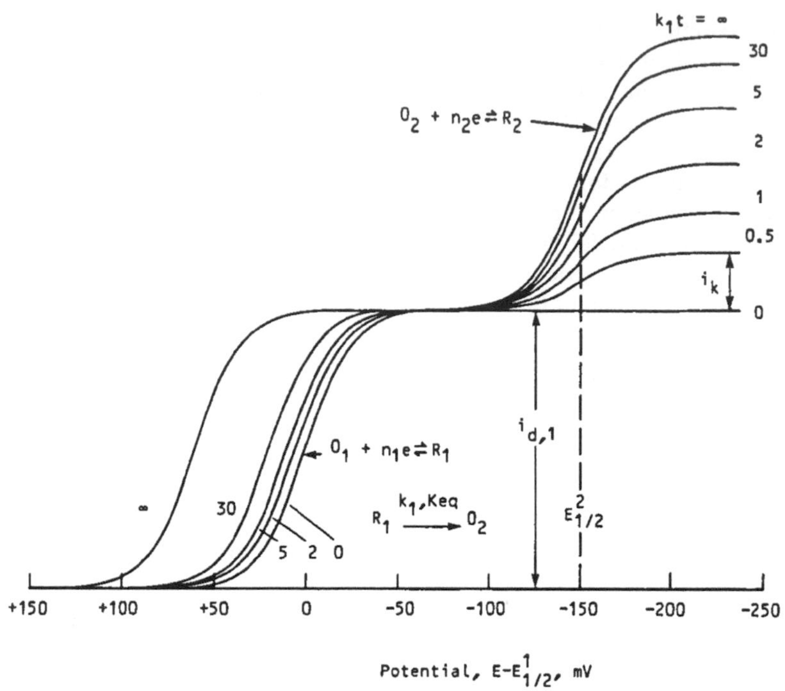

FIGURE 6. DC polarogram, ECE mechanism, reversible electron transfers.[44,45] Parameter values: $n_1 = 2$; $n_2 = 2$; $k_2 = 0$; $E^2_{1/2} - E^1_{1/2} = 150$ mV; k_1t, as indicated.

electron transfer steps are reversible with $E^1_{1/2} \gg E^2_{1/2}$ or from the formal potentials as indicated. The equilibrium constant and absolute magnitudes of k_3 and k_4 for this reaction determine its importance on the polarographic time scale.

The effects produced by the presence of the homogeneous redox reaction can be predicted by the limiting cases where the chemical reaction is assumed to be irreversible and the half-wave potentials are well separated. If a net forward reaction occurs ($k_3 \gg k_4$, $E^\circ_1 \gg E^\circ_2$), then the limiting current of the first wave will not be affected when $E^1_{1/2} \gg E^2_{1/2}$ (two waves) since R_2 is not produced until the second wave. The height of the second wave will be decreased since some electroactive O_1 (2n electrons) will be converted into equal amounts of O_2 (only n electrons) and the (temporarily) nonelectroactive R_1. When $E^1_{1/2} \ll E^2_{1/2}$ (one wave), a similar effect will occur on the height of the total wave for a net forward homogeneous redox reaction.

If the net homogeneous redox reaction occurs in the opposite direction ($k_3 \ll k_4$, $E^\circ_1 \ll E^\circ_2$, then the flux of O_1 will be increased and that of O_2 decreased at potentials where R_1 is formed. Thus, when $E^1_{1/2} \gg E^2_{1/2}$ (two waves), the height of the first wave will be increased at the expense of the second. When $E^1_{1/2} \ll E^2_{1/2}$ (one wave), the redox reaction will not affect the limiting current if all diffusion coefficients are equal.

The ECE mechanism is even more complex when the homogeneous redox reaction given in Equation 41 is reversible, the electrode reactions are rate controlled or the half-wave potentials are not well separated. The second electrode reaction in the ECE mechanism also can occur in the opposite direction from the first, e.g., electron transfer one can be a reduction and electron transfer two an oxidation. Because of the complex variations which can occur with the ECE mechanism, no further discussions of the ECE case will be made.

In summary, the ECE mechanism uncomplicated by the homogeneous redox reaction exhibits n_1 and $\leqslant n_2$ electron waves which may be separated or overlapped. The height of the second wave is increased as k_1 or K_{eq} is increased. The first electron transfer is made easier by a large value of K_{eq} similar to the EC mechanism; the second electron transfer becomes more reversible as K_{eq} increases ($K_{eq} < 1$), similar to the CE mechanism. If the homogeneous redox reaction is operational, the flux of the initial species (O_1) may be increased or decreased depending upon the direction of the homogeneous reaction (determined by the formal electrode potentials) and the polarographic half-wave potentials (determined by the formal potentials and the degree of charge transfer reversibility). Additional information is available for the simple ECE mechanism[24,34,44-46] and for the ECE mechanism complicated by the homogeneous chemical reaction.[13,47]

f. Catalytic Regeneration of the Reactant

The catalytic regeneration mechanism is a special case of the EC sequence in which the product of the chemical reaction is the starting material. This reaction sequence can be represented by

$$O + ne \rightleftharpoons R \qquad (42)$$

$$R + Z \underset{k_2}{\overset{k_1}{\rightleftharpoons}} O \qquad k_1' = k_1 C_Z^* \qquad K_{eq}' = k_1'/k_2 \qquad (42a)$$

where species Z in nonelectroactive and present in the bulk of the solution. When the substance Z is present in relatively high concentrations, the pseudo-first order chemical rate constant K_1' may be defined. If $C_Z \gg C_O$ and k_1 is large (i.e., k_1' is large), then the reduced species will react almost immediately upon leaving the electrode. Therefore, additional O is formed near the electrode, quickly diffuses to the electrode, and reacts to produce R which can be recycled. This sequence of reactions leads to a greatly increased current over that expected in the absence of the regeneration reaction.

One would expect that Z should be electroactive if it is a sufficiently powerful oxidant to regenerate O; however, this electrode reaction may not occur due to unfavorable kinetics. A catalytic regeneration mechanism is found with solutions of ferric ions containing either peroxide or hydroxylamine which react with the electrogenerated ferrous ions but which do not react at the electrode (at the potential required for the reduction of ferric ions). Analytically, the increased sensitivity produced by catalytic regeneration has been employed to measure trace concentrations of substances. The most notable of these methods is the Brdicka reaction of proteins, other sulfhydryl compounds, and certain nonsulfur containing compounds in the presence of Co(II) or Ni(II). A 100- to 500-fold increase in sensitivity is observed in the Brdicka assay.[41] Another example of catalytic currents is the catalytic hydrogen reduction, a frequent contributor to residual currents at negative potentials. The mechanism for the catalytic hydrogen generation reaction was presented in Chapter 3 of this volume, Coulometry (Equation 24).

When the electrode reaction is rapid, the ratio of the current observed in the presence of the catalytic regeneration to the equivalent diffusion controlled current is expressed by a mathematical series similar to that obtained for the rate controlled electron transfer. When the catalytic current is large, the ratio of the catalytic current averaged over the life of a mercury drop ($\bar{i}_{cat}$) to the normal average diffusion controlled current ($\bar{i}_d$) is given by

$$\bar{i}_{cat}/\bar{i}_d = 0.81\,[(k_1' + k_2)\,t]^{1/2} + 1.92\,[(k_1' + k_2)\,t]^{-7/6} \qquad (43)$$

This equation is valid when $(k_1' + k_2)\,t > 10$. Note that the first term predominates for very large values of $(k_1' + k_2)t$. When the electrode reaction is reversible, no shift in half-wave potentials occurs, and Equation 43 is valid for the current measured at any potential. When the electrode process is rate controlled, Equation 43 is valid only for the limiting current. The diffusion controlled current must be measured in the absence of Z or estimated from the Ilkovic equation in order to use Equation 43 to estimate $\bar{i}_{cat}$ or the kinetic parameters.

As stated above, the shape of the catalytic polarographic wave obtained with an electrochemical system which has a rapid electron transfer reaction is identical to that of the reversible electron transfer case. From Equation 43, it is apparent that the catalytic current is approximately proportional to the two-thirds power of the drop time ($\bar{i}_{cat} \propto \bar{i}_d\, t^{1/2} \propto t^{1/6}\, t^{1/2}$) when the first term predominates. Thus, the limiting current is proportional to $m^{2/3}\, t^{2/3}$ or independent of the mercury column height and directly proportional to the drop area for a naturally dropping DME. For a controlled drop time DME, the limiting current is proportional to the two-thirds power of the mercury height and the two-thirds power of the drop time. The current-time profile for an individual drop is related to the two-thirds power of time and the average current is equal to 0.6 times the current observed at the end of the drop. Equation 43 also predicts that the catalytic current should be proportional to the concentration of O when the kinetic current is given by the first term and the concentration of Z is large enough to maintain pseudo-first order kinetics.

Because of the nature of the catalytic regeneration mechanism, the limiting current may be sensitive to small changes in conditions, e.g., nature of the catalyst, temperature, species which enhance or decrease the catalyst activity, etc. However, the increased sensitivity obtainable for catalytic assays frequently is preferable over the problems which produce decreased precision. Other catalytic regeneration schemes are possible, e.g., half-regeneration or surface (heterogeneous) catalysis. The EE mechanism with disproportionation of the first reduction product could be considered a second order regeneration reaction. The catalytic regeneration mechanism is reviewed extensively by Mairanovskii[41] and by others.[10,11,20,24,34]

g. Adsorption

When additional molecules or ions are present in the solution layer next to the electrode compared to that expected on the basis of electrostatic attraction, the phenomena of adsorption has occurred. Adsorption of ions or dipoles may occur only over certain potential ranges. Anions are more likely to adsorb at positive potentials and cations at negative potentials referenced to the potential of zero charge (E_{pzc}). The extent of adsorption is also dependent upon the solution conditions, e.g., the presence of other species which are more strongly adsorbed, solvent polarity, solute solubility, ionic strength, etc. For example, large, nonpolar organic molecules are more likely to adsorb in aqueous solutions than in an organic solvent but may only adsorb in certain potential regions due to competitive adsorption from other species in the electrolyte, especially at highly positive or highly negative potentials vs. E_{pzc}.

The effect of adsorption on the charging current has already been described. However, the adsorption of the electroactive species causes other, more important changes in the polarogram. Specifically, adsorption produces a shift in the potential energy of the species undergoing adsorption. Thus, adsorption of the electrolysis product causes the electrode reaction to occur more easily (a positive shift in the half-wave potential of reductions), similar to that observed for a rapid following chemical reaction (EC mechanism). Conversely, adsorption of the electroactive species will tend to cause a negative shift in the half-wave potential for a reduction. Note, however, that a negative shift in the half-wave potential implies that a portion of the material diffusing to the electrode is adsorbed (without being reduced), and that it remains on the electrode surface at potentials where the nonadsorbed O is being reduced. Thus, the current will be less than the diffusion controlled limit until the electrode potential is sufficiently negative (on the postwave) to reduce all of the material diffusing to the electrode. The presence of this film may produce autoinhibition and cause distortion of the normal polarographic wave.[41] If both the reactant and the product are adsorbed, the effect observed will depend upon the relative strengths of each process (when O and R are equally adsorbed, an apparently unaffected polarographic wave is obtained).

Because of the complex nature of adsorption, the remaining description will be limited to the case which has a reversible electron transfer reaction and involves strong adsorption of the electrolysis product, the most common situation. It will be assumed that the adsorption process follows the Langmuir isotherm although the characteristics of other adsorption isotherms are available.[34,41] It also will be assumed that the adsorption equilibrium is rapid on the polarographic time scale so that no kinetic effects are involved.

Quantitatively, the height of the adsorption wave is limited by the maximum amount of material which can occupy the electrode surface. This quantity is frequently termed Γ_m (mole/cm²) while the amount of material at any specific potential is termed Γ, or the surface excess. These quantities are related by

$$\Gamma = \Gamma_m B C^* / (1 + BC^*) \tag{44}$$

when only one species is adsorbed, B is its adsorption coefficient (cm³/mole) and C* is the bulk concentration in moles/cm³ rather than the usual millimolar units.

When the surface of the electrode has been completely covered by adsorption of the electrolysis product, any additional material which is electrolyzed must diffuse into the solution. The polarogram, therefore, is divided into two separate waves. The first, or prewave, corresponds to the production of the adsorbed product at a more positive half-wave potential and the second corresponds to the production of the material which diffuses into the bulk of the solution. These chemical processes can be described by the reactions

$$O + ne \rightleftharpoons R_{ads} \qquad E_{1/2(ads)} > E_{1/2} \tag{45}$$

$$O + ne \rightleftharpoons R \qquad E_{1/2} \tag{45a}$$

$$R_{ads} \underset{k_2}{\overset{k_1}{\rightleftharpoons}} R \tag{45b}$$

where R_{ads} is the adsorbed electrolysis product in equilibrium with the soluble form of that species. The half-wave potential for the adsorption wave is shifted [(59/n) (B_R/B_O] mV positive when the electrode is completely covered and B_R/B_O is the adsorption coefficient ratio of the reduced to the oxidized species. Figure 7 illustrates the behavior for this situation for different concentrations of the electroactive species when the electrolysis product is strongly adsorbed.

When the solution concentration is sufficiently low, all of the electrolysis product will be adsorbed and only one wave will be observed. In this concentration region, the height of the adsorption prewave will be proportional to the concentration of the electroactive species. The concentration corresponding to a saturation coverage of the mercury drop is typically about 10^{-4} to 10^{-5} *M*, and it results in a limiting adsorption current (i_{ads}, microamps) measured at time (t) which is given by

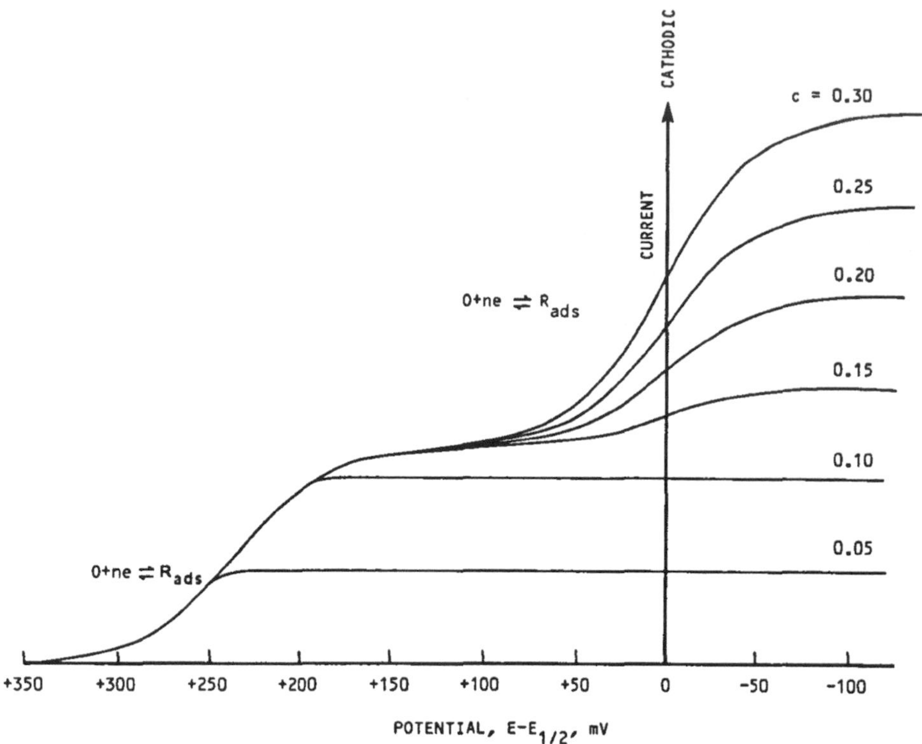

FIGURE 7. DC polarogram, strong adsorption of product, reversible electron transfer.[48] Parameter values; $n = 1$; $m = 2$ mg/sec; $t = 3$ sec; $B = 10^9$ cm³/mole; $\Gamma_m = 10^{-9}$ moles/cm²; C (m*M*), as indicated. (Data calculated from Laviron, E., *J. Electronal. Chem.*, 52, 355, 1974.)

$$i_{ads} = 9.10\, n\, m^{2/3} / a\, t^{1/3}$$

$$= 548 \times 10^6\, n\, m^{2/3}\, \Gamma_m / t^{1/3} \qquad (46)$$

where a is the area (square Angstroms) occupied per molecule on the electrode surface and Γ_m has already been defined. When a is equal to 100 Å² (observed for methylene blue[41]), corresponding to a Γ_m value of about 1.66×10^{-10} mole/cm², then the calculated maximum adsorption current is about 0.1 μA (m, 2 mg/sec; t, 3 sec). From Equation 46, the current in the adsorption wave is proportional to $m^{2/3}\, t^{-1/3}$ or to the rate of growth of the mercury drop (area divided by time). This relationship also predicts that the adsorption current should be directly proportional to the height of the mercury column for a natural dropping DME or to the two-thirds power of the height for a controlled drop time DME. The current-time profiles for individual drops observed in the adsorption wave will have a unique characteristic, the current will decrease during the life of the drop because of the $t^{-1/3}$ dependence. The time dependence also gives an average current over the life of the drop which is 1.5 times as large as the current measured at the end of the drop. The shape of the adsorption wave should be identical to that of a reversible electron transfer wave when the assumptions previously stated are valid.

When the solution concentration is above that limit which produces saturation coverage, the second wave appears. In this situation, neither wave height is proportional to concentration. However, the sum of the adsorption current and the current in the main wave will follow the Ilkovic equation for diffusion control since the total current is still controlled by the rate of diffusion of material to the electrode. The limiting current will also exhibit the normal diffusion controlled behavior with regard to mercury column height dependence. The half-wave potential of the second wave is unchanged from the value which would be observed in the absence of adsorption (for the assumptions stated above).

The extent of adsorption is dependent upon the physiochemical properties of the system. For example, adsorption is generally decreased as the temperature or electrolyte concentration increases. When adsorption is weak, the main wave may be observable before the prewave reaches its maximum value. Additional information can be found in other references.[10,11,13,20,34,41,49]

4. Summary of Electrode Mechanisms

From the above discussions, it is obvious that a multiplicity of mechanisms can result in a seemingly simple electrode reaction. Of course, possibilities of even greater complexity exist, such as mixed mechanisms, second order chemical reactions including the dimerization of the electrolysis product in either the EC or ECE mechanism, multiple homogeneous chemical equilibria for multiprotic species or several monoprotic species, etc. Electrode mechanisms can include heterogeneous chemical reactions which occur on the electrode surface (such as the hydrogen evolution reaction on Pt) and double-layer influences on coupled chemical reactions (i.e., modification of the homogeneous chemical reactions in the solution layer surrounding the electrode). Modern mathematical techniques are available to characterize the behavior of specific mechanisms using a combination of numerical analysis and analytical solutions in the equations which describe the system.[50-54] These characterizations allow the relative roles of diffusion, charge transfer, homogeneous chemical reactions, adsorption, etc. to be determined for a range of parameters associated with the system. In spite of the potential for unforseen problems introduced by complicated electrode mechanisms, an optimized analytical method should result from a knowledge of generalized electrode mechanisms combined with at least a limited understanding of the behavior of a specific compound (obtained during the basic screening of the polarographic characteristics). The limiting behavior of the uncomplicated mechanisms (discussed above) is summarized in Table 4. Additional information on electrode processes can be found in the current electrochemical literature, review articles, and monographs.[6,13,20,24,33,34,41,55-58]

5. Correlations Involving Half-Wave Potentials

From the previous discussions of the current-potential relationships, it is evident that the half-wave potential is related to the thermodynamic equilibrium between reactants and products in the case of reversible electron transfers, or to a combination of equilibrium and kinetic factors in other situations. Thus, within limitations, the variation of half-wave potentials can be used as a measure of the electrochemical reactivity of compounds. Furthermore, this reactivity can be correlated with structure or, sometimes, with other physiochemical properties (spectra, chemical reactivity, etc.) or molecular properties (such as pharmacological activity). In fact, half-wave potentials exhibit gross correlations to gas phase ionization potentials (oxidations) or electron affinities (reductions) and have provided convincing corroborations of molecular orbital theory.[6] In contrast to other chemical measurements, the polarographic half-wave potentials are much easier to obtain and exhibit a high accuracy when compared to rate or equilibrium constants for the corresponding homogeneous chemical reactions.

Quantitative electrochemical correlations of half-wave potentials or predictions of behavior of new species are complicated by several factors. The principal requirement for quantitative comparison of reactivities is that the electrode processes in question

Table 4
PARAMETERS AND BEHAVIOR OF SELECTED ELECTRON TRANSFER MECHANISMS FOR DC POLAROGRAPHY

Mechanism	Reactions	System parameters	Limiting current vs. bulk concentration	Current vs. h_{corr}, natural drop time DME	Current vs. time	Half-Wave potential	$E_{1/2} - E_{1/2}$ (rev)	Slope of log plot	$\bar{i}/i$
Reversible (diffusion controlled)	$Ox + ne \rightleftharpoons Red$	$i_d, E_{1/2}, E^\circ$	C_o	$h^{1/2}$	$t^{1/6}$	constant	0 (Defined)	59/n	6/7
Rate controlled Electron transfer (irreversible)	$O + ne \underset{k_b,\,1-\alpha}{\overset{k_f,\,\alpha}{\rightleftharpoons}} R$	$i_d, E_{1/2}, E^\circ$ k_s, α, t	C_o	Foot: h^0 Limiting: $h^{1/2}$	$t^{2/3}$ $t^{1/6}$	$k_s\downarrow$: $E_{1/2}\downarrow$ $t\downarrow$: $E_{1/2}\downarrow$	Negative	$\sim 54/\alpha n_a$	—
EE[a,b]	$O + n_1e \rightleftharpoons I$	$i_{d1}, E^1_{1/2}$	C^c_o	$h^{1/2}$	$t^{1/6}$	Constant	0	59/n	6/7
	$I + n_2e \rightleftharpoons R$	$i_{d2}, E^2_{1/2}$		$h^{1/2}$	$t^{1/6}$	Constant	0	59/n	6/7
CE[a]	$A \underset{k_2}{\overset{k_1}{\rightleftharpoons}} O$	$C^*_A, C^*_o, K_{eq}, k_1$	$C_A + C_o$			$t\uparrow$: $E_{1/2}\downarrow$			
					$t^{2/3}$	$K_{eq}\downarrow$: $E_{1/2}\downarrow$	Negative	Complex	3/5
	$O + ne \rightleftharpoons R$	$i_k, t, E_{1/2}$	$i_k\uparrow$ for $K_{eq}\uparrow$, $k_1\uparrow$, $t\uparrow$	h^0		$k_1\uparrow$: $E_{1/2}\downarrow$			
EC[a]	$O + ne \rightleftharpoons R$	$i_d, E_{1/2}$	C_o	$h^{1/2}$	$t^{1/6}$	$t\uparrow$: $E_{1/2}\uparrow$	Positive	Complex	6/7
	$R \underset{k_2}{\overset{k_1}{\rightleftharpoons}} Z$	k_1, K_{eq}				$K_{eq}\uparrow$: $E_{1/2}\uparrow$ $k_1\uparrow$: $E_{1/2}\uparrow$			
ECE[a,b]	$O_1 + n_1e \rightleftharpoons R_1$	$i_{d,1}, E^1_{1/2}$	C^c_o	$h^{1/2}$	$t^{1/6}$	$t\uparrow$, $K_{eq}\uparrow$, $k_1\uparrow$: $E^1_{1/2}\uparrow$	Positive	Complex	6/7
	$R_1 \underset{k_2}{\overset{k_1}{\rightleftharpoons}} O_2$	k_1, K_{eq}							
	$O_2 + n_2e \rightleftharpoons R_2$	$i_k, E^2_{1/2}$	$i_k\uparrow$ for $K_{eq}\uparrow$, $k_1\uparrow$, $t\uparrow$	Complex	Complex	$t\downarrow$, $K_{eq}\uparrow$, $k_1\downarrow$: $E^2_{1/2}\uparrow$	Negative	Complex	—

Catalytic Regeneration[a]	$O + ne \rightleftharpoons R$ $\overset{k'_1}{}$ $R + Z \rightleftharpoons O$ $\underset{k_2}{}$	$\bar{i}_{cat}$, $E_{1/2}$ k'_1	C_o when $C_o \ll C_z$ and $(k_1' + k_2)t \gg 1$	h^o	$t^{2/3}$	Constant	0	—	3/5
Adsorption of product[a]	$O + ne \rightleftharpoons R_{ads}$	i_{ads}, $E^{ads}_{1/2}$, Γ_m, B	—	Prewave: h^1	$t^{-1/3}$	B↑: $E_{1/2}$↑	Positive	Complex	3/2
	$O + ne \rightleftharpoons R$	i_d, $E_{1/2}$	C_o (total current)	Main: $h^{1/2}$ (total current)	$t^{1/6}$	Constant	0	59/n	6/7

[a] Reversible electron transfer process(es) assumed.

[b] $E^1_{1/2} \gg E^2_{1/2}$ (reduction): homogeneous redox reaction (disproportionation and/or recombination) assumed to be absent.

[c] See individual sections for exceptions and additional information.

exhibit the same overall mechanism, i.e., stoichiometry, number of electrons transferred, type of electrode mechanism, and its degree of reversibility. In general, comparisons of molecules with different electroactive groups are not possible. But comparisons of electrochemical behavior are possible if only the electron distribution (but not the bonds or atoms themselves) of the electroactive group is affected.[2,6,59] Functional group substitution at a site which does not appreciably affect the degree of conjugation of the electroactive group generally allows valid comparisons.

Because both equilibrium and kinetic effects may be important in an electrochemical reaction, substituents affecting either the equilibrium electronic states (polarities of substituents, etc.) or factors affecting the transition state (ionic atmosphere at the electrode, extent of adsorption, steric effects, etc.) can affect the half-wave potentials. These effects have been termed static and dynamic, respectively, by Zuman in his extensive, classic review of substituent effects.[59] Generally, the half-wave potential of a series of substituted compounds can be correlated using relationships of the Hammett or Taft form. Other reviews of this subject include that of Eberson and Nyberg[6] and Elving.[2]

Because the determination of the mechanism of electrochemical reactions is quite time consuming, frequently only partial investigations are undertaken. A minimum investigation should include the number of electrons transferred (relative heights of waves), numbers of waves observed, establishment of diffusion control for all waves of interest, the effect of pH, and the degree of reversibility (current-potential relationship). For at least one member of the series, an absolute determination of the number of electrons and an isolation and identification of the electrode products should be undertaken unless previous data are available for the identical conditions. Even minor variations in conditions (particularly solvation effects such as proton availability in aprotic solvents or adsorption) can cause changes in mechanisms which invalidate the attempted correlations. However, the comparison can be valid if any set of conditions can be found where the compounds react similarly.

6. Determination of n Values

Several methods are available to determine the number of electrons involved in the electrochemical reaction. Analogy with model compounds which have well defined electrochemical characteristics is one method. Analysis of the polarographic wave itself yields two other estimates of the n value. First, the diffusion-limited current can be used to directly estimate n through the Ilkovic equation (Equation 6) if the capillary constants are measured experimentally and if independent estimates of the diffusion coefficient and purity are available. Since n must be a small integer (in most cases), this technique will usually lead to a value which is sufficiently close to the true value to be useful. It should be noted in this regard that most organic compounds consume (or give up) an even number of electrons in aqueous solutions (or other proton donating electrolytes), two, four, or occasionally six being the most common numbers. In some aprotic electrolytes, formation of radical anions or cations by a one electron transfer may be observed. The common occurrence of diffusion controlled currents makes this technique a very useful estimate of the n value.

Second, the shape of the current rise of the polarographic wave is dependent upon the n value. However, an estimate obtained from the log plot (or wave shape) is influenced by the characteristics of the electron transfer step (e.g., reversibility or coupled chemical reactions) as described previously. These characteristics may be unknown or so complicated that they cannot easily be determined.

The use of controlled potential coulometry and its limitations in the determination of n values is discussed in Chapter 3 of this volume. Other methods for the determination of n include related electrochemical techniques such as differential pulse polarography or cyclic voltammetry. These techniques may involve different time scales as well as limitations such as those found in the use of the polarographic wave shape.

IV. RELATED POLAROGRAPHIC AND VOLTAMMETRIC TECHNIQUES

Although DME polarography is the oldest polarographic technique, it is not always the technique of choice. Some of the limitations previously mentioned include complex quantitative descriptions of the mass transport process, limited sensitivity, limited potential ranges, and limited ranges of experimental variables (such as time) for the study of electrode mechanisms. Depending upon the application involved, one or more of these limitations might be important. For example, the study of electrode mechanisms requires the widest possible range of experimental variables and the simplest possible mathematical descriptions (which generally include mass transport, kinetic, and equilibrium terms). In developing analytical methods, reproducibility, sensitivity, simple or inexpensive instrumentation, automation, or a large accessible potential range may be the most important criteria. Because of these limitations in DC polarography, a variety of related polarographic techniques have been developed. The related techniques include experiments performed at the DME such as pulse polarography, differential pulse polarography, and AC polarography; those which employ diffusion to a stationary electrode such as cyclic voltammetry; and those which employ convective mass transfer such as stripping analysis and hydrodynamic and amperometric techniques.

In most instances, the initial electrochemical approach to a problem should include the collection of some basic screening information (as described later) using either DC polarography or voltammetry (when mercury is not suitable as an electrode material). Then, if necessary, one of the related polarographic techniques can be employed to gain additional information or to perform an analysis. The basic characteristics of these related methods will be described in this section, and references will be given to more extensive discussions.

A. Pulse Polarography

Pulse polarography is usually performed at the DME using conditions which give mass transfer by diffusion. Pulse polarography is most often used when additional sensitivity is necessary when compared to the DC polarographic case. However, it is not used as frequently as differential pulse polarography which requires similar instrumentation and usually has greater sensitivity than either DC or pulse polarography. Pulse polarography has the advantage of a particularly simple theory while maintaining the advantages of a renewable electrode surface. Thus, it can also be applied to mechanistic investigations.

In pulse polarography, the potential of the DME is held constant at the initial potential (E_1, potential at the start of the experiment) throughout most of the drop life. At some time just before the end of the drop, a short duration (typically 1 to 100 msec) voltage pulse is applied. The magnitude of the voltage pulse is increased with each successive drop so that their height increases at the same rate as a normal DC polarographic scan, i.e., several millivolts per second. The current is measured near the end of the voltage pulse. The pulse polarogram consists of the measured current plotted vs. the final potential (E_2). The applied voltage, the instantaneous current, and the pulse polarogram are schematically illustrated in Figure 8. In the electrochemical literature, no standard convention exists for designating the time intervals. In this chapter, t_1 designates the time from the start of the drop to the current measurement and t_2 the time from the start of the applied voltage pulse to the current measurement (Figure 8).

The current consists of the faradaic current due to the electrode reaction which occurs at the new potential, the charging current due to the change in electrode potential, and other residual current processes due to the presence of impurities. If the width of the voltage pulse is short enough so that the electrode area can be considered constant and the electrode reaction is rapid, then the magnitude of the faradaic current will be given by Equation 2 multiplied by the fraction of material which is reduced at that potential (see Equations 17 and 18). Thus, the faradaic current decreases with time within an individual drop but increases as the final (pulse) potential becomes more negative (for a reduction). It remains constant when the diffusion limited current region of the DC polarogram is reached. The diffusion limited current for the pulse polarographic experiment (i_{pp}, microamps) is given by

$$i_{pp} = n F A C^* \sqrt{D/\pi t_2}$$

$$= 463\, n\, D^{1/2}\, C^*\, m^{2/3}\, t_1^{2/3} / t_2^{1/2} \qquad (47)$$

where t_1 is related to the electrode area, t_2 is related to the faradaic current decay and the other symbols have their usual definition and units. Comparing Equation 47 to the Ilkovic equation (Equation 6), the diffusion limited current measured in the pulse polarographic experiment will be larger than the equivalent current measured in the DC polarographic experiment by the ratio

$$i_{pp}/i_d = \sqrt{3/7}\ \sqrt{t_1/t_2} \qquad (48)$$

where the first term results from the DC polarographic compression of the diffusion layer and the second term results from the different measurement times. This ratio has a range of about 2 to 20 for typical values of t_1 and t_2.

The charging current in the pulse polarographic experiment is given in Equation 13 where the area varies in the same manner as in DC polarography and dE/dt is, in principle, infinite at the leading edge of the applied pulse. The charging current will return to the value observed in a DC polarographic experiment as soon as enough charge has passed to bring the electrode to its new potential. The charging time is controlled by the product of the uncompensated solution resistance times the electrode integral capacitance, i.e., the cell time constant, but is usually much less than the pulse width for moderately conducting solutions (supporting electrolyte ionic strengths greater than about 0.1 *M*). Therefore, the charging current at the time of the current measurement will have decayed to approximately that observed in a DC polarogram at the same potential. Thus, the greatest sensitivity (greatest faradaic to charging current ratio) should be obtained at long drop times with the current measurement made as soon as the charging current decays to the DC value. These conclusions are supported by the calculations of Christie and Osteryoung[60] and by the increase in sensitivity observed for practical pulse polarographic methods (about 10^{-6} *M*) compared to the DC polarographic experiment (about 10^{-5} *M*).

Because the diffusion limited current is proportional to the drop area, it will be proportional to the product $m^{2/3} (t_1)^{2/3}$ or independent of the corrected mercury column height when the delay time remains constant. For a controlled drop time DME, it will be proportional to the two-thirds power of the corrected height when both t_1 and t_2 are constant. The dependence of the current on the electrode area is typical of the pulse polarographic experiment since the $(1/\sqrt{t})$ dependence is effectively removed by holding t_2 constant. When the current-time behavior after an individual potential pulse is examined, it is apparent that the current decays with time after application of the pulse so that the product $(i/\sqrt{t_2})$ is constant in time regions where the charging

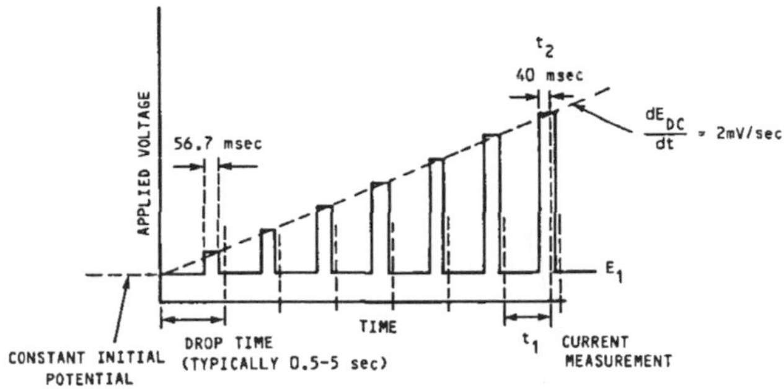

FIGURE 8A

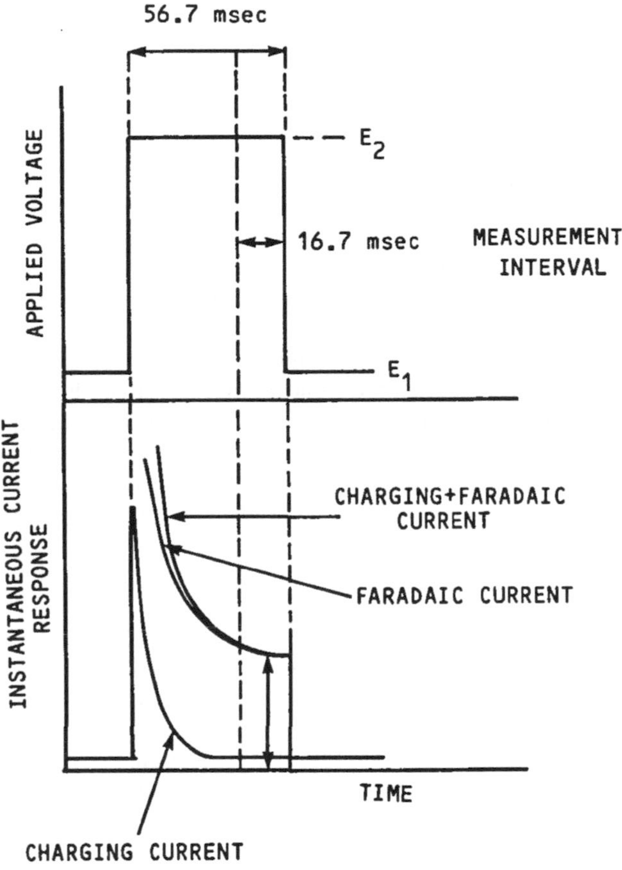

FIGURE 8B

FIGURE 8. Applied signal and response for pulse polarography, reversible electron transfer. (A) Applied signal indicating scan rate, E_1, E_2, drop time, t_1, and t_2. (B) Applied signal, faradaic, charging and total current for one pulse. (C) Pulse polarogram indicating limiting current and half-wave potential.

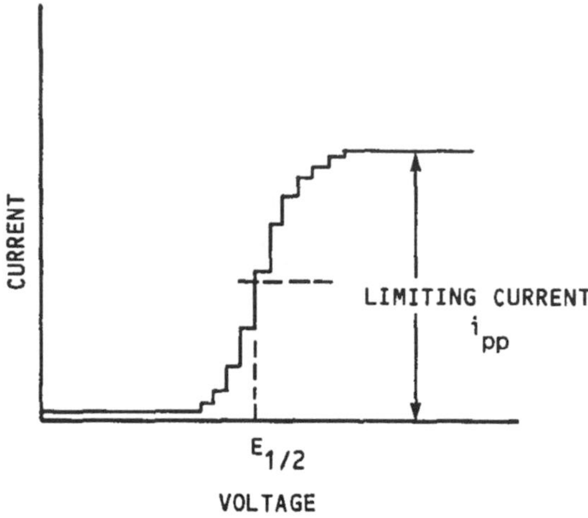

FIGURE 8C

current is insignificant. The current-time behavior is identical to that observed in a chronoamperometric experiment with application of a potential step, a technique with a well developed theory.[24,61] These current-height and current-time behaviors will occur with other mechanisms which exhibit diffusion limited behavior (e.g., at very negative potentials with the rate controlled electron transfer mechanism, with the EC mechanism, or with the first electron transfer of an ECE mechanism).

The shape and half-wave potential of the reversible polarogram obtained in pulse polarography is identical to that obtained in DC polarography. The resolution of overlapping waves will be identical to that of the DC case (for reversible electron transfers). When the electron transfer step is rate controlled, the effect will be similar to that observed in DC polarography, i.e., the half-wave potential will be dependent upon the delay time (t_2), the heterogeneous rate constant (k_s), and the electron transfer coefficient (alpha).[40]

The behavior of most mechanisms with coupled chemical reactions will qualitatively resemble those of the corresponding DC polarographic case. The most important differences between pulse polarography and DC polarography are: (1) the area of the electrode can be considered constant if t_2 is short, and (2) the diffusion layer remains in its initial state (no distortion of the depleted diffusion layer due to the growth of the mercury drop) until the application of the voltage pulse. Thus, the mathematical expressions describing the electrode behavior are simpler. The quantitative pulse polarographic response will be given by the behavior for the corresponding mechanism evaluated in a series of chronoamperometric experiments with a constant initial potential (E_1), variable final potential (E_2), and a single current measurement at a fixed time (t_2). The time scale for observing coupled chemical reactions in pulse polarography is the delay time, t_2.

The limiting currents for the EE mechanism will depart from the simple case only when the homogeneous redox reaction is important. The CE mechanism will exhibit wave heights controlled by the rate of conversion of the precursor to the electroactive species and the equilibrium concentration of the electroactive species. The EC mechanism will give a somewhat distorted wave shape but the height will be determined by the rate of diffusion. The simple ECE mechanism will give a total wave height corresponding to the range of n_1 to (n_1 + n_2) electrons depending upon t_2 and the rate constant of the chemical reaction. The current for the catalytic regeneration mechanism will depend upon the rate of the regeneration reaction and will decrease less slowly than the diffusion controlled case as t_2 is lengthened (eventually becoming constant if the concentration of Z is sufficiently large). Finally, the behavior of the adsorption wave will be somewhat different. When the reactant is adsorbed and the final potential is sufficiently negative to reduce the adsorbed material, a large current will flow immediately after the application of the voltage pulse.[61] This current will not be observed at the time of the current measurement (t_2) unless its duration is prolonged by instrumental limitations or high solution resistances. Thus, no postwave will be observed. If the product of the electrolysis reaction is adsorbed, an adsorption current will flow after the application of the pulse when E_2 corresponds to the DC polarographic prewave region. This current will exhibit the normal diffusion controlled decay until the electrode surface is covered; then, no more current will flow assuming that the electrode area is constant. This time (t_a, sec) is given by the relationship

$$t_a = (\pi/4) \times 10^{12}\, \Gamma_m^2 \,/\, (C^*)^2 D \qquad (49)$$

when the area is assumed to be constant. The value of t_a may be less than or greater than t_2. Thus, the magnitude (or appearance) of a prewave in pulse polarography depends upon the relative values of t_2 and t_a.

In all kinetic cases, the shifts of half-wave potentials and effects of disproportionation or reproportionation (recombination) will be similar to that previously described. These behaviors are summarized in Table 5. Additional information can be found in the electrochemical literature[40,57,60,62] or reviews.[11,63]

The implementation of pulse polarography requires several instrumental modifications of a DC polarographic instrument.[64-66] First, timing circuits are required to accurately reproduce the times t_1 and t_2. Because these times must be measured from the time of the fall of the previous mercury drop, the drop fall must be either sensed or (mechanically) controlled. A sample-and-hold amplifier generally is used to present a continuous signal to the recording device since the current measurement is made only once per drop. In addition to these requirements, the pulse polarographic instrument must be capable of responding to the more rapid potential changes and larger currents obtained with the application of the potential pulse. The theoretical pulse polarographic response assumes a constant pulse potential (E_2); however, some instruments produce a nonconstant voltage pulse which increases at the scan rate (several millivolts per second). The faradaic current recorded from these instruments will be essentially unaffected, but the charging current will contain a steady state contribution of the dE/dt term of Equation 13 (as well as the initial charging spike).

The sensitivity of the pulse polarographic experiment could be further increased if the charging current contribution could be eliminated, i.e., t_2 shortened without charging current contributions. One possible compensation method involves subtraction of a previously recorded polarogram obtained on a blank solution. Recently, an alternate drop mode was proposed to decrease the charging current.[67,68] In an alternate drop mode experiment, the waveform applied to the first drop is the normal pulse. Then, the final pulse potential (E_2) is held and applied to the second drop before the potential is returned to its initial value. At time t_1 for the second drop, the faradaic current will be equal to that of the DC polarographic experiment with the charging current contribution equal to that measured in the pulse polarographic experiment. Thus, if the two signals (measured at time t_2 on successive drops) are subtracted, the difference will be equal to the pulse polarographic faradaic current minus the DC polarographic faradaic current. Although the signal will be somewhat reduced in magnitude, the charging current will be almost completely eliminated and greater sensitivity will be observed. After the second drop, the potential is returned to its initial value (E_1) and the sequence repeated with a new final potential. Alternatively, the charging current measured be-

Table 5
PARAMETERS AND BEHAVIORS OF SELECTED ELECTRON TRANSFER MECHANISMS FOR PULSE POLAROGRAPHY, CYCLIC VOLTAMMETRY AND VOLTAMMETRY AT THE ROTATING DISK ELECTRODE

	Pulse polarography[a]	Cyclic voltammetry[b]	Voltammetry at RDE[c]
Experimental parameters	t_1, t_2, E_1	E_i, ν	ω
Response of reversible system (mass transport controlled)	$i_{pp} = n F A C^* \sqrt{D/\pi t_2}$	$i_p = 0.4463\, n F A C^* \sqrt{n F \nu D/RT}$	$i_L = 620\, n F A D^{2/3} C^* \nu^{-1/6} \omega^{1/2}$
	$E = E_{1/2} - (0.059/n) \log [i/(i_{pp} - i)]$	$i_{p,c} = i_{p,a}$	$E = E_{1/2} - (0.059/n) \log [i/(i_L - i)]$
	$E_{1/2} = E_{1/2,\,DC}$	$E_{p,c} = E_{1/2,DC} - 0.0285/n$	
	$i_{pp}/i_d = \sqrt{3/7}\ \sqrt{t_2/t_1}$	$i_p/i_d = 379\, (n^{1/2} A \nu^{1/2}) / (m^{2/3} t^{1/6})$	$i_L/i_d = 8.45 \times 10^4\, D^{1/6}(\nu^{-1/6}\omega^{1/2})/m^{2/3} t^{1/6})$
Rate controlled electron transfer	$i_{lim} = i_{pp}$	$i_{p,c} = 0.4958\, n F A C^* \sqrt{\alpha n_a F \nu D/RT}$	$i_{lim} = i_L$
	$E_{1/2} < (E_{1/2})_{rev}$	$E_{p,c} = E^o - (RT/\alpha n_a F)\,[0.780 + \ln \sqrt{\alpha n_a F \nu D/RT} - \ln k_s]$	$E_{1/2} < (E_{1/2})_{rev}$
EE[d]	(both waves): $i_{lim} = i_{pp}$	$i_{p,c} = i_p$ and $i_{p,c} = i_{p,a}$	$i_{lim} = i_L$
CE[d,e]	$i_{lim} = nFAC_o^* \sqrt{D_o/\pi t_2} + nFAC_A^* \sqrt{D_A k_1 K_{eq}} \exp(k_1 K_{eq} t_2) \operatorname{erfc}(\sqrt{k_1 K_{eq} t_2})$	$i_{p,c} < i_p$ and $i_{p,a} > i_{p,c}$	$i_{lim} = \dfrac{nFAD\,(C_o^* + C_A^*)}{1.61 D^{1/3} \nu^{1/6} \omega^{-1/2} + D^{1/2}/(K_{eq} \sqrt{k_1 + k_2})}$
		$E_{p,c} < (E_{p,c})_{rev}$	$E_{1/2} < (E_{1/2})_{rev}$
EC[d,e]	$i_{lim} = i_{pp}$	$i_p \leq i_{p,c} \leq (0.4958/0.4463)\, i_p$	$i_{lim} = i_L$
		$i_a < i_c$	
		$E_{p,c} > (E_{p,c})_{rev}$	$E_{1/2} = E^o + \frac{RT}{nF} \ln (1.61 D^{-1/6} \nu^{1/6} \omega^{1/2} k_1^{1/2})$
ECE[d,e]	$i^1_{lim} = i_{pp}$ for n_1	$i^1_{p,c} = i_p \quad i^1_{p,a} < i^1_{p,c}$	$i^1_{lim} = i_L$
	$i^2_{lim} = i_{pp} [1 - \exp(-k_1 t_2)]$	$i^2_{p,c} < i_p \quad i^2_{p,a} > i^2_{p,c}$	$i^2_{lim} = i_L [1 - \exp(-1.61^2 D^{-1/3} \nu^{1/3} k_1/\pi\omega)]$
Catalytic regeneration[d]	$i_{lim} = i_{pp} [\exp(-k'_1 t_2) + \sqrt{\pi k'_1 t_2} \operatorname{erf}(\sqrt{k'_1 t_2})$	$i_c = nFAC^* \sqrt{Dk'_1}$ for $(k'_1 RT/nF\nu) >> 1$	$i_{lim} = nFAC_o^* \sqrt{Dk'_1}$ for $k'_1 >> k_2$
	for $k'_1 >> k_2$	$i_a = i_c$	
Adsorption of product[d]	Prewave possible	Prepeak	(Not relevant for steady state experiments)

[a] $E_1 >> E_{1/2}$ (reduction); area assumed to be constant; i_{pp} is the diffusion limited current and i_{lim} is the observed limiting current.
[b] $E_i >> E_{1/2}$ (reduction); i_p is the diffusion limited current; c and a refer to observed values on cathodic and anodic scans, respectively.
[c] i_L is the mass transfer limited current; i_{lim} is the observed limiting current; Nernst diffusion layer current calculations.
[d] Reversible electron transfer process(es) assumed; $E^1_{1/2} >> E^2_{1/2}$; homogeneous redox reaction assumed to be absent.
[e] Irreversible chemical reaction assumed (except CE at RDE).

fore the application of the pulse in the normal pulse polarographic experiment can be subtracted. However, this approach is less effective since the charging current is obtained at a different electrode potential (E_1) than that used for the faradaic current measurement (E_2). Other reported techniques to reduce the charging current measurement (E_2). Other reported techniques to reduce the charging current contribution include the use of modified waveforms and/or current measurements made at multiple times[69,70] or the use of a constant final electrode potential with a variable initial potential (all current measurements are made at a fixed potential so that the charging current remains constant).[71]

In summary, the pulse polarographic experiment gives increased sensitivity compared to the DC polarographic experiment at the cost of some increase in complexity of instrumentation. Theoretical descriptions of behavior tend to be simplified compared to DC polarography. Compensation of the charging current can provide even greater sensitivity. Resolution and linearity are essentially unaffected from the DC experiment; precision at low concentrations may be improved due to the larger faradaic to charging current ratio.

B. Differential Pulse Polarography

Differential pulse polarography usually is performed at the DME, although theory and experimental data are available for differential pulse polarography (or differential pulse voltammetry) at a stationary, planar electrode.[72,73] These experiments are performed under conditions where mass transfer occurs by diffusion. Differential pulse polarography is most often employed for quantitative analysis, particularly at low levels. The waveform is well suited for easy measurement of the important parameters. It also can be combined with anodic stripping to produce a differential pulse anodic stripping technique.

In differential pulse polarography at the DME, the potential is scanned in the same manner as for DC polarography. However, a pulse is superimposed on the linear scan at a time just before the end of the drop. The magnitude of the pulse (ΔE) is typically between 5 and 100 mV and its width between 1 and 100 msec. The current is measured at two times: just before the application of the pulse (t_3) and near the end of the pulse (t_1). The differential pulse polarogram consists of the difference in current (current at time t_1 minus current at time t_3) displayed vs. the electrode potential before the voltage pulse is applied (E_1). The applied potential and the resulting instantaneous current and polarogram are schematically illustrated in Figure 9. Although no standard convention exists for specifying the time intervals, t_1 and t_2 are defined as before and t_3 is added to specify the time of the first current measurement (Figure 9).

The current at time t_3 is equal to the DC polarographic current which would be observed at potential E_1, i.e., the faradaic and residual currents normally observed at that time in the drop life. After the application of the pulse, the current consists of a new faradaic current and a new residual current for the pulse potential ($E_1 + \Delta E$).

For a reversible electrode reaction, the increase in faradaic current at the new potential ($E_1 + \Delta E$) can be approximated by the difference in currents which would be observed in two pulse polarographic experiments with initial potentials of E_1, current measurements at time t_2, and a final potential of ($E_1 + \Delta E$) for the first pulse and a final potential of E_1 for the second pulse (i.e., the current which would have been measured at time t_2 if no pulse were applied).[60] Intuitively, this current difference must be zero in the potential region where no electrochemical reaction occurs ($E_1 \gg E_{1/2}$) and in the potential region where all the material reaching the electrode is reduced ($E_1 \ll E_{1/2}$) such that an additional potential step does not change the current. The current (ΔI_{DPP}, microamps) for a diffusion controlled electrode reaction at a constant area, planar electrode is approximately given by

$$\Delta i_{DPP} = n F A C^* \sqrt{D/\pi t_2}\ [1/(1+\theta_1) - 1/(1+\theta_2)]$$

$$= 463\, n\, D^{1/2} C^* m^{2/3} (t_1^{2/3}/t_2^{1/2})(\theta_1 - \theta_2)/[(1+\theta_1)(1+\theta_2)] \quad (50)$$

where

$$\theta_i = \exp[(n F/RT)(E - E^i_{1/2})] \quad (51)$$

and the other parameters have their normal definitions and units. When the differential pulse polarographic current is differentiated with respect to potential and the result set equal to zero, the maximum current (Δi_{max}, microamps) is found to be

$$\Delta i_{max} = 463\, n\, D^{1/2} C^* m^{2/3} (t_1^{2/3}/t_2^{1/2}) [(1-\sigma)/(1+\sigma)] \quad (52)$$

where

$$\sigma = \sqrt{\theta_2/\theta_1} = \exp[(nF/RT)(\Delta E)/2] \quad (53)$$

and ΔE is a signed number (negative for a cathodic step). The differential pulse polarographic peak potential (E_p) is related to the DC polarographic half-wave potential by

$$E_p = E_{1/2} - (\Delta E)/2 \quad (54)$$

From these equations, it can be observed that the current in differential pulse polarography is equal to that in pulse polarography times the term containing the pulse height. This term exhibits values between zero and one so that the maximum current is normally larger than that for the DC polarographic diffusion limited current but less than that for the pulse polarographic diffusion limited current.

It was later realized[60,67] that Equations 50 and 52 should contain a correction term which corresponds to the increase in current which would be observed due to the growth in area of the mercury drop between times t_3 and t_1, i.e., the difference between the Ilkovic equation evaluated for times t_3 and t_1 at the potential E_1. This minor correction results in a negligible shift of the peak potential, a slight increase in the calculated peak current (less than 5%) and a baseline shift (towards the peak current) for the more negative baseline segment of a diffusion controlled reduction process when typical experimental parameters are assumed.

The charging current observed after the application of the voltage pulse is similar to that described for the pulse polarographic experiment, i.e., a large spike which decays to the normal DC polarographic residual current at a rate limited by the cell time constant. Thus, if the second current measurement is made at a time when the charging current spike has decayed, the difference in residual currents will consist of the difference of the two DC charging currents evaluated from Equation 13 for times t_2 and t_3 and the corresponding potentials ($E_1 + \Delta E$ and E_1, respectively). Since the area and potential changes are very small, the charging current contribution to the measured current will be very small. This characteristic is primarily responsible for the increased sensitivity of the differential pulse polarographic method compared to the pulse polarographic method. That is, the signal-to-noise ratio is increased by decreasing the charging current and a practical sensitivity limit of about 10^{-7} *M* is available with detectable signals for concentrations less than 10^{-9} *M*.

Because the approximate maximum current is proportional to the drop area, it will be independent of the mercury column height for a constant delay time (t_2). For a controlled drop time DME, it will be proportional to the two-thirds power of the cor-

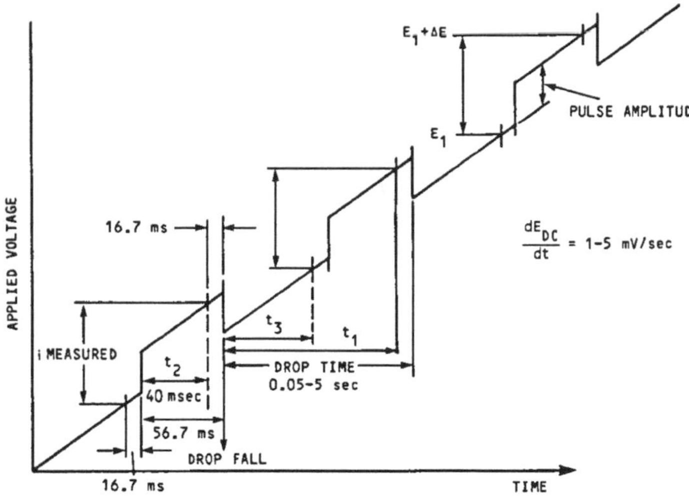

FIGURE 9A

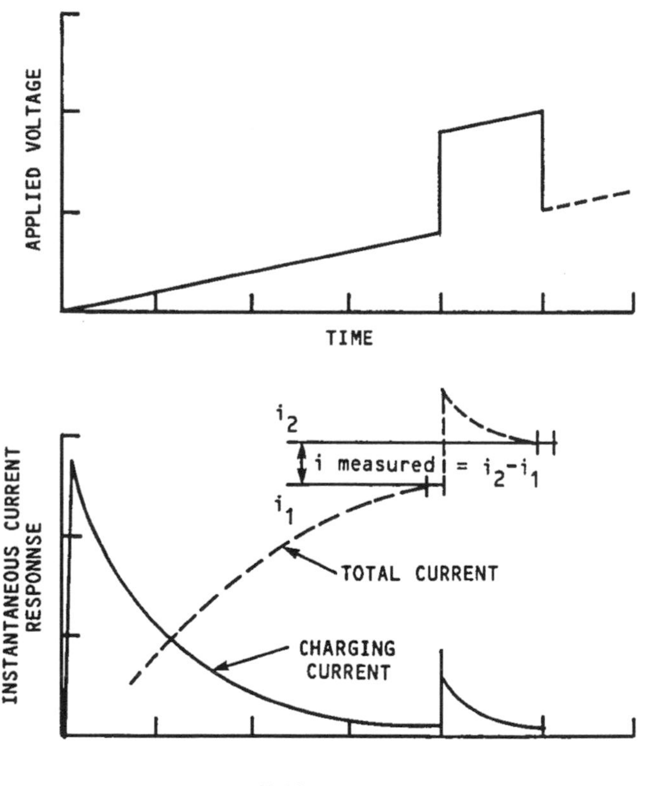

FIGURE 9B

FIGURE 9. Applied signal and response for differential pulse polarography, reversible electron transfer. (A) Applied signal indicating scan rate, E_1, ΔE, drop time, t_1, t_2, and t_3. (B) Applied signal, charging and total current for one pulse. (C) Differential pulse polarogram indicating peak current and peak potential with a DC polarogram for comparison of potentials.

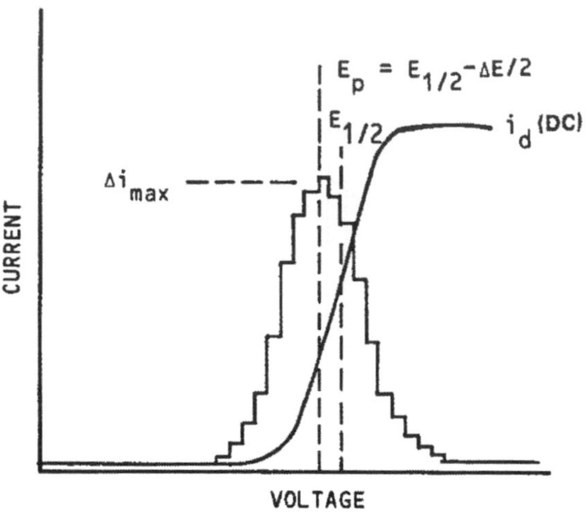

FIGURE 9C

rected height when both t_1 and t_2 are constant. The more exact form of the differential pulse polarographic current has a more complicated dependence on the capillary characteristics.

Equations 50 to 54 predict a symmetrical peak (for a reversible electrode reaction and equal diffusion coefficients) centered at potential E_p. E_p is shifted positively (for a negative potential pulse) from the corresponding DC polarographic half-wave potential. The peak width is also dependent upon the potential step size, with minimum half-widths of 90.4, 45.2, and 30.1 mV for 1, 2, or 3 electron steps, respectively, and an infinitesimal potential step.[62] For potential steps larger than about 10 to 30 mV (i.e., an appreciable fraction of the limiting half-width), the peaks are broadened considerably. The limiting peak width measured when the current is 10% of its maximum value is about 180/n mV and increases an additional 120/n mV for each decade decrease in current. The separation needed for no overlap at the 1% level is 300/n mV compared to only 240/n mV needed for separation in DC polarography. However, if baseline separation is not needed, but it is required that the first peak contribute less than 1% of its Δi_{DPP} at the potential of the second peak, then the minimum separation drops to 150/n mV for differential pulse polarography. As Parry and Osteryoung state,[62] differential pulse polarography cannot completely resolve peaks which are separated by amounts less than these values although such claims have been made.

The behavior of the rate controlled electron transfer mechanism in a differential pulse polarographic experiment has been described as part of a general digital simulation treatment.[74] As the standard heterogeneous rate constant (k_s) becomes smaller, the wave broadens, decreases in height and shifts negatively (for a reduction). The transition range from reversible behavior to almost completely irreversible behavior covers from about 10^{-1} to 10^{-4} cm/sec, respectively, for typical experimental parameters. The transfer coefficient affects the symmetry of the wave. The theoretical behavior for other mechanisms is not as well developed. Generally, the qualitative effects will be similar to those predicted for the difference of the pulse polarographic response evaluated at potential ($E_1 + \Delta E$) with time t_2 and area proportional to t_1 vs. that evaluated at potential E_1 with time t_3 and area proportional to time t_3. Adsorption, however, produces unusual behavior.[75] Differential pulse polarographic waves are frequently affected by adsorption. The peak heights may be either increased or decreased depending upon the electrode charge, the surfactant charge and the charge of the electroactive species. In addition, anomalous current peaks are observed in potential regions where adsorption or desorption of surfactants occur without any transfer of electrons across the electrode-solution interface (i.e., no DC polarographic wave is observed). These anomalous peaks are called tensammetric peaks. Surfactants also can affect the residual current in those potential regions where they are adsorbed. Additional information can be found in the electrochemical literature and reviews.[4,11,57,62,63]

The instrumental requirements for the implementation of differential pulse polarography are a mixture of those required for DC polarography and pulse polarography.[64-66] The time intervals, t_1, t_2, and t_3 must be controlled in a reproducible manner. The voltage pulse is normally stepwise selectable (e.g., 5, 10, 25, 50, or 100 mV) but remains constant throughout an experiment and is superimposed on the DC voltage ramp. The current measurements must be held, subtracted, and presented to a recorder. Because the charging currents are less in differential pulse polarography than in many other techniques, the inherent sensitivity is higher. In order to achieve the theoretical sensitivity limits, the instrumentation must be designed to minimize noise while still providing a signal which accurately reflects the electrode process. One common commercial instrument (Model 174A, EG&G Princeton Applied Research®, Princeton, N.J.) tends to distort differential pulse polarographic signals[64] because of the large data acquisition time constant employed to suppress noise. The modification of this instrument to achieve additional flexibility and new capabilities while removing the distortion was recently described by the same group.[65] There also is an apparently previously unreported gain change in the Model 174A when this instrument is used in the differential pulse polarographic mode (i.e., the actual currents are ten times smaller than the indicated currents).[76] Another paper[77] describes the use of an integrator to subtract the current measured before the pulse from that measured after the pulse to improve the signal-to-noise ratio. Other, more general instrumental techniques for noise reduction can be found in the instrumentation section.

The sensitivity of differential pulse polarography could be improved even more if the residual charging current could be reduced. As mentioned previously, the charging current remaining in the differential pulse polarographic experiment is due to the change in area and the change in potential of the electrode between the two measurements.[60] This residual charging current can be minimized by making the current measurements at long times (t_1 and t_3) when the area is changing less rapidly and by using a moderate value for the pulse potential. Large pulse potentials tend to increase the charging current more than the faradaic current, whereas pulse potentials that are too small result in currents which are low in absolute magnitude.[60] Other modifications proposed to minimize this problem include: (1) the use of alternate drop, differential pulse polarography in much the same manner as proposed for pulse polarography,[67,68] (2) measurement of the current at more than one time after the pulse has been applied and then mathematically correcting for the change in charging current,[70] and, (3) the use of multiple voltage pulses to increase the current, at least for reversible electrode reactions.[69]

In summary, the differential pulse polarographic experiment gives increased sensitivity compared to the DC and the pulse polarographic experiments. The instrumentation is somewhat more complicated and theoretical descriptions of behavior are considerably more complicated. The normal compensation of charging current and the special modifications available to further decrease this noise source are responsible for the great sensitivity of the technique. Resolution of overlapping waves is much better than for the DC polarographic experiment but not as good as has been claimed in the past. Linearity may be affected by adsorption of either the electroactive species or a surfactant present in the solution. Precision at low concentrations may be better than for other techniques due to the increased faradaic to charging current ratio.

C. Alternating Current (AC) Polarography

In AC polarography, a small amplitude alternating current signal is superimposed on a slowly changing DC potential applied to the DME. The applied signal produces a response which consists of both a DC and an AC current. The AC current signal contains both amplitude and phase (relative to the applied voltage) information. An AC polarogram is a plot of this information vs. the DC potential of the electrode. In some cases, a phase selected AC component will be measured.

The applied AC voltage is typically about 10 mV in amplitude with a frequency of about 100 Hz and the DC potential scan rate is several millivolts per second. The AC response to this signal can be predicted by a consideration of the electrode polarization process at a constant DC potential. As the applied AC signal becomes more negative, a greater fraction of the material is reduced than would be reduced at the nominal DC signal level so that the current flow is increased. As the AC signal reverses direction, the material which was previously reduced can be reoxidized when the electrode reaction is reversible and the electrogenerated material is still present (i.e., has not reacted via a coupled chemical reaction). The reoxidation produces an anodic current component superimposed on the DC current. Therefore, the applied AC voltage produces a corresponding AC modulation of the current. As one would expect, the magnitude of the AC response varies from zero at positive potentials where no electrode reaction occurs (for a reduction), to a maximum signal where the applied AC signal produces the greatest change in solution concentrations (measured at the electrode surface), and then returns to zero at sufficiently negative potentials where all the material reaching the electrode is reduced independent of the AC signal. If kinetic processes are present, the relative magnitude (and phase shift) of the response will vary with the frequency of the applied signal. The applied signal, the AC and DC responses, and the AC polarogram are illustrated in Figure 10. The AC polarogram is illustrated for a normal current mode, i.e., the AC current is measured continuously. When the sampled mode is used, the AC current will be measured only at the end of the drop life to produce a polarogram which does not contain the oscillations due to drop growth and fall.

The AC faradaic polarographic current-potential relationship for a diffusion controlled electrode reaction in response to a small amplitude AC applied voltage (ΔE, mV peak-to-peak) is given by

$$I(\omega t) = \frac{0.001\, n^2\, F^2\, A\, C^* \sqrt{\omega D}\, \Delta E \sin(\omega t + \pi/4)}{4RT \cosh^2 [(E_{DC} - E_{1/2})(nF/RT)]} \tag{55}$$

where $I(\omega t)$ is the instantaneous current (microamps) which is shifted 45° in phase from the applied voltage, ω is the angular frequency (radians per second), C^* is the bulk concentration (m*M*), A is the electrode area (cm²), E_{DC} is the DC component of the applied voltage, and $E_{1/2}$ is the DC polarographic half-wave potential. The maximum response occurs when the argument of the hyperbolic cosine function is zero, i.e., at the DC polarographic half-wave potential. The magnitude of the maximum AC current response ($I_{P,AC}$, microamps) is

$$\begin{aligned} I_{P,AC} &= 0.001\, n^2\, F^2\, A\, C^* \sqrt{\omega D}\, \Delta E/(4RT) \\ &= 7.99\, n^2\, D^{1/2}\, C^*\, m^{2/3}\, t^{2/3}\, \Delta E\, \omega^{1/2} \end{aligned} \tag{56}$$

where the terms have the same definitions as given above. Comparing Equation 56 to the Ilkovic equation, the ratio of the diffusion limited AC current to the diffusion limited DC current is given by

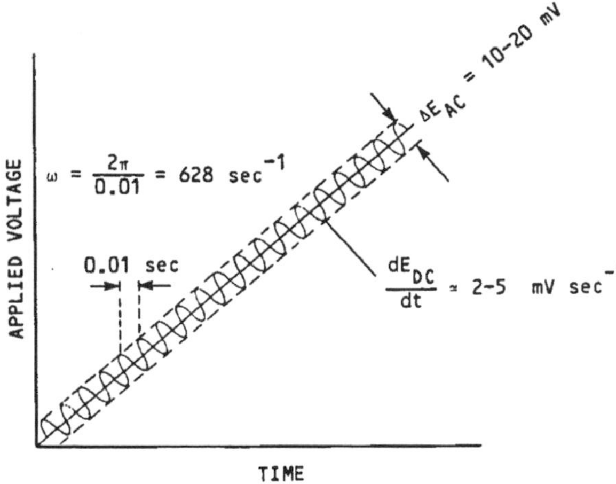

FIGURE 10A

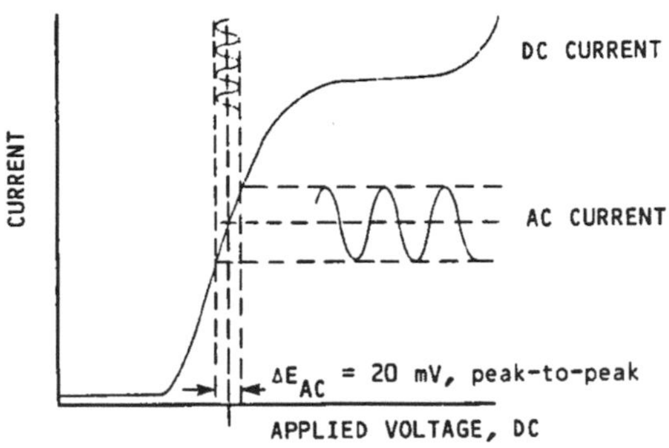

FIGURE 10B

FIGURE 10. Applied signal and response for AC polarography, reversible electron transfer. (A) Applied signal indicating DC and AC components. (B) AC current. (C) AC polarogram indicating peak potential and peak current.

$$I_{P,AC}/i_d = (0.001)\, n\, F\, \Delta E \omega^{1/2} \sqrt{3\pi t/7}\ /\ (4RT)$$

$$= 0.01128\, n\, (\Delta E \omega^{1/2})\, t^{1/2} \qquad (57)$$

This ratio is unity for typical values of the experimental parameters at a frequency of about 25 radians per second and increases at higher frequencies. Thus, the AC polarographic method has a comparable or slightly greater sensitivity (about 10^{-5} *M* for practical methods with a detection level of about 10^{-6} to 10^{-7} *M*) than DC polarography for the reversible electron transfer reaction.

The charging current in the AC polarographic experiment is a function of the drop growth (DC charging current) and of the applied voltage (AC charging current). The AC charging current is proportional to the frequency of the applied signal, so that the magnitude of the AC faradaic current decreases relative to the AC charging current as the frequency increases. However, the charging current is a capacitive current which will be shifted 90° in phase with respect to the applied AC voltage. Since the faradaic

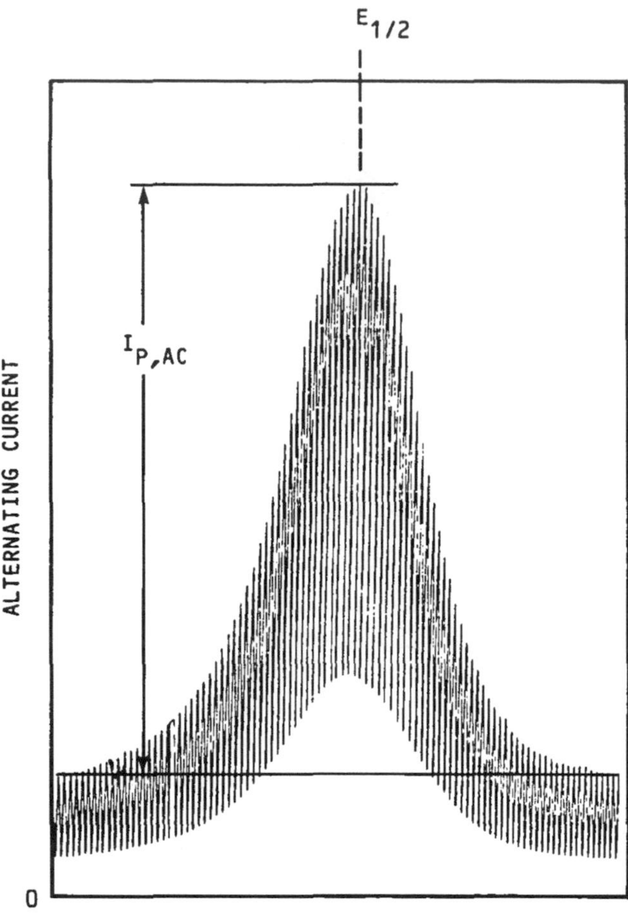

FIGURE 10C

current is shifted only 45° (for a reversible electrode reaction), it is possible to discriminate between these two signals with a phase sensitive detection system. When the electrode reaction is not diffusion controlled (high frequencies and/or slow kinetic steps), the phase shift of the faradaic signal will generally not be 45° and usually will be a function of the frequency. The complex phase behavior of the signal in these cases limits the effectiveness of phase sensitive detection. However, the phase information of the signal can be used to establish the kinetic characteristics of the electrode reaction.

Several behaviors are characteristic of a diffusion controlled electron-transfer reaction. First, the magnitude of the AC current is directly dependent upon the electrode area. Therefore, it will be independent of the mercury column height for a naturally dropping DME and vary with the two-thirds power of the corrected height and the two-thirds power of the drop time for a controlled drop time DME. Second, the peak potential will be equal to the DC polarographic half-wave potential and the peak width at half-height will be 90/n mV, both independent of the frequency of the applied signal. Third, the faradaic current will be proportional to the square root of the applied frequency with a zero intercept and its phase shift will be 45° independent of the frequency and of the DC potential. Finally, the log plot will be linear with a slope of 118/n mV with the wave shape given by the equation

$$E = E_{1/2} - (0.118/n) \log [\sqrt{I_P/I} - \sqrt{(I_P - I)/I}] \qquad (58)$$

Equation 58 can be used to predict the overlap of two waves. This function has almost the same shape as that for differential pulse polarography and results in a required peak potential separation of about 308/n mV to achieve a 1% baseline separation of the waves. If it is only required that the first peak contribute 1% of its current at the peak potential of the second wave, a separation of about 154/n mV is needed, somewhat better than the 240/n mV required for DC or pulse polarography and approximately equal to the 150/n mV required for differential pulse polarography.

In general, the AC polarographic current will be decreased in magnitude when either the initial reactant is not available for electrolysis (e.g., a CE mechanism) or when the electrolysis product cannot be reoxidized (e.g., an irreversible electron-transfer process or product removal by a coupled chemical reaction). The relevant time scale for the observation of these effects is dependent upon the frequency of the applied signal. Since the observation time scale can assume a wide range of values (frequencies from about 1 to 10^4 sec^{-1}), AC polarography is a versatile tool for the investigation of electrode mechanisms. For example, an electron transfer which appears to be reversible at low frequencies may appear to be totally irreversible at high frequencies. Although the predicted behaviors of various mechanisms seem simple, the mathematical descriptions are quite complex. Also, the experimentally observed behavior can be quite confusing due to the presence of charging current, adsorption processes (adsorption of electroactive species and tensammetric processes), instrumental limitations, etc. Thus, the reader is referred to more extensive reviews of the AC polarographic electron transfer mechanisms.[17,18,78,79] The electrochemical literature contains many recent papers and reviews[4,11,12,57] on AC polarography which include theory, digital simulation of behavior for specific mechanisms,[42,46,47,80] and analytical applications.

The analytical applications of AC polarography have usually involved cases where the selectivity (resolution) or sensitivity offered some advantages. In most other applications, the experimental difficulties tend to limit its routine application.

Because the instrument must respond to DC signals and AC signals without introducing distortion or destroying the phase information, the AC polarograph can be rather complicated. If phase selective detection or measurements of harmonic or derivative signals are used, the instrumental requirements are even more severe. The reviews by Smith[18] and Sluyters-Rehbach and Sluyters[79] discuss some of these problems and their solution. Smith also has presented the use of on-line digital computers for experimental control and data analysis in AC polarography.[81]

In summary, the large time scale accessible with AC polarography has made it a valuable tool in the investigation of electrode mechanisms. Alternating current polarography has potentially increased sensitivity over DC polarography and increased resolution, leading to some analytical applications. However, the theoretical descriptions of the process are complicated and usually must be solved by numerical or simulation techniques. In addition, AC polarography is subject to several experimental and instrumental complications.

D. Triangular Wave Polarography and Cyclic Voltammetry

A number of related techniques use a rapidly changing voltage ramp or two consecutive ramps in the form of a triangle as the applied voltage signal. When the signal is applied to a DME within the lifetime of a single drop, it is a polarographic technique; if the signal is applied to a stationary electrode, it is a voltammetric technique. These techniques have been named in various ways in the electrochemical literature, but the suggested names,[19] some of the variations in names and the description of these techniques follow.

1. Single sweep polarography, dropping electrode chronoamperometry with linear potential sweep, single sweep oscillographic polarography, cathode ray polarography: linear voltage scan applied to a single drop at the DME

2. Triangular wave polarography: same as single sweep polarography except that the applied signal is a complete triangle
3. Cyclic triangular wave polarography: same as single sweep polarography except that the applied signal consists of multiple triangular waves
4. Multisweep polarography: same as single sweep polarography except that the applied signal consists of multiple single sweeps separated by a nearly instantaneous return to the initial potential
5. Chronamperometry with linear potential sweep (LVP), linear sweep voltammetry (LSV), stationary electrode polarography (SEP): same as single sweep polarography except that a stationary electrode is used
6. Triangular wave voltammetry, cyclic voltammetry (CV): same as chronoamperometry with linear potential sweep except that the applied signal is a complete triangle
7. Cyclic triangular wave voltammetry: same as chronoamperometry with linear potential sweep except that the applied signal consists of multiple triangular waves
8. Staircase voltammetry: same as triangular wave voltammetry except that the applied signal is a series of discrete steps rather than a linear sweep

The principal difference in the measured response between the polarographic techniques and the voltammetric techniques is due to the change in the DME area with time. The response is quite similar qualitatively and becomes identical in the limit of sufficiently rapid signals so that the electrode area can be considered constant. When a complete triangular signal is used, the reverse scan measures the reoxidation of the product produced on the forward scan (for an initial reduction process). Multiple waveform methods are simple extensions of the triangular scan methods which may include the detection of species produced during the first scan which were not present in the original solution. Thus, single sweep polarography and triangular waveform methods are subsets of the multiple waveform methods and the polarographic techniques are closely related to the voltammetric techniques. In view of the many similarities, the following discussion will be limited to the technique commonly known as cyclic voltammetry (triangular wave voltammetry) which has a more thoroughly investigated theoretical basis than most of the other techniques.

Cyclic voltammetry can be performed at any stationary electrode, although the hanging mercury drop electrode is a common choice. The hanging mercury drop electrode is a single mercury drop (area typically 0.05 cm^2) suspended from a syringe type mechanism (Kemula electrode) or from a platinum wire sealed into the end of a glass tube. Characteristics of other materials which can be used to construct stationary electrodes are described in the next section (Section IV.E.).

In cyclic voltammetry, the potential of the electrode is varied from an initial potential (E_i) to a switching potential (E_r), then the direction of the potential scan is reversed and the electrode potential is scanned back to the initial potential. The current observed during this experiment (for a reduction process) is initially zero when $E_i \gg E_{1/2}$. The surface concentration ratio for a reversible or rapid electron transfer process is related to the electrode potential through the Nernst equation (Equation 15). Therefore, as the electrode potential becomes more negative, the surface concentration of the electroactive (oxidized) species is decreased and the current increases. Eventually, the surface concentration of the electroactive species reaches zero indicating that the solution layer surrounding the electrode has become depleted. At this point, the current becomes limited by the rate of diffusion of the electroactive material from the bulk of the solution. This sequence of events produces a maximum or peak current which then decays with the characteristic ($1 / \sqrt{t}$) dependence for diffusion limited processes. The applied voltage and observed voltammogram are schematically illustrated in Figure 11.

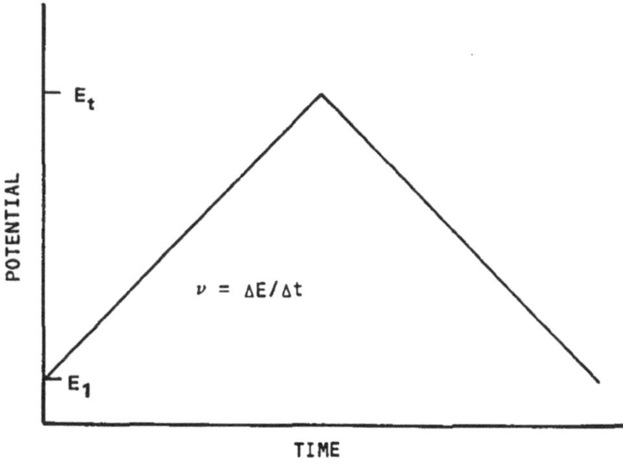

FIGURE 11A

FIGURE 11. Applied signal and response for cyclic voltammetry, reversible electron transfer. (A) Applied signal indicating initial (E_i) and switching (E_r) potentials. (B) Cyclic voltammogram indicating peak potentials ($E_{p,c}$ and $E_{p,a}$), half-wave potential ($E_{1/2}$), half-peak potential ($E_{p/2,c}$), and peak currents ($i_{p,c}$ and $i_{p,a}$) with $i_{p,a}$ measured from the extrapolated cathodic current baseline.

The faradaic current-potential relationship for this process is complex, but it can be described using finite difference methods or integral equations or digital simulation methods. The peak current (i_p, microamps) for the reversible electron transfer case can be given by the simple relationship

$$i_p = 0.4463 \text{ n F A C}^* \sqrt{\text{nF } \nu \text{D / RT}}$$
$$= 2.687 \times 10^5 \text{ n}^{3/2} \text{ A D}^{1/2} \text{ C}^* \nu^{1/2} \qquad (59)$$

where A is the electrode area (cm^2), C* is the concentration of the electroactive species (m*M*), ν is the potential scan rate (V/sec) and the other constants are defined in the normal manner. The constant 0.4463 is the value of a potential dependent current function at the peak potential, whereas the other terms are independent of the potential. Equation 59 predicts that the faradaic current is directly proportional to the square root of the scan rate and to the three-halves power of the number of electrons transferred.

Comparing Equation 59 with the Ilkovic equation (Equation 6), the diffusion limited peak current measured in the cyclic voltammetric experiment will be larger than the diffusion limited current measured for a DC polarographic experiment by the ratio

$$i_p / i_d = 379 \text{ n}^{1/2} (\text{A } \nu^{1/2}) / (\text{m}^{2/3} \text{ t}^{1/6}) \qquad (60)$$

where A and ν refer to the cyclic voltammetric experiment and m and t refer to the DC polarographic experiment. For normal values of the parameters (n, 2; A, 0.05 cm^2; m, 2 mg/sec; t, 3 sec), the peak current will be about ($14\sqrt{\nu}$) times the polarographic limiting current. The scan rates for cyclic voltammetry can vary from about 0.01 to 1000 V/sec, limited at low scan rates by the onset of convection and (at high scan rates) by charging currents and instrument characteristics. Thus, the cyclic voltammetric peak current will be slightly larger than the polarographic limiting current at low

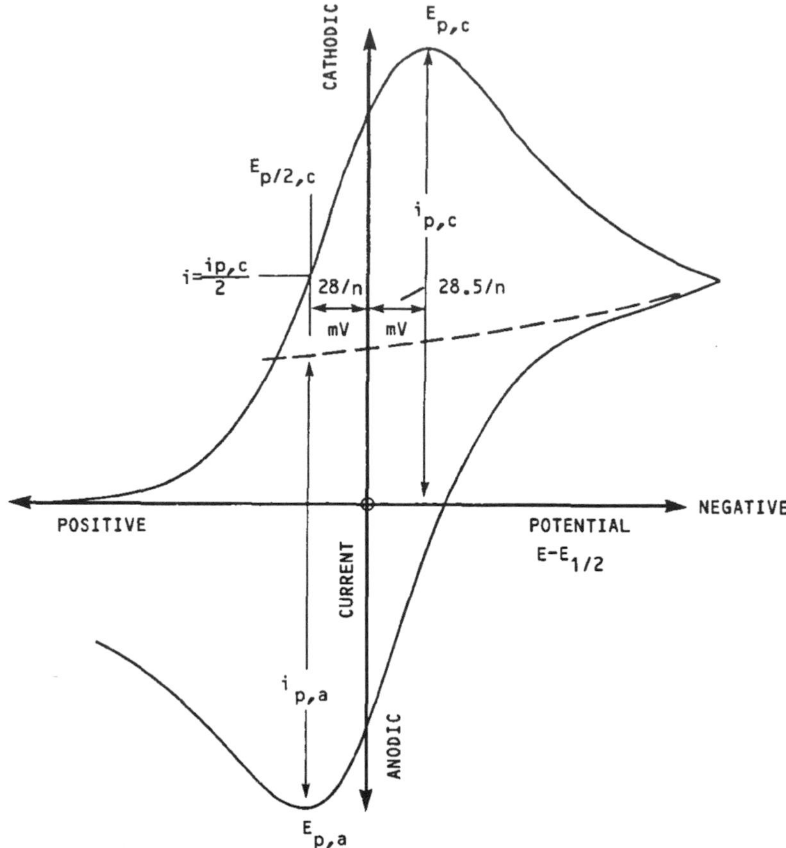

FIGURE 11B

scan rates and several hundred times larger at high scan rates. The practical sensitivity for analytical methods is about 10^{-6} *M* and the detection level is about 10^{-8} *M*.

When Equation 13 is applied to a constant area electrode, the charging current is predicted to be proportional to the rate of change of the potential. Thus, the faradaic current in a cyclic voltammetric experiment increases more slowly with scan rate ($\nu^{1/2}$) than does the charging current (ν). As the scan rate is increased, an optimum value will be reached when the signal is large enough to measure but not appreciably distorted by the charging current. The charging current in a cyclic voltammetric experiment produces two effects. One is the simple baseline offset which can be easily subtracted during the measurement of the current. The second charging current effect is more severe since the effective potential applied to the electrode is offset by the variable ohmic (resistive) potential drop. That is, the linear potential scan is distorted by a voltage drop proportional to the current. This distortion tends to shift the reduction peak to more negative potentials, lower the value of the observed peak current, and broaden or distort the shape of the wave.[82,83] This distortion is more severe in a cyclic voltammetric experiment than in DC polarography due to the larger currents, particularly at high scan rates. The distortion introduced by the charging current can be partially compensated by using various cell design or positive feedback techniques.

The reversible electron transfer in cyclic voltammetry will exhibit several other characteristics in addition to its dependence on the square root of scan rate. The peak current (and peak shape) for the anodic process (reverse scan) will be exactly equivalent to that of the cathodic peak when the extrapolated diffusion controlled decay from the cathodic peak is used as the baseline for the measurement of the anodic peak. The peak potential of the forward scan will be 28.50/n mV negative of the polarographic half-wave potential and 56.50/n mV negative of the half-peak potential (potential where the current is one-half of its maximum value), both independent of the scan rate. When the switching potential is more than about 300/n mV negative of the half-wave potential, the anodic peak potential will be 57/n mV positive of the cathodic peak potential or the anodic half-peak potential will be 56/n mV negative of the cathodic half-peak potential within an error of less than 1 mV.

The behavior of other electrode mechanisms in cyclic voltammetry can be qualitatively predicted by a consideration of the processes involved. The quantitative descriptions for these cases are available in review articles[3,4,11,12,17,24,57,84] as well as in the original research articles, particularly in a series of papers from Shain and associates (see for example, references 85, 86). Adams has extensively applied this technique to oxidation reactions at stationary solid electrodes.[14-16] Some of the characteristics of various electrode mechanisms are summarized in Table 5. For example, the peak current for a totally irreversible rate controlled electron transfer step is given by an equation which has the same form as Equation 59, but in which the current function constant is replaced by 0.4958 and the parameter ($n^{3/2}$) is replaced by ($\alpha n_a \sqrt{n}$). In the case of the preceding chemical reaction (CE), the cathodic current is decreased by an amount which depends upon the relative rates of the voltage scan and the chemical reaction and the peak potential is shifted negative of the normal value (for a reduction) analogous to the negative shift observed in the DC polarographic half-wave potential. In the reverse scan of a CE mechanism, the anodic peak current is larger than the cathodic peak current since some extra conversion occurs in the time interval required to scan back to the anodic peak (i.e., a larger chemical reaction time elapses for the material being reoxidized than for the material being reduced). The adsorption mechanism produces prepeaks when the electrolysis product is strongly adsorbed and postpeaks when the reactant is strongly adsorbed. When a system exhibits more than one wave and the waves are separated by more than about 180/n mV, the diffusion controlled decay of the first wave can be used as a baseline for the measurement of the second wave. If the potential is held at a constant value negative of the peak potential, then the current for the first peak will decay and allow the second peak to be more easily measured. This process was implemented with real time computer control in one series of investigations.[87] The extremely large accessible time scale (more than five orders of magnitude) allows discrimination between mechanisms and evaluation of kinetic parameters from the behavior at different scan rates.

The cyclic voltammetric peak current can be used for quantitative measurements. However, it is directly dependent upon the electrode area which is difficult to reproduce with extreme precision. In addition, the surface of a solid electrode can become contaminated (which produces distorted peaks) since it is renewed only periodically. Thus, the principal analytical application has been in trace analysis where lower precision can be tolerated for the advantage of increased sensitivity. Cyclic voltammetry also is used for quantitative measurements in the variant known as stripping analysis. Triangular wave polarographic techniques which have the same general characteristics as cyclic voltammetry but which are performed at the reproducible dropping mercury electrode have been used extensively for quantitation. The use of high scan rates leads to increased sensitivity compared to DC polarography; but the presence of charging current and the lack of accurate, high-speed recording devices (until digital data acquisition becomes more generally available) tends to limit precision. At large scan rates, unsuspected kinetic processes may become important and either decrease precision or introduce nonlinearity in the calibration curve.

The instrumentation for cyclic voltammetry is basically similar to that described for DC polarography except that higher scan rates are used and the measured currents are larger. In practice, this means that the potentiostat characteristics and cell design are much more important, especially for operation at high scan rates. Many results for investigations of instrument and electrochemical cell responses are given in the application section of this chapter. The staircase modification of the cyclic voltammetric experiment is easy to implement with digital logic circuits; it also tends to reduce the charging current problems. The theory for staircase voltammetry of reversible and irreversible electron-transfer mechanisms[88-90] and for multiple potential steps[73] at stationary electrodes is available.

In summary, cyclic voltammetry is an excellent tool for the investigation of electrode mechanisms, but it is not often used for routine quantitation. Triangular wave polarographic techniques, which use the DME, have been applied in analytical methods. Both types of techniques offer increased sensitivity and a much greater range of accessible time scales compared to DC polarography.

E. Hydrodynamic and Amperometric Techniques

Hydrodynamic electrochemical techniques utilize convective mass transfer to a stationary, solid electrode to determine the current-potential relationships. When the flow is laminar or nonturbulent, there is a quiescent layer of solution adjacent to the electrode. Therefore, these techniques really involve a combination of convection and diffusion. Hydrodynamic techniques include tubular electrodes in which the electrolysis solution flows down or around the tube, packed bed electrolysis cells in which the electrolysis solution flows over and around a porous, sometimes particulate electrode, and various geometric shapes of electrodes over which the electrolysis solution is passed. In hydrodynamic techniques with a homogeneous solution, the current assumes a steady state value which can easily be measured. In principle, a wide range of concentrations can be determined because the size of the electrode can be adjusted almost indefinitely and the solution flow rate can be varied. In practice, the processes which contribute to the residual current may vary in the same manner as the desired faradaic current so that gains in sensitivity may be limited to a certain range of experimental parameters. Also, the rate at which the current reaches its equilibrium value tends to decrease as the solution concentration decreases making it impossible to indefinitely extend the lower detection limit by increasing the size of the electrode. However, these types of electrolysis cells have found applications in certain areas, most notably in synthetic electrochemistry (due to the large amounts of material which can be accommodated) and in monitoring process control streams or trace environmental pollutants (due to their minimal instrumental requirements and ease of application). The electrode materials available, instrumentation, and practical problems are very similar to those discussed below for amperometric techniques. No further discussion of hydrodynamic techniques which use a stationary electrode will be presented in this chapter, but several reviews containing this information are available.[15,16,91,92]

Amperometric techniques which employ convective mass transfer to a solid, nonstationary electrode are frequently classified with hydrodynamic techniques. Like those techniques which utilize convective mass transfer to a stationary electrode, the limiting current of amperometric techniques is controlled by a combination of convection and diffusion when the system is described by a laminar flow model. Amperometric techniques include the rotating disk electrode (RDE), the rotating ring-disk electrode (RRDE), rotating wire electrodes, vibrating wire electrodes, etc. The remainder of this section will be limited to the RDE and RRDE geometries.

The RDE technique consists of measuring the current produced as a function of potential applied to a rotating electrode. The electrode configuration is most frequently a disk attached to a shaft which contains the electrical contacts and which can be rotated. The applied potential function is similar to that in DC polarography, i.e., a voltage scan at a rate of several millivolts per second. The information obtained from this experiment is analogous to that obtained from DC polarography. The principal difference consists in the choice of the electrode material and the changes which result from this choice. The RDE is generally limited to solid materials, although mercury plated metals (such as gold or platinum) and chemically modified surfaces (including the attachment of enzymes) have been reported.[93] The various electrode materials allow investigations of electrode processes which occur outside of the usable potential range of mercury. However, the solid electrode surface is continually exposed to the solution environment whereas the DME is constantly being renewed. Thus, contamination of solid electrodes and effects of surface films (oxides, adsorbed material, etc.) are much more important than in DC polarography.

In the RDE experiment, the current is independent of time if other parameters are kept constant and if the solution composition is not changed by electrolysis of appreciable fractions of the electroactive material. In this case, the mass transfer limited current (i_L, microamps) is given by the Levich equation[94]

$$\begin{aligned} i_L &= 620\, n\, F\, A\, D^{2/3}\, C^*\, \nu^{-1/6}\, \omega^{1/2} \\ &= 5.98 \times 10^7\, n\, A\, D^{2/3}\, C^*\, \nu^{-1/6}\, \omega^{1/2} \end{aligned} \qquad (61)$$

where A is the geometric area of the electrode (cm^2), D is the diffusion coefficient (cm^2/sec), C^* is the concentration of the electroactive species in the bulk of the solution (m*M*), ν is the kinematic viscosity (cm^2/sec), ω is the angular frequency (radians/sec) and the other terms have their usual definitions and units. The Levich equation assumes that the flow of solution at the edges of the electrode is not disturbed, i.e., no edge effects are present. This is usually achieved experimentally by surrounding the active electrode surface with an inert, nonconducting material which influences the mass transfer process but does not change the electrochemical process. Several corrections have been proposed for the Levich equation; applying these corrections to typical aqueous solutions leads to values of 602 and 5.81×10^7 for the constants in Equation 61.[16] Newman has provided a comprehensive theoretical description of the mass transport problem for this geometry.[91]

The kinematic viscosity has a value of about 0.01 cm^2/sec for most aqueous solutions.[16] The values of ω available are restricted to those which produce laminar flow. Laminar flow is obtained for rotational rates which have a Reynolds number ($r^2\omega/\nu$) less than about 10^5, which corresponds to ω values less than about 1000/sec for a 1 cm electrode radius (the radius, r, refers to the total surface of the rotating electrode, not just the active electrode). Comparing the Levich equation (Equation 61) with the Ilkovic equation (Equation 6), it is apparent that much higher currents (about 1000-fold greater for typical experimental parameters) are obtained at a RDE with an area equal to that of the mercury drop of a DME. This disparity results from the much greater efficiency of transport by convection compared to diffusion. The ratio of the mass transport limited current at the RDE to the diffusion limited current at the DME is given by

$$i_L/i_d = 720\, D^{1/6}\, (A\nu^{-1/6}\, \omega^{1/2}) / (0.008515\, m^{2/3}\, t^{2/3}) \qquad (62)$$

where A is the area of the RDE electrode and ($0.008515\ m^{2/3}\ t^{2/3}$) is the area of the DME. The practical concentration range for RDE voltammetry is about 10^{-2} to 10^{-6} *M* with a detection level of about 10^{-8} *M*. Thus, the increased efficiency of mass transport results in a sensitivity about the same as pulse polarography.

According to the Levich equation (Equation 61), the mass transfer limited current should be directly proportional to the electrode area and proportional to the square root of the rotational rate. Systematic variation of the rotational rate is the most commonly used test for mass transfer limited currents. When the current is controlled by the rate of mass transfer, the current-potential relationship for the RDE experiment is identical in shape to the polarogram obtained in DC polarography. Thus, a plot of the quantity log $[i/(i_L - i)]$ vs. potential should result in a straight line with a slope of 59/n mV. This behavior, tested at various concentrations of electroactive species and rotational rates, can also be applied to test for mass transfer limited currents. The half-wave potential should be independent of the rotation rate when the current is controlled by the mass transport process. Unfortunately, the half-wave potential observed in an RDE experiment cannot be simply related to the DC polarographic half-wave potential because of various thermodynamic and kinetic effects involving the electrode and its surface.

The behavior of the limiting current with various electron transfer mechanisms is quite similar to that for DC polarography. However, the time scale available for the observation of kinetic events is somewhat broadened by the larger range possible for rotation speeds compared to the range of drop times. For example, the current-potential relationship for the totally irreversible electron transfer consists of two terms, the first corresponds to mass transfer and a second term which predominates when the electron transfer process is slow. The equation for this wave is

$$i = n F A D C^* / [1.612 D^{1/3} \nu^{1/6} \omega^{-1/2} + (D/k_f)] \qquad (63)$$

The totally irreversible electron transfer ($k_s < 10^{-4}$ cm/sec) produces a constant current independent of the rotation rate, diffusion coefficient or solution viscosity. However, the rate of the electrode reaction (k_f) can be increased by changing the electrode potential to a more negative value (for a reduction). Values of the rate constant greater than about 0.1 cm/sec produce mass transfer limited behavior. In mechanisms which involve coupled chemical reactions, the current also generally consists of a mass transfer term which dominates at slow rotation rates or with rapid kinetics and a second term which dominates at fast rotation rates or with slow kinetics. The simplicity and steady state character of the RDE leads to relatively simple mathematical descriptions of the experimental behavior. The electrochemical literature[57,93] and several reviews[15,16,58,95-98] contain mathematical descriptions of many kinetic models of the RDE experiment as well as discussions of edge effects, nonuniform current flow and responses for electrode shapes other than the rotating disk. The behavior of the limiting current at the RDE with several electron transfer mechanisms is given in Table 5.

The practical aspects of the RDE experiment have been summarized by Adams,[15] Pierkarski and Adams,[16] and others.[10,97,98] Many important practical aspects are connected with the electrode itself (shape, material, rotation, etc.). The most common electrode shape is a circular disk attached to a connecting rod. This disk must be planar, perpendicular to, and concentric with the rotational axis to achieve laminar flow at the highest rotational rates. However, it has recently been demonstrated that a rather highly eccentric electrode produces currents which are almost identical to those achieved with a concentric electrode within those rotation rates where the flow remains laminar.[99]

Many electrode materials are available for the RDE experiment, some of which were previously mentioned. Other common materials include platinum, gold, graphite, pyrolytic graphite, carbon paste, and boron carbide. Platinum exhibits oxide film formation with a complicated behavior at positive potentials and adsorption of evolved hydrogen at negative potentials. Gold responds in a similar manner at positive potentials, but does not exhibit hydrogen adsorption. Untreated graphite electrodes are generally porous; thus, a wax impregnation technique may be used to prevent penetration of the electrolysis solution. Pyrolytic graphite is less porous, but is more difficult to obtain in a reproducible form. It also exhibits a narrower available potential range and is considerably more expensive. Carbon paste electrodes are easily constructed from inexpensive materials, powdered graphite, and a binding agent such as Nujol®. However, the binding agent can be removed by the solvent action of the electrolyte. Very low residual currents are reported for carbon paste electrodes. Boron carbide and similar materials have been investigated because of their chemical inertness and the wide potential range available; however, they have not been employed routinely in chemical analysis.

The precision and accuracy of the limiting currents obtained at a RDE are largely unreported. However, precision can be affected by processes which change the electrode surface in a significant manner (adsorption, oxide film formation, etc.). The limited information suggests relatively good precision when these interfering processes do not vary with time. Before the acceptance of the RDE, rotating wire electrodes were occasionally used for quantitation. The results generally were marginal to poor for rotating wire electrodes (mass transfer is not as reproducible as with the rotating disk) unless it was used as an indicator electrode in an amperometric titration or a similar application where a change in response was measured rather than the absolute response.

Instrumentally, the RDE experiment imposes very few requirements. The currents tend to be larger than those in the DC polarographic experiment so that a higher output potentiostat and greater attention to the potential control characteristics of the cell are required. Response times are generally limited by the electrode reaction and the solution hydrodynamics. Thus, additional filtering or averaging may easily be used to reduce the noise without degradation of the signal. The most important instrumental limitation concerns the reproducible rotation of the electrode. The rotation unit must be able to rotate the electrode without introducing wobble or eccentricity. It is also important to maintain a constant rotation speed, preferably variable over a rather wide range. Several commercial units are available and numerous custom constructed units have been reported. The units with the best characteristics incorporate a measurement of the rotation rate with a feedback signal designed to maintain the desired speed.

The rotating ring-disk electrode (RRDE) experiment is used primarily for investigations of electrode mechanisms. In this technique, the disk electrode is surrounded with an electrically isolated ring electrode which has separate potential control and current-measuring circuits. When the electrode assembly is rotated, the product of the electrochemical reaction at the disk electrode is hydrodynamically transported (radially and tangentially) across the electrode surface where it may encounter the ring electrode. Since some of the electrolysis product diffuses into the surrounding solution, the theoretical collection efficiency (ratio of ring current to disk current, usually between 0.1 and 0.5) must be calculated from the geometrical configuration (disk radius, ring radii, and separation distance) or experimentally determined with a known electrochemical system.

The ring electrode can be used to monitor the products of the disk electrolysis reaction. The information obtained at the ring may consist of a voltammogram to identify the species present and a quantitative determination of the amount of material reaching the ring compared to the amount expected. For example, if the ring is used to reoxidize the product of an EC reduction process which occurs at the disk, the decrease in the ring current is related to the amount of material consumed by the chemical reaction which involves the electrolysis product. The fraction of material which appears at the ring will be a function of the delay time between generation and detection. Since the rotation rate is easily variable, and the RRDE geometry can be controlled during electrode construction, a wide range of time scales is accessible, approaching the range exhibited by chronoamperometry and cyclic voltammetric experiments. Other mechanistic applications involve experiments in which the ring electrode potential is maintained at a constant value and the disk electrode potential is scanned. Additional information on the application of the RRDE is available.[16,57,84,93,97,98] Prater has reviewed the use of digital simulation for the calculation of the complicated theoretical response of the RRDE for various kinetic mechanisms.[100]

In summary, hydrodynamic techniques are available for a wide variety of conditions. They are especially useful when mercury is not a suitable electrode material, i.e., for many oxidation reactions. The RDE experiment is widely employed for the investigation of electrochemical behavior and can be used for quantitation. The RRDE is primarily used for investigations of electrochemical mechanisms and evaluations of kinetic parameters.

F. Stripping Analysis

Stripping analysis is a combination of several electrochemical technqiues. In this method, the substance to be determined is preconcentrated by deposition into or onto an electrode. Then, the deposited material is measured during a removal or stripping step. Stripping analysis is useful for determinations at concentrations so low that they cannot be directly measured with the techniques previously discussed. The lower detection limit can be extended to less than 10^{-9} *M* with appropriate experimental procedures.

Various electrode materials have been used for stripping analysis. A hanging mercury drop electrode is one of the most common choices. Inert electrodes such as platinum or gold as well as mercury plated electrodes have been used also. The material deposited may dissolve into a mercury electrode (such as zinc or lead), form intermetallic compounds, form surface deposits (metals deposited onto inert electrodes), or form precipitates (halides deposited onto mercury or silver electrodes). The first three processes usually involve reductions of metallic ions; the latter, oxidation of the electrode to form a salt with the anion in solution.

Several processes produce nonideal behavior during the stripping step. Intermetallic compound formation may occur between metals simultaneously deposited (e.g., Cu-Cd, Cu-Zn, Ni-Zn, or Ni-Sn) at high concentrations (>1 m*M* in an amalgam or surface deposits on a solid electrode) or between the deposited metal and the electrode material (e.g., mercury with Co, Fe, Ni, and Mn or platinum with Sb, Sn, or Zn). The stabilization due to intermetallic compound formation may be so great that the material cannot be removed from the electrode or it may exhibit several waves corresponding to different chemical species. The nature of the electrode may be significantly altered from that of the original surface when amalgam concentrations are very high, when intermetallic compounds containing mercury are formed or when surface films are formed. Thus, a variety of electrolysis behaviors are possible.

The preconcentration or deposition step is usually performed at a constant electrode potential chosen to achieve some degree of selectivity while being negative enough (for a reduction) to maintain the equilibrium between the concentrated amalgam and the dilute solution. A potential quite negative of the polarographic half-wave potential is required to maintain this equilibrium (see Equation 15 in which C_o is the amalgam concentration). The maximum degree of preconcentration (concentration ratio for Equation 15) is reached when all of the material is removed from the solution (large volume, low concentration) and dissolved in the electrode (small volume, large concentration).

During the deposition step, material is transported to the electrode by convection. If the stirring rate is uniform and the solution is not depleted in the electroactive species, then the amount of material deposited will be proportional to the deposition time. At very low concentrations, significant fractions of the electroactive material may be removed. In either situation, the most common experimental arrangement involves analysis of standard or spiked samples to determine the amount of material deposited (i.e., to obtain a calibration curve).

The removal or stripping step can consist of several controlled potential or controlled current methods; chronoamperometry with a linearly varying potential (LVP) and differential pulse polarography (DPP) are frequently used. If the deposited material forms an ideal solution with the electrode, then the removal is controlled by the rate of diffusion of the oxidized species in the mercury electrode. In this case, the peak current (LVP or DPP) is directly related to the original solution concentration. When intermetallic compounds, surface films, or surface deposits are involved, these relationships are more complex. In particular, when surface films of precipitated salts or deposited metals are encountered, it is necessary to integrate the total charge consumed during the stripping step. Additional experimental details and theoretical relationships[101-103] and examples[10,57,93] are available.

G. Miscellaneous Techniques

Several additional variations of the polarographic technique might be encountered occasionally. These techniques will be mentioned with sufficient description to recognize their mode of operation and primary usefulness.

Square Wave Polarography[10,18,104]— This technique is equivalent to differential pulse polarography with a symmetrical waveform; but square wave polarography preceded the development of differential pulse polarography. The current is measured after the charging current has decayed to the residual value. The polarogram has a similar peak shape as differential pulse polarography with an increased sensitivity compared to DC polarography. The detection level is about 10^{-8} *M*, approximately the same as that found for pulse polarography or cyclic voltammetry. Square-wave polarography has been successfully used in analytical methods, especially trace analysis where its sensitivity is an advantage.

Radio Frequency (RF) Polarography[18,104]— This technique utilizes a small amplitude RF signal (100 KHz to 6.4 MHz) modulated by a low frequency (225 Hz) square wave superimposed on the DC voltage scan. The low frequency DC rectification signal is measured. Detection levels of 10^{-7} to 10^{-8} *M* can be achieved and analytical applications have been reported. Sine wave modulation and large amplitude faradaic rectification methods have been investigated also.

Faradaic Impedance[18,105]— This technique measured the cell impedance (resistance and capacitance) as a function of the frequency of the applied AC signal and the magnitude of the DC potential. Complex plane analysis may be used to separate the individual current components. The information content of this experiment is essentially identical to that obtained in AC polarography. Faradaic impedance measurements have been used almost exclusively for mechanistic and theoretical investigations.

Oscillopolarography[10,11]— This technique is actually a controlled AC current method in which the rate of change of the electrode potential (dE/dt) is measured vs. the DC potential. Variation of the current magnitude and frequency is possible. Few analytical applications have been reported since the range of linear response with concentration is small and the precision is quite poor.

V. APPLICATION OF PRINCIPLES

Polarographic methods of analysis can be divided into direct and indirect methods. In a direct method, the substance being determined is electroactive itself and gives the wave being measured. In an indirect method, the substance to be determined can be converted into an electroactive species (or a species electroactive in a more convenient potential range), the substance can be reacted with an excess of an electroactive species (followed by polarographic determination of the excess amount of reactant), or the substance can be determined by an effect on an electroactive species (adsorption analysis or kinetic and catalytic methods). Alternatively, polarography can be used to detect the equivalence point of a titration (amperometric titration). Meites,[10] Zuman,[7] Eberson and Nyberg[6] and Mairanovskii[41] have reported some general considerations for indirect analysis and amperometric titrations. Siggia,[106] Eberson and Nyberg,[6] Brezina and Zuman,[107] Kolthoff and Lingane,[108] Elving,[2] Hoffmann and Volke,[5] Smyth[22] and Brooks[23] give specific examples for indirect analyses and amperometric titrations of several classes of organic compounds. The analysis of drugs in body fluids or other cases with special problems in sensitivity, specificity, lack of better methods, etc., might require the use of indirect methods.[9,23,25,107] However, most practical routine polarographic analyses of drug substances involve direct methods while avoiding extensive sample preparation or reactions. Therefore this review will concentrate on direct polarographic methods of analysis.

Like other analytical methods, polarographic analyses should exhibit high selectivity or specificity, high accuracy and good precision in order to achieve maximum usefulness. The parameters affecting these qualities should be defined during the assay development process. Selectivity and accuracy are dependent upon the impurities and degradation products present. Accuracy and precision are influenced by the relative shape of the polarographic wave(s).

The best polarographic assay results are obtained from a well defined polarographic wave. A well defined wave has a limiting current plateau which extends over a large enough potential range to be easily measured and which is nearly parallel to the residual current curve obtained in the same potential range. Generally, maxima should be absent or small enough so that the wave shape is not grossly distorted. Also, the wave of interest must be separated from other electrochemical processes involving impurities, the electrolyte or the electrode. The analytical chemist, therefore, must define or identify a suitable set of assay conditions by performing preliminary screening experiments.

A. Screening

Basic polarographic screening should elucidate solubility, stability of prepared solutions, effects of impurities or degradation products, and the characteristics of the wave in different electrolytes and under various conditions. As the supporting electrolyte is changed, the accessible potential range will vary as well as the characteristics of the wave of interest. Observations of the polarographic behavior should provide some basic information on the electrochemical processes which occur, the range of assay validity, and the chemical variables which must be controlled.

A solubility in the millimolar range is required for analysis of a substance, although lesser concentrations which give currents above about 1 microamp (a few tenths millimolar) can be used with only minor loss in precision. Frequently, the determination of an impurity does not need to be as precise as the determination of the main component in a sample, and thus an impurity concentration of about 0.01 m*M* is sufficient in many cases. Even at these impurity concentrations, the sample must exhibit solubility greater than 1 m*M* in order to measure low levels ($< 1\%$) of the impurity in it.

Several alternatives are available when the aqueous solubility is not sufficiently large. First, aqueous solubility usually is greatly dependent upon pH for compounds containing ionizable functional groups. Second, high concentrations of sulfuric, acetic, or formic acids will increase solubility of some organic compounds.[109] Finally, mixed solvents or nonaqueous solvents may be used to increase solubility. Simple modifications include the addition of solvents such as methanol or ethanol to aqueous solutions. Solubility is more likely to be increased with the use of acetonitrile, dimethylformamide, or dimethylsulfoxide.[1,10,28,109,110] These latter solvents also are used to extend the accessible negative potential range (as explained in the next section) and to investigate electrochemical mechanisms under aprotic conditions. However, it must be emphasized that it is extremely difficult to perform routine assays when highly purified solvents, rigidly dried solvents (water contents in the millimolar range), or highly volatile solvents are required.

As was previously stated, hydrogen ions may be involved in the electrode reaction, and the electrolysis mechanism may be dependent upon the supporting electrolyte composition. Thus, the initial investigations should use buffered solutions to avoid artifacts in wave shapes or half-wave potentials caused by pH variations in the solution near the electrode. Second, the general nature of the reduction or oxidation wave should be observed in strongly acidic (0.1 *M* sulfuric acid), weakly acidic (acetate, citrate or phosphate buffers), neutral (phosphate buffers), mildly basic (phosphate, carbonate, borate, or ammonia buffers), and strongly basic (0.1 *M* sodium hydroxide) solutions.

The number of waves, their relative heights, and their separation from each other and from electrolysis of the electrolyte or electrode should be observed. If a polarographic wave cannot be shifted away from the electrolyte decomposition wave, then other supporting electrolytes which give a larger accessible potential range should be investigated. Preliminary assay conditions can be optimized by investigating the behavior on a finer pH grid and by determination of the effect of minor variations in the electrolyte composition (e.g., change of cation or ionic strength), the effect of impurities or degradation products, the stability of prepared solutions, the factors which govern the limiting current, and the degree of reversibility of the electrode reaction.

To achieve selectivity, the polarographic waves for the main component and any impurities must be sufficiently separated to allow the measurement of the desired wave without interference from the other wave(s). When the presence of impurities are known or suspected, they should be screened simultaneously with the main compound. When the impurity has the same electroactive group as the main component, it may be impossible to find conditions which give sufficient resolution between the two species. In such cases, other analytical techniques may be better than polarography. When the impurity has a different electroactive group, changes in pH, proton donor availability, etc. may be sufficient to separate the waves. In the intermediate case, where the compounds have the same electroactive group but differ in structure sufficiently to cause a small shift in half-wave potentials, other polarographic techniques with better resolution (differential pulse polarography, AC polarography, or derivative techniques, for example) may be better than DC polarography.

To insure assay validity, the stability of prepared solutions must be investigated unless relevant data are available from other sources. Even then, it is prudent to test the sample under the specific conditions which will be used during the assay. Time and exposure to light are the most common variables which should be investigated after the supporting electrolyte is chosen. For example, samples prepared for automated instruments may be prepared many hours before the actual assay unless the apparatus[31] is designed to dissolve samples immediately prior to analysis.

When the limiting current is determined by diffusion controlled processes, it is directly related to concentration. Tests for diffusion control and the relationships between limiting currents and concentration for nondiffusion controlled mechanisms have been discussed previously. It is wise to test for a diffusion controlled limiting current at the minimum and maximum concentration which will be employed for the assay. In addition, the linearity of the limiting current-concentration plot should be directly checked over at least a tenfold concentration range. The temperature dependence of the limiting current will suggest if thermostated cells are required to minimize assay variability.

The shape of the polarographic wave and the effect of changes in concentration, mercury height, or drop time indicate whether the electron transfer mechanism is reversible and indicate the separation of half-wave potentials required to distinguish an impurity. The shift in half-wave potential with pH can be used to detect the involvement of hydrogen ions in the electrode reaction (see Equation 23). When maxima are observed, it may be useful to determine the effects of maxima suppressors on the shape of the polarographic waves. In extreme cases, the fundamental features of the wave (half-wave potential and limiting current) may not be measurable in the presence of maxima.

Finally, actual polarograms of solutions which contain the main component and various amounts of impurities should be obtained. This procedure will establish the limits within which the assay is valid (not affected by the impurity) and the range of concentrations over which the impurity can be measured. The accuracy and precision of the assay should be determined under the expected operating conditions (instrument, analyst, purity levels, etc.). The rather complex subject of precision and accuracy is discussed separately.

B. Supporting Electrolytes

In addition to providing adequate solubility for the compound being analyzed, the electrolyte must have characteristics or perform functions such as:

1. Provide a satisfactory potential range to observe and quantitate the desired electrochemical reactions
2. Carry the current through the solution, thereby minimizing the migration of electroactive species
3. Have sufficient conductivity to allow accurate potential control and measurement
4. Buffer nonelectroactive species, particularly hydrogen ions, involved in the electrode reaction

Thus, the electrolyte must consist of a solvent containing an inert salt which will prevent migration, give the necessary conductivity, and perhaps act as a buffer while not limiting the available potential range. At the same time, the solvent and electrolyte should not introduce unwanted impurities. Purified water (distilled is preferred over ion exchanged which may contain trace organic material), high quality solvents, and pure inert salts are required.

Electrolytes can be generally classified as aqueous or nonaqueous, although combinations of organic solvents and water are sometimes used. Within nonaqueous solvents, distinctions are made between protic and aprotic solvents and between those which have high, moderate or low dielectric constants.[109] Protic solvents include acids (sulfuric, acetic, or hydrofluoric), neutral solvents (alcohols and water), and basic solvents (ammonia and amines). Common aprotic solvents include acetonitrile, dimethylformamide, *N*-methylpyrrolidone, dimethylsulfoxide, pyridine, hexamethylphosphoramide, propylene carbonate, and sulfolane. Changes in hydrogen ion activity can cause dramatic changes in the electrochemical behavior including the relative ease of reduction of different electroactive groups, reactivity of protonated intermediates versus the nonprotonated species, changes in relative rates of competing chemical and electrochemical reactions, changes in relative amounts of tautomers, or changes in the extent of adsorption.[8,108]

Solvents with high dielectric constants (>60), such as water, sulfuric acid, *N*-methylamide, formamide and propylene carbonate dissolve and dissociate salts best to form the supporting electrolyte. However, those with moderate dielectric constants (>20) include some of the most frequently used organic solvents such as acetonitrile, dimethylformamide, and dimethylsulfoxide as well as methanol, ammonia, hexamethylphosphoramide, and sulfolane. Other good chemical solvents which have low dielectric constants, such as hydrocarbons, halogenated hydrocarbons, and ethers, are infrequently used in polarography because they fail to dissolve sufficient quantities of ionic species for the supporting electrolyte. Acetic acid, with a low dielectric constant, is more frequently used. Listings of solvent properties, salts sufficiently soluble for use as supporting electrolytes, compatible reference electrodes, potential ranges and purification procedures are given by Mann,[110] Lund and Iverson,[109] Eberson and Nyberg,[6] and Badoz-Lambling and Cauquis.[27] Some of the differences between aqueous and essentially water free, nonaqueous electrolytes include:[5,6,8,10,23,27,28,109,110]

1. Water is nontoxic and easily obtained in pure form. Nonaqueous solvents may be toxic, difficult to purify, volatile, etc. Residual impurities in nonaqueous solvents may limit sensitivity.
2. Most organic species, especially nonpolar or slightly polar compounds, exhibit much greater solubility in nonaqueous rather than aqueous solvents.
3. Many salts can be used to prepare aqueous electrolytes. In nonaqueous solvents, only a few salts are sufficiently soluble, and these may form ion pairs and/or change the electrode process.
4. Nonaqueous solvents tend to have a larger accessible potential range.
5. It is difficult to remove all traces of water (to millimolar or lower concentrations), to measure the residual water content (the Karl Fisher titration is reliable to only about millimolar water concentrations), and to prevent water pickup in completely nonaqueous solvents. Millimolar water content corresponds to only about 0.002% water on a weight basis.
6. Oxygen removal is more difficult in nonaqueous media, and oxygen is known to react with radical anions (radical anions are not found in aqueous solutions).
7. Many reference electrodes are suitable for direct use in aqueous systems (SCE, silver/silver chloride, etc.). However, the half-wave potentials obtained in nonaqueous systems may contain errors due to liquid junction potentials (when salt bridges to aqueous reference electrodes are used) or may not be easily extrapolated to the common reference electrodes.
8. The overall electron transfer process in nonaqueous, aprotic solvents is generally less complicated (frequently a reversible one-electron transfer to the radical anion or radical cation) and more easily correlated with theory. Electrochemical processes in aqueous solutions or protic solvents are complicated by rapid protonations of radical anions and cations. In aprotic solvents, the protonation can be controlled by the selective addition of known proton donors, usually weak organic acids such as phenol, benzoic acid, salicyclic acid, or acetic acid (in increasing proton donating strength).
9. Much of the polarographic data has been obtained in aqueous solutions. Thus, electrochemical data for related substances is more likely to be available for aqueous rather than nonaqueous media.
10. Buffered electrolytes are more easily prepared in water than in nonaqueous solvents.

Some of the problems associated with nonaqueous solvents are minimized when appreciable amounts of water are present (compared to the completely anhydrous system). As a general rule, aqueous solutions are preferred for routine assays if solubility permits and the potential range is sufficient. Otherwise, partially or totally nonaqueous electrolytes can be used.

The range of accessible potentials is determined by the electrode, the solvent, and by the salts used in the supporting electrolyte. Mercury is easily oxidized, especially in the presence of species which form complexes or precipitates with it (e.g., halides, pseudohalides, hydroxide, thiols, etc.). The use of other salts such as acetates, sulfates, nitrates, or perchlorates permits the greatest positive potential range. The maximum positive range is about +0.15 V vs. SCE. The hydrogen ion is more easily reduced than many other cations. Therefore, the use of basic solutions, when compound stability and solubility permit, extends the accessible negative potential range. Also, substitution of lithium or tetramethylammonium cations for the more common, but more easily reduced sodium and potassium ions extends the negative range to about −2.7 V vs. SCE in water. Alternatively, salts of these two species may be used when pH buffering is not needed. Organic solvents such as acetonitrile, dimethylformamide or dimethylsulfoxide are used to extend the negative potential range to about −3.5V vs. SCE. In these solvents, the less polar tetra (n-butyl) ammonium salts (halides, perchlorates and tetrafluoroborates) and lithium salts may be used.[110] However, the electrolyte may interact with the electrode (e.g., adsorption of iodide or tetra (n-butyl) ammonium) or cause ion pairing (e.g., lithium in many solvents or many other ions in low dielectric constant solvents). These phenomena may cause undesired artifacts or complications in the electrode process.

Migration of charged electroactive species, as previously stated, can be prevented by the presence of about 100-fold excess of ionic species. When migration occurs, linearity between the limiting current and concentration may not be observed. A minimum supporting electrolyte concentration to provide adequate conductivity for potential control and measurement is about 0.01 *M*, while 0.1 to 1 *M* is a more normal concentration. Low conductivity may result in waves shifted on the potential axis due to resistive potential drop, limiting currents determined by solution conductivity rather than the concentration of the electroactive species or other anomalous results. For example, the potential of the DME measured with the reference electrode will be in error by the product of the uncompensated resistance times the current. Because the currents in polarography are quite low (microamps compared to milliamps or more for controlled potential coulometry), this error term will be significant only when very low conductivity solutions are encountered. A more comprehensive discussion and references to other articles on potential control and measurement within electrochemical cells is given in the chapter on coulometry in this volume.

As stated previously, the polarographic half-wave potential and the electron transfer process itself may be highly dependent upon chemical species other than the electroactive species. The other chemical species may be in thermodynamic equilibria or involved in kinetically controlled reactions with the electroactive species or with the product of the electrode reaction. These chemical processes are experimentally apparent from shifts in the half-wave potential, changes in wave shape or height, etc. as the electrolyte is varied. Generally, homogeneous chemical reactions are most important in the solution layer surrounding the electrode (for chemical reactions with a half-life less than a few minutes). Thus, reacting nonelectroactive species such as hydrogen ions, electrophiles, nucleophiles, or ligands associated with metallic complexes should be present in sufficiently high concentrations so that no appreciable concentration changes occur, even in the solution near the electrode.[6] Generally, a 10- to 20-fold excess of the reacting species or least concentrated buffer component is sufficient. With unbuffered electrolytes, minor changes in the pH of the water used in solution preparation or in the concentration of the electroactive species may produce variable or complicated changes in the recorded polarograms.[20] For example, the totally irreversible reduction wave of a species in unbuffered neutral or alkaline (hypochlorite) or acidic (dimethylglyoxime) solution shifted negative and then became elongated or split into two partial waves, respectively, as the concentration of the electroactive species increased.[111,112] In other situations, the protonated conjugate acid is usually more easily reduced than the corresponding base, but the ratio will vary depending upon the local solution environment. The composition of some common pH buffers is given in Table 6.

Table 6
COMPOSITION OF BUFFER SOLUTIONS FOR USE IN POLAROGRAPHY

A. Constant-Ionic-Strength McIlvaine Buffers[a]

Approximate pH (25°C)	Grams/liter of Solution		Ionic strength of buffer system (*M*)	Grams KCl/liter of solution to produce ionic strength of	
	Na_2HPO_4 anhydrous	$H_3C_6H_5O_7 \cdot H_2O$		(0.5 *M*)	(1.0 *M*)
2.2	0.567	20.6	0.0108	37.2	74.5
2.4	1.76	19.7	0.0245	35.4	72.7
2.6	3.09	18.7	0.0410	34.2	71.5
2.8	4.50	17.7	0.0592	32.9	70.2
3.0	5.83	16.7	0.0771	31.4	68.7
3.2	7.02	15.8	0.0934	30.3	67.6
3.4	8.09	15.0	0.112	28.9	66.2
3.6	9.16	14.2	0.128	27.6	64.9
3.8	10.07	13.6	0.142	26.7	64.0
4.0	10.9	12.9	0.157	25.5	62.8
4.2	11.8	12.3	0.173	24.4	61.7
4.4	12.5	11.7	0.190	23.1	60.4
4.6	13.2	11.2	0.210	21.6	58.9
4.8	14.0	10.7	0.232	19.9	57.2
5.0	14.6	10.2	0.256	18.2	55.5
5.2	15.2	9.75	0.278	16.5	53.8
5.4	15.9	9.29	0.302	14.8	52.1
5.6	16.5	8.72	0.313	13.9	51.2
5.8	17.2	8.32	0.336	12.2	49.5
6.0	17.9	7.74	0.344	11.6	48.9
6.2	18.8	7.12	0.358	10.6	47.9
6.4	19.7	6.47	0.371	9.62	46.9
6.6	20.7	5.72	0.385	8.50	45.8
6.8	22.0	4.79	0.404	7.23	44.5
7.0	23.4	3.70	0.427	5.44	42.7
7.2	24.7	2.74	0.457	3.10	40.4
7.4	25.8	1.91	0.488	0.488	38.2
7.6	26.7	1.35	0.516	—	36.0
7.8	27.2	0.893	0.540	—	34.3
8.0	27.6	0.589	0.559	—	32.9

B. Phosphate Buffers

Composition of phosphate buffers of constant ionic strength for the pH range of 5.2 to 7.8 may be calculated from the nomogram found in Figure 6 of Volume I Chapter 2, "Separation of Drugs from Excipients".

C. Acetate Buffers

Composition of acetate buffers of constant ionic strength for the pH range of 3.5 to 6 may be calculated from the nomogram found in Figure 7 of Volume I Chapter 2, "Separation of Drugs from Excipients".

D. Bates and Bower's Alkaline Buffers

Composition and ionic strength of these buffers covering the pH range of 7 to 13 may be found in Table 3.B of Volume I Chapter 2, "Separation of Drugs from Excipients". Their composition may be adjusted to a constant ionic strength by the addition of KCl or other suitable ionic salts if desired.

Table 6 (continued)
COMPOSITION OF BUFFER SOLUTIONS FOR USE IN POLAROGRAPHY

E. Clark and Lub's Buffers

Composition of these buffers may be found in Table 3.A of Volume I Chapter 2, "Separation of Drugs from Excipients". Their ionic strength may be calculated from the appropriate dissociation constants and the solutions adjusted to a constant ionic strength if desired.

• Data from Elving, P. J., Markowitz, J. M., and Rosenthal, I., *Anal. Chem.*, 28, 1179, 1956.

The variation of half-wave potential with ionic strength (i.e., variation of activity coefficients) is often overlooked during the evaluation of electrochemical characteristics. This minor variation can be confusing when determining the shift of half-wave potentials as the pH is varied even though only the relative amounts of the buffer components are changed. In order to eliminate this artifact, the ionic strength of buffers can be adjusted to a convenient value, such as 0.5 or 1 *M*, with the use of inert salts. The total ionic strength (μ) of an electrolyte can be calculated by the equation

$$\mu = 0.5 \sum_i (z_i^2 C_i) \quad (64)$$

where the z_i are the charges on the ions and the C_i are the individual ionic concentrations. Of course, the distribution of the buffer between the anionic and variously protonated forms at the specific pH must be known or calculated from the equilibrium constants.

C. Deoxygenation or Deaeration

Oxygen exhibits two reduction waves at the dropping mercury electrode. The first corresponds to the reduction of oxygen to hydrogen peroxide as indicated in the following:

Acidic

$$O_2 + 2H^+ + 2e \rightleftharpoons H_2O_2 \quad (65)$$

Neutral or Basic

$$O_2 + 2H_2O + 2e \rightleftharpoons H_2O_2 + 2OH^- \quad (65a)$$

The half-wave potential for this wave is nearly pH independent at about −0.05 V vs. SCE in most solutions, but may occur as negative as −0.25 V in some electrolytes. The second reduction wave corresponds to the conversion of the peroxide to water given by the reaction:

Acidic

$$H_2O_2 + 2H^+ + 2e \rightleftharpoons 2H_2O \quad (66)$$

Neutral or Basic

$$H_2O_2 + 2e \rightleftharpoons 2OH^- \quad (66a)$$

The half-wave potential for this wave is about −0.9 to 1.0 V vs. SCE in most solutions. However, the wave tends to cover a large potential range between −0.5 and

−1.3 V vs. SCE. At the equilibrium concentration of oxygen in water (about 8 mg/ℓ or 0.25 m*M*), its diffusion current will be about 5 μA.[10] Therefore, it is usually desirable to remove oxygen from the solution being analyzed. Two methods are available for removing the oxygen: chemical and physical deoxygenation or deaeration.

Chemical removal of oxygen is accomplished by the addition of a substance which reacts rapidly with it. Sulfite ion, which removes oxygen in neutral or basic solutions, is the most common chemical reactant. However sulfite can also react with the substance being measured. Hydrazine has been used in basic solutions; but, hydrazine is moderately easy to oxidize and can interfere with polarographic waves which occur at potentials less negative than about −0.2 to −0.3 V vs. SCE. Because of these difficulties, chemical removal of oxygen is seldom practiced, especially in organic electrochemistry. It is of some use when the sample must be added directly to an oxygen free electrolyte, for long-term removal of oxygen or in nonlaboratory situations where physical deoxygenation techniques are impractical.

Physical deoxygenation or deaeration is accomplished by passing an oxygen free gas stream through the solution. The gas used should not react chemically with the solution, should not itself be reducible or oxidizable, and should be free of oxygen. The first two requirements can be met by using an inert gas such as nitrogen or argon. Other gases such as hydrogen and carbon dioxide have been used for this purpose, however, nitrogen is the least expensive, and it can be obtained in a "prepurified" grade with relatively low oxygen content (2 to 20 ppm). This grade of gas is suitable for many routine assays.

The maximum acceptable oxygen content depends upon the particular use. A few parts per million of oxygen introduced by typical flows of deoxygenation gas (perhaps 200 mℓ/min) can be sufficient to cause air oxidation of oxygen sensitive solutions (e.g., V(II) or Cr(II)) or to interfere with assays of trace components. For example, 20 ppp oxygen in deoxygenation gas flowing at a rate of 200 mℓ/min carries 0.002 m*M*/min oxygen into the cell and will result in an equilibrium oxygen concentration of about 25×10^{-9} *M* (decrease of 10,000 times). When necessary, final trace amounts of oxygen in the gas stream can be removed by several purification procedures.

The classical method for oxygen removal is passage of the gas through a heated (450 to 500°C) Pyrex® tube filled with copper wire. When the proper temperature is maintained, the oxygen reacts to form cupric oxide. At lower temperatures, the oxygen does not react and at higher temperatures the cupric oxide is decomposed yielding free oxygen. The copper is regenerated periodically by the passage of hydrogen gas over the copper. A catalyst coated with about 30% finely divided copper can be obtained also (R3-11®, BASF®, Wyandotte Corp., Parsippany, N.J.). Alternatively, the inert gas stream can be passed through a solution of vanadous sulfate or chromous chloride.[10] The vanadous solution is easier to prepare and regenerate. The vanadium is kept in the +2 oxidation state by the presence of amalgamated zinc in an acidic solution.[10] A particularly convenient, but expensive, form of chromium is available as a prepackaged cylinder of chromic oxide for insertion into the gas line (Oxisorb™, MG Scientific Gases, Kearny, N.J. or EG & G Princeton Applied Research, Princeton, N.J.).

Oxygen can contaminate the solution being deoxygenated in several ways. First, oxygen can penetrate rubber tubing and some types of plastics. Therefore, gas lines should be constructed of glass or metal if possible with the use of relatively impermeable tubing (such as Tygon®) when necessary. Lines which have not been recently used should be thoroughly flushed to remove oxygen before samples are deoxygenated. Second, oxygen can diffuse into the cell from auxiliary or reference electrode compartments (when used). This diffusion can be minimized by use of frits, agar plugs, etc. Third, oxygen can leak into the cell around openings (such as that for the DME). Leakage around openings is especially important when mechanically controlled drop timers are used, i.e., the DME opening must be sufficiently large to allow for the capillary movement as the drop is knocked off. The thoroughness of oxygen removal and elimination of means of recontamination depend upon the oxygen level which interferes with the particular assay. An experimental verification of the deoxygenation process may be justified. One method for verifying the absence of oxygen is the absence of oxygen reduction waves in a polarogram obtained on the supporting electrolyte itself. Observation of unusual phenomena at low concentrations or when oxygen is known to be present would also suggest re-examination of the deoxygenation process.

Just before the deoxygenation gas is introduced into the cell, it should be equilibrated with a solution identical in composition to the supporting electrolyte and maintained at approximately the same temperature as the cell. This procedure prevents the evaporation or dilution of the sample by the gas stream. Equilibration is normally accomplished using a gas washing bottle (coarse or medium porosity frit) containing the electrolyte.

The purified and equilibrated gas stream should be introduced into the cell in such a manner that fine, highly dispersed gas bubbles are produced. This can be accomplished by using a medium porosity sintered glass frit as the dispersion device (e.g., immersion tube with fritted disk available as catalog number 39535 from Corning® Glass Works, Corning, N.Y.). Coarse glass frits or small diameter glass tubes do not give fine enough bubbles for efficient deoxygenation. Fine glass frits may produce higher than desired back pressures, but are otherwise suitable. After the deoxygenation process is complete (2 to 20 min, depending upon its efficiency), the gas stream should be diverted to flow over the solution in the cell to prevent reabsorption of oxygen from the air. In extreme cases, the cell may need to be completely sealed from the atmosphere to prevent problems.

D. Measurement of Limiting Current

The diffusion limited current, proportional to concentration, must be extracted from the polarogram which also contains the residual current. The most satisfactory method involves first recording a polarogram under the same conditions as the assay without the presence of the sample. This polarogram will consist of the residual current except for those currents directly due to the presence of the sample (e.g., charging current due to adsorption of the electroactive species or faradaic residual current from impurities within the sample itself). Then, the polarogram for a solution containing the sample is recorded and corrected for the measured residual current at each potential. This process, unfortunately, is lengthy and tedious unless the polarograms can be stored and manipulated in a digital format. Also, it is still subject to the uncorrected residual currents mentioned above.

A second technique for residual current correction involves the extrapolation of the residual current baseline obtained at potentials where the major component is not electroactive. This commonly used correction procedure (see Figure 2) will be quite accurate unless the extrapolation region crosses the potential of zero charge. At that potential, the charging current changes significantly in magnitude so that the extrapolation will not be valid. When a preceding wave is present, this procedure may be used in a similar manner except that the diffusion current for the preceding wave is extrapolated for the baseline measurement (subject to similar errors in the extrapolation procedure).

In addition to establishment of the residual current baseline, the limiting current plateau must be measured. When the polarographic wave is sufficiently well separated from other current producing processes, this procedure is straightforward (see Figure 2). In other situations, a consistent, reproducible technique must be found. Some possible solutions were discussed by Meites.[19] Some digital data processing techniques designed to extract the electrochemical parameters are discussed later. If a well defined limiting current plateau is not present, precision and accuracy will be compromised. Therefore, the best possible wave shape should be one of the goals during assay development.

1. Effects of Multiple Waves

In general, qualitative as well as quantitative analysis becomes increasingly difficult as the number of electroactive species increases since the multiple waves may not be well resolved. Because the shape of the polarographic wave depends on the value of n and on the nature of the electron-transfer reaction, the difference in half-wave potentials needed for adequate separation varies. For reversible waves, however, the minimum separation is independent of the concentration ratio or capillary constants. Generally, it is stated that a separation of 180/n mV in the half-wave potentials gives adequate resolution for reversible waves. For qualitative work or cases where some error can be tolerated (e.g., determination of low level impurities), less separation (about 120/n mV) may be acceptable. However, a larger separation, perhaps 240/n to 300/n mV, is generally needed for high precision quantitative applications. For elongated waves, such as those due to heterogeneous electron transfer reactions, an even greater separation will be required. In most cases, an experimental examination of polarographic waves obtained with various ratios of components will be required to confirm the predicted behavior of a mixture.

The electrolysis of the supporting electrolyte is another interference in the measurement of the diffusion current from the species of interest. Because components of the supporting electrolyte are present in much larger amounts than the sample, the current involved is very much greater than that for the sample. The separation from the large currents due to electrolysis of the electrolyte must be approximately the same as stated above when the potential of the electrolyte decomposition is measured at a value of the current comparable to that for the sample (instead of at the half-wave potential for the electrolyte decomposition).

Considerable effort has been expended in developing methods to improve resolution between overlapping waves. Proposed modifications have included changes in supporting electrolyte composition to shift one wave relative to the other, variation of instrumental parameters, mathematical processing and actual changes in the type of experiment. The improvement of sensitivity has been a related goal in many of these investigations.

Equations 23 and 26 indicate the typical variation of the half-wave potentials observed with changes in supporting electrolyte composition. When more than one wave is present, it may be possible to achieve the desired separation by varying the concentration of a nonelectroactive species involved in one of the half-reactions. For instance, the half-wave potential of many metallic ions can be varied by the addition of ligands which form complexes with the ions. Depending upon the relative stability constants and stoichiometry of the metal-ligand complexes, the reduction waves frequently can be separated. In other cases, an electrochemical process involving hydrogen ions may be shifted relative to the electrolyte decomposition as the pH is varied.

In some polarograms that have multiple waves, a variation in instrumental settings may be useful. The instrumental parameters, of course, cannot change the true separation between waves, but it may allow scale expansion or an offset in current in order to more easily detect the wave of interest. The use of scale expansions or offsets is most beneficial when measuring a small wave due to a minor impurity that is more difficult to reduce (or oxidize) than a major component (i.e., the major component limiting current is the baseline for the impurity wave).

Mathematical separation of overlapping data has received much attention in spectroscopy, chromatography, and titrimetry as well as in electrochemistry. Meites and Lampugnani have written a series of papers on multiparametric curve fitting including one paper specifically concerning polarographic applications.[113] Barker and Gardner[114] recommended a generalized procedure for combining the basic wave with different derivatives of the wave to enhance resolution. Other mathematical transformations, such as logarithmic analysis, have been applied.[115] Generally, these mathematical treatments are feasible only if the data are collected in a digital format and computer facilities are available.

The development of related polarographic techniques which have greater resolution and/or sensitivity is another effort to extend the application of polarographic analyses. These techniques range from simple derivative experiments to those of a more complex nature. For example, differential pulse polarography, square wave polarography, AC polarography, derivative DC polarography, and higher harmonic AC polarography all offer some increase in both resolution and sensitivity at a cost of increasing complexity and sometimes poorer precision. They also tend to minimize the effect of a more easily reduced (or oxidized) major component.

2. Effects of Maxima

Maxima or streaming maxima have been classified into three categories depending upon the mechanism which causes the mass transport: maxima of the first, second, and third kind.[10,20,27,41,116] Generally, all three types result in a movement (streaming) of the solution immediately adjacent to the electrode. The convective mass transport brings extra material to the electrode compared to the diffusion controlled case and, therefore, causes an increase in the measured current. These convective processes usually occur only within a limited potential range.

The distortion of the current-potential curve causes difficulty when quantitative analyses are performed. In addition, the wave distortion makes it difficult to elucidate the nature of the electrochemical processes involved in the electrode reactions. Thus, suppression or elimination of the convective processes may be desirable. Usually, these solution movements can be inhibited by the use of surfactants (materials which are adsorbed onto the electrode surface). In fact, the impurities contained within typical analytical grade reagent chemicals and purified water (i.e., solutions of normal purity) may be sufficient to decrease or prevent the maxima.

Maxima occur within specific potential ranges and depend upon the experimental parameters. As the experimental parameters change, a wave may exhibit a different type of maxima, a combination of first and second kind of maxima or change between a strictly diffusion controlled behavior and a wave which exhibits a maxima. When a maxima is present, the solution movement causes a nonuniform concentration of the electroactive substance from one area of the diffusion layer (solution adjacent to the electrode) to another. This effect, termed the streaming induced depolarizer effect by Bauer,[116] may enhance (higher magnitudes, occurrence over broader potential ranges, etc.) or limit the maxima.

a. Maxima of the First Kind

Maxima of the first kind are caused by differences in the surface tension over the surface of the mercury drop.[116] Surface tension differences are believed to arise from nonsymmetrical primary current distribution within the solution. Distortions in the current distribution are primarily due to geometric causes (e.g., shielding of the drop by the capillary) with minor contributions from the compression of the diffusion layer at the bottom of the drop (nonconstant drop center during the growth of the drop) and nonuniform polarization of the double layer (very dilute solutions). The solution movement is downward for a reduction process which occurs at a potential positive of the E_{pzc} and upward when the potential is negative of the E_{pzc}. For oxidation processes, the solution movement is in the opposite direction.[116,117]

Maxima of the first kind are usually observed on the rising portion of the wave.

The appearance has frequently been reported to be very sharp, but their actual shape is more likely to be rounded. The sharp maximum is an artifact resulting from Ohm's law resistive control of the current when the uncompensated solution resistance is high enough to limit the current. The behavior of maxima of the first kind and their dependence on experimental variables are listed as follows:[10,20,116]

1. The magnitude of the current rise may be 5 to 50 times the diffusion limited current.
2. The magnitude of the current maxima will be smallest near E_{pzc}, but not necessarily absent as sometimes claimed.
3. The ratio of current to concentration increases as the concentration increases and maxima of the first kind are most often observed at high concentrations of the electroactive species.
4. Increasing the conductivity (concentration of inert salt) of the electrolyte tends to decrease the maxima. They are most often observed in dilute electrolytes.
5. The current is dependent upon $m^{2/3}t^{1/6}$ (diffusion controlled) at low m values and $m^{1.15-1.4}t^{1/6}$ at high m values. The dependence on height is the same as diffusion controlled currents at low m values and greater at high m values. The current-time curves for individual drops show a maximum current at times shorter than the drop time for solutions of normal purity; they show a discontinuity with a maximum current at the end of the drop for highly purified solutions.
6. The streaming induced depolarizer effect tends to enhance maxima of the first kind at potentials positive of the E_{pzc} for reductions and negative of the E_{pzc} for oxidations. The electrochemical processes at oppositely charged electrodes tend to be self-limiting.

b. Maxima of the Second Kind

Maxima of the second kind are caused by nonradial expansion of the drop. That is, they appear when the mercury streams from the capillary tip to the bottom of the drop and then flows upward at the outside drop surface. This drop movement induces movement within the solution and convective mass transport. As the mercury flow rate becomes larger (greater mercury column height or smaller capillary radius), the tendency for these phenomena increases.

Maxima of the second kind are typically observed in the same potential region as the limiting current. The maxima may appear as a broad peak or mimic an extra wave. Sometimes it may be detected only by the fact that the current plateau is distinctly higher than one would predict from the Ilkovic equation or that the current is nonlinear with concentration. Some other observations and dependence on experimental variables include:[20,116,118]

1. High mercury flow rates (m) cause maxima of the second kind. They are infrequently observed at m values less than 0.5 mg/sec.
2. The current maxima will be greatest near the E_{pzc}.
3. The ratio of current to concentration decreases as concentration increases (high concentrations of the electroactive species favor maxima of the first kind over maxima of the second kind).
4. The maxima is independent of the supporting electrolyte (low electrolyte concentrations favor maxima of the first kind over maxima of the second kind).
5. The current is dependent upon $m^{2/3}t^{1/6}$ (diffusion controlled) at low m values and $m^{1.15-1.4}$ $t^{1/6}$ at high m values. The dependence on height is the same as diffusion controlled currents at low m values and greater at high m values. The current-time curves for individual drops show a maximum current at times shorter than the drop time for solutions of normal purity; they show a discontinuity with a maximum current at the end of the drop for highly purified solutions.

6. The streaming induced depolarizer effect tends to enhance maxima at potentials positive of the E_{pzc} for reductions and negative of the E_{pzc} for oxidations while limiting processes of the second kind at oppositely charged electrodes (same as for maxima of the first kind).
7. The temperature dependence is approximately the same as that for diffusion limited currents.

c. Maxima of the Third Kind

Maxima of the third kind are produced by nonuniform adsorption of slightly water soluble compounds.[116,119,120] The surface tension is reduced in the area of adsorption causing movement of the mercury towards these areas and the resulting convection processes. It has been described for only a limited number of organic compounds (surfactants) which form closely packed condensed layers on the electrode surface, apparently having a two-dimensional structure. Frequently, these compounds exhibit abrupt transitions between adsorbed and nonadsorbed states as the electrode potential is changed. Some of the compounds which exhibit maxima of the third kind include camphor, borneol, adamantanol, 2-oxoadamantane, guanine, and pelargic acid.

Maxima of the third kind are observed in potential regions where the surface coverage by the adsorbed material is approximately 20 to 50%. The maximum current is observed at the E_{pzc} for low concentrations of the material; it splits in two peaks which move away from the E_{pzc} (positively and negatively) and change in height as the amount of surfactant increases. The potential regions including the maxima correspond to the region where the change in surface coverage with change in surfactant concentration is the largest. However, these currents are not desorption currents (due to changes in the differential capacitance of the electrode). Maxima of the third kind also are not associated with electric field gradients. Because maxima of the third kind do not directly involve the electroactive species, no induced depolarizer effect is observed.

d. Detection and Elimination of Maxima

Maxima can be detected by several observations:

1. Waves with characteristic shapes associated with one or more types of maxima
2. Changes in wave shape with changes in concentration of the electroactive species or supporting electrolyte
3. Changes in wave shape with the addition of typical maxima suppressors
4. Dependence of the current upon the height of the mercury column (square root of corrected height dependence not found)
5. Currents greater than expected for a given concentration
6. Nonlinear variation of limiting current with concentration of the electroactive species

Of course, one type of maxima will not exhibit all types of behavior. Also, observations of more than one behavior will tend to differentiate maxima from electron-transfer mechanisms which may give somewhat similar behavior (adsorption processes, catalytic regeneration, etc.).

Most types of maxima can be reduced or eliminated by the addition of low amounts of surfactants. Typical surfactants include Triton X-100® (Rohm and Haas Co., Philadelphia, Pa.), gelatin, and other large organic species (frequently dyes such as methyl red, methylene blue, or fuchsin; polypeptides; and cationic or anionic soaps). Triton X-100® is a nonionic detergent which can easily be prepared in dilute solutions of good stability. A typical concentration of maxima suppressor might be 10^{-3}% (10 μg/mℓ) Triton® or 5×10^{-3}% (50 μg/mℓ) gelatin. However, one should use the *minimum* amount of suppressor needed to achieve the desired effects since excessive amounts can distort the normal electrochemical behavior of the species which is being quantitated. These distortions include changes in the current-time profile of the drop and a decrease in the limiting current. The distortions may arise because of a physical blocking of the electrode surface, changes in the electrical double layer, etc. Obviously, the quantitative characteristics of the wave (linearity, diffusion control of currents, etc.) should be checked thoroughly in the presence of the maxima suppressor.

In addition to the use of maxima suppressors, several changes in experimental conditions can be made to eliminate a specific type of maximum. Maxima of the first kind can be minimized by using a low concentration of the electroactive species or a high concentration of supporting electrolyte. They can also be reduced by making the primary current distribution more symmetrical, e.g., use of a shield at the bottom of the electrode similar to the blocking effect of the DME capillary at the top of the electrode.[121] Maxima of the second kind can be controlled by using a smaller mercury flow rate or larger concentrations of the electroactive species. Because of the complementary nature of maxima of the first and second kinds, changing conditions to eliminate one type may induce the other type of maxima or a mixed maxima. Maxima of the third kind can be eliminated by avoiding specific classes of compounds which produce this phenomenon.

E. Calculation of Concentration

1. Direct Calculation

In principle, the concentration of the electroactive species can be directly calculated from the Ilkovic equation (Equation 6) or from a previously measured diffusion current constant (Equation 7). Calculation with the Ilkovic equation requires measurements of the diffusion coefficient (D) and the capillary constants (m and t) as well as an independent evaluation of n. Use of the diffusion current constant requires only the measurement of the capillary constants. However, accurate experimental evaluation of these quantities is difficult and time consuming, while the use of previously acquired values for D and I assumes that the Ilkovic equation is followed exactly for data from two different capillaries. Thus it is usually not feasible to perform a direct calculation of the concentration. Instead, one of three comparative calculational techniques (standard curve, standard addition or pilot ion) is routinely used.[10]

2. Standard Curve

The standard curve method is essentially identical to procedures commonly used in many other analytical techniques. First, the response (limiting current) for several different concentrations of the electroactive species is obtained. Then, the limiting current of the sample is obtained under identical conditions, and the sample concentration calculated from the calibration curve of the standards. In some instances, the intercept of the standard curve is assumed to be zero and the average limiting currents for the standards are normalized for their concentration. These procedures assume that all parameters affecting the limiting current remain the same between recording the standard and the sample polarograms. That is, the capillary constants and the solution conditions affecting drop time or diffusion coefficient must be controlled and identical.

3. Standard Addition

The standard addition method is more time consuming than the standard curve method, but it eliminates most errors due to different matrix effects in standard and sample solutions. In the standard addition method, the polarogram of the sample solution is obtained first. Then, a known amount of standard is added to the solution and an additional polarogram recorded. Several additions are normally made. When the response (Y-axis, corrected for dilution) is plotted vs. volume of standard added

(X-axis), a straight line should be obtained. The response corrected for dilution (i_{corr}) is given by:

$$i_{corr} = i_d (V + v) / V \quad (67)$$

where i_d is the measured diffusion limited current, V is the volume of the sample, and v is the volume of the standard solution added. The original concentration of the sample is obtained by dividing the y-intercept of the plot by its slope. If only one addition is made, the sample concentration (C_u) can be calculated from the standard concentration (C_s), the volumes (defined above), and the measured limiting currents of the sample (i_1), and the sample with added standard (i_2) using:

$$C_u = i_1 \, v \, C_s / i_2 \, (V + v) - i_1 \, V \quad (68)$$

A relatively concentrated solution of the standard should be added to minimize dilution effects and changes in the matrix. Usually the amount of standard added is adjusted to give approximately twice the response of the unknown.

4. Pilot Ion (Internal Standard)

The pilot ion or internal standard method of calculation uses the relative diffusion currents of two species. The second species is added at a known concentration during sample preparation and the polarogram recorded. Then, the concentration of the unknown is calculated from the ratio of the two diffusion limited currents and the previously established ratio for standards and samples. The basis for this method is the fact that the limiting current ratio will be nearly independent of experimental variables such as mercury flow rate, drop time, temperature, supporting electrolyte composition, and matrix effects arising from the sample. Unfortunately, it usually is difficult to find a second species which has a well-defined wave that is sufficiently separated from the sample wave, from waves due to possible impurities, and from the wave due to electrolysis of the supporting electrolyte. Thus, the pilot ion method is not normally used.

F. Precision and Accuracy

Many factors determine the precision and accuracy of an analytical method. Precision can be measured rather easily; many evaluations of precision for specific assays and several investigations of factors which limit precision in polarographic assays are available.[7,10,20,30,122] However, it is more difficult to determine the accuracy of an assay since evaluation of accuracy requires comparison to an independent assay that itself has been characterized for precision and accuracy. Alternatively, indirect estimates of accuracy can be made based on linearity, intercept, material balance, etc.

Precision in polarography depends upon:

1. Proper choice of experimental conditions; e.g., concentration of electroactive species, supporting electrolyte composition, current measurement technique, and instrumentation used
2. Minimization of processes which produce nonideal behavior, e.g., migration, convection, residual current, and suppression of maxima
3. Control of experimental variables which affect the current, e.g., electrode potential, mercury flow rate, drop time, and temperature

On the other hand, accuracy depends upon specificity (freedom from interferences due to impurities) and lack of determinate errors (during sample or standard preparation, current measurements, etc.). Many of these subjects have already been presented; instrumentation will be discussed later. In this section, the results of several investigations of the factors which limit precision will be reviewed.

Reported values for the precision range from about 1 to about 3% for assays of the main component.[2,7,10,20,30,122] Precision is poorer for the assay of an impurity and the precision decreases as the wave approaches the detection level. The best precision is obtained with well-defined waves and when the waves are separated from the limits of the potential range with proper control of the experimental variables.

Many investigators have reported the effect of long-term (hours to days) variation in the reproducibility of the DME. These variations are basically due to changes in flow rate, capillary diameter (dependent upon temperature), drop time (dependent upon surface tension, solution creep inside the capillary, impurities, physical blockages, etc.), or nonconstant mercury column height.[10] Because these numerous variables are difficult to control, the usual analytical procedure involves a comparison of standard and sample polarograms obtained within a short period of time. Preferably, the standards should be intermixed with the samples to detect any drift in response. Using these sequential, single cell determinations, Taylor[122] reported that a coefficient of variation of 2% (or 1% in favorable cases) could be achieved by controlling the mercury height to 0.3% (about 0.2 cm), mercury flow rate to 1% (about 0.02 mg/sec), temperature to 0.5°C, and drop time to 2% (about 0.1 sec). When the experimental variables are more rigidly controlled, an even better precision is obtained. However, control of temperature to better than 0.5°C is not feasible in many cases and it may be difficult to achieve even this level of control. Also, convenient methods for maintaining (or measuring) constant mercury flow rates are not available for most polarographic instruments.

Fisher[30] reported an interesting series of observations concerning the control of experimental variables. In the first experiment (first derivative DC polarography was employed), the peak current was read with a five digit digital voltmeter, the temperature was controlled to 0.05°C, the drop time (0.5 sec) was controlled to 0.03% and the mercury height was maintained constant. Although a short-term precision (several minutes, same solution, $n = 7$) of 0.03% was obtained, the long-term precision (several hours, replicate solutions) was about 1%. Even when standards and synthetic samples were alternated, the precision of samples was about 1%. The principal cause of imprecision was found to be variations in the mercury flow rate. When a positive displacement, mechanical device was used to control the mercury flow (to 0.1% at 1 mg/sec) and the temperature control was increased to 0.02°C, a long-term precision (several days, $n = 28$) of 0.15% was found for either the first derivative peak height or the normal DC polarographic diffusion current.

An alternate approach for the improvement of precision is the use of difference (sometimes called subtractive or differential) polarography. In difference polarography, the currents obtained in two matched cells are subtracted and the difference signal is displayed. This approach allows the direct comparison of a standard and sample or correction for residual currents or matrix effects. The latter two corrections are important in the case of trace level analyses or with complex matrices such as urine. Several investigators[30] have achieved precision levels of about 0.1% using these methods. However, this approach is experimentally complicated because the capillary constants, cell geometries, electrode potentials, and current measuring circuits must be carefully matched.

Digital processing of data, as previously mentioned, can improve the signal-to-noise ratio, correct for residual current, improve timing of current measurements, etc. Therefore, digital data processing will improve precision and lower the detection level. However, digital control of the most important experimental variables (mercury flow rate and temperature) with suitable designed instrumentation would permit an even greater improvement in precision.

VI. APPARATUS

A. Polarographic Cell

Many designs for polarographic cells have been reported in the literature[2,7,10,21,27,30] or made available commercially (Table 7). In most cases, the performance requirements for the cell are minimal so that the cell may be designed for convenience of use (size, cleaning, deoxygenation inlets, etc.) or for specific practical problems (exclusion of oxygen or water, temperature control, etc.) dependent upon the situation. The choice of reasonable cell designs may be limited in certain applications (e.g., when sample size is limited or when leaching of metal ions from glass cells occurs during a trace level metal assay). The various cell designs differ with respect to size, temperature control capabilities, two or three electrode designs, number of cell compartments, and internal or external reference electrode configurations.

For quantitative work, the sample preparation should involve large enough weights and volumes so that the errors associated with these steps are on the order of 0.1 to 0.5% or less. For normal analytical balances (0.1 mg resolution), this would require sample weights of about 50 mg; for semimicrobalances (0.005 to 0.01 mg resolution), 10 to 20 mg of sample is reasonable. However, the minimum amount of sample taken must be sufficiently large to assure that no appreciable sampling error is involved.[123] Volumes greater than about 10 m*ℓ* can be measured with reasonable accuracy. This volume is appropriate for filling a typical polarographic cell which holds about 10 to 50 m*ℓ*. However, additional solution is required to rinse the cell so that contamination from the previous sample is removed. Small cell designs (several milliliters) or microcells (0.1 to 1 m*ℓ*) have been used when samples are more limited. For example, Zuman[7] has described a simple polarographic cell constructed from a small beaker.

Because the diffusion current and the potential of the reference electrode are both temperature dependent, it is necessary to maintain controlled temperatures when either accurate current or potential measurements are made. Accurate temperature control is most easily achieved by immersing or jacketing the cell in a constant temperature bath. For controlled temperatures near room temperature, a small immersion heater or a heater combined with a cold water coil will suffice. In semiquantitative or qualitative work, a well-thermostated room is sufficient.

1. Cell Design

The cell current may flow through the reference electrode (two-electrode cell) or through a separate auxiliary electrode (three-electrode cell). In the earlier two-electrode designs, the potential was simply applied between the reference electrode and the DME. There are two problems with this type of design. First, the applied potential is divided between the DME-solution interface and the undesired ohmic losses in the solution. Second, the current flow through the reference electrode produces an electrochemical reaction (usually oxidation) at the reference electrode opposite of that occurring at the DME. To maintain a constant reference electrode potential under these conditions, the surface area of the reference electrode must be large (to obtain a low current density) and the electrochemical reaction of the reference electrode must be rapid (to remain in equilibrium). Generally, the two-electrode design is used now only in special cases. For example, it may be more practical to have only the two electrodes of this simplified design when a microcell is used.[21,30]

The three-electrode cell consists of the DME, the reference electrode, and a separate auxiliary or counter electrode. The three-electrode design was developed to minimize the problems associated with two-electrode designs. In a three-electrode cell, the reference electrode is used to measure the actual potential difference between the DME and the solution. Then, the electronic feedback circuits of the potentiostat adjust the

Table 7
SELECTED POLAROGRAPHIC SYSTEMS AND COMPONENTS

Manufacturer	Address	Components[a]
Beckman Instruments®, Inc.	2500 Harbor Blvd. Fullerton, CA 92634	Cell, Pol
Bioanalytical Systems®, Inc.	P.O. Box 2206 W. Lafayette, IN 47906	Cell, MFP, HPLC (carbon paste, glassy carbon, gold)
Brinkmann Instruments®, Inc.	Cantiague Road Westbury, N.Y. 11590	Cell, Pol, HPLC (glassy carbon)
Chromatix®	560 Oakmead Parkway Sunnyvale, CA 94086	HPLC
Dionex Corp.®	1228 Titan Way Sunnyvale, CA 94086	HPLC
ECO®, Inc.	56 Rogers St. Cambridge, MA 02142	Cell, CV, MFP, Pol, Syn
EG&G Princeton Applied Research®	P.O. Box 2565 Princeton, N.J. 08540	AC, Cell, CV, MFP, Pol, SA, Auto, HPLC (static mercury drop)
Environmental Science & Engineering Corp.®	1776 May's Chapel Rd. Mt. Juliet, Tenn. 37122	MFP, SA
Environmental Sciences Associates, Inc.®	45 Wiggens Ave. Bedford, MA 01730	Cell, CV, HPLC, MFP, SV
Harrick Scientific®	P.O. Box 867 Ossining, N.Y. 10562	Cell, RS
IBM Instruments, Inc.®	P.O. Box 3020 Wallingford, CT 06492	Cell, CV, MFP, SA
Pacific Photometric Instruments®, McKee-Pedersen Instruments Division	5675 Landregan St. Emeryville, CA 94608	Cell, Mod
Pine Instrument Co.	P.O. Box 429 Grove City, PA 16127	RDE
Sargent Welch Scientific Co.®	7300 N. Linder Ave. Skokie, IL 60077	Cell, MFP
Tacussel Electronics® (distributed by Astra Scientific International, Inc.®)	P.O. Box 2004 Santa Clara, CA 95051	Cell, CV, MFP, RDE, HPLC

[a] Components are AC = AC polarograph, Auto = automated polarograph, Cell = polarographic cells, CV = cyclic voltammetric instrumentation, HPLC = electrochemical detectors for HPLC, MFP = multifunctional polarograph, Mod = modular components, Pol = DC polarograph, RDE = rotator assembly and electrodes for RDE voltammetry, RS = rapid scan spectrophotometer for spectroscopic monitoring of electrochemical reactions, SA = instrumentation for stripping analysis, and Syn = synthetic (preparative) scale instrumentation.

voltage applied to the auxiliary electrode to maintain the desired potential. This feature gives better potential control in cases where high resistance components are present in the cell (low conductivity solutions, diaphragms separating cell compartments, etc.). With the three-electrode design, a greater variety of reference electrodes is available since its only function is the maintenance of a constant potential.

The DME, reference, and auxiliary electrodes may be placed in the same cell compartment or isolated from each other. When separate cell compartments are used, a fine or ultrafine porosity sintered glass disk (or equivalent) may be used to prevent mixing of the solutions. Sometimes an agar plug (4% solution of agar dissolved in water saturated with potassium chloride or potassium nitrate) may be added to further reduce mixing. The electrical resistance added by these components causes no problem with the three electrode design and the small currents obtained at the DME. However, the frits must be cleaned periodically and/or the agar replaced when it becomes contaminated.

2. Dropping Mercury Electrode

The DME is the single most important aspect of the cell. The precision and accuracy of the entire quantitative experiment depends upon reproducible behavior of the DME. Because of its importance, much attention has been given to its characterization, to its care, and to modifications which attempt to improve its performance.[7,10,20,21,30]

A glass capillary DME has the following typical properties: length, 10 to 15 cm; internal diameter, about 0.05 to 0.1 m; mercury flow rate, 1 to 2 mg/sec for mercury column heights of 30 to 50 cm; drop diameter, about 1 mm; drop area, about 0.03 to 0.05 cm^2; and ohmic resistance of about 100 Ω. In the naturally dropping mode, a typical drop time might be 3 to 6 sec. Of course, the drop time is related to the parameters above (flow rate and drop size) and to the electrode potential and solution composition as previously discussed.

The capillary length is selected to give a reasonable drop time. When it is necessary to break a capillary, a clean, right-angle break can be achieved by scoring the capillary with a sharp file and then breaking it using a cork borer which just fits over the capillary. When the new DME is first used, the capillary constants, m and t, should be determined under specific conditions. These constants are useful in evaluating diffusion control of the limiting current, in determination of n values and in detecting future changes in the behavior of the particular DME which might indicate need for replacement or cleaning. Generally, replacement or cleaning is indicated when erratic currents or changes in drop times are observed.

Cleaning of the DME can sometimes be accomplished by breaking off the dirty end of the capillary or by washing with organic solvents or with dilute nitric acid and water while mercury is still flowing through the capillary. More drastic cleaning involves washing the inside of the capillary with organic solvents or water, followed by nitric acid, a final water rinse and thorough drying. However, prevention of conditions which lead to dirty or clogged capillaries or replacement of bad capillaries are generally better approaches. For example, the DME should never be inserted into a solution unless the mercury is flowing. When an empty DME is placed in solution, the solution enters the capillary and then leaves a film of dissolved salts and dirt when it evaporates. During normal use, the solution naturally wets the glass and very slightly enters the capillary at the end of each drop as the mercury column withdraws up the capillary (due to its surface tension). This material can be removed by a final water rinse at the end of the day before the mercury flow is stopped. The DME also should never be allowed to stand in a dry state. It can be left in a beaker of distilled water (either dropping or at its equilibrium height) or rinsed and capped with a rubber policeman.

A simple dropping mercury electrode assembly can be made by attaching the capillary to a mercury containing leveling bulb using Tygon® or rubber tubing. When rubber tubing is used, it should be cleaned by boiling 30 to 60 min in 2 *M* sodium hydroxide to remove sulfur compounds.[10] The leveling bulb can be adjusted to keep a constant mercury height. A stopcock may be used to clos off the mercury flow when the DME is not being used. A stand tube which incorporates a meter stick allows convenient measurement of the mercury height. Electrical contact to the DME can be made by means of a platinum wire inserted into the leveling bulb or sealed into the stand tube.

The mercury used in the DME should be clean and free of oxide films. The oxide film will quickly form on the surface of the mercury in the leveling bulb, but should not be allowed to enter the DME where it will quickly produce clogging. The mercury itself is most conveniently obtained from commercial sources, generally as a triple distilled grade. The used mercury that accumulates may be returned for reprocessing at a nominal cost. Cleaning of one's own mercury is generally feasible only when large amounts are used. Washing with acid and base to remove active metals and/or distillation to remove base metals is necessary.

The most common and most important modification of the DME is the use of controlled drop time instrumentation. The controlled drop time gives more reproducible behavior and is less sensitive to mechanical disturbances. The effect of controlled drop times on tests for diffusion controlled mass transfer and other parameters is described above. Controlled drop times (or sensing the fall of the drop) is required for application of related techniques such as sampled DC (Tast) polarography, pulse polarography and differential pulse polarography. Other modifications of the DME include those with mechanically driven syringes to control the size of the drop, those with horizontal capillaries, and the use of a Teflon® DME. The syringe or Kemula capillaries can be driven with a stepping motor to deliver extremely reproducible mercury drops as stated in the section on precision and accuracy. The horizontal capillary was proposed to reduce the tendency of the DME to stream (form a series of very small drops with a short drop time) at very negative electrode potentials. This electrode design has not been accepted for general use. The Teflon® DME has been used for acidic fluoride solutions which attack glass.

A novel version of the DME has recently been introduced commercially (EG & G Princeton Applied Research, Princeton, N.J.). This device, Model 303, is termed a static mercury drop electrode. It operates by squeezing a variable sized mercury drop (using a stepping motor driven syringe) at the beginning of a drop cycle (in the first 50, 100, or 200 msec, depending on the drop size selected). Then the mercury flow is stopped and the mercury drop holds a constant area for the remainder of the drop life (drop times are controlled at 0.5, 1, 2, or 5 sec). This operation changes the nature of the experiment in several ways. First, the individual current-time curves consist of a rapid increase in current (as expected for a rapidly growing drop at a normal DME), then the current begins to decrease with an approximate square root of time dependence. This current decay is similar to that expected for diffusion to a stationary spherical electrode (current given by Equation 2 plus a small spherical correction term) except that the initial diffusion layer concentrations are distorted due to the early, rapid drop growth. Thus, the current which is sampled in a sampled DC polarographic experiment is really that of a current sampled chronoamperometric experiment with a very slow potential scan and a distorted diffusion layer. In differential pulse polarography, the diffusion layer also is distorted compared to the normal techniques so that the response will no longer be given by the corresponding theoretical relationships. In pulse polarography, no distortion of the diffusion layer will occur when the initial potential is selected in a region where the species is nonelectroactive. Thus, the pulse polarographic response will be equal to that expected for a constant area electrode. The second major difference with this electrode is that the portion of charging current due to the change in area is eliminated leaving only the minor contribution from the change in potential. Thus, sensitivity should be increased compared to the corresponding normal polarographic technique. In differential pulse polarography, the charging current elimination will be very similar to that described by Vassos and Osteryoung[65] for the alternate drop differential pulse mode since both measurements will be obtained on the same size drop (but not at the same potential when used with most polarographs). The reproducibility of polarograms obtained with this instrument (currently unknown) depends mainly upon the reproducibility of the drop area in successive drops.

3. Reference Electrode

In either the two-electrode or the three-electrode design, the reference electrode may be placed internal or external to the cell. Examples of reference electrodes which could be used internally (in the same cell compartment as the DME) are the mercury pool electrode and the silver/silver halide electrodes. However, the potential of a reference electrode used in this manner is dependent upon the identity and concentration of the anion involved in the electrochemical couple. Thus, the electrolyte must contain a known amount of the anion and cannot contain other ions which would form competing electrochemical couples (sulfides, amines, etc.) or react chemically (oxidize the metal or reduce the cation). For these reasons, most people prefer to use an external reference electrode separated from the cell compartment.

The most common external reference electrodes are the aqueous SCE and the silver/silver chloride electrodes. The mercury/mercurous sulfate/sulfate electrode can be substituted for the SCE when chloride must be eliminated from the cell. These reference electrodes can be directly inserted into the cell when the reference electrode filling solution is compatible with the supporting electrolyte. For example the silver/silver halide electrodes can be modified for use in many nonaqueous solvents by replacing water with the sample solvent when constructing the electrode. However, acetonitrile solutions present some special problems since the silver halide salts are appreciably soluble, and calomel electrodes cannot be used due to disproportionation of mercurous ions. Sometimes, the most acceptable reference electrode is simply a platinum wire inserted into the electrolysis cell. When this procedure is adopted, the reference electrode really does not assume a known potential and its potential may not even be constant, but it does permit the operation of normal three-electrode potential control circuits.

Any of these electrodes can be connected to the test solution by means of a salt bridge or Luggin capillary. The salt bridge provides better separation of incompatible species (e.g., leakage of potassium ions from a SCE causes precipitation in solutions containing perchlorate) or of a nonaqueous solution from an aqueous reference electrode. The only disadvantage of the external reference electrode is the possibility of introducing liquid junction potentials between the two compartments. The liquid junction potential can be quite large (perhaps as much as several hundred millivolts) and variable when an aqueous reference electrode is used with a nonaqueous electrolyte. The liquid junction potential causes a direct offset error of the same magnitude in the apparent reference electrode potential measurement which cannot easily be detected or corrected. Thus, liquid junction potentials make it more difficult to compare values for half-wave potentials obtained in different solvents or obtained by different investigators, but they do not affect the cell operation or stability.

The connection of the reference electrode to the instrument must be made properly. For example, the lead to a SCE should make contact with the metallic mercury phase (only). Otherwise, the reference electrode may function as a platinum electrode in a potassium chloride/mercurous chloride solution. This consideration is especially important when the internal mercury pool reference electrodes are used.

4. Auxiliary or Counter Electrode

The auxiliary or counter electrode functions as the source of current which flows to the DME in the three-electrode cell design. In this capacity, an electrochemical reaction of opposite polarity (usually oxidation) occurs at the auxiliary electrode. A simple platinum wire or foil electrode with about a 1 cm^2 surface area is generally used. The area should be appreciably greater than the DME drop area (about 0.05 cm^2) so that the current is not limited by the reaction at the auxiliary electrode. If the solution contains species which might be removed by electrolysis at the auxiliary electrode or if the auxiliary electrode reaction might produce a chemical species which would react with the sample, the auxiliary electrode can be removed from the cell and placed in a separate compartment.

B. Laboratory Design and Safety

A description of a polarographic laboratory was given by Meites.[10] Many of his ideas have already been mentioned in the appropriate section of this chapter. Others deserve a brief mention here. All connections on tubing carrying mercury between the mercury reservoir and the DME must be made secure with clamps or wire because of the heavy weight of mercury. Also, there are several ways to remove spilled mercury. It can be mechanically picked up with the help of a copper wire or brush (to which it clings and forms an amalgam), by using a vacuum aspirator and small diameter glass tube, or by chemical removal using sulfur. Spillage can be kept localized by the use of stainless steel, fiberglass or plastic trays or pans placed under the equipment to catch spilled mercury.

When one works in an electrochemical laboratory, he is exposed to mercury in the form of vapor and in the form of contamination of equipment from spillage. Special safety precautions are necessary due to its high toxicity. Precautions might include good laboratory cleanup procedures, adequate ventilation, periodic blood or urine tests for mercury, mercury vapor monitors, good personal hygiene before eating or smoking, etc. Other laboratory hazards might arise for those working with large quantities of organic solvents. Many of the common electrochemical solvents (such as acetonitrile, dimethylformamide and dimethylsulfoxide) exhibit some toxic properties.

C. Instrumentation

The instrumentation needed to conduct a polarographic experiment is very simple. There are many commercial instruments available (Table 7) as well as a variety of instrument designs reported in the literature.[11,24,30,124] Modern instruments are almost always based on the three-electrode cell-potentiostat design and incorporate either discrete or integrated operational amplifier circuits. With only a moderate increase in complexity, multifunctional instrument designs can incorporate the circuits needed to perform many other related electrochemical techniques. The most versatile instruments incorporate design features such as control of the DME drop time, other timing circuits for pulse applications or sampling times, sample-and-hold amplifiers, digital control circuits, counters, and digital voltage or ramp generation circuits. In some cases, analog-to-digital converters, digital-to-analog converters, microprocessors, and digital memory may be included to provide complete instrument control, data storage, and data manipulation (scaling, subtraction, smoothing, etc.) capabilities. This tendency towards more complex instrumentation produces more meaningful data and presents data in a better format for the chemist with a minimum increase in complexity of operation. However, the rapid advances in instrumentation mean that most of the books on electrochemical or polarographic techniques are inadequate or out-of-date in their coverage of instrumentation and automation. Recent improvements in instrument design may be found in relevant journals or review articles.[57]

The general-purpose polarograph consists of a potentiostat section to control the voltage applied to the cell, a signal generator section incorporating a constant initial potential and a scan (ramp) generator, and a current measurement section. Some of the desirable features for a polarograph include:

1. An initial potential variable from about −3 to + 3 V vs. the reference electrode with a resolution of 1 mV is required.

2. Variable scan rates including zero (constant potential operation) and perhaps 0.2 to 5 mV/sec in 1-2-5 steps are required. Additional ranges would be needed for some related electrochemical techniques. A provision for recording on an expanded potential scale is helpful for accurate half-wave potential measurements or to examine individual current-time curves.
3. A variable current sensitivity in 1-2-5 or smaller steps from perhaps 0.1 to 50 μA fullscale is required. Additional ranges would be needed for some related electrochemical techniques (e.g., cyclic voltammetry or hydrodynamic voltammetry) or when polarographic instrumentation is used for special applications (e.g., HPLC detectors or amperometric titrations).
4. A variable current offset, up to several times the full scale sensitivity, may be used to offset a preceding wave.
5. Damping of current output, if used, should have variable time constants and an off position.
6. Controlled drop times, if used, should have several selections.
7. An internal dummy cell and well-placed test points are extremely helpful for trouble shooting problems.
8. When used, high level logic and software must be wisely designed to avoid bias while giving the user flexibility and control over parameters. For example, sample-and-hold amplifiers should accurately follow normal changes in signal (current) levels and software baseline selection techniques must handle a variety of peak shapes, overlapped peaks, etc.

Conventional instrument designs will be discussed next; specific instruments designed for automation or which incorporate digital computer control will be discussed separately.

1. Potentiostat

The design concept for the potentiostat section of most electrochemical instrumentation is essentially identical for all electrochemical techniques. The only important variations in potentiostat design are those instruments which feature direct digital control of the potential (those which use a computer in the feedback circuit of the potentiostat) or those which are designed for fast response, low noise, external digital control, automation, etc. DC polarography places very few limitations on the design that may be chosen, although an evolution has occurred as general potential control concepts have matured. Multifunctional instruments generally include a DC polarographic mode.

The older two-electrode instrument designs usually recorded the current by measuring the voltage drop across a load resistor inserted in series with the cell. This load resistor introduced a voltage error proportional to the current since a portion of the applied potential appeared across the resistor rather than across the cell. The voltage drop across the load resistor typically was used as the Y-axis input of a 1 to 100 mV full scale recorder. Time was usually used for the X-axis which could be synchronized (more or less accurately) to the voltage ramp. The measured potentials can be corrected for ohmic losses if the value of the load resistor or the full scale sensitivity of the chart recorder is known and the ohmic losses in solution are estimated. However, this tedious process is rarely applied except with digital systems.

In contrast, almost all three-electrode circuits are based on potentiostat designs which eliminate the use of an uncompensated load resistor for current measurement. In addition, the three-electrode potentiostat circuits are designed to compensate for the cell resistance between the auxiliary electrode and the solution phase near the DME (specifically between the auxiliary electrode and the equipotential line intersecting the reference electrode or its point of physical connection to the cell). This procedure allows a more accurate potential control across the DME-solution interface, and, at the same time, diverts the current flow from a reference electrode-DME path to an auxiliary electrode-DME path.

The remaining (uncompensated) solution resistance is due to the portion of the solution between the reference electrode and the DME, i.e., the resistive drop is equivalent to the difference in equipotential lines through the reference electrode and that through the outer boundary of the diffuse double layer. This resistance is important only in very low conductivity electrolytes (those with low dielectric constant solvents or those having very low concentrations of added ionic salts) or at high current densities (not usually obtained in DC polarography). It is somewhat more important in techniques with a more rapid change in signal (pulse polarography, differential pulse polarography, oscillographic polarography and cyclic voltammetry) and/or larger currents (hydrodynamic or amperometric voltammetry, potential step chronoamperometry, and controlled potential coulometry).

If necessary, these residual potential errors due to ohmic losses can be further reduced or eliminated by careful cell design or positive feedback techniques. When the cell is designed to minimize uncompensated resistance, the reference electrode probe is placed as closely as possible to the DME (working) electrode while maintaining a symmetrical DME-auxiliary electrode geometry conducive to uniform current distribution. Further information and references on cell design can be found in the controlled potential coulometry section of the following chapter.

Positive feedback refers to a procedure in which the voltage applied to the cell is increased by an amount equal to the uncompensated potential drop. Positive feedback in DC polarograph is complicated by the dynamic change in uncompensated solution resistance as the drop grows and falls, since the solution resistance (for a fixed geometry) is mainly dependent upon the drop size.[125-127] Thus a dynamic change in the amount of positive feedback applied to the cell is required. Several schemes have been proposed to accomplish this task, usually based on a dynamic measurement of the cell resistance[128,129] or on a current interruption technique.[130,131] in which the cell current is periodically halted and the true reference electrode potential is sensed. Several general treatments[30,124,132] of the uncompensated resistance problem are available also.

2. Signal Generator

The type of excitation (potential, current, or charge) and its waveform determines the nature of the response and the information content of an electrochemical experiment. Thus, various electrochemical experiments require different types of signal sources. In controlled potential applications, constant potentials, linear potential ramps or scans, potential steps or pulses, and various AC signals may be needed. In polarography, some instruments are designed for DC polarography only while many others are of a multifunctional design.

The initial potential section of a DC polarograph is extremely simple, consisting of a user selectable constant voltage of either polarity. High accuracy in the potential settings is not necessary for routine measurements; but, accurate measurements of half-wave potentials and their shift as a function of electrolyte composition is helpful during method development or in generating data for publication. These applications require an initial potential setting accurate to about 1 mV.

The DC polarographic scan or ramp generator must be linear and reproducible to obtain good polarograms. In many cases, the scan generator and the recorder X-axis will both operate with respect to time so that accurate measurement of potentials will depend on knowing the scan rate accurately. In other cases, the current can be recorded directly as a function of potential using an X-Y recorder.

The polarographic theory assumes that the potential during the life of any one drop is constant. That is, the diffusion layer changes only due to the drop growth and the time since the beginning of the drop and not because of changes in the fraction of material which reacts at the electrode. With typical drop times of 1 to 5 sec and allowing a 1 mV change in potential, the voltage scan rates should be less than 1 to 0.2 mV/sec. In practice, little change is observed even at scan rates of 5 mV/sec. Faster scans, however, will not give a sufficient number of drops to adequately define the wave without a corresponding decrease in drop time. Behavior at shorter drop times is not entirely consistent with normal DC polarographic theory, but still may be usable for quantitative polarography.[133]

In older instruments, the voltage ramp was generally produced from a constant or stepwise selectable voltage source and a mechanically driven potentiometer. In more recent instrumentation designs, the voltage ramp is generally produced by analog integration of a stepwise selectable input signal or integration of a constant signal with a variable integrator time constant. Many signal generators of this type have been proposed, especially for use in cyclic voltammetry which also requires a voltage ramp(s). It is convenient to control the start and duration of the ramps by means of digital logic circuits, especially in the case of cyclic voltammetry where sweep-and-hold or triangular waveforms are required.

The voltage ramp can be generated also by using a digital-to-analog converter (DAC) controlled by clocks and counters or other digital logic. When a DAC converter is used, the size of the potential steps is determined by its resolution (range and number of input bits). Although 10- to 16-bit DACs have been used, about 12 bits are required for adequate resolution (12 bits give 4096 steps). Since the normal DAC range is ± 10 V, a scaling amplifier may be connected to its output to improve resolution while reducing the range to more usable values. If the DAC is used in conjunction with drop timing circuits, the voltage can be held constant during the life of any one drop so that the charging current due to a change in the voltage is eliminated. However, the main contribution to the charging current is the change in drop area with time as stated earlier. If the voltage is varied in steps (with a DAC) but is not synchronized with drop time, the size of any voltage steps which occur during the life of a single drop must be small (probably less than 1 mV) to keep from changing the nature of the experiment.

An extremely flexible type of signal generator has been used in some investigations. This apparatus consists of a DAC controlled by a minicomputer or microprocessor to generate any arbitrary waveform with or without monitoring the actual cell potentials. Real-time modification of the waveform (e.g., to correct for uncompensated ohmic potential losses, to allow decay of current from preceding peaks in cyclic voltammetry, etc.) is possible with a design of this type.

3. Current Measurement

The design of current measuring circuits is dependent upon the type of polarographic experiment. For instance, pulse experiments (pulse polarography and differential pulse polarography) require sampling of the current at specific times in relation to the application of the pulse and at a specific point in the drop time, while AC polarography may require a lock-in amplifier tuned to the frequency of the applied signal or one of its harmonics.

Older DC polarographic instrumentation generally measured the voltage drop across a calibrated resistor as previously described. In new instrumentation, the most common circuits use operational amplifiers to achieve the same result while eliminating the voltage drop in the potential control circuits. Operational amplifier circuits for current-to-voltage conversion in DC polarography are essentially the same as those described for controlled potential coulometry except that smaller current ranges (i.e., greater signal amplifications) are needed.

Because polarography is a comparative technique, extreme accuracy is not needed in the current-to-voltage conversion. However, proportionality between two or more current ranges should be maintained (to 0.1%) so that range changes will not introduce errors. Likewise, zero offsets in the current measuring section are not as critical as in controlled potential coulometry. Some calibration suggestions may be found in the calibration and trouble shooting section of this chapter.

Damping of the current response was routinely used in the past to observe average rather than instantaneous currents during the life of the mercury drop. This damping was generally dictated by slow instrument, galvanometer, or recorder response times. With the advent of modern polarographs and recorders, this type of signal processing is no longer necessary from an instrumental perspective. From an electrochemical perspective, it also is unnecessary and will obscure desired information about current-time response on individual drops. Therefore the use of damping is not advised unless the signal is noisy (near the detection level). Even in this case, digital signal manipulation (averaging of multiple drops, smoothing, Fourier transform processing, etc.) or a better instrument design may provide more reliable signal enhancement.[65,72,134-136] As an alternative to digital processing after the experiment, multiple data points can be acquired with a fast ADC and digitally averaged in real-time to give an average signal with less noise.[137]

One of the most effective devices[65] for minimizing moderate- to high-frequency noise is the use of an integrating analog-to-digital converter (ADC) with an integration time which is a multiple of the line period (1/60 second in the U.S.). This technique averages out line noise and also acts as a low pass filter with a time constant equal to the sampling time but without introducing the distortion that an analog filter with the same time constant would introduce. Analog filters must have a time constant less than about one-seventh of the sampling time to prevent distortion, i.e., to accurately track the signal. One commonly used, commercially available instrument (Model 174A from EG & G Princeton Applied Research, Princeton, N.J.) suffers from distortion of this type,[64] apparently in an effort to suppress noise at low signal levels.

Current offsets may be used to suppress the current due to a major component when looking at the current due to a minor component. This process may tend to increase the detection limit for the small impurity whose wave occurs on the diffusion plateau of the more abundant species. However, the ultimate detection limit is determined by the signal-to-noise ratio where the noise component is composed of electronic noise, residual current, and fluctuations in the current of any more easily electrolyzed species (due to drop-to-drop variations in area, convection contributions, etc.). When the latter noise source predominates, which might occur for a very low impurity concentration with a more easily electrolyzed major component, then the use of current offsets will not enhance sensitivity or improve precision.

Residual current compensation, in principle, should also extend the useful range of DC polarography to lower concentration by permitting the measurement of smaller limiting currents. In the past, some proposals were made to null out the residual current with a current offset proportional to the applied potential. Although the charging current portion of the residual current is proportional to the difference between the applied potential and the E_{pzc}, this procedure does not work well experimentally. Most likely, this is due to the change in slope of the charging current as the applied potential crosses the E_{pzc}, contributions to the residual current other than charging current and the experimental difficulty of selecting the correct slope for the compensation current.

Digital processing techniques seem more promising for elimination of residual current effects.[30] For example, recording a separate residual current curve and then subtracting it from the analytical data on a point-by-point basis is possible. Of course, subtraction of residual current will not correct for those changes in charging current or impurity waves caused by the introduction of the sample into the electrolyte. One recent paper proposed three *in-situ* methods for charging current compensation.[138] All were based on a superimposed AC potential with analog processing of the response to yield a quantity proportional to the DC charging current. Although these techniques correct for dynamic changes in charging current due to changes in potential and drop size, amount of adsorption, etc., they add considerable complexity to the instrumentation and do not correct for other contributions to the residual current. The related techniques of sampled DC (Tast) polarography, pulse polarography, differential pulse polarography, and AC polarography tend to improve the signal-to-noise ratio as discussed previously. Stripping analysis uses an electrochemical deposition step to concentrate the electroactive material (usually a metal) into or onto the electrode surface in order to increase the signal level. Thus a change in the type of polarographic experiment may be required at very low concentrations to improve the signal-to-noise ratio.

Digital measurements of the current are usually made with the aid of an ADC. If the analog signal is changing fast relative to the conversion time of the ADC, a sample-and-hold amplifier (SHA) should be used to provide a constant signal to the ADC. The SHA should be able to track the signal accurately and then hold the value for the time necessary for the ADC conversion. The ADC resolution must be sufficient to measure the current accurately over the range of the residual current to the limiting current. Generally, about 10 to 12 bits covering the appropriate current-to-voltage converter output range would be sufficient. If ADC's with smaller resolution are used (e.g., an 8-bit ADC with an 8-bit microprocessor), programmable gain amplifiers or multiplexed fixed gain amplifiers may be used to effectively extend the dynamic range of the ADC. The use of an integrating ADC for noise reduction has already been mentioned. Integrating ADCs do not require a SHA.

Current measurements in sampled DC polarography and pulse polarography are essentially the same as in DC polarography except that the current is measured at a specific point in the drop time and the output to the recorder is generally held constant until the next sampling time. For differential pulse polarography, the difference in currents measured at two times is required. This may be accomplished by using two SHAs and taking the analog difference or by taking the arithmetic difference between two digitized values. A subtle error can arise, particularly when analog differences are used, if either of the two amplifiers reach their full scale value even though the difference in current remains within the output range. AC polarography requires circuits designed specifically for that technique. They are generally frequency and/or phase sensitive.

4. Calibration and Trouble Shooting

Generally, very little calibration is needed when using modern instrumentation. The primary calibration step will probably be an adjustment of the zero output settings of the various amplifiers as described by the manufacturer. The applied potential calibration can be checked by measuring (three- or four-digit voltmeter) the voltage between the reference and DME electrode connections for a specific applied potential. This voltage may also be available internally in the instrument. Linearity of the voltage scan can be observed by recording the current across an external (or internal) resistor substituted for the cell. The voltage scan rate can be checked simultaneously by measuring the change in potential for a measured time interval. For instruments which display current vs. potential (rather than current vs. time), an exactly calibrated scan rate is not needed for polarography.

The current range may be calibrated by substituting a calibrated resistor (0.1% or better, 100 K to 1 $M\Omega$) for the cell and checking the output for a specific applied potential. The calibration resistor is directly substituted for the cell with a two-electrode polarograph. With a three-electrode polarograph, the calibration resistor should be connected between the reference and DME terminals. The auxiliary electrode can be connected directly to the reference electrode junction or connected through a separate resistor (1 to 10 KΩ) to the reference electrode terminal.

To a beginner, it may seem as if problems arise fairly frequently with a polarograph. After the elimination of operator error (incorrect initial potential, scan rate, or current range settings; incorrect sample preparation; etc.), the most common source of problems is the cell. Substitution of an internal or external dummy cell (constructed in the same manner as for the calibration) will quickly clarify the situation. If the external dummy cell gives the correct response (e.g., 10 μA for 1 V applied to a 100 K Ω resistor and a linear current response for a voltage scan), then the problem is in the cell. If the external dummy cell does not respond, but the internal dummy cell does, check the connections and cables leading to the cell. If the internal dummy cell does not respond, check for trouble shooting aids in the instrument manual. Generally, the trouble can be traced to the ramp generation section (no voltage scan), the current measuring section (no current response but the correct potential is applied to the cell), the power supply (no supply voltages and multiple failure modes), or other specific sections with only minor difficulties. Further refinement of the problem and its repair depend upon the skill of the chemist, the documentation available, and the alternatives.

Within the cell, several problems may occur. In the three-electrode mode, a broken connection in the auxiliary electrode circuit will prevent the application of the applied voltage to the cell. The auxiliary electrode lead (before the break) will be driven to the voltage limit of the output amplifier which may be indicated by a voltage limit indicator on the instrument. Within the cell, the platinum wire or foil used for the auxiliary electrode can be broken. In either case, the indicated current will be zero.

A broken connection in the reference electrode circuit in the three-electrode mode also will lead to voltage limiting since the reference electrode will not be sensing the applied voltage. However, a large current will also be obtained since the auxiliary electrode-DME circuit is still intact and the auxiliary electrode is being driven to its potential limit. A two-electrode instrument will simply not respond, i.e., no current will flow but the reference electrode lead (before the circuit discontinuity) will be at its proper voltage. In many cases, the connection to the reference electrode will be a platinum wire. This wire should be checked to see that it is intact and in proper position.

A broken connection in the DME circuit in either the two- or three-electrode modes will give the correct applied potential, but no current. Since the DME circuit includes a portion of the mercury column and the DME itself, an air bubble in the mercury column or a break in the mercury in the capillary can be the source of this problem. Of course, if the capillary is clogged, no drops will form and no current will be obtained.

Other sources of problems within the cell include an erratic dropping DME which will give noisy or variable currents. The DME should be cleaned (as explained in that section) or replaced in this case. When multiple cell compartments are used, the diaphragms separating the compartments may become clogged (leading to high resistances) or an air bubble may cause a circuit break (e.g., in the Luggin capillary from the reference electrode). Aged agar can exhibit similar high resistances. Excessively high resistances will give noisy polarograms (due to partial loss of voltage control) or may cause voltage limiting in the control circuits.

Mechanical vibration of the cell also can lead to polarograms which have erratic limiting currents. The primary effect for a natural drop time DME is a low erratic current due to the premature fall of the drop. However, higher currents could result from vibrational induced convection. Mechanically controlled drop time DMEs can give similar behavior complicated by action of the drop dislodging process itself. Premature fall of the drop causes a lower current while a larger drop and higher currents are observed when the previous drop was not dislodged. Any of these phenomena (dirty DME, vibration, or irreproducible mechanical drop dislodgment) can usually be detected by inspection of the drop times and current-time profiles for individual drops after recording a normal DC polarogram. Use of a sampled DC mode tends to obscure the cause of the problem.

The remaining problem which may be associated with the cell is leakage of oxygen. As discussed in the section on deoxygenation, oxygen gives two reduction waves which will be superimposed on the polarogram of the sample. Depending upon the relative heights and half-wave potentials, oxygen may cause a distorted wave shape or an abnormally high wave. The presence of oxygen in the cell can be confirmed by recording the polarogram of the supporting electrolyte itself and looking for the oxygen reduction waves. Impurities within the supporting electrolyte can be detected in a similar manner.

The recorder is another source of possible problems. Almost all modern recorders have high impedance inputs so that no appreciable current is drawn from the polarograph. However, the recorder calibration should be checked with known input voltages and times (time-based strip chart recorders) when accurate, absolute current and potential values are needed. The linearity of the recorder can be tested in a similar manner as the linearity of the potential scan on the polarograph. The recorder gain and damping should be adjusted (as described in the operating manual) to values which allow the recorder to accurately follow the input signal. A dirty slidewire or worn contacts on the recorder may give "noisy" signals at a particular spot on the X- or Y-axis. Other recorder malfunctions can be diagnosed using the instrument manual and known input signals.

One electrochemical phenomena may be mistaken for an instrumental problem. This situation arises when a species present in solution is sufficiently reactive to oxidize the mercury electrode at open circuit conditions (no electrodes connected). Thus, the polarogram observed (as the potential is scanned from positive to negative values) consists of a very large anodic current (due to the oxidation of mercury) which rises immediately into a normal cathodic current (due to the reduction of the species in solution) without any residual current separation between the two processes. The diffusion current measurement in these cases must be made from a previously recorded residual current curve or from the zero current line (if the diffusion current can be measured near the potential where the residual current is zero).

D. Digital Processing and Control

The increasing availability of computers and logic circuits have given flexibility in the choice of components for a particular system. These could be classified as analog, hardware logic circuits, microprocessors, and micro-, or minicomputers. Each has its own strengths and weaknesses. For example, analog circuits respond rapidly to changes in signal levels, are inexpensive and easy to implement in simple circuits, are capable of high accuracy (0.1% or better), and circuit descriptions and operating theory for instruments using these devices are readily available.

On the other hand, digital logic circuits are outstanding for their precision in timing and for their switching and control capabilities. However, they are not well suited to generation of waveforms, control of cell potentials, etc. Both analog and digital circuits, due to their hardwired interconnections, are not easily changed as experimental needs change. Microprocessors or microcomputers combined with external logic, however, are more easily modified and allow at least some degree of decision making and data storage. Finally, a small or minicomputer can offer the capability for extensive software programs with variations necessary for many types of experiments, the capability to acquire and store relatively large amounts of data and to process that data, perhaps even making real-time modifications of the experiment dependent upon the data. However, computers do not respond as quickly to changes in data as analog devices and, thus, do not make good, fast response potentiostats. Neither do they make good timing circuits because of their software overhead which causes variable delays in response dependent upon other programs or upon the internal structure of the computer. Thus, the most versatile instrument might consist of a computer to provide flexibility, to control the experiment and to analyze data; digital circuits for timing, switching, and digitizing; and analog circuits for potential control and current measurement. Several instruments have been proposed within this general context, many of which have been reviewed by Fisher[30] and Harrar.[132] In addition, Osteryoung[136] has reviewed the use of on-line computers in electrochemistry and Smith[81] has reviewed specific applications for AC polarography. Several specific examples which illustrate these concepts applied to polarography are discussed below. The author anticipates that future system design will incorporate many of the ideas which are pioneered in these examples.

Ramaley[139] has described a digitally controlled potentiostat for electrochemical studies in polarography, sampled DC polarography, staircase voltammetry (cyclic voltammetry with a staircase voltage scan rather than a linear scan), and square wave polarography. The initial, final, and scan reversal potentials; the initial sweep direction; sweep rate; potential step size; and the time of cell current measurement are digitally selectable either manually or with a computer. The sweep rate was selected using 12 control bits, the sweep range with 2 control bits (0.5, 1, 2, or 3 V), the square wave generator by 5 bits, and the output current configuration by 3 bits. The potentiostat section was a conventional three-electrode analog design. The signal generator consisted of a 12-bit, clock driven DAC with the scaled sweep range to provide less than a 1 mV step size (synchronized with drop dislodgement) and a sweep accuracy of 0.02%. A variable initial potential and a square wave generator were included in the signal generator. The current measurement section consisted of the current-to-voltage converter, two SHAs and a differential amplifier which could be digitally switched into various configurations for maximum flexibility. The drop fall was detected by an optical drop detector for accurate timing of current measurements. When the data were brought into the computer, a linear least squares procedure was used to determine the half-wave potentials (± 0.1 to 0.2 mV), n values (± about 1%), and diffusion coefficients (± about 1%) for a series of five experiments each on 0.1 m*M* Pb(II) and 0.1 m*M* Cd(II) in potassium chloride in the sampled DC polarographic mode.

Vassos and Osteryoung[65] described a low-noise pulse polarograph configured for automation. It was suitable for sampled DC polarography, pulse polarography, differential pulse polarography, and alternate drop differential pulse polarography. The design objectives for this instrument included very low background noise, good noise rejection, internal timing circuits (not dependent upon the external computer), low cost, simplicity, and manual or remote (computer) controlled operation.

Clem and Goldsworthy have described two general purpose instruments. The first[140] consisted of a collection of function modules which could be interconnected to perform almost any pulse or DC controlled potential electrochemical experiment. The potentiostat design was a conventional analog, three-electrode design. The signal generator provided a linear ramp, stairsteps, and constant voltages. The output section consisted of a current-to-voltage converter and a multichannel analyzer for storage and digital manipulation of data (differentiation, etc.) to obtain a current measuring precision of 0.02%. Their second instrument[141] was termed a bipolar digipotentiograter. It functions as a potentiostat through the pulsed injection of constant charge increments to maintain the correct electrode potential. Thus, the current is proportional to the frequency of the charge injection and could be measured to a precision of 0.01%.

Two instruments which include the computer in the feedback loop for polarography have been described. The first, by Pomernacki and Harrar,[142] has a system response time of about 8 msec, sufficient for polarography or other electrochemical techniques with slow response. The cell potential was generated by a 12-bit DAC, sensed by a 12-bit ADC (using a voltage follower with gain) at a 1 kHz rate, and a feedback control alogrithm of the proportional-plus-integral type was applied using the calculational powers of the computer. Thus, the applied potential was dynamically calculated to adjust for changes in solution resistance or other factors affecting the voltage at the reference electrode. In practice, it was possible to incorporate positive feedback so as to minimize the rise time of the signal within the limits imposed by the time response of the control system and the accuracy of the control alogrithm.

The second instrument which incorporates direct digital control of the cell was described by Bos.[143] In this case, the auxiliary electrode was driven by a 12-bit DAC synchronized to the drop dislodger (also driven by a DAC). The scan rate, initial potential, and drop time were variable in 1 mV/drop, 1 mV, and 1 sec increments, respectively. A load resistor between the working electrode and ground and two 11-bit ADCs were used to measure the current and potential. After the data were obtained (only the sampled DC mode is described), the residual current was determined by a linear least-squares line through 20 points selected 40 points before the start of the wave. Then a multiparametric curve-fitting alogrithm was used to determine the diffusion limited current (±1 or 2%), half-wave potential (±1 or 2 mV), and slope of the log plot (±0.5 or 2 mV) for Cd(II) in potassium chloride (a system with a well defined wave) or for K(I) in an ethanol/tetraethylammonium perchlorate electrolyte (a very marginal system with the potassium wave nearly merged with the wave due to the reduction of the electrolyte), respectively.

A general purpose instrument available commercially (Environmental Science and Engineering Corp., Mt. Juliet, Tenn.) was recently described by Matson et al.[144] This instrument, although not digitally configured, is novel in design philosophy. It was assumed that background corrections, especially at low concentration levels, are difficult to achieve by normal procedures due to differences in hydrogen evolution, impurities, electrode area, or history (for solid electrodes) between the blank and sample runs. Thus, the design provided a variety of methods to correct for electrode and solution background during acquisition of the signal. This approach, although logically conceived, has no convenient theoretical justification for the corrections implemented. Thus, its success will depend upon its ability to perform in typical analytical assays and upon its acceptance by users.

E. Automation

Automated methods have been developed to increase the productivity of workers and, in some cases, to increase the reliability of the results. Productivity is increased by freeing the analyst from the routine details in the analysis of the sample and in operation of the equipment over extended intervals of time. Reliability can be increased by the reproducible handling of multiple samples, although it can also be decreased if key steps require judgmental decisions. Within this context, polarographic techniques are well suited to automation from the sample preparation stage through at least the data acquisition portion of the assay. Much less work has been devoted to the automated data analysis, partially because it tends to involve judgment decisions.

An automated polarographic unit will include the normal potentiostat, signal generator, and current measuring components, as well as devices for sample preparation and transport of material to the electrochemical cell. In addition, automated equipment could involve data processing techniques such as those used by Ramaley[139] or Bos[143] which are noted above.

The electrochemical cell on automated equipment can be designed for either flowing solutions or static solutions. The DME has been used with either configuration. Tubular solid electrodes of platinum or graphite also have been used for flowing systems. Efficient deoxygenation has been difficult to achieve with many of the proposed cell designs.

A practical problem exists within the sample measurement step. The current required for the calculation of concentration is actually the difference between the observed limiting current and the residual current. However, the simplest experimental arrangement is the monitoring of the current at a single potential. Thus, the residual current must be obtained from a separate measurement on a blank solution and the appropriate corrections made to the sample current. This procedure is viable when the sample itself does not change the residual current (due to impurities within the sample or adsorption of sample components). Alternatively, the current can be measured at multiple potentials either once for the residual current and once for the limiting current, or by recording the entire polarogram.

The instrument configuration, data handling, and cell design must be considered when selecting an automated system. Only one instrument company (EG & G Princeton Applied Research, Princeton, N.J.) is currently marketing a system designed specifically for automated polarography. It requires prior dissolution of the sample, i.e., the prepared solutions must be stable in solution for the system to be useful. This section will briefly describe several approaches to automation with the realization that one has a limited choice for buying equipment or must construct equipment to perform the required tasks. Other approaches to these problems are summarized in the review by Foreman and Stockwell.[29]

Over 10 years ago, Lento[145] presented a design for an automated polarographic unit based on an AutoAnalyzer® system (Technicon Instrument Corp., Tarrytown, N.Y.) and a flow cell. This system used a two-electrode cell design with a DME and a mercury pool auxiliary electrode. Deoxygenation was achieved by using nitrogen as the separator gas. The oxygen dissolved in the sample equilibrated with the nitrogen bubbles which were later removed by the debubbler. The linearity of response appeared good although statistical evaluations were not presented. The analysis rate was 20 samples per hour.

Cullen et al.[146] later used the Lento cell design with a better deoxygenation system and a modern, three-electrode polarograph. In this case, the entire polarogram was recorded in the sampled DC mode (some data were presented for the differential pulse mode) using the autoanalyzer to control the polarograph. A SolidPrep® unit for dissolution of tablets and capsules and a filter unit (both from Technicon Instrument Corp., Tarrytown, N.Y.) were added to increase the types of samples which could be accommodated. In one case, a sample rate of 15 /hr was used while achieving a precision of 1.4% (CV) and a 99% recovery.

Lund and Opheim, in a series of papers,[147-149] have described another system based on the autoanalyzer. In this case, a revised flow cell was designed using a DME with a platinum auxiliary electrode located on opposite sides of the flowing stream and a downstream reference electrode (Ag/AgCl). The current was measured at a constant applied potential. They extensively studied the effects of cell design, solution flow, deoxygenation techniques, position of the debubbler, and the use of an expanded recorder scale to increase precision. Better precision (but longer response times) was found when the pump was stopped between aspirations of adjacent samples in order to avoid the introduction of air into the system. With the uncovered samples, argon was found to give better deoxygenation than nitrogen, presumably because of its blanketing effect. A precision of 0.3 to 0.6% (CV) was found for several assays performed under the optimum conditions.

The design proposed by Flann[150] is different in that the electrodes are lowered into the sample container of the autoanalyzer. Natural drop times, monitoring of average current at a constant potential, and correction for blanks were used. This system suffered from poor temperature control, carryover between samples, and poor data processing, but achieved a precision of 1.1% (CV) at a sample rate of 70 /hr for a nitroglycerin assay.

A microprocessor (Model 8080, MMD-1 System®, E & L Instruments Inc., Derby, Conn.) controlled automated unit designed in the author's laboratory has recently been described.[31] It includes a 25-position solid sampler unit which dissolves the samples in variable amounts of a two-component electrolyte with stirring after each addition. The dry samples are weighed directly into the sample tube to reduce transfer errors and the dissolution step is performed just prior to analysis to minimize decomposition of aged sample solutions. The dissolved sample is then displaced into the cell, with a provision for two rinses before the final fill. The rinses prevent carryover between samples. The solution is deoxygenated with nitrogen dispersed through a sintered glass frit fitted into the bottom of the cell. Finally, the polarogram (variable scan length and number of scans per sample) is obtained. At the present time, the microprocessor controls all of the functions described above as well as starting a strip chart recorder. The initial potential, scan rate, and current range are manually set on the commercial polarograph (Model 174A, EG & G Princeton Applied Research, Princeton, N.J.) and data are measured manually. A microcomputer system is currently being added to provide data acquisition, data storage, and data processing capabilities. The precision of the automated system was 1.1% (CV for sample vs. standard, single scan, n = 26) for the sampled DC mode and 0.8% (CV for sample vs. standard, single scan, n = 24) for the differential pulse polarographic mode for the assay of cefamandole nafate, a cephalosporin antibiotic. Approximately 20 min are required per sample and more than 40 samples can be assayed per day including an overnight run.

VII. PHARMACEUTICAL APPLICATIONS

Many practical pharmaceutical applications of polarographic and voltammetric techniques have been reported. Electrochemical techniques are suitable for both inorganic and organic species. Some of the organic functional groups which are directly electroactive have been listed (Tables 1 and 2); other compounds can be analyzed after derivatization. The use of solid electrodes is possible for the investigation of those compounds which are electroactive in a potential region where mercury is oxidized. Polarographic and voltammetric techniques can be used for quantitation of the drug substance, its formulation, or its impurities and degradation products; to monitor reactions or determine kinetic parameters;[151] and as detectors for other analytical techniques such as HPLC or titrations. Sensitivity and specificity can be increased by use of techniques such as differential pulse polarography when required for the analysis of dilute solutions or complex mixtures, such as many biological samples. Depending upon the application, separation by conventional techniques (typically chromatography or extraction) may be required prior to analysis. This section will discuss some generalized electrode processes and guide the reader to literature containing specific examples.

A. Inorganic Compounds

Electrochemical determinations of inorganic compounds have an extensive history and many applications have been reported. For example, Meites[10] has reported the half-wave potentials for metallic ions in a variety of supporting electrolytes. Others such as Brezina and Zuman[107] and Kolthoff and Lingane[108] have reported applications specific for pharmaceutical compounds.

Many organometallic compounds (e.g., thimerosal) or salts can be reduced directly at the DME. In other cases, the free metal must be liberated prior to analysis by techniques such as ignition or acidic decomposition. Unfortunately, destruction of the compound prevents differentiation between the compound of interest and metal containing impurities or degradation products. Additional examples of applications can be found in several review series.[152,153]

B. Organic Compounds

Organic electrode reactions have been classified by Eberson and Nyberg[6] according to the final mode of product formation as: (1) pure electron transfer, (2) conversion of functional groups, (3) addition, (4) coupling, (5) substitution (6) elimination, or (7) cleavage. Some functionalities will tend to react almost exclusively by one of these modes; other functional groups will give products which vary in composition depending upon factors such as molecular structure, pH or proton availability, electrolyte composition, relative rates (kinetics) of possible reaction paths, stereochemistry of the reactant, extent of adsorption, electrode potential, or electrode material.

The electrochemistry of organic compounds includes reactions of a majority of the organic functional groups. Thus, a complete discussion of organic electrochemistry is obviously beyond the capabilities of this review chapter. A few of the reaction types will be discussed with references to more extensive investigations of electrode reactions and pharmaceutical applications. Thus, two modes of entry into the electrochemical literature will be available; reviews of specific applications and the electrochemistry of functional groups.

Pure electron transfer reactions without coupled chemical reactions are much less common in organic electrochemistry than in inorganic electrochemistry. For example, radical ions (or neutral radicals) are usually unstable in aqueous solutions. Radical anions will react rapidly with hydrogen ions, unless the solvent is aprotic and proton donors have been removed. Radical anions will react with other electrophiles also. Radical cations will react with water or other nucleophiles to give a loss of hydrogen ion or addition of the nucleophile. Either radical species can disproportionate, couple (dimerize), or undergo a second electron transfer step (usually more difficult than the first). Products of a two-electron transfer (dianions or dications) tend to have even greater reactivity. On the other hand, highly conjugated species or neutral radicals may tend to have lower reactivities. Although the products of a simple-electron transfer may be stable enough to obtain spectra or to investigate their chemical behaviors, they are rarely stable indefinitely. In fact, Elving[8] considers the single-electron addition as the first step in his formal mechanism for a generalized reduction. Of course, these distinctions are purely arbitrary on a time scale which includes the continuum of behaviors from an essentially instantaneous chemical reaction following the electron transfer (EC; heterogeneous chemical reaction on the electrode surface or homogeneous chemical reaction in the solution double layer) to the quasi-stable electrolysis product.

Products formed by functional group conversion refer to electrochemical reactions in which the initial functional group is reduced or oxidized to another functional group. This formalism would include the more-or-less related couples of nitro-nitroso-hydroxylamine-amine, azoxy-azo-hydrazine, thiol-disulfide, and quinone-hydroquinone.

Functional groups which can be reduced to give a variety of products dependent upon conditions include carbonyls, imines, oximes, hydrazones, carbazones, N-oxides, sulfoxides, and sulfones. It is apparent that this classification of reactions would include many pharmaceutical compounds as well as a variety of chemical types.

Generally, nitro compounds are relatively easily reduced (4e) to the hydroxylamine in neutral to slightly acidic solutions, as are the corresponding non-tautomerized nitroso compounds (2e). In certain situations, the hydroxylamine may be further reduced (2e) to the amine. However, the reverse oxidations are more complex and occur at potentials which require the use of solid electrodes. For example, oxidation of aliphatic amines tends to give substitution or dimerization while oxidation products of aromatic amines tend to couple to give a more easily oxidizable intermediate (ECE mechanism), although both examples are influenced by substitution at the nitrogen and on the adjacent atom. Azoxy compounds, which can be produced by anodic oxidation of hydroxylamines, can be reduced (4e) to the hydrazine. The intermediate azo compound may exhibit a less difficult (and sometimes nearly reversible) 2e reduction to the hydrazine. The thiol-disulfide couple is unique in that its behavior is intimately linked to the choice of electrode materials. At a platinum electrode, the thiol-disulfide couple is interconvertible although the disulfide may tend to be insoluble (in aqueous solutions) and form a film on the electrode. At the mercury electrode, the oxidation product of the thiol has the form of the mercurous salt of the thiol [$(RS)_2Hg_2$] while the free thiol is the reduction product of the disulfide. In aprotic solvents, the thiol or disulfide may be oxidized to the corresponding sulfoxide or sulfone. The quinone-hydroquinone couple tends to be one of the most nearly reversible couples in organic electrochemistry in aqueous solutions. Numerous examples of these reactions are available in the literature. The behavior of other groups is more variable depending upon the conditions previously mentioned. In all cases, the reader is directed to the references below for further information and exceptions to the general behavior.

Addition reactions are closely related to those which interconvert functional groups. For example, the reductions of imines, nitriles (to amines), hydrazones, or carbazones could be assigned to either classification. The most important "pure" addition reaction is the addition of hydrogen to unsaturated molecules, particularly aromatic hydrocarbons, alkynes, and aryl-conjugated olefins. Unsaturated compounds can undergo anodic addition reactions with other nucleophiles to give alkoxylation, acyloxylation, hydroxylation, acetamidation, cyanation, fluorination, or similar reactions when the electrolysis is conducted in solutions containing the corresponding nucleophile.

Coupling reactions likewise could be considered a special form of the addition reaction in which the substrate and reactant are identical (except, perhaps, with respect to electronic configuration). These reactions are important synthetically, particularly in producing bifunctional molecules from monofunctional monomers. The most important cathodic (or reductive) coupling is the electrohydrodimerization reaction in which coupling of two activated olefins occurs with the addition of two hydrogen ions. Baizer pioneered the preparative aspect of this mechanism with the reaction for the production of adiponitril directly from acrylonitrile. Hydrodimerization also can occur with carbonyls, especially α,β-unsaturated species. Anodic coupling reactions include the Kolbe reaction for the production of R−R from RCOOH. The anodic coupling reactions of aromatic amines and thiols (at platinum electrodes) have already been mentioned. Analytically, these coupling reactions may form insoluble polymeric films on the electrode surface which can then cause problems with precision, especially at solid electrodes.

Substitution reactions also are interesting synthetically because a carbon-hydrogen bond can be transformed into a carbon-heteroatom bond and vice-versa. Reductive substitution frequently involves substitution of the electrophilic hydrogen ion for a leaving nucleophilic group on the substrate molecule. Anodic substitution reactions can occur with the same types of nucleophiles that participate in addition reactions. Substitutions are most likely to occur in activated molecules (aromatic compounds or compounds with carbon atoms activated by aryl, vinyl, or similar groups), although it is not limited to those cases. Pure substitution reactions are not generally suitable for quantitative analysis methods.

Elimination reactions could be considered to be the reverse of addition reactions. However, only a formal distinction exists between elimination and cleavage reactions and these will be considered together. Cleavage is obviously involved in substitution reactions also. Cleavage can occur with carbon-carbon, carbon-oxygen, carbon-nitrogen, carbon-halogen, oxygen-oxygen, nitrogen-oxygen, nitrogen-nitrogen, and various other bond types. Carbon-carbon bonds can be reductively cleaved as in substituted aromatic nitriles; or oxidatively cleaved (the Kolbe reaction involves cleavage of a carbon-carbon bond). Anodic decarboxylation with the formation of a carbonium ion and subsequent reaction with the solvent (such as water, which produces the alcohol) or another species is another example of the carbon-carbon cleavage reaction. Although carbon-oxygen bonds can be either reductively or oxidatively cleaved, the most important example is the oxidation of phenolic compounds. Only a few examples are known for the cleavage of the carbon-nitrogen bond, although it has been reported in heterocyclic compounds.[154-156] In general, the ease of reductive cleavage for carbon-halogen bonds is in the same order as the stability constant for the formation of the mercury halide (I>Br>Cl>>F). Also, allyl halides are easier to reduce than alkyl halides which are easier to reduce than vinyl halides. The carbon-halogen reduction is facilitated by the presence of electron withdrawing groups. For example, 5-bromo-5-nitro-*m*-dioxane is spontaneously reduced at a mercury electrode whereas the unsubstituted compound would be reduced only at very negative potentials.[76] Oxygen-oxygen bonds are easily cleaved reductively in a number of peroxy compounds. Reductive cleavage of nitrogen-nitrogen bonds may occur with hydrazones, hydrazides, azides, triazines, nitrosoamines, and in nitrogen heterocyclics. Other bond types can also be cleaved including: nitrogen-oxygen as in reduction of amine oxides, sulfur-oxygen as in the reduction of sulfoxides or sulfones, carbon-sulfur as in the reduction of thioethers or sulfonium; sulfur-sulfur as in the previously discussed reduction of disulfides, and quaternary compounds. Quaternary compounds which can be reductively cleaved include quaternary arsonium, phosphonium, stilbonium, and ammonium in order of reduction half-wave potentials from the most easily reduced to the least easily reduced.

1. Electrochemical Reviews of Specific Compounds

Several review series contain references to recently published articles on polarographic methods developed for pharmaceutical and biochemical applications.[152,153] Meites and Zuman[157] have published the most extensive compilation of electrochemical data on specific compounds. Others[10,21] have included less extensive tabulations of data as part of a discussion of polarography. Many research articles report data on single components such as a specific vitamin, mineral, or drug. Reviews concerned specifically with pharmaceutical and biological compounds include:

1. Brezina and Zuman:[107] metallic ions, anions, oxygen, peroxides, sulfur compounds, quinones, halogenated compounds, unsaturated hydrocarbons and acids, aldehydes, ketones, sugars, oxygen heterocyclics, nitrogen-containing compounds, nitrogen heterocyclics, alkaloids, vitamins, hormones, steroids, various electrooxidations, proteins, and enzymes; this source also discusses indirect determinations (derivatizations) and amperometric titrations
2. Brezina and Volke:[158] metallic ions, oxygen, nucleic acids and related compounds, and thiols and proteins; this reference also discusses derivatization reactions, analyses of formulations, and the use of polarography to monitor biologically important reactions

3. Hoffmann and Volke:[5] several specific examples are presented as well as sample pretreatments for solutions, tablets, ointments, creams, suppositories and similar preparations, and plant samples; preliminary chromatographic separations, indirect determinations, electrodes, solvents, and brief descriptions of polarographic methods are included
4. Underwood and Burnett:[159] pyridine coenzymes and nicotinamide model compounds, purine and pyrimidine derivatives, flavins, quinones, and iron porphyrins; a discussion of biological redox systems involved in metabolism is also included
5. Dryhurst:[11] pyrimidines, purine and pyrimidine nucleosides and nucleotides, polyribonucleotides and nucleic acids, pteridines, isoalloxazines, flavins and flavin nucleotides, pyrroles and porphyrins, and pyridines and pyridine nucleotides; a brief description of polarographic techniques and instrumentation is included
6. Zuman:[7] specific pharmaceutical, medicinal and biochemical applications, amperometric titrations, indirect methods, and separation techniques
7. Siegerman[160] and Unterman and Weissbuch:[161] antibiotics
8. Lindquist and Farroha:[162] differential pulse polarography of vitamins
9. Woodson and Smith:[163] DC and AC polarography of various pharmaceuticals in aprotic solvents (barbiturates, salicylates, corticosteroids, alkaloids, sulfa drugs, estrogens, etc.)
10. Brdicka, Brezina, and Kalous:[164] proteins

2. Electrochemical Reviews of Functional Groups

Some reviews are oriented towards functional groups or molecular types. Such information can be extremely helpful in predicting or interpreting behavior of compounds which have similar functional groups. These reviews include:

1. Siggia:[106] hydroxyl compounds, carbonyls, carboxylic acids and their derivatives, peroxides, unsaturated compounds, amines, hydrazines and hydrazides, diazonium salts, thiols, disulfides, and dialkyl thioethers
2. Kolthoff and Lingane:[108] inorganic compounds, unsaturated hydrocarbons, halogenated organics, carbonyls, quinones, organic acids and their derivatives, nitro and related compounds, sulfur compounds, oxygen heterocyclics, nitrogen heterocyclics and proteins; this reference also contains information on amperometric titrations and some derivatization reactions
3. Baizer (oriented towards synthetic aspects): hydrocarbons,[165,166] organohalides,[167] nitro compounds,[168] saturated carbonyls,[169] α,β-unsaturated carbonyls,[170] carboxylic acids and their derivatives,[171,172] onium compounds,[173] amines,[174] oxidation of oxygen containing compounds,[175] oxidation of sulfur-containing compounds,[176] heterocyclics,[156] and organometallics;[177] this comprehensive text also includes discussions of reductive coupling,[178] oxidative coupling,[179] cleavage,[180] anodic substitution,[181] the stereochemistry of electrode processes,[182] and a number of other topics
4. Mann and Barnes (nonaqueous electrochemistry):[26] hydrocarbons, decarboxylation, anodic substitution, carbonyls, carbon-halogen bonds, carbon-oxygen single bonds, amines, amides and ammonium salts, heterocyclic aromatic and other nitrogen compounds, nitro and nitroso compounds, organosulfur compounds, and organometallics and inorganic compounds
5. Adams:[15] anodic oxidation of hydrocarbons, amines and amides, acids, aldehydes, alcohols, halides, sulfur compounds, and heterocyclic compounds

6. Volke[154] and Lund:[155] nitrogen, oxygen, and sulfur heterocyclic compounds
7. Lund:[183] aliphatic compounds

Certain other reviews are oriented towards a broader area of organic electrochemistry. For example, Eberson and Nyberg[6] discuss electron-transfer mechanisms, the role of adsorption, and the relationship of structure and electrochemical activity. Elving[2] includes a general treatment of qualitative and quantitative analysis.

3. Electrochemical Methods in The U.S. Pharmacopeia and *The National Formulary*

Polarographic methods are described in *The U.S. Pharmacopeia*[184] for the following compounds:

1. Acetazolamide tablets: assay
2. Azathioprine tablets: assay
3. Chlorothiazide tablets: identification and assay
4. Dichlorophenamide: identification and assay
5. Bis(2-ethylhexyl)maleate impurity in dioctyl sodium sulfosuccinate
6. Nitrilotriaceticacid (NTA) impurity in Edetate disodium (disodium EDTA)
7. Ethacrynic acid tablets and ethacrynate sodium for injection: assay
8. Methazolamide tablets: identification and assay
9. Nitrofurantoin tablets: identification and assay
10. Procarbazine hydrochloride capsules: identification and assay
11. Thimerosal: assay
12. Phenylmercuric nitrate: assay

Polarographic methods included in The *National Formulary*[185] are:

1. Dienestrol cream and tablets: assay
2. Bis(2-ethylhexyl)maleate impurity in Dioctyl calcium sulfosuccinate
3. Methylclothiazide tablets: identification
4. Xylometazoline hydrochloride: assay

REFERENCES

1. Baizer, M. M., *J. Electrochem. Soc.*, 124, 185C, 1977.
2. Elving, P. J., Voltammetry in organic analysis, in *Electroanalytical Chemistry*, Nürnberg, H. W., Ed., Wiley-Interscience, New York, 1974, chap. 3.
3. Laitinen, H. A., *J. Electrochem. Soc.*, 125, 250C, 1978.
4. Flato, J., *Anal. Chem.*, 44(11), 75A, 1972.
5. Hoffmann, H. and Volke, J., Polarographic analysis in pharmacy, in *Electroanalytical Chemistry*, Nürnberg, H. W., Ed., Wiley-Interscience, New York, 1974, chap. 4.
6. Eberson, L. and Nyberg, K., Structure and mechanism in organic electrochemistry, in *Adv. Phys. Org. Chem.*, Vol. 12, Gold, V. and Bethell, D., Eds., Academic Press, New York, 1976, 2.
7. Zuman, P., *Organic Polarographic Analysis*, Pergamon Press, New York, 1964.
8. Elving, P. J., *Can. J. Chem.*, 55, 3392, 1977.
9. Smyth, W. F., *Proc. Soc. Anal. Chem.*, 10, 9, 1973.
10. Meites, L., *Polarographic Techniques*, 2nd ed., Wiley-Interscience, New York, 1965.
11. Dryhurst, G., *Electrochemistry of Biological Molecules*, Academic Press, New York, 1977.
12. Heitbaum, J. and Vielstich, W., *Angew. Chem. Int. Ed.*, 13, 683, 1974.
13. Cauquis, G., Basic concepts, in *Organic Electrochemistry*, Baizer, M. M., Ed., Marcel Dekker, New York, 1973, chap. 1.
14. Adams, R. N., *J. Pharm. Sci.*, 58, 1171, 1969.
15. Adams, R. N., *Electrochemistry at Solid Electrodes*, Marcel Dekker, New York, 1969.
16. Piekarski, S. and Adams, R. N., Voltammetry with stationary and rotating electrodes, in *Physical Methods of Chemistry*, Vol. 1, Part IIA, Weissberger, A. and Rossiter, B. W., Eds., Wiley-Interscience, New York, 1971, chap. 7.
17. Brown, E. R. and Large, R. F., Cyclic voltammetry, AC polarography and related techniques, in *Physical Methods of Chemistry*, Vol. 1, Part IIA, Weissberger, A. and Rossiter, B. W., Eds., Wiley-Interscience, New York, 1971, chap. 6.
18. Smith, D. E., AC polarography and related techniques: theory and practice, in *Electroanalytical Chemistry*, Vol. 1, Bard, A. J., Ed., Marcel Dekker, New York, 1966, 1.
19. Bates, R. G., Chm., Analytical Chemistry Division, Committee on Electroanalytical Chemistry, *Pure Appl. Chem.*, 45, 81, 1976.
20. Muller, O. H., Polarography, in *Physical Methods of Chemistry*, Vol. 1, Part IIA, Weissberger, A. and Rossiter, B. W., Eds., Wiley-Interscience, New York, 1971, chap. 5.
21. Heyrovsky, J. and Zuman, P., *Practical Polarography*, Academic Press, New York, 1968.
22. Smyth, W. F., *Proc. Anal. Div. Chem. Soc.*, 1975, 187.
23. Brooks, M. A., Polarographic Analysis in the Development of a Drug, presented at 169th American Chemical Society National Meeting, Philadelphia, April, 1975, available from the American Chemical Society Lab Professional Audiovisual Dept.
24. Shain, I. and Evans, D. H., Electroanalytical chemistry, American Chemical Society Short Course, American Chemical Society, Washington, D.C., 1969.
25. Brooks, M. A., de Silva, J. A. F. and Hackman, M. R., *Am. Lab.*, 5(9), 23, 1973.
26. Mann, C. K., and Barnes, K. K., *Electrochemical Reactions in Nonaqueous Systems*, Marcel Dekker, New York, 1970.
27. Badoz-Lambling, J., and Cauquis, G., Analytical aspects of voltammetry in non-aqueous solvents and melts, in *Electroanalytical Chemistry*, Nürnberg, H. W., Ed., Wiley-Interscience, New York, 1974, chap. 5.
28. Zuman, P., *Electrochim. Acta*, 21, 687, 1976.
29. Foreman, J. K. and Stockwell, P. B., *Automatic Chemical Analysis*, Wiley-Halstead Press, New York, 1975, chap. 2.
30. Fisher, D. J., Advances in instrumentation for DC polarography and coulometry, in *Electroanalytical Chemistry*, Nürnberg, H. W., Ed., Wiley-Interscience New York, 1974, chap. 1.

31a. Cooley, R. E. and Stevenson, C. E., presented at 30th Pittsburgh Conference on Analytical Chemistry and Applied Spectroscopy, Cleveland, March, 1979, Abstract 651.

31b. Cooley, R. E., Stevenson, C. E., and Rickard, E. C., A microprocessor controlled scanning polarograph for solution labile compounds", *J. Automated Chemistry*, 2, 60, 1980.

32. Bates, R. G., Chm., Analytical Chemistry Division, Committee on Electroanalytical Chemistry, *Pure Appl. Chem.*, 45, 131, 1976.
33. Koryta, J., Dvorak, J., and Bohackova, V., *Electrochemistry*, Methuen and Co., London, 1970.
34. Guidelli, R., Chemical reactions in polarography, in *Electroanalytical Chemistry*, Vol. 5, Bard, A. J., Ed., Marcel Dekker, New York, 1971, 149.
35. O'Brien, R. N. and Dieken, F. P., *J. Electroanal. Chem.*, 42, 25, 1973.
36. O'Brien, R. N. and Dieken, F. P., *J. Electroanal. Chem.*, 42, 37, 1973.

37. Mohilner, D. M., The electrical double layer. Part I: Elements of double-layer theory, in *Electroanalytical Chemistry*, Vol. 1, Bard, A. J., Ed., Marcel Dekker, New York, 1966, 241.
38. Kubicek, P., *J. Electroanal. Chem.*, 78, 161, 1977.
39. Nishihara, C. and Matsuda, H., *J. Electroanal. Chem.*, 73, 261, 1976.
40. Oldham, K. B. and Parry, E. P., *Anal. Chem.*, 40, 65, 1968.
41. Mairanovskii, S. G., *Catalytic and Kinetic Waves in Polarography*, Plenum Press, New York, 1968.
42. Ruzic, I. and Smith, D. E., *J. Electroanal. Chem.*, 58, 145, 1975.
43. Nishihara, C. and Matsuda, H., *J. Electroanal. Chem.*, 51, 287, 1974.
44. Matsuda, H., *J. Electroanal. Chem.*, 56, 165, 1974.
45. Nicholson, R. S., Wilson, J. M., and Olmstead, M. L., *Anal. Chem.*, 38, 542, 1966.
46. Ruzic, I., Sobel, H. R., and Smith, D. E., *J. Electroanal. Chem.*, 65, 21, 1975.
47. Ruzic, I., Schwall, R. J., and Smith, D. E., *Croatica Chem. Acta*, 48, 651, 1976.
48. Laviron, E., *J. Electroanal. Chem.*, 52, 355, 1974.
49. Damaskin, B. B., Petrii, O. A., and Batrakov, V. V., *Adsorption of Organic Compounds on Electrodes*, Plenum Press, New York, 1971.
50. Nicholson, R. S. and Olmstead, M. L., Numerical solution of integral equations, in *Electrochemistry: Calculations, Simulation and Instrumentation*, Mattson, J. S., Mark, H. B., Jr. and MacDonald, H. C., Jr., Eds., Marcel Dekker, New York, 1972, chap. 5.
51. Feldberg, S. W., Digital simulation: a general method for solving electrochemical diffusion-kinetic problems, in *Electroanalytical Chemistry*, Vol. 3, Bard, A. J., Ed., Marcel Dekker, New York, 1969, 199.
52. Feldberg, S. W., Digital simulation of electrochemical surface boundary phenomena: multiple electron transfer and adsorption, in *Electrochemistry: Calculations, Simulation and Instrumentation*, Mattson, J. S., Mark, H. B., Jr., and MacDonald, H. C., Jr., Eds., Marcel Dekker, New York, 1972, chap. 7.
53. Herman, H. B., Application of computers to solution of organic electrode reaction mechanisms, in *Electrochemistry: Calculations, Simulation and Instrumentation*, Mattson, J. S., Mark, H. B., Jr., and MacDonald, H. C., Jr., Eds., Marcel Dekker, New York, 1972, chap. 3.
54. Ruzic, I. and Feldberg, S. W., *J. Electroanal. Chem.*, 63, 1, 1975.
55. Vetter, K. J., *Electrochemical Kinetics*, Academic Press, New York, 1967.
56. Thirsk, H. R. and Harrison, J. A., *A Guide to the Study of Electrode Kinetics*, Academic Press, New York, 1972.
57. Roe, D. K., *Anal. Chem.*, 50, 9R, 1978 and previous biennial reviews of Analytical Electrochemistry: Theory and Instrumentation of Dynamic Techniques.
58. Albery, J., *Electrode Kinetics*, Clarendon Press, Oxford, 1975.
59. Zuman, P., *Substitutent Effects in Organic Polarography*, Plenum Press, New York, 1967.
60. Christie, J. H. and Osteryoung, R. A., *J. Electroanal. Chem.*, 49, 301, 1974.
61. Murray, R. W., Chronoamperometry, chronocoulometry and chronopotentiometry, in *Physical Methods of Chemistry*, Vol. 1, Part IIA, Weissberger, A. and Rossiter, B. W., Eds., Wiley-Interscience, New York, 1971, chap. 8.
62. Parry, E. P. and Osteryoung, R. A., *Anal. Chem.*, 37, 1634, 1965.
63. Osteryoung, J. and Hasebe, K., *Rev. Polarog.*, 22, 1, 1976.
64. Christie, J. H., Osteryoung, J., and Osteryoung, R. A., *Anal. Chem.*, 45, 210, 1973.
65. Vassos, B. H. and Osteryoung, R. A., *Chem. Instrum.*, 5, 257, 1974.
66. Abel, R. H., Christie, J. H., Jackson, L. L., Osteryoung, J., and Osteryoung, R. A., *Chem. Instrum.*, 7, 123, 1976.
67. Christie, J. H., Jackson, L. L., and Osteryoung, R. A., *Anal. Chem.*, 48, 242, 1976.
68. Turner, J. A. and Osteryoung, R. A., *Anal. Chem.*, 50, 1496, 1978.
69. Klein, N. and Yarnitzky, C., *J. Electroanal. Chem.*, 61, 1, 1975.
70. van Bennekom, W. P. and Schute, J. B., *Anal. Chim. Acta*, 89, 71, 1977.
71. Christie, J. H., Jackson, L. L., and Osteryoung, R. A., *Anal. Chem.*, 48, 561, 1976.
72. Keller, H. E. and Osteryoung, R. A., *Anal. Chem.*, 43, 342, 1971.
73. Rifkin, S. C. and Evans, D. H., *Anal. Chem.*, 48, 1616, 1976.
74. Dillard, J. W. and Hanck, K. W., *Anal. Chem.*, 48, 218, 1976.
75. Jacobsen, E. and Lindseth, H., *Anal. Chim. Acta*, 86, 123, 1976.
76. Rickard, E. C., Eli Lilly and Co., unpublished data.
77. Kalvoda, R. and Trojanek, A., *J. Electroanal. Chem.*, 75, 151, 1977.
78. Breyer, B. and Bauer, H. H., Alternating current polarography and tensammetry, in *Chemical Analysis*, Vol. 13, Elving, P. J. and Kolthoff, I. M., Eds., Wiley-Interscience, New York, 1963.
79. Sluyters-Rehbach, M. and Sluyters, J. H., Sine wave methods in the study of electrode processes, in *Electroanalytical Chemistry*, Vol. 4, Bard, A. J., Ed., Marcel Dekker, New York, 1970, 1.
80. Sobel, H. R. and Smith, D. E., *J. Electroanal. Chem.*, 26, 271, 1970.

81. Smith, D. E., Applications of on-line digital computers in AC polarography and related techniques, in *Electrochemistry: Calculations, Simulation and Instrumentation*, Mattson, J. S., Mark, H. B., Jr., and MacDonald, H. C., Jr., Eds., Marcel Dekker, New York, 1972, chap. 12.
82. Nicholson, R. S., *Anal. Chem.*, 37, 667, 1965.
83. de Vries, W. T. and van Dalen, E., *J. Electroanal. Chem.*, 10, 183, 1965.
84. Cauquis, G. and Parker, V. D., Methods for the elucidation of organic electrochemical reactions, in *Organic Electrochemistry*, Baizer, M. M., Ed., Marcel Dekker, New York, 1973, chap. 2.
85. Nicholson, R. S. and Shain, I., *Anal. Chem.*, 36, 706, 1964.
86. Nicholson, R. S. and Shain, I., *Anal. Chem.*, 37, 178, 1965.
87. Perone, S. P., Enhancement of electroanalytical measurement techniques by real-time computer interaction, in *Electrochemistry: Calculations, Simulation and Instrumentation*, Mattson, J. S., Mark, H. B., Jr., and MacDonald, H. C., Jr., Eds., Marcel Dekker, New York, 1972, chap. 13.
88. Christie, J. H. and Lingane, P. J., *J. Electroanal. Chem.*, 10, 176, 1965.
89. Ferrier, D. R. and Schroeder, R. R., *J. Electroanal. Chem.*, 45, 343, 1973.
90. Zipper, J. J. and Perone, S. P., *Anal. Chem.*, 45, 452, 1973.
91. Newman, J., The fundamental principles of current distribution and mass transport in electrochemical cells, in *Electroanalytical Chemistry*, Vol. 6, Bard, A. J., Ed., Marcel Dekker, New York, 1973, 187.
92. Stock, J. T., *Anal. Chem.*, 50, 1R, 1978 and previous biennial reviews of Amperometric, Bipotentiometric and Coulometric Titrations.
93. Heineman, W. R. and Kissinger, P. T., *Anal. Chem.*, 50, 166R, 1978 and previous biennial reviews of Analytical Electrochemistry: Methodology and Application of Dynamic Techniques.
94. Levich, V. G., *Physicochemical Hydrodynamics*, Prentice-Hall, Englewood Cliffs, N.J., 1962.
95. Riddiford, A. C., Rotating disk system, in *Advances in Electrochemistry and Electrochemical Engineering*, Vol. 4, Delahay, P., Ed., Wiley-Interscience, New York, 1966, 47.
96. Albery, W. J. and Hitchman, M. L., *Ring-Disk Electrodes*, Oxford University Press, New York, 1971.
97. Opekar, F. and Beran, P., *J. Electroanal. Chem.*, 69, 1, 1976.
98. Bruckenstein, S. and Miller, B., *Acc. Chem. Res.*, 10, 54, 1977.
99. Mohr, C. M. and Newman, J., *J. Electrochem. Soc.*, 122, 928, 1975.
100. Prater, K. B., Digital simulation of the rotating ring-disk electrode, in *Electrochemistry: Calculations, Simulation and Instrumentation*, Mattson, J. S., Mark, H. B., Jr., and MacDonald, H. C., Jr., Eds., Marcel Dekker, New York, 1972, chap. 8.
101. Shain, I., Stripping analysis, in *Treatise on Analytical Chemistry*, Part 1, Vol. 4, Section D-2, Kolthoff, I. M., Elving, P. J., and Sandell, E. B., Eds., Wiley-Interscience, New York, 1963, chap. 50.
102. Barendrecht, E., Stripping voltammetry, in *Electroanalytical Chemistry*, Vol. 2, Bard, A. J., Ed., Marcel Dekker, New York, 1967, chap. 2.
103. Brainina, Kh. Z., *Stripping Voltammetry in Chemical Analysis*, Wiley-Halsted Press, New York, 1974.
104. Barker, G. C., *Proc. Anal. Div. Chem. Soc.*, 12, 179, 1975.
105. Agarwal, H. P., Faradaic rectification method and its applications in the study of electrode processes, in *Electroanalytical Chemistry*, Vol. 7, Bard, A. J., Ed., Marcel Dekker, New York, 1974, 161.
106. Siggia, S., Ed., *Instrumental Methods of Organic Functional Group Analysis*, Wiley-Interscience, New York, 1972.
107. Brezina, M. and Zuman, P., *Polarography in Medicine, Biochemistry and Pharmacy*, rev. English ed., Wawzonek, S., transl. ed., Wiley-Interscience, New York, 1958.
108. Kolthoff, I. M. and Lingane, J. J., Polarography, *2nd ed., Vol. 2, Wiley-Interscience*, New York, 1952.
109. Lund, H. and Iverson, P., Practical problems in electrolysis, in *Organic Electrochemistry*, Baizer, M. M., Ed., Marcel Dekker, New York, 1973, chap. 4.
110. Mann, C. K., Nonaqueous solvents for electrochemical use, in *Electroanalytical Chemistry*, Vol. 3, Bard, A. J., Ed., Marcel Dekker, New York, 1969, 57.
111. Guidelli, R., Pezzatini, G., and Foresti, M. L., *J. Electroanal. Chem.*, 43, 83, 1973.
112. Guidelli, R., Pezzatini, G., and Foresti, M. L., *J. Electroanal. Chem.*, 43, 95, 1973.
113. Meites, L. and Lampugnani, L., *Anal. Chem.* 45, 1317, 1973.
114. Barker, G. C. and Gardner, A. W., *J. Electroanal. Chem.*, 46, 150, 1973.
115. Cipak, J., Ruzic, I., and Jeftic, L., *J. Electroanal. Chem.*, 75, 9, 1977.
116. Bauer, H. H., Streaming maxima in polarography, in *Electroanalytical Chemistry*, Vol. 8, Bard, A. J., Ed., Marcel Dekker, New York, 1975, 170.
117. O'Brien, R. N. and Dieken, F. P., *Can. J. Chem.*, 54, 402, 1976.
118. Lal, S. and Bauer, H. H., *Anal. Lett.*, 9, 13, 1976.
119. Frumkin, A. N., Fedorovich, N. V., Damaskin, B. B., Stenina, E. V., and Krylov, V. S., *J. Electroanal. Chem.*, 50, 103, 1974.

120. Stenina, E. V., Frumkin, A. N., Nikolaeva-Fedorovich, N. V., and Osipov, I. V., *J. Electroanal. Chem.*, 62, 11, 1975.
121. Bauer, H. H. and Shallal, A. K., *J. Electroanal. Chem.*, 73, 367, 1976.
122. Taylor, J. K., *J. Assoc. Off. Anal. Chem.*, 47, 21, 1964.
123. Laitinen, H. A., *Chemical Analysis, An Advanced Text and Reference,* McGraw-Hill, New York, 1960, chap. 27.
124. Schroeder, R. R., Operational amplifier instruments for electrochemistry, in *Electrochemistry: Calculations, Simulation and Instrumentation,* Mattson, J. S., Mark, H. B., Jr., and MacDonald, H. C., Jr., Eds., Marcel Dekker, New York, 1972, chap. 10.
125. Britz, D. and Bauer, H. H., *J. Electroanal. Chem.*, 18, 1, 1968.
126. Taylor, D. F. and Barradas, R. G., *J. Electroanal. Chem.*, 23, 166, 1969.
127. Bauer, H. H. and Britz, D., *J. Electroanal. Chem.* 23, 167, 1969.
128. Yarnitzky, C. and Friedman, Y., *Anal. Chem.*, 47, 876, 1975.
129. Yarnitzky, C. and Klein, N., *Anal. Chem.*, 47, 880, 1975.
130. Bezman, R., *Anal. Chem.*, 44, 1781, 1972.
131. Britz, D. and Brocke, W. A., *J. Electroanal. Chem.*, 58, 301, 1975.
132. Harrar, J. E., Techniques, apparatus and analytical applications of controlled potential coulometry, in *Electroanalytical Chemistry,* Vol. 8, Bard, A. J., Ed., Marcel Dekker, New York, 1975, 1.
133. Canterford, D. R., *J. Electroanal. Chem.*, 77, 113, 1977.
134. Hayes, J. W., Glover, D. E., Smith, D. E., and Overton, M. W., *Anal. Chem.*, 45, 277, 1973.
135. Horlick, G., *Anal. Chem.*, 47, 352, 1975.
136. Osteryoung, R. A., Introduction to the on-line use of computers in electrochemistry, in *Electrochemistry: Calculations, Simulation and Instrumentation,* Mattson, J. S., Mark, H. B., Jr., and MacDonald, H. C., Jr., Eds., Marcel Dekker, New York, 1972, chap. 11.
137. Zynger, J. and Rickard, E. C., presented at 28th Pittsburgh Conference on Analytical Chemistry and Applied Spectroscopy, Cleveland, Feb., 1977, Abstract 479.
138. Rajagopalan, S. R., Poojary, A., and Rangarajan, S. K., *J. Electroanal. Chem.*, 75, 135, 1977.
139. Ramaley, L., *Chem. Instrum.*, 6, 119, 1975.
140. Clem, R. G. and Goldsworthy, W. W., *Anal. Chem.*, 43, 918, 1971.
141. Goldsworthy, W. W. and Clem, R. G., *Anal. Chem.*, 44, 1360, 1972.
142. Pomernacki, C. L. and Harrar, J. E., *Anal. Chem.*, 47, 1894, 1975.
143. Bos, M., *Anal. Chim. Acta,* 81, 21, 1976.
144. Matson, W. R., Zink, E., and Vitukevitch, R., *Am. Lab.*, 9(7), 59, 1977.
145. Lento, H. G., Automation in analytical chemistry, *Technicon Symposia, 1966,* Vol. 1, Mediad, White Plains, N.Y., 1967, 598.
146. Cullen, L. F., Brindle, M. P., and Papariello, G. J., *J. Pharm. Sci.*, 62, 1708, 1973.
147. Lund, W. and Opheim, L., *Anal. Chim. Acta,* 79, 35, 1975.
148. Lund, W. and Opheim, L., *Anal. Chim. Acta,* 82, 245, 1976.
149. Lund, W. and Opheim, L., *Anal. Chim. Acta,* 88, 275, 1977.
150. Flann, B., *Chem. Instrum.*, 7, 241, 1976.
151. Evans, D. H., *Acc. Chem. Res.*, 10, 313, 1977.
152. Gilpin, R. K., *Anal. Chem.*, 51, 257R, 1979 and previous biennial reviews of Pharmaceuticals and Related Drugs.
153. Evenson, M. A. and Carmack, G. D., *Anal. Chem.*, 51, 35R, 1979 and previous biennial reviews of Clinical Chemistry.
154. Volke, J., *Talanta,* 12, 1081, 1965.
155. Lund, H., *Adv. Heterocycl. Chem.*, 12, 213, 1970.
156. Lund, H., Electrolysis of heterocyclic compounds, in *Organic Electrochemistry,* Baizer, M. M., Ed., Marcel Dekker, New York, 1973, chap 17.
157. Meites, L. and Zuman, P., *Electrochemical Data. Part I: Organic, Organometallic and Biochemical Substances,* Wiley-Interscience, New York, 1974. (This information is also available as CRC Handbook Series in Organic Electrochemistry, CRC Press, Boca Raton, FL, 1977.)
158. Brezina, M. and Volke, J., Polarography in biochemistry, pharmacology and toxicology, in *Progress in Medicinal Chemistry,* Vol. 12, Ellis, G. P. and West, G. B., Eds., Elsevier-North Holland, New York, 1975, 247.
159. Underwood, A. L. and Burnett, R. W., Electrochemistry of biological compounds, in *Electroanalytical Chemistry,* Vol. 6, Bard, A. J., Ed., Marcel Dekker, New York, 1973, 1.
160. Siegerman, H., Differential pulse polarography of antibiotics, in *Methods in Enzymology,* Vol. 43, Hash, J. H., Ed., Academic Press, New York, 1975, 373.
161. Unterman, H. W. and Weissbuch, S., *Pharmazie,* 29, 752, 1974.
162. Lindquist, J. and Farroha, S. M., *Analyst,* 100, 377, 1975.
163. Woodson, A. L. and Smith, D. E., *Anal. Chem.*, 42, 242, 1970.
164. Brdicka, R., Brezina, M., and Kalous, V., *Talanta,* 12, 1149, 1965.

165. Dietz, R., Cathodic reactions of hydrocarbons, in *Organic Electrochemistry*, Baizer, M. M., Ed., Marcel Dekker, New York, 1973, chap. 5.
166. Eberson, L., Hydrocarbons, in *Organic Electrochemistry*, Baizer, M. M., Ed., Marcel Dekker, New York, 1973, chap. 12.
167. Rifi, M. R., Electrochemical reduction of organic halides, in *Organic Electrochemistry*, Baizer, M. M., Ed., Marcel Dekker, New York, 1973, chap. 6.
168. Lund, H., Cathodic reduction of nitro compounds, in *Organic Electrochemistry*, Baizer, M. M., Ed., Marcel Dekker, New York, 1973, chap. 7.
169. Feoktistov, L. G. and Lund, H., Saturated carbonyl compounds and derivatives, in *Organic Electrochemistry*, Baizer, M. M., Ed., Marcel Dekker, New York, 1973, chap. 8.
170. Baizer, M. M., α,β-Unsaturated carbonyls, in *Organic Electrochemistry*, Baizer, M. M., Ed., Marcel Dekker, New York, 1973, chap. 9.
171. Eberson, L., Carboxylic acids and derivatives, in *Organic Electrochemistry*, Baizer, M. M., Ed., Marcel Dekker, New York, 1973, chap. 10.
172. Eberson, L., Carboxylic acids, in *Organic Electrochemistry*, Baizer, M. M., Ed., Marcel Dekker, New York, 1973, chap. 13.
173. Horner, L., Onium compounds, in *Organic Electrochemistry*, Baizer, M. M., Ed., Marcel Dekker, New York, 1973, chap. 11.
174. Parker, V. D., Anodic oxidation of amines, in *Organic Electrochemistry*, Baizer, M. M., Ed., Marcel Dekker, New York, 1973, chap. 14.
175. Parker, V. D., Anodic oxidation of oxygen-containing compounds, in *Organic Electrochemistry*, Baizer, M. M., Ed., Marcel Dekker, New York, 1973, chap. 15.
176. Parker, V. D., Anodic oxidation of sulfur-containing compounds, in *Organic Electrochemistry*, Baizer, M. M., Ed., Marcel Dekker, New York, 1973, chap. 16.
177. Lehmkuhl, H., Organometallic syntheses, in *Organic Electrochemistry*, Baizer, M. M., Ed., Marcel Dekker, New York, 1973, chap. 18.
178. Baizer, M. M., Electrolytic reductive coupling, in *Organic Electrochemistry*, Baizer, M. M., Ed., Marcel Dekker, New York, 1973, chap. 19.
179. Nyberg, K., Oxidative coupling, in *Organic Electrochemistry*, Baizer, M. M., Ed., Marcel Dekker, New York, 1973, chap. 20.
180. Horner, L., and Lund, H., Cleavages, in *Organic Electrochemistry*, Baizer, M. M., Ed., Marcel Dekker, New York, 1973, chap. 21.
181. Eberson, L., Anodic substitution, in *Organic Electrochemistry*, Baizer, M. M., Ed., Marcel Dekker, New York, 1973, chap. 22.
182. Eberson, L. and Horner, L., Stereochemistry of organic electrode processes, in *Organic Electrochemistry*, Baizer, M. M., Ed., Marcel Dekker, New York, 1973, chap. 27.
183. Lund, H., *Talanta*, 12, 1065, 1965.
184. *U.S. Pharmacopeia*, 19th rev., United States Pharmacopeial Convention, Rockville, Md., 1975.
185. *National Formulary*, 14th ed., American Pharmaceutical Association, Washington, D.C., 1975.

Chapter 2

COULOMETRY

E. C. Rickard

TABLE OF CONTENTS

I. INTRODUCTION

Coulometry is an electrochemical assay technique involving complete reduction or oxidation of a chemical species. The titrant employed, directly or indirectly, is the electron. The total titrant transferred is equal to the amount of charge consumed, that is, the integral of current vs. time. Since charge can be directly measured, coulometry ranks as one of only six techniques that qualify as absolute methods of analysis. When combined with the ease of charge generation, the high accuracy of its measurement, and the elimination of the need to prepare, store, or standardize titrants, coulometry becomes a particularly valuable technique for quantitative analysis.

There are two major classifications of coulometric techniques. Primary coulometric techniques involve electron transfer to (reduction) or from (oxidation) the species of interest at an appropriate electrode. Assays using this technique are generally performed at a constant electrode potential and are called controlled potential coulometry. Secondary coulometric techniques involve the generation of a chemical reactant at the electrode which then undergoes a homogeneous chemical reaction with the species of interest. These secondary techniques are usually performed at constant current and are called coulometric titrations.

The choice of electrolysis conditions defines the selectivity of the assay. Some factors which must be considered include instrumentation, cell design (reference electrode, counter electrode, and cell geometry), and electrochemical parameters (electrode potential or current, working electrode material, sample solvent, supporting electrolyte, and electroactive species concentration). Preliminary conditions for controlled potential coulometric assays are generally selected from polarographic or other electrochemical data. Those for coulometric titrations are selected from a combination of chemical and electrochemical data. An endpoint detection technique must also be chosen for coulometric titrations.

In addition to quantitative analysis, controlled potential electrolysis techniques also have been used for electrosynthetic purposes[1-3] and for elucidation of the reaction mechanisms of electron transfer reactions.[4,5] Selection of electrochemical parameters, cell design and instrumentation will be greatly influenced by the specific purpose. This chapter will primarily discuss the use of coulometry in analytical assays of pharmaceutical compounds.

II. CONTROLLED POTENTIAL COULOMETRY

A. Advantages and Limitations

The advantages of controlled potential coulometry include:[7-11]

1. It is an absolute technique which requires no comparison to a chemical standard. In reality, it is a preferred method for evaluation of standard materials.
2. There is no reagent standardization nor storage and no time spent in their preparation.
3. Sample size is small, typically 1 μmol to 1 mmol.
4. It is a high precision technique. Routine coulometric assays in the author's laboratory exhibit coefficients of variation (CV) of 0.1 to 1%. High precision controlled potential coulometry has been reported to a standard deviation (σ) of about 20 ppm (CV = 0.002%).
5. Electronic circuits for measuring current and charge are simple to construct and calibrate. The measurements depend only upon time, mass [Standard International (SI) base units], resistance, and voltage (SI derived units).
6. Controlled potential coulometry is more selective than coulometric titrations.

Several limitations to controlled potential coulometric methods include:

1. Processes which lead to less than 100% current efficiency must be absent or appropriate corrections made.
2. The species must accept or give up electrons at an electrode, at a reasonable rate and in an accessible potential region.
3. Controlled potential coulometry does not indicate the presence of impurities. If present, impurities must exhibit different electrochemical characteristics or they must be determined by a separate method and an appropriate correction made.
4. Precision frequently decreases at low concentrations or if a minor component is being determined.
5. Controlled potential coulometry is not as precise as coulometric titrations.

B. Principle

In controlled potential coulometry, it is necessary to obtain complete conversion (reduction or oxidation) of the electroactive species at the electrode. The substance must be dissolved in an appropriate electrolyte and brought to the electrode surface. At suitable electrode potentials, the desired electron transfer process can then occur. Transfer of electrons to the electroactive species is called a reduction process and gives a cathodic current. Removal of electrons is an oxidation and gives an anodic current. As the working electrode becomes more negative with respect to the reference electrode, reduction processes are more favored. At some negative potential, reduction of water or a supporting electrolyte constituent will occur. Similarly, at some positive potential, oxidation of an electrolyte constituent or of the electrode itself will occur. These potentials define the useful voltage range for the electrode-supporting electrolyte combination.

The term electrode or working electrode refers to the electrode where the desired electron transfer occurs. Its potential with respect to the bulk of the solution is measured using a nonpolarizable reference electrode, frequently a saturated calomel electrode (SCE). Of course, the current flowing through the working electrode must also flow through an auxiliary or counter electrode added for that purpose. Occasionally, the functions of the reference and counter electrodes are combined.

For most substances, it is desirable that the resulting electrolysis product also be soluble in the electrolyte. If it is not soluble, it may form a film on the working electrode surface which prevents mass or electron transfer. In the electrolysis of metallic ions, one species may be dissolved into the electrode (to form an amalgam at a mercury electrode) or may be plated onto the electrode surface (for a platinum or other solid electrode). If the change in weight of the electrode is measured instead of the amount of charge transferred, the technique is called electrogravimetry.

The supporting electrolyte transports the flow of current through the solution in addition to acting as a solvent. Thus, aqueous or other high dielectric constant solvents containing dissolved ionic salts are preferred. The sample solvent must be identical to the supporting electrolyte or compatible with it.

1. Purity Calculation

The charge consumed by the electroactive species is related to the number of equivalents of material by Faraday's law and to the weight of material by the equivalent weight of the compound being electrolyzed. These relationships are expressed in the following equation:

$$Q = \int I \times dt = n \times F \times W/MW \qquad (1)$$

where Q is the total charge in millicoulombs, I is the current in milliamperes, t is the time in seconds, n is the number of electrons transferred per molecule of electroactive material, F is Faraday's constant, 96486.7 mC/milliequivalent, W is the sample weight in milligrams, and MW is the molecular weight of the compound. If the material is composed of only the major component and substances which are nonelectroactive under the electrolysis conditions, this relation may be restated in terms of the percent purity calculated from the measured charge.

$$\text{Percent Purity} = Q \times MW \times 100/(n \times F \times W) \qquad (2)$$

2. Potential Selection

The electrode potential can be selected from polarographic or other voltammetric data obtained under similar conditions — electrode type, solvent, and supporting electrolyte. However, voltammetric data obtained at microelectrodes may not be representative of the processes occurring during bulk electrolysis.[12] For example, relatively slow coupled chemical reactions may occur during a long controlled potential electrolysis (5 to 60 min) which would not occur during a polarographic experiment (1 to 5 sec drop time). The higher solution concentrations used in controlled potential coulometry will increase the importance of higher order chemical reactions (e.g., dimerization). In controlled potential coulometry, the electrodes are larger and it is more difficult to maintain pH control and electrode potential control. Therefore, voltammetric data are useful only for preliminary experimental design. The final design should be based on a current-voltage scan obtained under the coulometric conditions using the large area electrode. Agreement between the voltammetric and coulometric scan data qualitatively (number of waves) and quantitatively (half-wave potential and wave shape) give preliminary confirmation of the choice for experimental parameters.

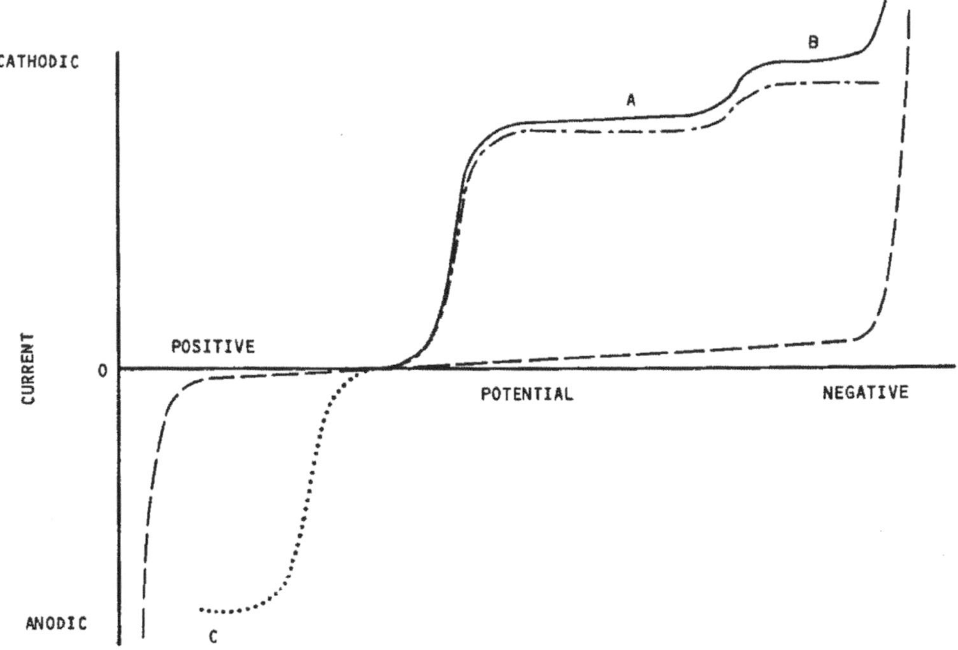

FIGURE 1. Current-potential curve; stirred mercury pool electrode. Residual current (——), faradaic reduction current(-·-·-·), faradiac oxidation current (for reduced form of main component (……), and total current (———).

A simulated curve and its components are illustrated in Figure 1 for a reduction process at a stirred mercury pool. At very positive potentials, the residual current is due to the oxidation of mercury. At very negative potentials, it is due to the electrogeneration of hydrogen. The chosen electrolysis potential (Region A) must avoid both of these regions as well as any potential region where other components are electroactive (Region B). Otherwise, residual currents will be high or other species will electrolyze. Residual currents tend to decrease precision and/or accuracy. Electrolysis of other sample components will give high results.

The oxidation wave of the reduced form of the main component (Region C) may be observed depending upon the nature of the electrochemical mechanism. If it is observed, it can be used to preoxidize the material and thereby assure that all material is initially in the same oxidation state. Preoxidation (or prereduction preceding an oxidation) is most useful in the case of metal ion electrolysis.

3. *Reversible Electrochemical Reactions*

The electrochemical reaction or electron transfer process which occurs at the working electrode can be a complex function of equilibria and kinetics. It is composed of the actual electron transfer(s) and any related chemical reaction(s). These chemical reactions may involve the electrochemical reactant (preceding chemical reaction), the electrochemical product (following chemical reaction) or both. For a simple, reversible reduction (cathodic) process (Equation 3), the concentration of oxidized and reduced species at equilibrium with the electrode is related to its potential by the Nernst relationship (Equation 4). A reversible reduction process, by this definition, does not involve coupled chemical reactions. In addition, the Nernst relationship is valid when the rate of the electrode process is fast relative to the rate of mass transfer. Thus, electrode processes are always in equilibrium for reversible electron transfers.

$$Ox + ne \rightleftarrows Red \tag{3}$$

$$E = E_{1/2} - (RT/nF) \ln [Red]/[Ox] \tag{4}$$

In the Nernst equation, E is the electrode potential in volts, $E_{1/2}$ is the polarographic half-wave potential, (RT/F) ln 10 has the value of 0.05915 at 25°C, and n is the number of electrons transferred per molecule of electroactive material. Thus, 99.9% of the oxidized species is converted to the reduced form when it is brought into contact with an electrode maintained 180/n mV negative (cathodic) of the half-wave potential. As might be expected, the rate of mass transfer will determine the time required for this conversion.

The polarographic half-wave potential is defined as the potential where the concentrations of oxidized and reduced species measured at the electrode surface are equal. It is a function of the standard potential for the half-reaction, the diffusion coefficients, the activity coefficients and chemical equilibria. When hydrogen ions are involved in the electrochemical reaction, the half-wave potential will be a function of the ratio of hydrogen ions transferred to electrons transferred. For example if p hydrogen ions are consumed by the product of an organic reduction process, the half-wave potential will shift negative by 59 × (p/n) millivolts for each unit increase in pH.[13] For a reversible electron transfer, the half-wave potential for the reduction of species Ox is the same as the half-wave potential for the oxidation of species Red for a given set of conditions. The polarographic and coulometric half-wave potentials for reversible processes are essentially equal. However, their relationship is more complex when kinetic processes are involved.

If the rate of material transport to the electrode is the limiting factor, the current at time t, I(t), decays in an exponential manner from its initial value, I_o, as given by

$$I(t) = I_o \exp(-s_o \times t) \tag{5}$$

where s_o is a mass transport constant defined by

$$s_o = D_o A/V\delta \tag{6}$$

In this Nernst diffusion-layer approach which generally holds with a constant stirring rate, D_o is the diffusion coefficient of the electroactive material in cm^2/sec, A is the electrode area in cm^2, V is the solution volume in cm^3, and δ is the width (in cm) of the solution layer next to the electrode where mass transport occurs by diffusion and across which the concentration gradient occurs. Delta is an empirical quantity which reflects undefined hydrodynamic conditions such as laminar, turbulent or other flow conditions and solution viscosity.

The exponential current decay leads to an exponential growth in the measured charge at time t, Q(t),

$$\begin{aligned} Q(t) &= [I_o/s_o] \times [1 - \exp(-s_o \times t)] \\ &= Q_\infty \times [1 - \exp(-s_o \times t)] \\ &= [I_o - I(t)]/s_o \end{aligned} \tag{7}$$

If the charge obtained when the electrolysis is carried to completion (infinite time) is termed Q_∞, then it is given by

$$Q_\infty = I_o/s_o \tag{8}$$

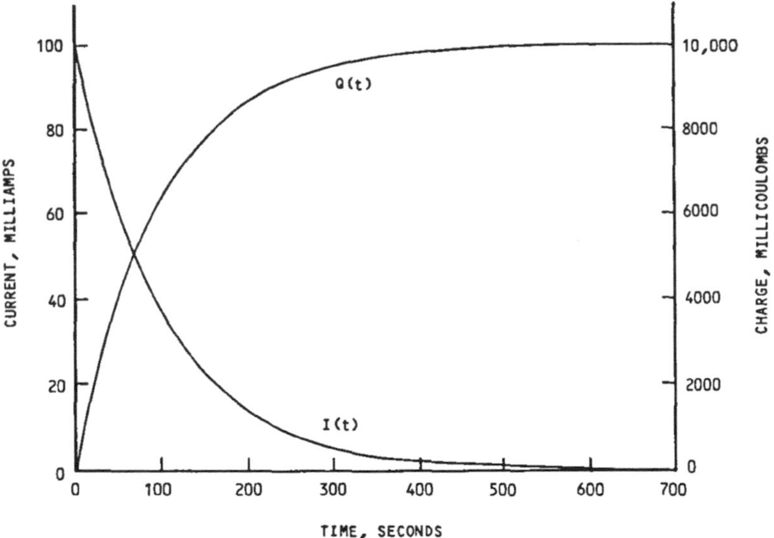

FIGURE 2. Current-charge-time relationships. Reversible, diffusion controlled case in absence of residual current. $Q_\infty = 10^4$ mC; $s_o = 0.01$ sec^{-1}; $I_o = s_o \times Q_\infty = 100$ mA.

It can be seen from Equation 7 that infinite time in practice is that time required for the current to decrease to 0.01 to 0.001 of its original value or until the quantity ($s_o \times t$) is 4.6 to 6.9 (99 to 99.9% conversion, respectively for each case). Because of the relationship of the mass transport constant to the electrolysis time, it is sometimes called the cell constant. These current-charge-time relations are illustrated in Figure 2.

a. Data Analysis

Experimental data can be more conveniently examined for reversible, mass transfer limited behavior if Equation 5 is rewritten in a linear form as

$$\text{Log } I(t) = \text{Log } I_o - s_o \times t \times \text{Log } 10 \tag{9}$$

Thus, a plot of log current versus time (Figure 3) is linear with an intercept of I_o and a slope of $-s_o/$ Log 10 or $-0.434 \times s_o$. This relationship should be experimentally valid between the 5 and 95% electrolysis points when the residual current is very small or absent. In addition, the total charge can be estimated from the ratio of the intercept to the slope. However, better methods for estimating the total charge under nonideal conditions are discussed in the section on charge measurement.

Examination of the magnitude of the mass transfer constant during the electrolysis is a more sensitive test of mass transfer limited behavior.[7] This is facilitated by using the relationship

$$Q(t) = Q_\infty - Q(r) \tag{10}$$

where Q (r) is the charge remaining to be accumulated. Upon substituting Equations 8 and 10 into 7 and taking the natural logarithm,

$$s_o = -(1/t) \times \ln [Q(r) / Q_\infty] \tag{11}$$

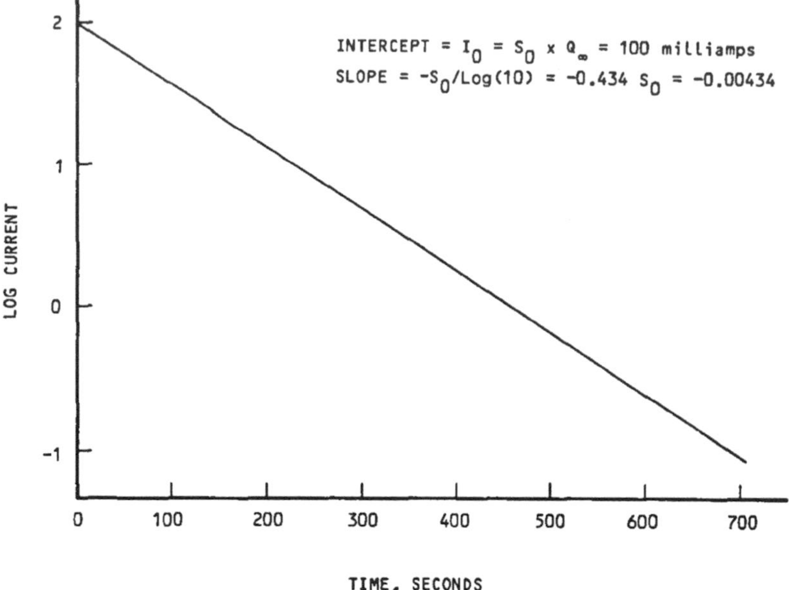

FIGURE 3. Plot of log current vs. time. Reversible, diffusion controlled case in absence of residual current. $Q_\infty = 10^4$ mC; $s_o = 0.01$ sec^{-1}

Now s_o can be independently calculated from any set of charge-time values provided that the total charge is known. An even more sensitive calculation involves the comparison of two sets of charge-time values obtained at times t_1 and t_2 using[5,14]

$$s_o = [1 / (t_2 - t_1)] \times \ln[Q(r_1) / Q(r_2)] \qquad (12)$$

Plots of s_o vs. time from Equations 11 or 12 are more sensitive tests of reversible, mass transfer limited coulometric behavior than log current versus time plots. However, these equations are valid only if all values of the charge have been corrected for the residual current.

If the current consists of the electrolysis current and a constant residual current, i(b), it will be given by

$$I(t) = I_o \times \exp(-s_o \times t) + i(b) \qquad (13)$$

instead of Equation 5. The accumulated measured charge, Q_{meas}, will now be the sum of Q(t) from Equation 7 and that due to the residual current or

$$Q_{meas} = Q(t) + i(b) \times t \qquad (14)$$

It can readily be seen that a constant residual current leads to an always increasing measured charge (Figure 4) and a more difficult data analysis. Correction procedures will be described in the residual current section.

The electrochemical processes involving pharmaceutical compounds are usually more complicated than the simple, reversible process described. Heterogeneous kinetics, multiple electron transfer paths, or coupled chemical reactions give more complicated current-time relationships.[4,5] Several of these processes and their effects on con-

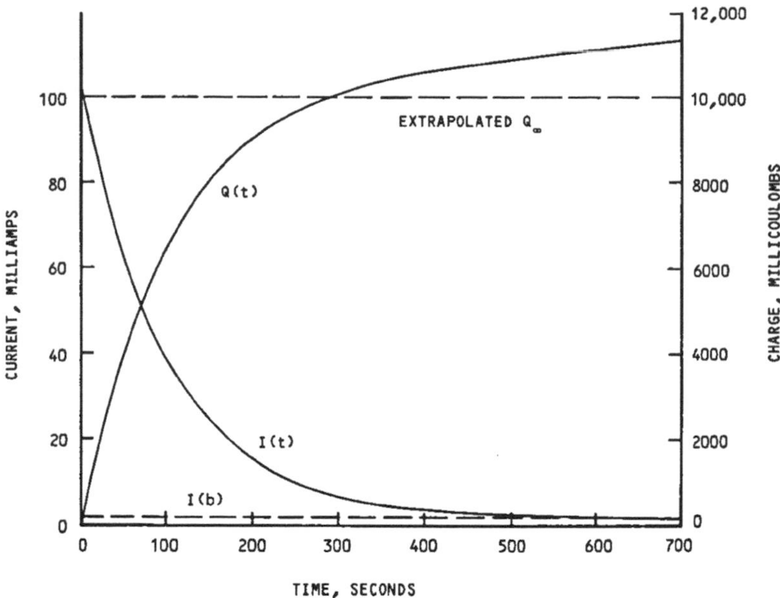

FIGURE 4. Current-charge-time relationships. Reversible, diffusion controlled case in presence of constant residual current. $Q_\infty = 10^4$ mC; $s_o = 0.01$ sec^{-1}; $I(b)/I_o = 0.02$; $I_o = s_o \times Q_\infty + i_b = 102$ mA.

trolled potential coulometry are described below. They will be described as reduction processes, but the description is equally valid for oxidations.

4. Other Electrochemical Mechanisms

a. Heterogeneous Kinetics

An electron transfer process can be so slow that the electrode is not in equilibrium with the solution. This phenomena is called a rate-controlled electron transfer and exhibits heterogeneous kinetics. The half-wave potentials for reduction and oxidation are not equal and the polarographic wave is elongated along the potential axis. It can be described by a standard heterogeneous rate constant, k_s, and an electron transfer parameter, α.

$$Ox + ne \underset{\alpha}{\overset{k_s}{\rightarrow}} Red \tag{15}$$

The relationship between the polarographic half-wave potential and the coulometric half-wave potential depends upon the polarographic drop time, s_o, V, A, D_o, α and n.[5] Controlling the electrode potential more than 180/n mV from the coulometric half-wave potential will generally increase the electron transfer rate and allow the electrolysis to reach completion. When the potential is in the limiting current region, no difference in coulometric behavior versus the reversible case will be observed. The log current vs. time plot is always linear. The true integer n value will be observed.

b. Multiple Electron Transfer Reactions

Several simultaneous electrochemical processes may occur as illustrated.

$$Ox + n_1 e \rightleftharpoons Red_1 \qquad E^1_{1/2}$$

$$Ox + n_2 e \rightleftharpoons Red_2 \qquad E^2_{1/2} \tag{16}$$

This poorly defined analytical situation results in a mixture of products and an average noninteger n value (unless $n_1 = n_2$). Again, the log current vs. time plot will be linear assuming each electron transfer is reversible. Adjustment of solution conditions (pH, solvent, etc.) and electrode potential can sometimes favor one path at the expense of the second.

c. Coupled Chemical Reactions

Homogeneous chemical reactions involving the reactant or product of the initial electron transfer reaction are called coupled chemical reactions.[4,5,15] These reactions may result in a net gain or loss of the measured charge. Several common examples of mechanisms involving coupled chemical reactions will be described. Generally, the magnitude of the effects depend upon the rates of the chemical reaction relative to the time scale of the coulometric experiment. If the time scales overlap, the log current vs. time plot will usually not be linear and the apparent n value may be nonintegral.

When a chemical reaction precedes the electron transfer (CE mechanism), the electroactive species is generated from a non-electroactive species, A, via the homogeneous chemical reaction.

$$A \underset{k_b}{\overset{k_f}{\rightleftharpoons}} Ox \qquad K_{eq} = k_f/k_b$$

$$Ox + ne \rightleftharpoons Red \tag{17}$$

If the rate of the chemical reaction is slow, then the current will be dependent upon the chemical reaction rather than controlled solely by the rate of mass transfer to the electrode. For a slow chemical reaction rate and an equilibrium constant favoring species A ($K_{eq} \ll 1$), a linear log plot having an apparent mass transport constant smaller than expected will result. If both species A and Ox are initially present ($K_{eq} \approx 1$), then a plot of log current vs. time will consist of two linear segments — one corresponding to the reduction of free Ox and one due to the rate of chemical production of species Ox.[5] In either situation, the electrolysis will require several chemical half-lives for completion. Shorter electrolysis times will give incomplete electrolysis and a smaller than expected measured charge. In contrast, the chemical reaction will have no effect and cannot even be detected if it occurs fast relative to the electrolysis. Examples of the CE mechanism include recombination of an organic acid anion to give a more easily reducible neutral species, the dissociation of a coordination complex to give a more easily reducible hydrated metal ion and dehydration of a hydrated, less easily reduced form of an aldehyde or ketone. Modification of the chemical conditions can frequently prevent these problems.

In the regeneration (catalytic) mechanism, additional reactant is generated through reaction of the product with a nonelectroactive species, Z.

$$Ox + ne \rightleftharpoons Red$$

$$Red + Z \xrightarrow{k} Ox \tag{18}$$

This species could be an oxidant such as hydrogen ion, water, perchlorate, or dissolved oxygen. A higher than expected amount of charge is consumed and the current at the end of the electrolysis is always greater than that due to the electrolyte alone unless the oxidant is completely consumed. Alternatively, a soluble reduced species can be reoxidized at the counter electrode (if it is not separated from the cell by a diaphragm) and transported back to the working electrode to give a very similar behavior. The log current vs. time plot is nonlinear and the apparent n value is larger than the true n.

The final steady state current is reached when the rate of chemical production of species Ox is equal to its rate of reduction. This current will be dependent upon chemical parameters, stirring rate, etc. Many examples of the regeneration mechanism involve inorganic ions. But catalytic hydrogen waves involving organic bases are very common. Examples of these will be presented in the section on residual current.

The following chemical reaction (EC mechanism) is one of the most frequently observed coupled chemical reactions in organic electrochemistry.[15] It is important because the product of the electron transfer is removed via the chemical reaction.

$$\begin{aligned} &Ox + ne \rightleftharpoons Red \\ &Red \xrightarrow{k} Z \end{aligned} \qquad (19)$$

Thus, the oxidized and reduced forms are never in equilibrium at the electrode and the electron transfer process is nonreversible. This may be beneficial in that very complete electrolytic reductions can be performed at potentials more positive than normal as long as appreciable current flows (e.g., near the half-wave potential). This may avoid supporting electrolyte decomposition. However, it also means that a finite, net electrolysis occurs in potential ranges where a reversible electrochemical system with the same half-wave potential would give essentially no conversion. This restricts the ability to discriminate by potential between a desired reduction and a second, undesired but more difficult reduction when the second process occurs by an EC mechanism. Examples of the EC mechanism include reduction of organic compounds in proton donor solvents where irreversible protonation of the product occurs or where dimerization of the product occurs. The chemical reaction following the electron transfer has no effect on the measured charge or n value or on the log plot. It does result in a shift of the half-wave potential and may result in the isolation of a nonobvious electrolysis product. One special case of the EC mechanism is the coupling reaction where species Red reacts with Ox to give a nonelectroactive product. In the coupling reaction, the measured charge, n value, and log plot are obviously affected. Both the dimerization and the coupling reaction are important in synthetic organic electrochemistry.

The ECE mechanism is a more complicated mixture of chemical and electrochemical processes.

$$\begin{aligned} &Ox_1 + n_1 e \rightleftharpoons Red_1 && E^1_{1/2} \\ &Red_1 \underset{k_b}{\overset{k_f}{\rightleftharpoons}} Ox_2 && K_{eq} \\ &Ox_2 + n_2 e \rightleftharpoons Red_2 && E^2_{1/2} \end{aligned} \qquad (20)$$

The rate of the intermediate chemical reaction determines the rate of formation of the second electrochemical reactant Ox_2. In addition, chemical reactions such as disproportionation of Ox_2 or reaction of Ox_1 with Red_2 may be important depending upon the relative values of the half-wave potentials and the chemical rate constants. If Red_1 and Ox_2 are the same species (no chemical reaction), this is an EE mechanism. Thus, a variety of behaviors are possible with limits for the number of electrons transferred being n_1 to $(n_1 + n_2)$. Most electron transfer mechanisms with n values greater than two and many of those with n values of two are examples of the ECE mechanism. Analytically, these processes are generally useful only if the overall n value is either n_1 or $(n_1 + n_2)$. A n value of n_1 results when the chemical reaction is slow compared to the electrolysis time or when the second reduction is much more difficult (half-wave potential considerably more negative) than the first. Then, the second electron transfer does not appreciably occur at the electrolysis potential. A n value of ($n_1 + n_2$) occurs with a fast chemical reaction and an electrode potential sufficient to cause both reduction processes.

5. Criteria for Controlled Potential Coulometry

From this description of electrochemical mechanisms, one can formulate a set of criteria which must usually be met to perform successful quantitative controlled potential coulometry. These include formation of a single electrolysis product, an integer n value and no gain or loss of reactant through chemical reactions. A change in electrochemical mechanism is indicated if the n value and the product obtained are not independent of reasonable changes in stirring rates (or electrolysis times), the initial concentration of the electroactive species or changes in the electrode potential (as long as it is maintained in the limiting current region). Obviously a complete investigation of the electrochemical mechanism of that compound or a suitable model compound leads to a better understanding of the actual processes.

C. Application of Principles

Controlled potential coulometry can easily achieve levels of precision (0.1 to 1.0%) consistent with most laboratory needs. However, the discussion of the purity calculation (Equation 2) and the various electrochemical mechanisms emphasize that several factors determine the accuracy and precision. These include:

1. Complete transfer of charge to the electroactive species in a well defined manner so that the number of electrons transferred per molecule is known
2. Correction for residual current contributions to total charge
3. Accurate measurement of the charge transferred
4. Elimination of any impurities electroactive at the assay conditions
5. The number of coulombs per equivalent (Faraday's constant)
6. The molecular weight of the compound

Conversely, any other single factor (Faraday's constant, number of electrons transferred per molecule, or molecular weight) may be determined if the purity is known and the charge is accurately measured. The implications of these statements will be discussed in detail.

1. Number of Electrons Transferred

Frequently the number of electrons transferred can be determined by analogy with known compounds. For example, nitro groups in acidic or mildly alkaline solutions usually give a four electron reduction to the hydroxylamine followed by an additional two electron reduction to the amine at more negative potentials. In more basic solutions, only the four electron reduction is observed.[16] When data on reasonable model compounds are not available, several alternative methods to estimate n are available. First, the polarographic diffusion current (I_d, microamperes) is proportional to the number of electrons transferred (n), the concentration (C, millimolar) and to the square root of the diffusion coefficient (D_o) as given by the Ilkovic equation

$$I_d = 708\, n\, C\, D_o^{1/2}\, m^{2/3}\, t^{1/6} \tag{21}$$

where m is the mercury flow rate (mg/sec) and t is the drop time (sec). If an estimate of the purity is available (to calculate concentration), and a reasonable guess is made of the diffusion coefficient (species of similar molecular weight and ionic charge generally have nearly equal diffusion coefficients), then n may be estimated from polarograms of other materials obtained on the same dropping mercury electrode capillary. This estimate is valid only when the limiting current is diffusion controlled — a typical situation. Polarographic diffusion limited currents are proportional to concentration and to the square root of the corrected mercury height of pressure.

Second, the shape of a reversible polarographic wave is directly related to the n value by the slope of the "log plot" when potential is plotted vs. the log term in Equation 22.

$$E = E_{1/2} - (0.05916/n) \times \log [I/(I_d - I)] \quad (22)$$

The shape of cyclic or other voltammograms contain similar information. Estimates based on wave shape are best applied only when the electrochemical mechanism is reversible — a fairly stringent and uncommon situation. Third, the product of the controlled potential coulometric process may be identified and the number of electrons transferred inferred. In some cases, a complete elucidation of the controlled potential coulometric mechanism may be desired. Chronocoulometry and other electrochemical or chemical techniques may be used to obtain this information. The postulated n value should be confirmed by a direct controlled potential coulometric experiment on a sample of pure material. If these numbers do not agree, it is likely that differences in time scales of the experiment, concentration, etc. have caused a change in the mechanism.

2. *Residual Current*

Residual current is the term for that current which is consumed by undesired electrode processes. If the residual current were zero, then the current efficiency would be 100%. That is, all charge passing through the cell would be consumed by the electrochemical reaction of interest. In practice, the measured charge (Q_{meas}) almost always has to be corrected for that consumed by other processes ($Q_{residual}$). The total corrected charge (Q) is that used for the purity calculation by Equation 2.

$$Q = Q_{meas} - Q_{residual} \quad (23)$$

The residual current is composed of four components: charging current, faradaic currents, kinetic currents, and induced currents.[17] Charging current and faradaic residual currents are always present. Kinetic and induced residual currents may be present.

The charging current contributes an amount of charge necessary to change the electrode potential from its initial value to its final value. This charge is usually small (typically <0.1 mC) compared to the typical charge for the electrolysis (>100 mC) and can be ignored. When corrections are necessary, it can be measured in a blank experiment with only the electrolyte if the product of the electrochemical reaction does not change the electrode's integral capacitance. Alternatively, it may be essentially eliminated by maintaining the electrode at the electrolysis potential while adding the sample.

Faradaic residual current is due to the electrolysis of impurities or a major constituent of the electrolyte. For example, species such as water, hydrogen ion, alkali metal ions, and oxygen are reducible. If the electrode potential is sufficiently close to their half-wave potentials, then a significant current flow may result. The reduction of even 0.002% of the water would not appreciably affect the solution, but it would contribute the same amount of charge as the one electron reduction of a 10^{-3} *M* solution. Oxygen is another easily reducible species which is difficult to completely exclude from the electrochemical cell. The usual technique involves deoxygenation by bubbling oxygen free nitrogen or argon through the solution before and during the electrolysis. However, if the gas contains only 10 ppm oxygen, its reduction could account for about 0.02% of the total charge for electrolysis of a 10^{-3}*M* solution or 0.15% for a 10^{-4}*M* solution, assuming typical conditions.[5]

The preceding examples illustrate the large errors which can arise from uncorrected faradaic currents. If the faradaic residual current is due to a major component (water) or to a steady state concentration of a minor component (oxygen in nitrogen used for deoxygenation), it can be assumed to be constant through the electrolysis. The residual current observed at the end of the electrolysis should be equal to that observed at the end of the preelectrolysis. The faradaic residual current correction is then equal to the residual current multiplied by the electrolysis time.

The catalytic hydrogen mechanism (Equation 24) is a specific example of a regeneration mechanism giving a kinetic type of residual current. It involves protonation of the electrolysis product (R) followed by electrogeneration of hydrogen.

$$R + H^+ \rightleftharpoons RH^+$$

$$RH^+ + e \rightleftharpoons R + \tfrac{1}{2}H_2 \qquad (24)$$

The hydrogen reduction is effectively shifted to a less negative potential due to the presence of species R. Because this reaction is dependent upon the concentration of the reduction product, the magnitude of the kinetic current changes during the electrolysis. This results in a residual current which is larger at the end of the electrolysis than was observed after the preelectrolysis. Thus, use of the residual current after the preelectrolysis to calculate the corrected charge will produce an undercorrection and high results. Use of the residual current that was observed at the end of the electrolysis will give an overcorrection and low results. The latter will generally be more nearly correct since the product is rapidly formed at the start of the electrolysis and changes concentration less rapidly towards the end.

Induced residual currents are those due to an extraneous electrochemical reaction which would not normally occur. It is induced by the reduction or oxidation of the substance being electrolyzed. The induced current varies during the electrolysis, usually somewhat proportional to the current for the major process. Only a few specific examples have been cited. Thus, its contribution to the residual current will not be considered further.

Comparison of the residual current during preelectrolysis to that at the end of electrolysis serves to differentiate faradaic and kinetic or induced residual currents. It also gives an estimate of the possible error in assuming a constant residual current.

The residual current can be directly measured at the end of the preelectrolysis or electrolysis as appropriate (Figure 4). However, its fluctuation with stirring generally makes it easier to measure a partial residual charge by integration of the residual current over a predetermined time. Then, the total residual charge used for correcting the observed electrolysis charge is the average residual current multiplied by the electrolysis time or the integrated partial residual charge multiplied by the ratio of the electrolysis time to the partial integration time. These approaches are equivalent to extrapolation of the long time charge-time behavior back to zero time (Figure 4).

Another correction alternative is electronic compensation of the residual current by an amount equal to the average residual current observed at the end of the preelectrolysis.[11] This works, of course, only when faradaic contributions dominate the residual current. Any of the processing techniques described for prediction of the final charge in the presence of constant residual current are applicable also.

Generally, one cannot measure the small residual current nearly as accurately as one can measure the larger electrolysis current. Changes in the residual current during the electrolysis contribute to additional inaccuracy. The numerous possible situations prevent any generalizations about its maximum acceptable value. However, if an overall accuracy of 0.1% is desired and one decides that the possible error in measuring the residual current is 50%, then the total charge contributed by the residual current must be less than 0.2%. For example, if the electrolysis consumed about 10,000 mC (5 mℓ of 10 mM solution of a compound with n = 2), the maximum allowed residual current for a 15 min electrolysis would be 0.022 mA (900 seconds × 0.022 mA = 20 mC). The estimated maximum error of 10 mC due to uncertainty in the measured residual current is 0.1% of the total charge.

3. Charge Measurement

There have been many approaches to the measurement of charge. Historically, charge has been measured with chemical and mechanical coulometers or by manual integration of current-time curves (planimeters, graphical, etc.).[4,5,7,10] More recently, electronic analog integration of the current has been most widely used. The increasing use of laboratory computers provides an additional option of recording digital current-time curves followed by data processing techniques such as numerical integration or curve fitting. Fundamentally, the precision of charge measurement is limited by uncertainty in measuring current and time. This limitation is estimated to be several parts per million.[18] In practice, precision is usually limited by the characteristics of the analog circuit elements or data processing techniques.[7,10,11,14,17] Instrument calibration and nonideal electrochemical behavior also often contribute to imprecision and/or inaccuracy.[7,8,17,19,20] Calibration procedures will be discussed in the instrumentation section.

Several procedures have been proposed to estimate the total charge prior to completion of the electrolysis. These predictions can be used by on-line computer data systems to shorten the analysis time with only a minimal decrease in accuracy.[7,14] If three charge measurements (Q_1, Q_2, and Q_3) are taken at equally spaced time intervals (t_1, t_2, and t_3 with $t_2 - t_1 = t_3 - t_2$), then the total charge is estimated to be[5,14]

$$Q_\infty = [Q_2^2 - Q_1 \times Q_3] / [2Q_2 - Q_1 - Q_3] \qquad (25)$$

A more useful form of this equation for on-line computer calculation is[14]

$$Q_\infty = Q_3 + (Q_2 - Q_3)^2 / (2Q_2 - Q_1 - Q_3) \qquad (26)$$

Of course, these predictions are valid only in the absence of appreciable residual current. Unfortunately, the necessary corrections are somewhat complicated even with the assumption of constant residual current.[7,14]

Two additional procedures allow real time correction of the observed charge for a constant (faradaic) residual current, thus they can transform the data into a form suitable for the prediction of Q_∞. The first involves subtracting an amount of charge proportional to the electrolysis time until the log current vs. time plot (Equation 9) is linear.[10,19] Then Equations 25 and 26 may be applied to the corrected data. The amount of charge to be subtracted can be judged subjectively from plots or obtained from digital iterations to the best fit for a straight line. This has a possible disadvantage of confusing a mechanistic nonlinear log plot with one nonlinear due to residual currents.

The second processing method uses recording of successive coulometric runs.[7] The first is obtained on a blank only and second on the sample. This procedure is practical only for digital processing techniques and extremely reproducible electrolysis conditions (stirring rate, etc.). It will allow correction for charging current and any constant faradaic residual current but will not work with kinetic or induced residual currents related to the presence of the samples. Again Equations 25 and 26 may be applied to the corrected data.

4. Impurities

Impurities may be present in the supporting electrolyte, in the sample solvent, or in the sample being assayed. Solvents may be purified by distillation, molecular sieves, etc.[15] Several texts are available on chemical purifications.[21,22] Electroactive impurities in the supporting electrolyte are routinely removed by a preelectrolysis before sample addition. The preelectrolysis also tends to bring the electrode surface to an equilibrium state equivalent to that which will be present during the electrolysis.[7] The preelectrolysis is continued until the current decays to a low, constant value, the residual current described above. The sample solvent can also be preelectrolyzed to remove impurities. The preelectrolysis is generally performed at the potential that will be used in the controlled potential coulometric experiment. Preelectrolysis is most important when low sample concentrations must be used or when extended potential ranges are used. For example, alkali metal ions must be removed from quaternary ammonium salts if very negative potentials will be used.

Impurities present in the sample frequently interfere with quantitative assays of pharmaceutical compounds. That is, their structure and reactivity are often very similar to that of the main component. For aged samples, degradation products may be present also. Assays with higher selectivity (e.g., chromatography) may sometimes be preferred. Assay limitations can best be defined by a comprehensive screen followed by isolation and identification of impurities and degradation products. Then, each assay can be tested for interference using authentic samples of the impurities. The separation of half-wave potentials necessary to prevent interference in controlled potential coulometry depends upon the electrochemical mechanism. For example, if both electrochemical processes are reversible, a 240/n mV separation of half-wave potentials is required to reduce 99% of one material and leave 99% of the second. When nonreversible electrochemical mechanisms are involved, the separation required may be even larger, as was described for the EC mechanism.

Thus, large variations in the half-wave potentials are required for good selectivity in controlled potential coulometry. This usually means that good selectivity can be achieved only when the electroactive group itself is modified or if a very substantial molecular rearrangement occurs. These situations are most likely to occur if the last step of the synthesis creates or modifies the electroactive group or if the degradation mode involves a redox reaction.

5. Coulombs Per Equivalent

Faraday's constant, the number of coulombs per equivalent, is a well defined value.[18] The best electrochemical determination obtained by stripping silver anodes into perchloric acid gives 96,487.2 ± 1.8 mC/meq (CV = 18 ppm). When calculated from the electronic charge and Avogadro's constant, the value obtained is 96,486.7 ± 0.5 mC/meq (CV = 5 ppm). All values are based on the carbon-12 atomic scale. These values are essentially identical except that the electrochemical value has a higher standard deviation partly due to a 10 ppm coefficient of variation in the atomic weight of silver. Recent and continuing work on both electrochemical and physical evaluations of Faraday's constant have been described by Koch et al.[23] and Koch and Diehl.[23a] It should be noted that the precision of Faraday's constant is better than the precision of the relative atomic masses for 70 elements and better than such common analytical standards as sodium carbonate, benzoic acid, potassium dichromate, zinc, copper, and mercury.[18]

6. Molecular Weight

Generally the molecular weight of the compound being analyzed is known. However, if its purity and the n value can be determined by other means, it is possible to determine molecular weight by controlled potential coulometry. Reported examples include an atomic weight of zinc of 65.378 ± 0.0015 (23 ppm).[24,25] One example in the author's laboratory involved the determination of purity by differential stepping calorimetry[26] followed by the determination of equivalent weight by controlled potential coulometry to confirm the molecular ion postulated from high resolution mass spectrometry.[27]

D. Cell Design

The electrochemical cell design required is greatly influenced by the desired electrolysis time and by the nature of the electrolysis reaction. Many different cell designs have been proposed for different applications.[4,5,7] Basically, these may be considered in terms of their important characteristics such as size, mass transport rate, cell geometry and electrode-electrolyte selection. Several sources of electrochemical cells are listed in Table 1.

1. Size

Most analytical applications of controlled potential coulometry employ 0.001 to 1 millimole (100 n to 100,000 n mC) of material at a concentration of 1 to 10 m*M*. Normal cell volumes are about 1 to 25 mℓ. Micro cells or thin-layer cells can be used when samples are limited or if spectroelectrochemical techniques (e.g., ultraviolet, infrared, electron spin resonance, or nuclear magnetic resonance) are used for mechanistic investigations.[28-32] Larger volume cells would usually be used only for poorly soluble samples or semi-preparative work. Temperature regulation may be required for interpretation of electrochemical mechanisms from charge-time behavior but is not usually necessary for quantitative applications. Except in larger scale electrosynthesis, resistive heating effects are not important.

2. Mass Transport Rate

In the absence of slow preceding chemical reactions, the progress of an electrolysis at constant potential depends upon the mass transport rate as given in Equation 5. Therefore, many electrochemists have attempted to design cells with high mass transport rates and correspondingly short analysis times.[4,5,10,11,33-35] As indicated by Equation 6, the mass transport rate is directly proportional to the area/volume ratio and inversely proportional to the diffusion layer thickeness (δ). The diffusion layer is the nonstirred solution layer adjacent to the electrode where convectional transport is replaced by the slower or less efficient diffusional transport. Values of the mass transport constant, s_o, for typical cell designs are about 0.1 to 0.002/sec.[5,7,10,11,34,35] Thus, complete electrolysis requires about 2 to 60 min, respectively. More efficient mass transfer decreases the electrolysis time which increases the ratio of electrolysis charge to background charge and decreases the effects of instrumental drift or other errors.[7,11] It is estimated that halving the time of the electrolysis increases the accuracy and/or precision by a factor of ten.[7]

The diffusion layer thickness and electrolysis time are reduced with efficient mixing. Magnetic stirrers are convenient and do not require a physical inlet to the cell, but they are not as efficient, reproducible, or constant with time as mechanical stirrers (propeller or paddle blades) driven by synchronous motors. Mercury pool or other liquid working electrodes should be stirred at the electrode-solution interface as vigorously as possible without causing electrode breakup. However, the mercury surface itself (actually the distance between the working and reference electrodes) must not be disturbed if positive feedback voltage compensation techniques are used. With solid electrodes, any efficient stirring method can be used.

Clem[10,34,35] has designed an unique rotating cell which provides efficient stirring and

Table 1
SELECTED SYSTEMS AND COMPONENTS FOR CONTROLLED POTENTIAL COULOMETRY

Manufacturer	Address	Components*
Anteck Instruments, Inc.®	6005 N. Freeway Houston, Tex. 77076	Coul
Beckman Instruments, Inc.®	2500 Harbor Blvd. Fullerton, Cal. 92634	Cell, MP
Bioanalytical Systems, Inc.®	P.O. Box 2206 W. Lafayette, Ind. 47906	Cell, Mod, Pot
Brinkmann Instruments, Inc.®	Cantiague Road Westbury, N.Y. 11590	Cell, Int. Pot
Coulometrics, Inc.®	P.O. Box 350 Wheat Ridge, Col. 80033	Coul
Environmental Science & Engineering Corp.®	1776 May's Chapel Rd. Mt. Juliet, Tenn. 37122	Int, Pot, Sy
Hi-Tek® (distributed by) Coastal Associates, Inc.®	6122 Dryden Ave. Cincinnati, Ohio 45239	Int, Pot
Koslow Scientific Co.®	7800 River Rd. N. Bergen, N.J. 07047	Coul, Int
Pacific Photometric Instruments, McKee-Pedersen Instruments Division®	5675 Landregan St. Emeryville, Cal. 94608	Cell, Int, Mod, Pot, Sy
Petrolite Instruments®	4411 Bluebonnet Dr. Stafford, Tex. 77477	Coul
EG & G Princeton Applied Research®	P.O. Box 2565 Princeton, N.J. 08540	Cell, Pot, Sy
Sargent-Welch Scientific Co.®	7300 N. Linder Ave. Skokie, Ill. 60077	Cell, Coul
Tacussel Electronics® (distributed by Astra Scientific International, Inc.®)	P.O. Box 2004 Santa Clara, CA 95051	Int, Cell, Mod, Pot
United Technical Corp.®	81 Keystone Dr. Leominster, Mass. 01453	Int, Pot
ECO — Control, Inc.®	56 Rogers St. Cambridge, MA 02142	Cell, Coul, Int, Pot
Environmental Sciences Associates, Inc.®	45 Wiggins Ave. Bedford, MA 01730	MP

* Components are Cell = electrolysis cell, Int = integrator, MP = multipurpose instrument, Sy = system (cell, integrator and potentiostat), Coul = coulometer (usually integrator and potentiostat), Mod = modular components, and Pot = potentiostat.

deoxygenation with either mercury or platinum electrodes. Instead of stirring the solution, his design rotates the cell causing both the mercury and the solution to be forced onto the cell wall by centrifugal force. This creates a thin film of solution and gives rapid mass transport.

Another approach to efficient mass transport is the use of flow cells where the solution is forced through a porous electrode at a rate which gives complete electrolysis.[36] Flow cells have been used most often for studies of the electrochemical mechanism[37] or identification of electrolysis products.[28-31b] The rapid conversion time suggests another approach for rapid controlled potential coulometric analysis. Many other cell designs have been described as well.[7,15,32,38-40]

Nonuniform mass transport or momentary fluctuations in the electrode area (liquid electrodes) produce stirring "noise" or variations in the current. Electronic integrators can accurately follow these fluctuations. However, manual integrations of current-time curves or digitally sampled current-time curves could be adversely affected. According to Clem et al.,[34] smaller background or residual currents as well as smaller stirring noise artifacts are observed if constant electrode area is maintained during the stirring process.

3. Cell Geometry

Cell geometry affects potential control capability and counter electrode reactions as well as the mass transport characteristics. In certain cases, the cell must be designed to exclude environmental contaminants.

If the electrode potential temporarily or locally exceeds the control potential (more negative than the control potential for a reduction or more positive for an oxidation), unwanted electrochemical reactions such as solvent electrolysis may occur. If these electrochemical reactions are irreversible, the charge consumed is not recovered when the electrode potential returns to its control value and an analytical error results. Potentials which are less than the control potential may cause incomplete electrolysis or a change in the electrochemical mechanism. Thus, accurate control of the working electrode potential is desirable. Potential control is determined by the cell design and by the potentiostat characteristics (discussed later).

The potential difference between the working electrode and the solution is sensed by the reference electrode. There are two effects which interfere with this measurement. The most important effect is due to the potential and current distributions which, in turn, depend upon the counter-working electrode geometry.[7,41,42] In the limiting current region, the working electrode surface is equipotential, but the local current density is determined by the rate of mass transport to each region of the electrode. Therefore, the equipotential lines within the solution are not parallel to the electrode surface. This means that the measured potential difference depends upon the relative magnitudes of the mass transfer resistance to the electrolyte resistance and that it will vary depending upon the location of the reference electrode. The greatest potential difference is observed at the point of closest approach of the counter and working electrodes. Placing the reference electrode on the line of their minimum separation and on an equipotential line which just intersects the working electrode is the best arrangement. If the equipotential line at the reference electrode intersects the working electrode away from the line of minimum separation, then the potential will exceed the control value. If the equipotential line never intersects the working electrode, then the potential will be less than the control value due to uncompensated solution resistance. The error in potential due to uncompensated resistance can be minimized or eliminated by using positive feedback techniques. The electrode arrangement also influences the cell transfer function which describes the ability of the cell to respond to the potential imposed by the potentiostat. These problems are minimized by high conductivity solutions and by a cell design incorporating a symmetric arrangement of the counter and working electrodes. A more detailed discussion has been given by Harrar.[7]

The second reason for poor potential control within the cell arises if the product of the electrolysis current times the working electrode internal resistance is large (greater than a few millivolts). In this case, the electrode surface will not have a uniform potential due to its internal resistive potential drop. This should not be a problem with most metallic electrodes unless currents are extremely high. However, it may be a problem with semiconductor (tin oxide, etc.) or thin film electrodes.

Electrochemical reactions which can occur at the counter electrode may interfere with the coulometric experiment. This may be due to simple regenerations of the starting material (e.g., ferric reduced to ferrous at the working electrode which is convectively carried to the counter electrode and reoxidized) or to unwanted changes in the starting material (e.g., electrochemical reaction of the starting material at the counter electrode or its chemical reaction with electrogenerated chlorine). Separation of the counter electrode from the bulk of the solution prevents these phenomena. The separating diaphragm must prevent appreciable chemical mixing[43,44] while allowing good potential distribution with a minimum cell resistance.[4,7] Although these requirements are somewhat mutually exclusive, a separate counter electrode compartment arranged symmetrically with the working electrode is a reasonable configuration. Medium or fine porosity sintered glass disks, porous vycor, ion exchange membranes, or similar materials can be used. Porous vycor has a smaller electrical resistance and lower solution flow rates than a corresponding fritted glass separator, but it requires a few simple precautions for its use.[7] Ion exchange membranes are particularly effective in preventing migration or diffusion from the cell of ions with the opposite charge type as the electrode charge. If mixing is a particular problem, a double diaphragm cell with a center compartment can be used.[5] In general, the counter electrode and diaphragm should be as large as practical relative to the working electrode.

Some species such as Fe(II), V(II) or Cr(II) are susceptible to oxidation by air. Oxygen, as mentioned previously, is itself easily reduced at a mercury electrode and contributes to the residual current. Other species such as radical anions react extremely fast with proton donors such as water. If one is investigating these types of species, the cell design must necessarily be capable of excluding them. Vacuum techniques and various purge techniques have been used for these purposes.[4,15,38] The most common arrangement for excluding oxygen consists of a nitrogen or argon purge.[4,5,7] The purge gas is passed through an oxygen scrubber and then presaturated with the electrolyte solution to prevent evaporation within the cell. The oxygen scrubber may be a vanadous sulfate or chromous chloride solution over amalgamated zinc or a hot (450 to 500°C) quartz tube filled with copper turnings. The purge gas can be bubbled through the electrolysis solution using a fritted glass gas dispersion tube (being careful not to splash solution onto the cell top) or passed over the top of the solution. The purge should be continued during the electrolysis.

4. Electrode and Electrolyte Selection

The choice of materials for the working electrode depends upon the electrochemical characteristics of the compound. Mercury is a frequent choice for reductions while platinum is the most frequent choice for oxidations. The effects of electrode purity and surface states have been discussed by Harrar.[7] Glassy carbon (pyrolytic graphite), wax impregnated graphite, carbon paste, gold, silver, tin oxide, and other similar materials have also been used.[7,15] Available potential regions may be limited by the type of working electrode.[15,21] For example, mercury is easily oxidized but has a high overpotential for hydrogen evolution in aqueous solutions. Conversely, platinum has a low overpotential for hydrogen evolution making it unsuitable for work at moderately to highly negative potentials.

The counter electrode is usually constructed from platinum. In aqueous solutions, the most likely major product at the counter electrode is either oxygen or hydrogen depending upon the direction of current flow. However, moderately concentrated chloride solutions will give chlorine if the counter electrode functions as an anode. In these situations, a silver counter electrode may be used if the counter electrode cannot be separated from the electrolysis solution (silver chloride is electrodeposited).

The most common reference electrode is the aqueous saturated calomel electrode (S.C.E.) which is +0.2412 V vs. the normal hydrogen electrode. If the supporting electrolyte contains perchlorate which would cause precipitation of potassium perchlorate at the S.C.E. junction, the S.C.E. may be modified to use a sodium chloride solution or be separated by a sodium chloride salt bridge. Alternatively, the silver/silver chloride electrode may be used. The silver/silver chloride electrode is also more easily adapted to nonaqueous solutions. Numerous other combinations of electrolytes with reference electrode have been described.[21] Although it is convenient to know the absolute voltage of a particular reference electrode, stability and solution compatibility are more important criteria. Occasionally, a bare platinum wire inserted into the solution is the best choice for an electrode even though its absolute potential (versus S.C.E.) will be undefined.

Water is the most frequently used solvent in the supporting electrolyte. It has a high dielectric constant permitting easy dissolution of ionic species to give high conductivity solutions. Other frequently used solvents are dimethylformamide, acetonitrile, dimethylsulfoxide, alcohols or aqueous alcohol mixtures, propylene carbonate, hexamethylphosphoramide, tetrahydrofuran, and acetic or formic acid.[21] The choice of solvent also greatly affects the choice of the ionic species added to complete the supporting electrolyte. Salts, acids, and bases have all been used in aqueous solutions. Tetralkylammonium hydroxide or its salts give a larger negative range than Na, K, or Li salts. Generally, if the electrolysis involves uptake or release of hydrogen ions, a buffered electrolyte is advisable. The buffer capacity should be several times the concentration of the electroactive species. In nonaqueous media, species such as lithium chloride or the tetralkylammonium halides, perchlorates, or tetrafluoroborates are preferred.[21] The solution conductivity must be sufficient to transport the current (frequently as high as 100 mA at the start of the electrolysis) without voltage limiting of the potentiostat. High supporting electrolyte concentrations also minimize migration (ion transport due to electric field gradients) of ionic sample components into the counter electrode compartment. Thus, solvents with relatively high dielectric constants (>10) and concentrated supporting electrolytes (>0.1 *M*) are necessary.

E. Instrumentation

Extensive discussion of controlled potential coulometry instrumentation and its requirements, design and availability have been published.[4,7,10,11] The review by Harrar[7] includes an extensive summary of current theory on potential control circuits and the interactions of cell with instrumentation. Basically, the instrumentation for controlled potential coulometry consists of a potentiostat, a current to voltage converter and an integrator. Several sources of instrumentation are listed in Table 1.

1. Potentiostat

The potentiostat controls the electrolysis potential (working electrode to reference electrode potential difference) compared to the externally applied control potential. It must perform the control function accurately without voltage oscillations or other losses of control greater than a few millivolts, even with sudden changes in applied voltage. This function is usually implemented by analog amplifiers. However, digital circuits or computer controlled instrumentation is possible.[7,45] An automated system has also been described.[46]

The potentiostat needs sufficient current and voltage output capabilities to maintain its control functions. The current at time zero is the maximum demand. It will consist of I_o as well as the charging current. Typical values will be 50 to 200 mA. A capability of approximately 1A should be sufficient for most cases except perhaps semipreparative work. The control potential difference rarely exceeds 2V. However, the voltage output must overcome the ohmic losses due to the solution and diaphragm resistances between the counter electrode and the reference electrode. For even 200 mA currents and 100 ohm resistances, this amounts to an additional 20 V. Generally 50 to 100 V output capability is preferred, especially if work is performed in nonaqueous solvents where solution resistances are typically larger. The output voltage offset should be adjustable to zero. However, this is easily obtained within the required limits and only rarely is it necessary to readjust it with modern instrumentation.

2. Current to Voltage Converter

The current to voltage converter changes the current into a signal which can be measured. It is usually incorporated into and considered a part of the potentiostat. Since it is very important to accuracy and/or precision, it will be discussed separately.

The current to voltage transformation also is achieved with analog instrumentation. It may be as simple as measuring the voltage drop across a calibrated resistor in series with the cell. Usually it is better to use an amplifier circuit to achieve a similar result. The amplifiers must have the same current output capability and frequency response as the potentiostat. The voltage output required depends upon the integrator to be used as well as upon any current monitoring devices used. Frequently several ranges will be provided from about 1 μA full scale to about 1A full scale. Purchased instruments should perform within stated accuracy tolerances. However, calibration may be required for high precision work or after electronic repairs or following circuit modifications.

The current is almost always measured in a circuit employing one or more precision resistors. These resistors should be calibrated to an accuracy of at least 0.05 to 0.1% for high quality work. When electronic circuits are used, the zero output should be stable, easily adjustable and checked periodically since any offset is added directly to the apparent current. The calibration of the precision resistors needs to be checked only infrequently (e.g., after instrument repair or following prolonged current overload conditions). Direct measurement of the precision resistor(s) for calibration of the current to voltage conversion circuit may be possible. More frequently, a separate precision resistor (to be substituted for the electrochemical cell) and digital voltmeter (4½ digits preferred) are required for calibration. The measured converter output can be compared to that calculated from the voltage drop across the resistor and its resistance. The voltmeter and ohmmeter used for this calibration must be calibrated independently.

Electrolysis current (in milliamperes) measured with commercial instruments is usually obtained from the converter output (E, volts) using the relationship

$$I = \text{Sens} \times E \qquad (27)$$

where Sens is the current to voltage converter sensitivity in milliamperes per volt.

3. Integrator

The integrator performs the final data manipulation. However, it is more subject to errors than the other instrumentation. Integration may be performed by various combinations of analog circuits, digital circuits and numerical processing. The most common designs use the output of the current to voltage converter as the input to an analog integrator having a resistor-capacitor circuit. The output voltage (in volts) is given by

$$V = [1/RC] \int_0^t E \times dt \qquad (28)$$

where R is the integrator input resistance in ohms, C is the integrator capacitance in farads, E is the voltage being integrated (the output of the current to voltage circuit), and t is the time in seconds. The product of resistance times capacitance (RC) is called the integrator time constant and has units of seconds (equivalent to millicoulomb per milliampere). Circuits for integration are subject to capacitor related errors as well as to amplifier drift and offset errors.[7] The capacitor must have time stability, low leakage, low memory (hysteresis), and a low temperature coefficient. A polystyrene capacitor of one to ten microfarads is usually preferred. The amplifier must have a low time- and temperature-stable bias current which can be adjusted to zero. The integrator's bias current including drift must contribute insignificantly to the total charge on the capacitor. With a current to voltage converter output of 10 mV (common near the end of the electrolysis) and an integrator input resistor of one megohm, the faradaic current would be 10 nA. Therefore, the bias current should be no more than about 10 pA (10^{-11} A). This would contribute an additional 1 mV to the integrator output for a 1000 sec integration. These specifications can be obtained with high quality, discrete amplifiers and well designed circuits. Recently a circuit was described for the automatic compensation of drift and offset errors in the current to voltage converter and integrator circuits.[20]

Using Equation 1 and substituting Equations 27 and 28 gives the charge in millicoulombs from the integrator output voltage (V) as

$$Q = \text{Sens} \times \int_0^t E \times dt = \text{Sens} \times V \times (R\,C) \tag{29}$$

The resistance can be varied (using several precision resistors) to provide multiple charge ranges. For example, a 10 μF capacitor with a 1 Mohm resistor gives a time constant of 10 sec and a 10 V output signal for 10,000 mC if a 100 mA/V current range is used. Switching to either a 100 kilohm input resistor or to a 10 mA/V current range would give a 10 V output signal with only 1000 mC.

The integrator time constant is most easily calibrated by a single measurement rather than separate measurements of R and C. This calibration usually requires a voltage source for the input, timer, and digital voltmeter. The integrator calibration is performed by measuring the change in its output after integrating a constant voltage signal for a specified interval. Care must be taken not to introduce switching transients at the start or end of the timed interval.[7] The voltmeter is used to measure the constant input voltage. The time measurement must be accurate to at least 0.2 sec for a 7 min integration time to give 0.05% precision. Alternatively, the current to voltage converter and integrator may be calibrated as 1 unit. An overall accuracy of 0.1% or better is desirable.

When voltage to frequency conversion is used, the current to voltage circuit output is converted to a square wave whose frequency is proportional to the input voltage.[11] Then, a digital counter on the voltage to frequency converter output performs as an integrator. These circuits are commonly used in high quality digital voltmeters and it is expected that bipolar, low drift, high accuracy voltage to frequency circuits will gradually displace analog integration.[7,11]

The digipotentiogrator[7,45] acts as a potentiostat, current monitor, and integrator in one digital unit. Essentially it injects constant value charge pulses into the cell at a variable rate to maintain the appropriate control potential. If these charge pulses are sufficiently small, the response will be nearly that of analog instrumentation. However, the current is proportional to the rate of delivery (frequency) and the charge to the total number of pulses delivered (total count). Both of these can be easily measured very accurately.

When digital acquisition and processing systems are used, the number of data points and data processing algorithm must be adequate to insure a precision on the order of 0.1%. Whenever analog-to-digital conversion is used, its errors,[47-48a] and those due to 60 Hz noise pickup must be included in the error analysis. When the current is digitized, there is at least a 3 decade change in its magnitude. This requires a large dynamic range in the analog-to-digital converter to maintain adequate resolution at low current values. The digitization error resulting from the discrete output values is equal to ½ of the least significant bit. Thus, a 13 to 14 bit, 10 V bipolar analog-to-digital converter would be required for initial current values giving output signals of several volts. If less precise analog-to-digital converters are available (e.g., 8 to 10 bits), using a programmable gain preamplifier or multiplexing between parallel fixed gain amplifiers (e.g., relative gains of 1, 4, 16, and 64) will effectively extend the dynamic range. At the low signal levels which will be encountered near the end of the electrolysis, 60 Hz noise will almost always be a problem. The error due to 60 Hz pickup may be minimized by averaging several data points over a specific number of 60 Hz waveforms or by using analog or digital filters. A 50 msec time period (three 60 Hz waves) is a convenient digital averaging time.[49]

A check of the overall instrumental calibration accuracy may be obtained by replacing the electrochemical cell with a precision resistor. The measured charge can then be compared to that calculated for the known resistance, applied voltage and integration time. For an applied voltage (millivolts) with the precision (dummy) resistor in ohms and the time in seconds, the calculated relationship is

$$V = E(\text{applied}) \times t \,/\, R(\text{dummy}) \times \text{Sens} \times R\,C \tag{30}$$

If the integrator output is presented directly in millicoulombs, then the calculated charge is

$$Q = E(\text{applied}) \times t \,/\, R(\text{dummy}) \tag{31}$$

F. Analytical Applications

Several recent articles review applications of controlled potential coulometry to the study of organic electrochemical reactions.[15,22,40,50,51] However, only a few examples of controlled potential coulometric assays of pharmaceutical compounds have been published (Table 2). Experimental conditions and precision data are given when available. Additional results obtained in the author's laboratory are described below and included in Table 2.

New instrumentation and cell designs have reduced the complexity and increased the accuracy and speed of this technique. It is expected that many additional examples will be reported in the future.

The cephalosporin antibiotics cefamandole nafate (I, Table 3) and cefamandole sodium (II) undergo a two electron reductive cleavage of the thioether linkage.[52,53] The assay is performed in an aqueous, pH 2.3 McIlvaine buffer at a sample concentration of approximately 5 mM (0.05 mmol). The 12.6 cm^2 stirred mercury pool working electrode is in contact with approximately 10 mℓ of solution. The 4 cm^2 platinum counter electrode is separated from the solution by a medium porosity sintered glass frit, 1 cm in diameter. The analog integrator has a time constant of 10.73 sec (resistance about 1 megohm, capacitance about 10 μF). Corrections for residual current (believed to be due to a catalytic hydrogen mechanism) must be made. Other experimental details and the isolation of the coulometric reduction products have been reported.[52]

Cinoxacin (III), an antibacterial, undergoes a two electron reduction of the imine in a variety of acidic aqueous solvents.[27,54] Aqueous formic acid solutions have been used to provide increased sample solubility for the assay (5 mM, 0.05 mmol). Experimental details are similar to those for the cephalosporins.

Thimerosal (IV), used in topical antiseptics and as a preservative, undergoes a one electron reductive cleavage of the ethylmercury substituent at a stirred mercury pool.[55] The sample concentrations used were approximately 5 mM (0.05 mmol). Potassium chloride was added to the electrolyte to improve conductivity.

Penicillamine (V), used for treating rheumatoid arthritis and as a chelating agent, can be oxidized. Like most thiols, the one electron oxidation product is usually the disulfide at a platinum electrode and the mercurous mercaptide at a mercury electrode.[56,57] A concentration of 10 mM (0.1 mmol) was used. The platinum screen working electrode area was approximately 10 cm^2. The other cell and instrumental characteristics were similar to those described for the cephalosporins.

Sodium omadine (VI), a thiol derivative of pyridine N-oxide, also can undergo a one-electron oxidation at a platinum electrode.[58,59] Ethanol is added to increase solubility.

The bromo-nitro dioxanes nibroxane (VII) and compound VIII, used as antibacterials, are very easily reduced. At a mercury electrode, they undergo a spontaneous chemical reaction forming mercuric bromide. At a platinum electrode, they undergo a two electron reduction leading to quantitative production of the des-bromo compound.[53] A variety of solution conditions produce equivalent results. Methanol can be added to the solution to increase sample solubility. Normally, the residual current is very low for this assay. Sample concentration was 5 mM (0.05 mmol).

An experimental compound (IX), undergoes a four electron reduction.[53,60] Solution conditions were determined by sample solubility and the rather negative electrode potentials needed to produce the reduction. Sample concentration was approximately 2.5 mM (0.025 mmol). Potassium chloride was added to the supporting electrolyte to improve solution conductivity.

III. COULOMETRIC TITRATIONS

A. Advantages and Limitations

Coulometric titrations have many of the same advantages as controlled potential coulometry such as being an absolute technique. Specifically, the first three advantages listed under controlled potential coulometry and those listed below apply to coulometric titrations.[8-10,13,58,64-66]

1. It is a high precision technique. High precision constant current coulometry has been reported to a σ of about 10 ppm (CV = 0.001%).
2. Substances which cannot be electrolyzed directly (kinetically slow electron transfer processes) can be titrated using a chemical redox mediator or buffer.
3. The concentration of the redox buffer can be much larger than the concentration of the electroactive species. This permits higher currents and gives shorter analysis times and higher current efficiencies than controlled potential coulometry.
4. Instrumentation for coulometric titrations is usually less expensive, less complicated, and less subject to errors than that for controlled potential coulometry. Coulometric titrations require a current generator and timer, whereas controlled potential coulometry requires a potentiostat and integrator.
5. Coulometric titrations are not limited to redox reactions. Acid-base neutralizations, precipitation, or complexometric titrations can be performed with electrogenerated titrants.
6. It is possible to electrochemically generate titrants which are difficult to prepare, store, and standardize chemically: iodine, bromine, chlorine, Ag(II), Mn(III), Cu(I), Cr(II), V(II), Sn(II), Ti(III), etc. This is in addition to common redox couples such as Ce(III)/Ce(IV), Fe(II)/Fe(III), Hg(I)/Hg(II), and Cr(III)/Cr(VI). Generation of hydrogen ions, hydroxide ions, and the Karl Fischer reagent are also possible.
7. In some cases with a slow chemical reaction, an excess of titrant can be generated and then back titrated.

Table 2
PHARMACEUTICAL APPLICATIONS OF CONTROLLED POTENTIAL COULOMETRY[a]

Compound[b]	Working electrode	Supporting electrolyte	Control potential, V vs. SCE	Reaction	n Value	Sample weight, mg	Precision[c]	Replicates	Ref.
Ascorbic acid	Pt	0.2 *M* phthalate buffer, pH 6.0	+1.09	Oxidn	2	15—100	±0.7 mg	7	61
Atropine N-oxide	Hg	H_2SO_4/K_2SO_4 buffer, pH 1.57	−1.00	Redn	—	20	1.4%	>4	62
3-Bromopyridine N-oxide	Hg	H_2SO_4/K_2SO_4 buffer, pH 0.88	−0.80	Redn	—	10	0.4%	—	62
Brucine N-oxide	Hg	Acetate buffer, pH 4.63	−1.05	Redn	—	25	2.5%	—	62
Cefamandole nafate	Hg	pH 2.3 McIlvaine buffer + KCl ($\mu \approx 0.5$ *M*)	−0.875	Redn	2	25	0.95%	67[d]	52,53
Cefamandole sodium	Hg	pH 2.3 McIlvaine buffer + KCl ($\mu \approx 0.5$ *M*)	−0.875	Redn	2	25	0.84%	83[d]	52, 53
Chlordiazepoxide	Hg	H_2SO_4/K_2SO_4 buffer, pH 0.40	−0.30	Redn	—	20	0.3%	—	62
Chlorpromazine	Pt	12 *N* H_2SO_4	+0.60	Oxidn	1	10—35	±0.16 mg	7[d]	63
Chlorpromazine	Pt	1 *N* H_2SO_4	+0.95	Oxidn	2	35	±0.4 mg	3	63
Cinoxacin	Hg	Formic acid: H_2O at 1:1 (v/v)	−0.65	Redn	2	15	0.7%	16	27, 53, 54
Compound VIII (Table 3)	Pt	pH 7 McIlvaine buffer + KCl ($\mu \approx 0.5$ *M*)	−0.45	Redn	2	10	1.5%	8	53
Compound IX (Table 3)	Hg	pH 8.5 borate buffer + KCl ($\mu \approx 0.5$ *M*)	−1.65	Redn	4	5	0.8%	3	53
Nibroxane	Pt	pH 7 McIlvaine buffer + KCl ($\mu \approx 0.5$ *M*)	−0.50	Redn	2	10	0.5%	9[d]	53
Penicillamine	Hg	1 *M* NH_4OH/1M NH_4Cl	−0.30	Oxidn	1	15	0.7%	5	53
Prochlorperazine	Pt	12 *N* H_2SO_4	+0.60	Oxidn	1	10—56	±0.16 mg	6[d]	63
Prochlorperazine	Pt	1 *N* H_2SO_4	+1.00	Oxidn	2	56	±0.2 mg	3	63
Promazine	Pt	12 *N* H_2SO_4	+0.60	Oxidn	1	10—32	±0.1 mg	6[d]	63
Promazine	Pt	1 *N* H_2SO_4	+1.05	Oxidn	2	32	±0.2 mg	3	63
Promethazine	Pt	14 *N* H_2SO_4	+0.70	Oxidn	1	10—32	±0.15 mg	7[d]	63
Promethazine	Pt	1 *N* H_2SO_4	+0.75	Oxidn	2	32	—	—	63

Scopolamine N-oxide	Hg	Acetate buffer, pH 4.63	−0.75	Redn	—	20	3.0%	—	62
Sodium Omadine	Hg	pH 5.8 acetate buffer + ethanol	−0.15	Oxidn	1	15	0.7%	4	53
Strychnine N-oxide	Hg	H_2SO_4/K_2SO_4 buffer, pH 0.88	−0.90	Redn	—	20	0.7%	—	62
Thimerosal	Hg	pH 8 phosphate buffer, + KCl ($\mu \approx 0.5$ *M*)	−1.60	Redn	1	20	1.3%	4[d]	53
Thioridazine	Pt	12 *N* H_2SO_4 in 30% (v/v) ethanol	+0.55	Oxidn	1	10—40	±0.17 mg	3[d]	63
Thioridazine	Pt	1 *N* H_2SO_4	+0.75	Oxidn	2	40	±0.4 mg	3	63
Trifluoperazine	Pt	12 *N* H_2SO_4	+0.70	Oxidn	1	10—48	±0.22 mg	6[d]	63
Trifluoperazine	Pt	1 *N* H_2SO_4	+1.00	Oxidn	2	48	±0.2 mg	3	63
Triflupromazine	Pt	12 *N* H_2SO_4	+0.65	Oxidn	1	10—40	±0.1 mg	7[d]	63
Triflupromazine	Pt	1 *N* H_2SO_4	+1.05	Oxidn	2	40	±0.4 mg	3	63

[a] Data not given in references is indicated by a line.

[b] Names are titles assigned in the *Merck Index*, 9th ed., where applicable.

[c] Precision given as standard deviation (± mg) or coefficient of variation (%) as reported.

[d] Indicates data from more than one sample or from different weight ranges were pooled to obtain the standard deviation

$$\sigma_p = \left[\frac{\sum_i (n_i - 1)\, \sigma_i^2}{(\sum_i n_i) - i} \right]^{1/2}$$

where σ_p is the pooled deviation and the σ_i are individual standard deviations calculated from n_i measurements within the group.

Table 3
CHEMICAL STRUCTURES

	Structure	Name
I.	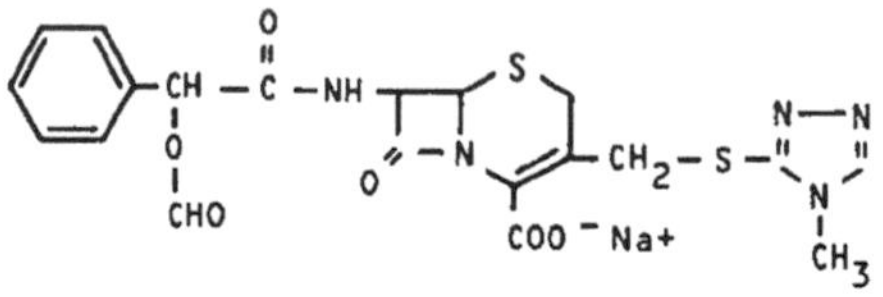	Cefamandole nafate
II.	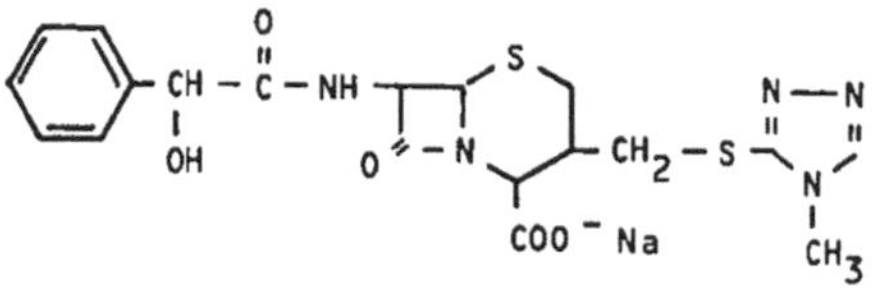	Cefamandole sodium
III.	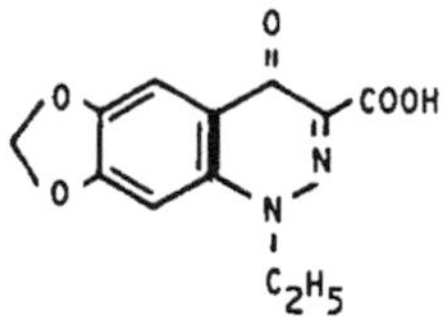	Cinoxacin
IV.	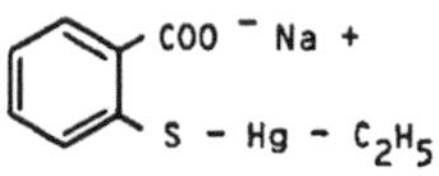	Thimerosal
V.	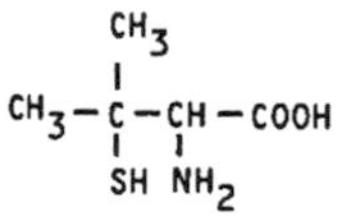	Penicillamine
VI.	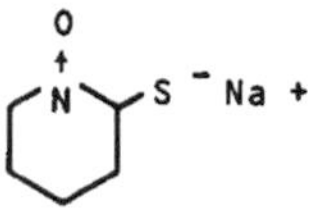	Sodium omadine
VII.	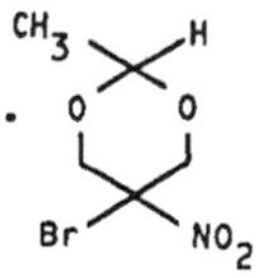	Nibroxane
VIII.	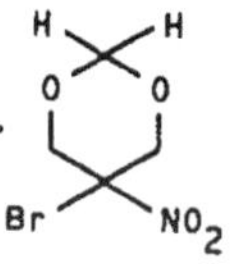	5-Bromo-5-nitro-m-dioxane
IX.	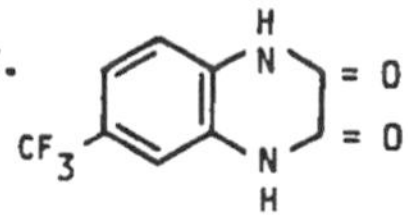	1,4-dihydro-6-trifluoro-methylquinox-aline-2,3-dione

However, coulometric titrations have certain disadvantages. In addition to the requirement of 100% current efficiency (for titrant and/or substance being titrated), other disadvantages include:

1. The electrogenerated titrant must react rapidly and quantitatively with the substance being titrated in a stoichiometrically well defined reaction.
2. Coulometric titrations are not as selective as controlled potential coulometry. In fact, they are no more selective than any other chemical titration. Impurities must be sufficiently different to avoid reaction and erroneous results.
3. An endpoint detection method is required.

B. Principle

Coulometric titrations combine the techniques of coulometry and titrations. That is, they depend upon the homogeneous reaction of a titrant with a substrate (the substance being titrated). The titrant, however, is generated electrochemically instead of being prepared from reagents. As such, it shares both the advantages and disadvantages of controlled potential coulometry and volumetric titrations as listed above. This section will discuss the differences between coulometric titrations and controlled potential coulometry and the application of coulometric titrations to pharmaceutical analysis.

Two fundamental differences exist between coulometric titrations and controlled potential coulometry. First, the controlled parameter for coulometric titrations is current; that for controlled potential coulometry is potential. Second, there is a discrete endpoint which must be detected for coulometric titrations, whereas the controlled potential coulometry endpoint is automatically approached asymptotically.

Coulometric titrations were developed before controlled potential coulometry because they were easy to implement in a high precision mode with a minimum of instrumental design problems. Generally, coulometric titrations are more accurate than controlled potential coulometry, but they are not as selective. The historical development[67] and many applications[68] of coulometric titrations have been sketched by Lingane.

1. Purity Calculation

In coulometric titrations, a relationship analogous to Equation 2 exists between charge and purity. However, the charge for constant current coulometric titrations is now given by the product of current (I) times electrolysis time (t). Thus, percent purity is given by

$$\text{Percent Purity} = I \times t \times MW \times 100 / (n \times F \times W) \qquad (32)$$

where the other parameters are defined as before. Integration is performed by measuring current and time directly rather than by analog integration of current. Nonconstant current variations will be discussed in the instrumentation section.

2. Types of Titrant Generation Reactions

For purposes of discussion, the electrochemical generation of titrants can be broken into several categories.[64,67] Direct coulometric titrations refer to those where the electrode actively participates in the titration reaction such as the titration of halides using electrogenerated Ag(I) from a silver electrode. Indirect coulometric titrations refer to those where the electrode does not participate except as a source of electrogenerated chemical titrants. Coulometric titrations also have been used for determination of excess reagent in back titrations with either chemical or electrochemical generation of the original titrant. Finally, electrogenerated reagents can be generated externally to the cell when sample interferences prohibit internal generation. The endpoint detection technique must monitor the homogeneous chemical reaction for any of these classes of coulometric titrations.

a. Direct Coulometric Titrations

Direct coulometric titrations are generally used for precipitation or complexation reactions with easily generated metallic ions such as silver(I), mercury(I), or mercury(II). The electrode consists of the corresponding metal. Thus, these titrations are generally applied to the determination of ionic halides, pseudohalides, or mercaptans. The selectivity is dependent upon differences in solubility products for precipitation or formation constants for complexation.

b. Indirect Coulometric Titrations

Indirect coulometric titrations with electrogenerated titrant (Species B) comprise the largest class. This sequence of reactions may be represented by

$$A + ne \rightleftharpoons B \qquad E_{1/2}$$

$$B + C \underset{}{\overset{K_{eq}}{\rightleftharpoons}} \text{Product} \qquad E_{eq} = [\text{Product}] \,/\, (B)\,(C) \qquad (33)$$

where species A is the titrant precursor and species C is the substrate or substance being titrated. In some cases such as the titration of Cr(VI) with electrogenerated Fe(II), the titrant precursor may be regenerated in the chemical reaction. In cases such as the addition of electrogenerated bromine to cyclohexene, the titrant is irreversibly consumed.

Both the titrant precursor (A) and the substrate (C) may react at the electrode depending upon their relative half-wave potentials and electron transfer rates. In the titration of Fe(II) with electrogenerated Ce(IV) in aqueous 1 *M* sulfuric acid, both Fe(II) and Ce(III) react directly at the platinum electrode.[69]

$$\text{Ce(III)} \rightleftharpoons \text{Ce(IV)} + e \qquad E^1_{1/2} = 1.35$$

$$\text{Fe(II)} \rightleftharpoons \text{Fe(III)} + e \qquad E^2_{1/2} = 0.60$$

$$\text{Ce(IV)} + \text{Fe(II)} \rightleftharpoons \text{Ce(III)} + \text{Fe(III)} \qquad K' \cong 10^{-12} \qquad (34)$$

Since Fe(II) is more easily oxidized than Ce(III) and readily transfers electrons at a platinum electrode, it will react at the electrode as well as undergoing the homogeneous chemical reaction with electrogenerated Ce(IV). A similar phenomena occurs during the titration of acids with electrogenerated hydroxide. The hydrogen ion can be directly reduced or water can be reduced to give hydroxide which will then react with the acid in solution. The relative importance of each of these reactions depends upon the magnitude of the generating current and the concentration of the relevant species (Ce(III) and Fe(II) or hydrogen ion and water).

The electrogenerated titrant is usually inorganic. Thus, complicated electrochemical mechanisms (heterogeneous kinetics or coupled chemical reactions) are less likely to be important than in controlled potential coulometry. Competing electrochemical reactions, such as those types included in residual current processes, are likely to occur and may contribute significantly to error.

As mentioned previously the chemical reaction must be quantitative ($K_{eq} \gg 1$) within the same restrictions as analogous volumetric titrations. Similarly, the stoichiometry must be well defined and reproducible to calculate the quantity of substrate present.

Competing homogeneous chemical reactions of the electrogenerated titrant are frequently the most important sources of error in coulometric titrations. Chemical reactions of the electrogenerated titrant with the supporting electrolyte, dissolved oxygen, or impurities consume charge and cause analytical errors.

If the rate of the chemical reaction is not rapid, there will be a tendency to halt the titration before the equivalence point because the titrant B has not had time to react with the substrate C. This situation produces a drifting endpoint, higher charge consumption when titration times are lengthened by using lower currents, and lower than expected charge consumption for titrations of standards. If solution conditions (pH, concentration, nature of titrant, etc.) cannot be modified to give a favorable reaction rate, then alternative approaches must be adopted. These could include periodic interruption in the current generation to allow time for the reaction to occur, a decrease in the current (titration rate) near the equivalence point, or a generation of excess titrant followed by back titration. The procedure adopted may affect the precision and accuracy of the method. Periodic interruption of the current generation is simple to implement and used most frequently when additional time must be allowed for chemical equilibrium. Decreasing the magnitude of the current near the equivalence point is effective, but the experiment is now a hybrid between controlled potential coulometry and coulometric titrations with a nonlinear or stepwise linear relationship between time and charge. Back titrations are discussed in the following section.

c. Back Titrations

Back titrations are most frequently used with halogen titrations requiring an extended reaction time and excess reagent to react quantitatively. A known amount of titrant, say bromine, is added chemically or electrochemically and allowed to react for a specified time. The amount of excess titrant and length of reaction time must be determined for each situation. The excess bromine can then be determined by back titration with thiosulfate, arsenic(III), or copper(I). Species such as arsenic(III) or copper(I) can be added volumetrically or by electrogeneration. Alternatively, the bromide produced can be determined by titration with electrogenerated silver(I). Obviously, a variety of titrants and back titrants combined with volumetric addition or electrogeneration have been used.

d. External Generation of Titrant

If the substrate being titrated does react directly at the electrode, it must undergo a reaction sequence identical to its homogeneous chemical reaction with the electrogenerated titrant to maintain assay validity. That is, with the reaction sequence illustrated in Equation 33, a reaction of the type

$$C + n'e \rightleftharpoons \text{Product} \qquad (35)$$

is not permitted unless $n' = n$. This could happen, for example, if analine were oxidized at the electrode during its titration with electrogenerated bromine. In this situation, the titrant can be electrogenerated externally to the reaction mixture.[67] External generation can also be used if the optimum conditions for electrogeneration of the titrant are not compatible with the conditions necessary for the occurrence of the homogeneous chemical reaction.[64] The same procedure may be used when an impurity reacts at the electrode but not with the electrogenerated titrant. External generation, however, is a considerable inconvenience.

3. Titration Efficiency and Current Efficiency

The term titration efficiency refers to that percentage of the total current which leads to the desired titration.[67] In the titration of Fe(II) with Ce(IV), this would be the ratio of the sum of the current for the direct oxidation of Fe(II) plus the current for the generation of Ce(IV) which reacts with Fe(II) to the sum of all processes which consume current. The titration efficiency must be nearly 100% for a good analytical method. The titration efficiency is generally increased by using a high ratio of titrant precursor to substrate. This also allows higher generating currents and shorter analysis times within the limitations described below. The titration efficiency can be estimated by comparing the residual current for the supporting electrolyte alone to the constant current used for the titration.[67] The residual current must be measured (using controlled potential electrolysis) at the electrode potential which the working electrode would assume during an actual constant current titration.

Current efficiency has frequently been used interchangeably with titration efficiency.[67] However, current efficiency should refer to the current for a specific electrode process [such as the Ce(III)/Ce(IV) couple] compared to the total current. If the substance being titrated can react directly at the electrode [as does Fe(III)], the current efficiency for Ce(IV) generation will be less than 100%. For reversible electrode reactions with both substrate and titrant precursor, the current efficiency for electrogeneration of titrant will be dependent upon the solution concentration ratio of titrant precursor to substrate.

4. Current Selection

The selection of the current to be used involves many of the same considerations as the selection of potential in controlled potential coulometry.[67] In both cases, the potential and current are related through the half-wave potential for the electrochemical reaction (Equation 4) and the mass transfer (Equations 5 to 8). Figure 5 indicates the relationships between current and potential at the start of a Ce(IV) titration of Fe(II) at the platinum electrode assuming a constant mass transfer rate.

The electrochemical reactions which occur depend upon the magnitude of the current. If the current were forced to assume a low value (a), almost all of it would be consumed by the residual current processes and a low titration efficiency would result. The residual current processes are the same as those discussed in controlled potential coulometry.

If current level b were chosen instead, a much lower percentage of the current would be consumed by residual current processes. At this level, the principal reaction would initially be the direct electrooxidation of Fe(II). As the titration proceeds, the Fe(II) would be consumed and its limiting current (c) would decrease. Eventually, some and then most of the current would be consumed by the oxidation of Ce(III). It can be seen that the Ce(III)/Ce(IV) couple acts as a redox buffer. That is, the electrolysis current is consumed by a redox couple which prevents the electrode potential from changing to a value where an undesired side reaction (oxidation of water) will occur. At the same time, almost all of the charge is effectively used to oxidize the ferrous ion.

If the current is increased to the value d, it can be seen that both Fe(II) and Ce(III) would initially react at the electrode. Again, the concentration of Fe(II) will decrease during the titration and a proportionately higher percentage of the current will be consumed by the Ce(III) oxidation. However, if the current level selected (e) was above that which could be furnished by the total Fe(II) and Ce(III) oxidations at any time during the titration, then some current would be consumed by the oxidation of water and titration efficiency would decrease. Note that the total current at the end of the titration will be composed only of the residual current and the current due to the nonconsumed titrant precursor [Ce(III)]. Thus, a current with value f would be satisfactory for the initial electrolysis, but would cause the oxidation of water later in the experi-

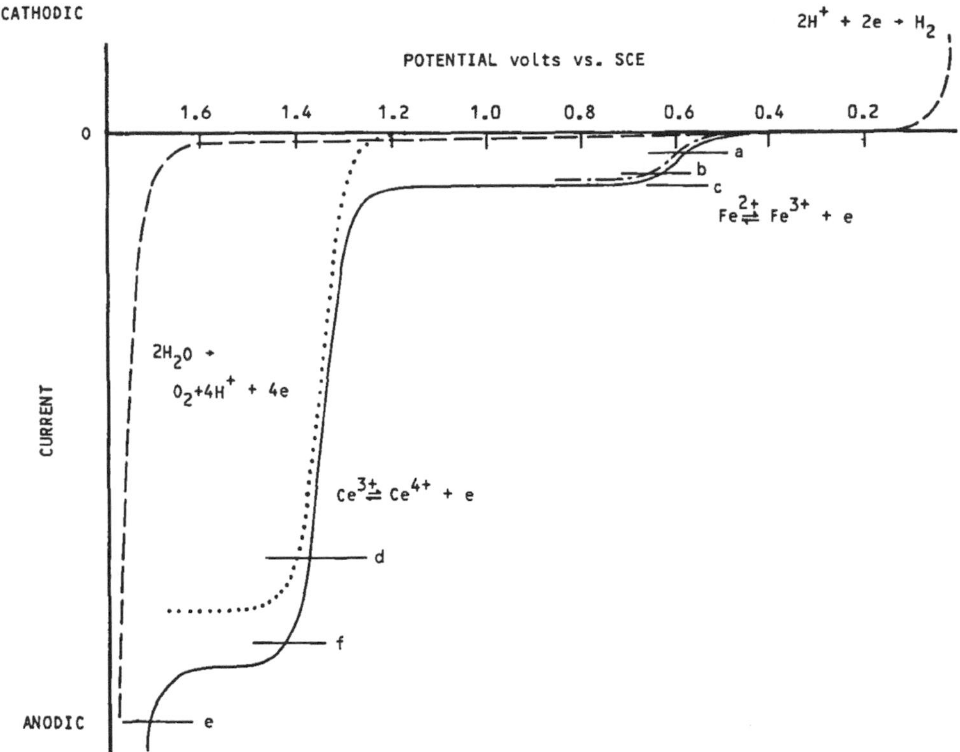

FIGURE 5. Current-potential relationships for coulometric titration of Fe^{+2} with electrogenerated Ce^{4+}. Residual Current (——), oxidation of Fe^{2+} (-·-·-·), oxidation of Ce^{3+} (......), and total current (———),electrogeneration current levels a-f discussed in text.

ment when the Fe(II) concentration had decreased. It also is important to allow for reasonable variations in mass transport rates so that a decrease in stirring rate does not cause an undesired electrochemical reaction to occur.

If the substance being titrated is not electroactive, similar experimental current-voltage curves will be obtained except that the first wave in Figure 5 will be absent. The logic in selecting an appropriate current will be analogous to that described for the Ce(IV) titration of Fe(II). In either case, the current selected should result in as near to 100% titration efficiency as possible. That is, the residual current should be only a small fraction of the total current while undesired side reactions such as the oxidation of water should not be allowed to occur.

The actual magnitude of the current to be used depends upon the electrode area, the solution concentrations, mass transport parameters, and the desired electrolysis time. The optimum current is directly proportional to the concentration of the titrant precursor (plus the substrate if it is electroactive) and to the electrode area. The optimum current is frequently found to be about 0.5 $mA/cm^2/mM$. This will, however, depend upon the other mass transfer parameters D_o and δ included in the mass transport constant, s_o. Thus, higher titrant precursor concentration, larger electrode area or better stirring will increase the absolute magnitude of the current value which should be chosen. Increasing the concentration of titrant precursor and increasing the current will decrease the ratio of residual current to total current. Changing the electrode area or the stirring generally affects the residual current and the titration current equally.

The charge passed through the cell is directly proportional to the time the current flows. Thus, the titration time is inversely proportional to the current. The titration time must be long enough to be accurately measured (200 sec with 0.2 sec accuracy gives 0.1% precision). It also must be long enough to allow mixing of the electrogenerated titrant with the solution and long enough to allow equilibration of the chemical reaction. However, excessive electrolysis times may allow loss of substrate or titrant by diffusion through cell diaphragms[43,44] and decreased efficiency from residual current losses.[10,67] If 0.5 mmol of a compound with a n value of 2 were titrated using a 100 mA constant current (4 cm^2 electrode, 50 m*M* titrant precursor), then the electrolysis time would be about 965 sec as calculated from Equation 32.

5. Criteria for Coulometric Titrations

Several fundamental criteria usually must be met to insure a high titration efficiency and a resulting successful application of the coulometric titration technique. These include a well defined electrode reaction for the titrant precursor and/or substrate and a moderately rapid, quantitative, homogeneous chemical reaction having a well defined stoichiometry. Competing electrochemical reactions of the residual current type may be minimized by appropriate selection of titrant precursor concentration and titration current. Other competing electrochemical and chemical reactions should be absent. Variations in the total charge consumed per mole of substrate as a result of reasonable variations in the current or titration time, or with variations in titrant precursor concentration or substrate concentration, are indicative of some types of competing electrochemical reactions, competing homogeneous reactions, unfavorable reaction rates, or variable stoichiometry for the homogeneous chemical reaction.

C. Application of Principles

The practical usefulness of the coulometric titration technique depends upon several factors other than the overall titration efficiency. These include constraints discussed under controlled potential coulometry such as a well characterized electrode reaction, an accurate value for Faraday's constant, and a known molecular weight for the substrate in addition to the following parameters:

1. Types of homogeneous chemical reactions
2. Titrant and supporting electrolyte selection
3. Endpoint detection technique
4. Pretitration
5. Effects of impurities in the sample
6. Other factors affecting precision and accuracy

1. Types of Homogeneous Chemical Reactions

Coulometric titrations can involve acid-base neutralizations, precipitation, complexation, or redox reactions. These reactions are completely analogous to common volumetric titrations except that the titrant is generated at an electrode. This allows a wide variety of titrants to be used, including many which cannot be conveniently prepared, stored, or standardized for volumetric titrations. Of course, some chemical reactions which depend upon colorimetric detection, complex organic titrants, or special conditions may be inconvenient or impossible to handle using coulometric titration techniques. Generalized reactions and references for the most common types of coulometric titration reactions are given in Table 4. More extensive discussion of these reactions with emphasis on pharmaceutical applications is given below.

2. Titrant and Supporting Electrolyte Selection

The selection of the titrant depends upon the selectivity required and previously reported applications. Selection of the most appropriate titrant is basically a chemical decision. For example, silver (I) is preferred over mercury (II) for the titration of mercaptans in proteins or amino acids since primary amines will complex with mercury (II).[65] In oxidation reactions, the oxidation strength and chemical reactivity can be selected. For example, monounsaturated fatty acids give equivalent titrations with chlorine and iodine, but highly unsaturated fatty acids give lower results for titration with chlorine compared to bromine.[65]

The greatest number of applications of coulometric titrations involve electrogeneration of halogens, especially bromine. Electrogeneration of bromine with nearly 100% current efficiency (low residual and extraneous currents) is easy because the potential of the bromine/bromide couple (+1.05 V vs. NHE) is more than 0.5 V below the potential at which water is oxidized at the platinum electrode.[67] In addition to being a fairly strong oxidant, bromine participates in a variety of substitution reactions. Iodine is also easily generated with nearly 100% current efficiency.

The supporting electrolyte performs a dual role in coulometric titrations. First, it must be compatible with the electrode reaction in terms of ionic conductivity and titrant precursor concentration. Second, it must act as sample solvent and reaction medium. These roles may conflict since a highly concentrated, high dielectric constant solvent will be preferred for conductivity, but sample solvation and reaction may be better in an organic solvent. Usually an aqueous or a mixed aqueous-organic solvent system containing the titrant precursor is used. The titrant precursor concentration is usually at least 0.1 *M* and may be much larger in order to maintain high titration efficiency and high current densities for rapid analysis as described above. Frequently, pH control is achieved by addition of strong acids or bases or the use of buffers. The relatively high concentrations of titrant precursor and buffers are usually sufficient to give adequate conductivity. However, neutral salts may be added for this purpose in some cases.

Bromocoulometric titrations (coulometric titrations with electrogenerated bromine) are usually performed in strongly acidic solutions. Organic solvents such as methanol are frequently added. Iodocoulometric titrations can be performed from strong acid up to pH 8 to 8.5 where disproportionation of iodine begins to occur.[68] It is more difficult to find a supporting electrolyte for chlorocoulometric titrations. Highly acidic chloride solutions tend to attack the platinum anode and less concentrated solutions may give low titration efficiencies.[68] Cerium (IV) titrations must be performed in strongly acidic sulfuric acid solutions (> 2 *M*) with current densities between 1 and 10 mA/cm^2 to maintain $>99\%$ titration efficiency.[68] Silver (II) titrations must be performed at low temperatures ($<-10°C$) to prevent disproportionation and reactions with water.[68,70] The ferrocyanide/ferricyanide couple is useful over a wide pH range.[68] Titrations involving metallic ions such as silver (I), mercury (II), or mercury (I) must generally be acidic enough to prevent hydrolysis and/or precipitation of the metallic ions. Karl Fischer titrations require very specific conditions in order to properly perform.[68] Neutralization titrations are an exception to most limitations in that a wide combination of electrolytes is possible. Electrolytes for other reactions may be found from the references in Table 4.

3. Endpoint Detection Techniques

Precision in a titration depends upon how well the endpoint can be located.[82a] Accuracy is determined by the correspondence between the true equivalence point and the located endpoint.[13] Since the endpoint being measured relates to the homogeneous chemical reaction, any of the commonly available techniques may be used. These include colorimetric, potentiometric, amperometric and conductometric methods. Because of the electrochemical nature of coulometric titrations, there is some bias towards usage of electrochemical detection techniques.

Table 4
TYPES OF COULOMETRIC TITRATIONS

Application	Titrant generated	Electrode	Electrode reaction	Typical supporting electrolyte	Type compounds titrated	Typical analytical reaction	Ref.
Acid base neutralization	OH^-	Pt	$2H_2O + 2e \rightleftharpoons 2OH^- + H_2\uparrow$	Na_2SO_4 or other electrolyte in water	Acid	$OH^- + HA \rightleftharpoons H_2O + A^-$	13,65,68, 71,72
	H^+	Pt	$H_2O \rightleftharpoons 2H^+ + \frac{1}{2}O_2\uparrow + 2e$	DMSO, ACN, isopropanol, DMF, etc.	Base	$H^+ + B \rightleftharpoons HB^+$	13,65,68, 71,73
	H^+ consumed	Pt	$H^+ + e \rightleftharpoons \frac{1}{2}H_2\uparrow$		Acid	–	68
Precipitation	Ag^+	Ag	$Ag \rightleftharpoons Ag^+ + e$	KNO_3, pH < 10	Halide	$Ag^+ + X^- \rightarrow \underline{AgX}$	64-66,68
					Mercaptan	$Ag^+ + RSH \rightarrow \underline{AgSR} + H^+$	65,66,68,72
	Hg_2^{++}	Hg	$2Hg \rightleftharpoons Hg_2^{++} + 2e$	0.5*M* $NaClO_4$ + 0.02*M* $HClO_4$	Halide	$Hg_2^{++} + 2X^- \rightarrow \underline{Hg_2X_2}$	65,68
	$Fe(CN)_6^{4-}$	Pt	$Fe(CN)_6^{3-} + e \rightleftharpoons Fe(CN)_6^{4-}$	0.2*M* $K_3Fe(CN)_6$	Metals	$2Fe(CN)_6^{4-} + 2K^+ + 3Zn^{++} \rightarrow \underline{K_2Zn_3[Fe(CN)_6]_2}$	65,66,68
Complexation	CN^-	Ag	$Ag(CN)_2^- + e \rightleftharpoons Ag + 2CN^-$	0.25*M* $KAg(CN)_2$ in 0.01*M* NaOH/0.1*M* Na_2SO_3	Au, Ni, Ag	$j\,CN^- + M^{+i} \rightleftharpoons M(CN)_j^{i-j}$	65,66
	EDTA (H_4Y)	Hg	$HgNH_3Y^= + NH_4^+ + 2e \rightleftharpoons Hg + 2NH_3 + HY^{3-}$	0.02*M* $HgNH_3Y^=$ in pH8.6 NH_4OH/NH_4NO_3 buffer	Metals	$HY^{3-} + Ca^{++} \rightleftharpoons CaY^= + H^+$	65,68
	Hg^{++}	Hg	$Hg \rightleftharpoons Hg^{++} + 2e$	0.5*M* $NaClO_4$ + 0.02*M* $HClO_4$	CN^-, mercaptans amines,	$Hg^{++} + jX^- \rightleftharpoons HgX_j^{2-j}$	65,66, 68

Oxidation[a]	Br_2	Pt	$2Br^- \rightleftharpoons Br_2 + 2e$	0.2*M* NaBr in 0.1*M* H_2SO_4	Unsaturation (addition)	$>C=C< + Br_2 \rightarrow -CBr-CBr-$	65,66, 68
				or 0.1*M* KBr in 0.3*M* HCl	Aromatic amines, phenols and sulfonamides (substitution)	$C_6H_5OH + 3Br_2 \rightarrow C_6H_2Br_3OH$ (2,4,6-Br) $+ 3HBr$	65,68, 72,74, 75
				or 0.1*M* KBr in 0.3*M* HCl in 85% methanol	Hydroquinones	$HOC_6H_4OH + Br_2 \rightarrow O{=}C_6H_4{=}O + 2HBr$	65,74
				or 60% glacial acetic acid, 26% methanol 14% aqueous 1*M* KBr	Hydrazides	$RCONHNH_2 + 2Br_2 + H_2O \rightarrow RCOOH + N_2\uparrow + 4HBr$	65,68, 72,74
					Mercaptan	$RSH + 3Br_2 + 2H_2O \rightarrow RSO_2Br + 5HBr$	72,74, 76
				or 0.1–1.2*M* KBr in 60% glacial acetic acid	Dialkyl sulfide	$RSR' + Br_2 + H_2O \rightarrow R{-}SO{-}R' + 2HBr$ $RSR' + 2Br + 2H_2O \rightarrow RSO_2R' + 4HBr$	72
					Disulfide	$RSSR + 5Br_2 + 4H_2O \rightarrow 2RSO_2Br + 8HBr$	72,74

Table 4 (continued)
TYPES OF COULOMETRIC TITRATIONS

Application	Titrant generated	Electrode	Electrode reaction	Typical supporting electrolyte	Type compounds titrated	Typical analytical reaction	Ref.
					α, β-Unsaturated lactone (ascorbic acid)	(HO, HO, O, O ring; $CHOH{-}CH_2OH$) + Br_2 → (O, O, O, O ring; $CHOH{-}CH_2OH$) + 2HBr	72,74
	OBr^-	Pt	$Br^- + OH^- \rightleftharpoons OBr^- + \frac{1}{2}H_2 + e$	1*M* NaBr in pH 8.6 borate buffer	Ammonia	$2NH_3 + 3OBr^- \rightarrow N_2\uparrow + 3H_2O + 3Br^-$	66,68
	Cl_2	Pt	$2Cl^- \rightleftharpoons Cl_2 + 2e$	0.2*M* NaCl in 1*M* H_2SO_4	Unsaturation	$>C{=}C< + Cl_2 \rightarrow -\overset{Cl}{\underset{\mid}{\overset{\mid}{C}}}-\overset{Cl}{\underset{\mid}{\overset{\mid}{C}}}-$	65,68
	I_2	Pt	$2I^- \rightleftharpoons I_2 + 2e$	0.1*M* KI, pH3–8	Peroxides	$ROOR' + 2HI \rightleftharpoons I_2 + ROH + R'OH$	77,78
					α, β-Unsaturated lactones	See Br_2	79
					Mercaptan	$2RSH + I_2 \rightarrow RSSR + 2HI$	74
					Karl Fisher	$H_2O + I_2 + SO_2 \xrightarrow[\text{methanol}]{\text{pyridine}} 2HI + SO_3$	58, 80-82

	Ag^{++}	Au	$Ag^+ \rightleftharpoons Ag^{++} + e$	0.1*M* $AgNO_3$ in 5*M* HNO_3 at −10°C	Hydrazides	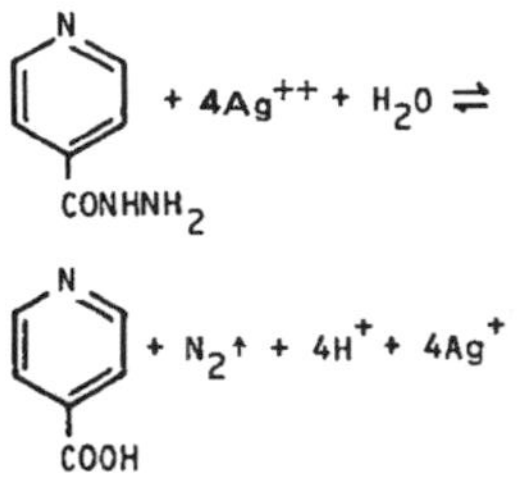 (N-pyridine ring)–$CONHNH_2$ $+ 4Ag^{++} + H_2O \rightleftharpoons$ (N-pyridine ring)–$COOH$ $+ N_2\uparrow + 4H^+ + 4Ag^+$	58,70
	Ce^{4+}	Pt	$Ce^{3+} \rightleftharpoons Ce^{4+} + e$	0.1*M* $Ce_2(SO_4)_3$ in 2–3*M* H_2SO_4		Varied	58,64,65
Reductions[b]	Fe^{++}	Pt	$Fe^{3+} + e \rightleftharpoons Fe^{++}$	Acidic $Fe(NH_4)(SO_4)_2$		Varied	58,65
	Sn^{++}	Pt	$Sn^{4+} + 2e \rightleftharpoons Sn^{++}$	0.2*M* $Sn(NO_3)_4$ in 0.2*M* HCl/3–4*M* NaBr		Varied	65,66
	Ti^{3+}	Pt	$TiO^{++} + 2H^+ + e \rightleftharpoons Ti^{3+} + H_2O$	3.6*M* $TiCl_4$ in 7*M* HCl		Varied	58,65,66

[a] Other titrants for oxidation are: Cr(VI), $Fe(CN)_6^{3-}$, Mn(III), Mo(V), $S_2O_6^{=}$, V(V).[58,65,66]

[b] Other titrants for reduction are: Cr(II), Cu(I), V(IV).[65,66]

Tackett[83] has compared the use of three endpoint detection techniques applied by students to the titration of arsenious oxide with electrogenerated iodine. These were visual observance of the starch-iodine complex, amperometry with two indicating electrodes, and potentiometric. The precision of the visual endpoint was significantly poorer ($\sigma \pm 0.9\%$) except for those students with exceptional color perception ($\sigma \pm 0.1$ to 0.2%), but could have been improved by use of a spectrophotometer. The precision of the other two methods were comparable ($\sigma \pm 0.1\%$) depending upon the equipment used (galvanometer sensitivity and regular or expanded scale pH meter). In a general sense, no one endpoint detection method can be described as being better, more precise or more accurate. For a specific application, one may be better than the others.

Colorimetric endpoint detection techniques are perhaps the most universal and easiest to apply. Phenolphthalein or other pH indicators can be used for neutralization titrations.[65] For iodometric titrations, the iodine color at 342 nm or the color of the starch-iodine complex can be monitored.[8,83] For bromine, the excess bromine can be monitored directly[65] or the decolorization of an indicator such as methyl orange can be followed. However, this requires that the substrate react faster with the bromine than does the methyl orange so that the indicator is not decolorized until an excess of bromine is present.[67] Either the color of the ferrous *o*-phenanthroline complex or the normal absorbance of cerium ion (400 nm) can be used to follow a cerium (IV) titration.[65,67] Generally, the use of a spectrophotometer gives more precise results than visual determinations. The use of fiber optics (accessories available with autotitrators from Brinkmann®, Fisher Scientific®, etc.) or flow cells facilitate this application. The titration curves and endpoint detection are identical to those for homogeneous titration reactions.

Potentiometric endpoint detection techniques are perhaps the most widely used.[8] These employ an indicator electrode, a reference electrode, and a potentiometer. The indicator electrode may be a pH electrode for neutralization titrations, platinum or gold for redox titrations, metallic electrodes such as silver for complexation or precipitation reactions, or specific ion electrodes for applicable species. The reference electrode may be any of the conventional ones available such as the SCE, mercury/mercurous sulfate/sulfate or silver/silver chloride. The titration curves (potential vs. time) are similar to volumetric titrations. However, they tend to be more asymmetric and exhibit smaller potential changes near the equivalence point since the titrant precursor is present in much greater amounts than the substrate. Usually this will cause only minor errors from any reasonable endpoint determination.[67] The use of expanded scale pH meters or potentiometers will generally give better precision.

An amperometric endpoint detection technique usually refers to a method using two indicator electrodes. The correct terminology is "amperometry with two indicator electrodes", but it is frequently called biamperometry.[84] A small, constant voltage (usually 20 to 500 mV) is applied between the two indicator electrodes and the resulting current is measured. The indicator electrodes are usually small area (0.1 to 2 cm^2) platinum or other inert metal composition. This technique is used only for redox titrations, usually those involving the halogens. It is applicable if there is a change in the electrochemical behavior at the indicator electrodes which affects the current flow during the titration. In order for a current to flow, a reversible electron transfer must occur at each electrode (oxidation at the anode and reduction at the cathode). In the titration of arsenious oxide with electrogenerated bromine, the arsenic (III)/arsenic (V) couple is not reversible but the bromine/bromide couple is reversible at the platinum electrodes. However, only bromide and the two arsenic species are present (in appreciable amounts) in the solution before the equivalence point. Thus, no (or little) current flows before the equivalence point and the current steadily increases after the equivalence point. The titration can be stopped when a small, but definite current flows (5 to 20 μA) due to the presence of a small excess of bromine. Other combinations of reversible and irreversible couples will give different behavior.[66,67] Generally, the indicator electrodes should be located away from the generating electrodes to avoid potential gradients caused by the flow of the generating current. This effect can be evident as a change in indicator current when the generating current is interrupted.

One variation in amperometric endpoint detection results when the current (rather than the potential) between the two indicator electrodes is controlled. This technique should actually be called "controlled current potentiometry with two indicator electrodes" rather than the common name of bipotentiometry.[84] The magnitude of the detection current must be small compared to the generating current for the titrant to avoid titration errors unless the detection reactions are cyclic (i.e., bromine to bromide at the indicator cathode and bromide to bromine at the indicator anode). The detection current is typically 1 to 10 μA. The electrode composition and behavior is much like that described for amperometry with two indicator electrodes.

Conductiometric endpoint detection techniques are used rather infrequently. They are best applied to titration of strong acids or bases where changes in conductivity are large. With weak acids or bases, a neutral salt may have to be added to get a measurable conductance. However, addition of the salt decreases the relative change in conductance during the titration.[65,67]

4. Pretitration

The most important purpose of a pretitration is the removal of impurities present in the supporting electrolyte (similar to the preelectrolysis in controlled potential coulometry). Since the coulometric titration sample size tends to be quite small compared to the amount of electrolyte, even very minor amounts of titratable impurities can cause significant error. Commonly, titrant is generated until the equivalence point color, potential, or current is attained. If necessary, a small portion of the compound to be titrated can be added before the pretitration. Then, the sample is titrated to the same point. There are some cases where this approach is incorrect, but usually the error due to pretitration is smaller than the error from impurities in an electrolyte which has not been pretitrated.[67]

Successive samples can usually be titrated in the same portion of the electrolyte using the previous titration as a pretitration for the current titration. However, it should be noted that the true equivalence potential in potentiometric titrations varies with the concentration of species present in the solution. These concentrations change with each successive titration. If the equilibrium constant for the chemical reaction is large, the potential change near the equivalence point will be large and no significant errors will result.

Other important functions of pretitration, especially with small sample sizes, concern the electrode surface and double layer. First, platinum and many other types of solid electrodes may form surface oxide films. The pretitration insures that the surface state will be similar before and after the titration so that no titration current will be consumed to produce net changes in the electrode surface. Second, double layer charging also requires current. A pretitration to the final electrode condition will tend to minimize changes in electrode potential or capacitance which could cause errors of the charging current type.

5. Effects of Impurities in the Sample

The lack of specificity is a major disadvantage of coulometric titrations.[66] In neutralization, precipitation, and complexation reactions, all compounds with similar functional groups will titrate. In redox coulometric titrations, there is a much less direct and, consequently, less discriminatory control over the strength of the redox couple than in controlled potential coulometry. These problems are essentially the same as those involved in volumetric titrations. Impurities must be detected by other means, isolated and characterized, and tested for interference, and conditions must be chosen to minimize any errors due to their presence.

6. Factors Specifically Affecting Precision and Accuracy

Regulation and measurement of the generating current and measurement of time limit the precision of constant current coulometric titrations in the absence of residual current errors, endpoint detection errors, and other electrochemical limitations. Limitations due to current and time measurements are estimated to be several parts per million.[18] The IUPAC committee report[18] also reviewed many results for high precision coulometric titrations, some with reported precisions as good as 10 ppm. Other high precision coulometric titrations have been reported for the titration of 4-aminopyridine with acid (8 ppm),[23] for sulfamic acid titrated with hydroxide (150 ppm),[85] and for dichromate titrated with ferrous ion (150 ppm).[86,86a]

Time traditionally has been measured using manual or power line based timers. Manual time measurements can be no better than approximately 0.1 sec, electric stopclocks may obtain a precision of 0.01 sec.[67] However, highly precise work (10 to 100 ppm) with stopclocks requires relatively long electrolysis times (minutes to hours). The accuracy of the power line frequency should be known when it determines the accuracy of the timer. Harrar[7] recently reported no deviations greater than 0.05% even for short periods. However, Lingane[67] had reported deviations up to 0.15% at a much earlier time in his laboratory. Accurate and highly stable crystal based timers are readily available alternatives for more precise work or equally precise work at shorter electrolysis times. With crystal based timers, the minimum titration time would be limited by the rates of mixing and chemical reactions rather than the errors in time measurement.

It should be realized that techniques which include periodic interruption of current flow are actually summing many electrolysis periods to arrive at the total electrolysis time. Therefore, overall precision and accuracy will include the accumulated errors of these multiple time period measurements.

The constant current is almost always generated by passing a constant voltage through a precision resistor. Thus accuracy depends upon accurate measurement of the voltage source and precise calibration of the resistance. Precision is affected by drift in the voltage source with time or by change in the value of the resistor with time or temperature. In addition, errors from auxiliary sources such as bias currents and offset voltages in operational amplifier circuits contribute to inaccuracy and/or imprecision.

D. Instrumentation

The basic instrumentation for coulometric titrations consists of a timer and a constant current generator (galvanostat). However, many variations in instrumentation exist including charge injection and variable current generation (sometimes coupled with integration of charge). The endpoint detection circuit is usually considered to be separate and distinct, but modern instruments tend to incorporate electronic comparator circuits into the current generator control section. Several sources of instrumentation are given in Table 5. Comparators are particularly effective in apparatus designed for automated coulometric titrations. One titration system applied to neutralization reactions[87] even reverses current direction if the endpoint is passed. Specialized instrumentation will be briefly discussed.

Table 5
SELECTED COULOMETRIC TITRATION SYSTEMS AND COMPONENTS

Manufacturer	Address	Components[a]
American Instrument Co.® Division of Travenol Labs®, Inc.	8030 Georgia Ave. Silver Spring, Md. 20910	Cl, Coul
Aminco Corp.®, Scientific Systems Division	21 Hartwell Ave. Lexington, Mass. 02173	Cl
Beckman Instruments, Inc.®	2500 Harbor Blvd. Fullerton, CA. 92634	Cell, Cl, Galv
Bioanalytical Systems, Inc.®	P.O. Box 2206 W. Lafayette, In. 47906	Mod
Brinkmann Instruments, Inc.®	Cantiague Road Westbury, N.Y. 11590	Cell, Sy
Buchler Instruments® Division of Searle Diagnostics, Inc.®	1327 16th St. Fort Lee, N.J. 07024	Cl
Dohrmann-Envirotech®	3240 Scott Blvd. Santa Clara, CA 95050	Cl, Sy
Koslow Scientific Co.®	7800 River Rd. N. Bergen, N.J. 07047	Coul
Pacific Photometric Instruments, McKee-Pedersen Instruments Division®	5675 Landregan St. Emeryville, CA 94608	Cell, Galv, Mod
Petrolite Instruments®	4411 Bluebonnet Dr. Stafford, Texas 77477	Coul
EG & G Princeton Applied Research®	P.O. Box 2565 Princeton, N.J. 08540	Cell, Galv
Process Analyzers, Inc.®	1101 State Road Princeton, N.J. 08540	Coul
Radiometer A/S®	Not Available	Cl
Sargent-Welch Scientific Co.®	7300 N. Linder Ave. Skokie, Ill. 60077	Cell, Sy
Tacussel Electronics® (distributed by Astra Scientific International, Inc.®)	P.O. Box 2004 Santa Clara, CA 95051	Cell, Galv, Sy, Mod
United Technical Corp.®	81 Keystone Dr. Leominster, Mass. 01453	Coul

[a] Components are Cell = electrolysis cell, Cl = chloride titrator, Coul = coulometer (usually galvanostat and timer or integrator), Galv = galvanostat, Mod = modular components, and Sy = system (cell, galvanostat and timer or integrator).

1. Timers

A recent article[88] included a review of time measuring techniques. Conventional timers include manual and electronic stopclocks. Many recently reported coulometric titrators incorporate digital timers.[88-91] Muha[90] also synchronized the start and end of current generation with the pulse from a digital timer. Thus, the maximum inaccuracy from any number of electrolysis cycles is one clock pulse (0.1 sec for that instrument design). Digital circuits are readily available for generating, counting, and controlling timing signals to a microsecond or less. New instrument designs should achieve more accurate time measurements by digital control of fast analog current switches (for stopping and starting current flow) synchronized with 1 kHz or higher frequency clock pulses. Timing accuracy is easy to maintain with digital clock pulses which can be calibrated by a measurement of frequency. Many of these new ideas have been incorporated in specific instruments,[19,88-91] but most commercially available instruments incorporate electric timers.

2. Current Generation

The history of current generation is long and varied.[8,10,65,67,87,92,93] Initially, current generation was limited to constant current sources, usually a high voltage battery connected in series with a large load resistor and the electrolysis cell. As long as the load resistance was much greater than the cell resistance and the battery drain was minimal, these circuits provided a highly stable but nonflexible current source. Next, regulated power supplies usually incorporating tube or transistor circuits became available.

Many current circuit designs, especially those constructed by chemists, use solid state operational amplifier circuits. The most frequent configuration employs a constant voltage source and precision resistor to generate a known current which is then forced through the titration cell placed in the feedback loop of the amplifier. If either the voltage source or precision resistor is variable, several current values may be obtained. Direction of current flow is determined by voltage source polarity. In these operational amplifier designs, the amplifier offset current must be small compared to the electrolysis current since the offset current will also pass through the cell.

In constant current designs, the electrolysis current can be accurately calculated from the measured potential drop across a precision resistor substituted for the cell or placed in series with the cell. With modern circuit designs, drift in current output is typically quite small and will usually be a problem only if very small electrolysis currents (<0.1 mA) are used.

Variable current techniques may be stepwise constant or continuously variable. Actually, a step reduction in current near the endpoint has been a rather common technique to increase precision. However, some newer current generator designs have incorporated a continuously variable current inversely related to the rate of change of a potentiometric indicator system[92] and a constant[88] or variable[19,45] charge injection. Charge injection refers to the use of current pulses which are usually constant in magnitude but variable in frequency or duration. In any nonconstant current generation technique, the total charge must be obtained by integration or calculation.

Calculation is usually used with stepwise constant current generation. It may be implemented and calibrated the same as a single constant current. Continuously variable current techniques usually use integration to determine the total charge consumed. Integration techniques and errors were discussed in the controlled potential coulometry section. One variable current instrument[92] has a precision resistor incorporated into its design to calibrate either current or the integrator.

Charge injection techniques may use integration or calculation to determine the total charge. If integration is used, the integrator must be able to handle the high frequency components of the current pulses. When calculation is used, the charge contained in each current pulse must be calibrated. This is usually accomplished by multiplying the measured current magnitude by the pulse width. Distorted pulse shapes (rise and fall times) and inaccuracy in measuring the width are the principal contributions to inaccuracy and imprecision. The analog switch behavior, high off resistance and low on resistance, is usually satisfactory.[45,88] However, if extremely high frequency pulses are used, capacitative coupling losses may be significant.[7]

3. Specialized Instrumentation

Automation of the titration process is the most common type of specialized instrumentation. Earl and Fletcher[89] have described a microprocessor controlled titrator which uses a variable current generator (±10 mA at ±150 V). They use a highly reproducible, solenoid-activated, long-life (millions of cycles) ceramic slide valve for volumetric sampling of flowing streams. The detection circuit output (any device whose signal can be converted to a voltage) is compared to an appropriate setpoint previously determined. The charge is calculated from the known current and (digital) time data.

Precision and accuracy are reported for single endpoint neutralization titrations (±0.2%, > 100 samples), consecutive endpoint titrations (±0.5%, > 300,000 samples), double endpoint titrations (±1%, 100 samples), and redox titrations (± 0.2%, 200 samples).

Jaycox and Curran[88] developed a coulometric titrator using digital logic for timing and control. It also included a comparator for detecting the endpoint based on a sensor output and a control value. Digital circuits also were used to slow the rate of titration near the endpoint. The sensor output at the end of the titration must be known from prior experiments. The instrument can generate a 10 μsec current pulse containing picocoulombs of charge (10^{-17} equivalents) with an accuracy of 10% or a larger pulse with better accuracy. Precision and accuracy of a few tenths of a percent were reported for microgram amounts of V (V) and Mn (VII). This decreased to 2% at the nanogram level.

A modular, automated constant current titration system for ascorbic acid has been reported.[94] A 50-position turntable, electrolyte dispenser, stirrer, and electrode station enabled samples to be prepared just prior to analysis. Integration of charge combined with interrupted current generation (180 sec delay) produced a standard deviation of 0.1%.

Miyake and Suto[91] have described an automatic titrator for the Karl Fischer determination of water. It features a digital counter, potentiometric detection, and rapid assay (2 min for 1 mg of water). Precision was reported at 5 μg (σ) for less than 3 mg of water.

Adams et al.[95] reported a coulometric pH-stat based on operational amplifier circuits. The titrator "delivered" titrant proportional to the rate of change in pH. A time constant of 2 to 3 sec prevented overshooting and ringing while being able to follow chemical reactions with a half-life greater than about 15 sec. Reported current efficiencies were high (> 99%) for a wide range of rates for electrogeneration of base (0.1 to 20 μmol/min). Precision was approximately 0.5%.

Harrar and Pomernacki[95a] described a programmable constant current source for use in automated coulometric titrations. The 0 to 500 mA output current range has a resolution of 30 μA when digitally controlled with a 14 bit digital to analog converter. Maximum error due to nonlinearity is 10 μA. Long term stability is about 0.03%/year. Sensitivity to temperature changes and line voltage changes are given. Compliance voltage is 45 V. A 0.01%, 1 ohm internal resistor and 4½-digit panel meter are used for calibration. A mercury wetted contact relay is used for switching the current source between a standby circuit and the electrolysis cell.

Hendler et al.[95b] recently described a microcomputer controlled coulometric titrator. It is a portion of a larger system designed to obtain spectrophotometric and redox behavior of biological compounds. An Intel® 8080 microprocessor with a 12-bit digital to analog converter is interfaced to an analog voltage controlled current source. The ±5 mA output current range has a resolution of 1.25 μA. Compliance voltage is about 22 V. The microprocessor also controls the titration to a present potentiometric endpoint, controls the spectrophotometer, and acquires the optical spectra.

Lindberg[95c] described a photometric endpoint detection system designed to minimize errors caused by lamp drift, gas bubbles, and turbidity. The light beam (generated by a single lamp) passing through the cell is split and sent to two monochromators with photodiode detectors. One monochromator is set to the wavelength of maximum absorbance of an indicator added to the electrolysis cell; the other is set to a wavelength of minimum absorbance to compensate for changes in light intensity due to physical processes not directly related to the titration. After correction for the relative photodiode response at the two wavelengths, the difference in intensities is compared to a preset value to determine the equivalence point.

Other specialized instrumentation includes automatic chloride analyzers[65,96,97] and sulfur analyzers.[98] Dougherty[98a] has summarized the properties of a coulometric titrator with potentiometric end point detection which can be operated in either a constant current or controlled current mode. Organic samples are generally combusted before these determinations.

E. Cell Design

Typical cell designs for coulometric titrations resemble those described for controlled potential coulometry in size and function. As in controlled potential coulometry, it is generally necessary to separate the counter electrode from the reacting solution using a sintered glass disk or other separator. If this is not done, many electroactive titrants generated at the working electrode (bromine, Fe(II), etc.) will be removed or transformed back to the titrant precursor species (bromide or Fe(III)) at the counter electrode. Occasionally, dual isolation is needed to prevent leakage.[8] Several sources of electrochemical cells for coulometric titrations are given in Table 5.

The working electrode is generally placed directly in the mixture of supporting electrolyte, titrant precursor, and substrate. However, if the titrant must be generated externally, then special cell designs will be needed.[64,67,99]

The control of current rather than potential simplifies the geometric design restrictions for coulometric titrations compared to controlled potential coulometry. The high titrant precursor concentration amplifies these design differences. First, a reference electrode is not needed except for preliminary investigations to select the correct current density. Second, the redox buffering of the titrant precursor/titrant couple prevents loss of potential control with resulting undesired electrochemical reactions occurring due to poor electrode geometry except in the most extreme cases. Third, the high titrant precursor concentration usually causes solution mixing and/or the rate of the chemical reaction to be the rate-limiting step rather than mass transfer to the electrode. In cases where rapid mixing is desirable, special cell designs may be needed.[88,89]

Over-titrations can be avoided by two types of cell designs.[8] First, the chemical sensors can usually be located within the local excess of titrant generated at the working electrode. With this electrode arrangement and a current interruption or variable current circuit, the sensor acts as a "chemical anticipator" signaling the approach of the endpoint. Second, the cell design itself can cause an over-titration within a selected portion of the solution containing the electrolysis sensors. This signal can also be used as an endpoint predictor.

Several special cell types have been reported. These include a cell for Karl Fischer reactions in a back titration mode[100] and a normal Karl Fischer titration.[101] Feher and Pungor[102] have described a flow cell for rapid mixing of an injected sample with a flowing stream of electrolyte. Constant potential electrolysis was used to determine the original substance or its reaction product in the flowing stream.

F. Electrode Selection

The combination of supporting electrolyte and working electrode should be chosen to give a low residual current combined with rapid electron transfer to the titrant precursor. This behavior is easiest to achieve when the electrode reaction involves a major supporting electrolyte species (e.g., electrogeneration of hydrogen ions from water) or the electrode itself (e.g., Ag (I) from silver). Silver and mercury electrodes are called active electrodes when their corresponding ions participate in the titration reaction.

It is more difficult to achieve ideal electrode behavior when the working electrode acts as an inert electrode for electrogeneration of titrant. In this situation, the electrode-electrolyte combination must maintain a potential (determined by the redox buffering action of the titrant precursor-titrant) where the residual current is low. This requires rapid electrode kinetics as well as appropriate selection of chemical conditions. Platinum is the most common electrode for oxidation reactions although gold, glassy carbon, and carbon fiber electrodes have been used.[10,15] With porous electrodes such as carbon fiber, absorption of electrogenerated species such as iodine may be a problem. For reduction, mercury may be preferred because of its higher hydrogen overvoltage.

G. Analytical Applications

Many applications of coulometric titrations are available from periodic reviews[50,66,103] and texts.[9,13,65,72] Patriarche[74] has reviewed the use of coulometric titrations in pharmaceutical analysis. Girard et al.[104,105] discussed the application of constant current coulometric titrations in the assay of aromatic amines and sulfur-containing pharmaceutical compounds. Unterman and Weissbach[106] reviewed electrochemical methods for the determination of antibiotics. The range of applications cover the generation of virtually any titrant that has been used in a conventional manner as well as many that cannot be prepared chemically.

Selected examples of coulometric titrations are listed in Table 6. The important experimental details and precision are given when available. These specific examples illustrate the general reactions previously discussed and presented in Table 4, e.g., neutralization, precipitation, complexation, and redox reactions. In addition, the specific application of coulometric titrations to the Karl Fischer reaction is briefly discussed.

1. Neutralization Reactions

Acid-base neutralizations are usually conducted using platinum electrodes. In solutions containing water, hydroxide ions can be generated at the cathode and hydrogen ions at the anode. In many cases, the undissociated acid or hydronium ions produced by dissociation may be reduced directly at the cathode. A variety of solvents are available depending upon acid or base strength of the substrate.

2. Precipitation Reactions

Precipitation titrations occur with the generation of ionic species such as silver (I) or mercury (I) from the corresponding metallic electrode. Other species such as ferricyanide can be generated at an inert electrode such as platinum. Precipitation titrations are used for ionic halides or pseudohalides, mercaptans, etc. Organohalides or organosulfur compounds can be determined after combustion. High selectivity, determined by differences in solubility products, may be difficult to achieve in precipitation titrations using either volumetric or coulometric techniques.

A general coulometric titration procedure for hydrochloride salts using electrogenerated silver (I) was reported by Lemahieu et al.[107] van Oort et al.[108] developed a microtitration of chloride with a CV of 1.3% (32 degrees of freedom) for 2 to 20 neq (0.07 to 0.7 μg) and a CV of 3.2% (18 degrees of freedom) for 0.5 to 1 neq (17 to 35 ng). Linnet[97] described an autotitrator designed for microdeterminations of chloride (0.1 to 5 μeq) in serum samples at the 1% precision level.

3. Complexation Reactions

Complexation titrations can involve an anion such as cyanide or the ethylenediaminetetracetic acid (EDTA) anion or a metallic cation such as mercury (II). Mercury (II) can be used for amines or mercaptans, whereas most other applications involve complexation of metallic cations.

4. Redox Reactions

Redox reactions are the most common type of coulometric titrations for pharmaceu-

Table 6
PHARMACEUTICAL APPLICATIONS OF COULOMETRIC TITRATIONS[a]

Name[b]	Electroactive group[c]	Titrants[d]	Reactions[e]	Supporting electrolyte	Working electrode	Detection[f]	Sample size (mg)	Current density[g] (mA/cm^2)	Precision[h]	Replicates	Ref.
1. N-Acetyl methionine	Dialkyl sulfide	Br_2	$CH_3-S-R + H_2O + Br_2 \rightarrow CH_3-SO-R + 2HBr$	0.42 *M* KBr, 0.2 *M* NH_4Cl in 10% HCl	–	Amp(15)	0.3–1	(< 12 mA)	0.7% CV	7[i]	74
2. Allobarbital	Allyl	Cl_2	$(CH_2{=}CH-CH_2)_2-R + 2Cl_2 \rightarrow (CH_2Cl-CHCl-CH_2)_2-R$	4% HCl	–	–	0.07–0.14	(< 12 mA)	0.9% SE	–	110
3. Aniline	Aromatic amine	Br_2/Cu^+	$C_6H_5NH_2 + 3Br_2 \rightarrow C_6H_2Br_3NH_2 + 3HBr$	0.1 *M* NaBr, 0.02 *M* $CuSO_4$ in 1 *M* HCl	Pt	Amp(200)	0.012–0.2	(10.4 mA)	0.2 μg AE	–	67
4. Antipyrine	–	Br_2	$C_{11}H_{12}N_2O + Br_2 \rightarrow C_{11}H_{11}BrN_2O + HBr$	0.1N KBr in glacial acetic acid	–	Amp(15)	0.1–0.5	(< 12 mA)	0.8% CV	6[i]	74
5. Antipyrine	–	Br_2	Same	0.2N KBr in 1N HCl	–	–	0.5–4	–	0.5% AE	–	111
6. Antipyrine	Base	H^+	$R_3N + H^+ \rightleftharpoons R_3NH^+$	0.1 *M* $NaClO_4$ in 95:5 acetic anhydride: acetic acid	Pt	Amp(80) (Bi/Bi)	0.6	(15 mA)	0.9% AE	7	112
7. Ascorbic acid	Unsaturated lactone	Br_2	$-O-\overset{O}{C}-\overset{OH}{C}{=}\overset{OH}{C}- + Br_2 \rightarrow -O-\overset{O}{C}-\overset{O}{C}-\overset{O}{C}- + 2HBr$	0.15 *M* KBr in 1:1 acetic acid:water	Pt	Amp(250)	50–500 (tablets)	(20 mA)	1.4% CV	14[i]	113
8. Ascorbic acid	Same	Br_2	Same	Same as number 1	–	Amp(10)	0.2–0.5	(< 12 mA)	1.2% CV	6[i]	74
9. Ascorbic acid or sodium ascorbate	Same	I_2	$-O-\overset{O}{C}-\overset{OH}{C}{=}\overset{OH}{C}- + I_2 \rightarrow -O-\overset{O}{C}-\overset{O}{C}-\overset{O}{C}- + 2HI$	0.5 *M* KI in 0.4 m*M* H_2SO_4	Pt	CCP(10)	30 30	5 5	0.2% CV 0.3% CV	22[i] 50[i]	94
10. Ascorbic acid	Same	I_2	Same	KI solution	Pt	–	< 1	4	0.3% AE	–	114
11. Ascorbic acid	Same	I_2/Ag^+	Same, then $Ag^+ + I^- \rightarrow AgI\downarrow$	40% Acetic acid, pH 1.6	Ag	CCP(1)	0.7	0.25	0.2% CV	6	115
12 Ascorbic acid	Same	I_2/I^-	Same, then $I_2 + 2e \rightleftharpoons 2I^-$ (constant potential)	1*M* NaI in 1*M* Na_2SO_4, pH 3	Pt	Int	0.002–9	–	~0.1% CV	–	79
13. Ascorbic acid	Same	$Mo(CN)_8^{3-}$	$-O-\overset{O}{C}-\overset{OH}{C}{=}\overset{OH}{C}- + 2Mo(CN)_8^{3-} \rightarrow -O-\overset{O}{C}-\overset{O}{C}-\overset{O}{C}- + 2Mo(CN)_8^{4-} + 2H^+$	pH 5 buffer	Pt	Pot-1	0.05–1	2	0.3% CV	–	116
14. Atropine sulfate	R_3N	$(C_6H_5)_4BH/Ag^+$	$R_3N + (C_6H_5)_4BH \rightarrow (C_6H_5)_4BH{\cdot}R_3N\downarrow$ $(C_6H_5)_4B^- + Ag^+ \rightarrow (C_6H_5)_4BAg\downarrow$	0.4 *M* $NaNO_3$ in acetone	Ag	Pot-2	10–20	–	0.1% CV	3	117
15. Brucine sulfate	R_3N	$(C_6H_5)_4BH/Ag^+$	Same	Same	Ag	Pot-2	5–10	–	0.1% CV	3[i]	117
16. Caffeine	–	Excess Cl_2/?	$C_8H_{10}N_4O_2 + 2Cl_2 \rightarrow$ products	5% HCl	–	–	0.2–1	–	0.2–0.6% CV	–	118
17. Carbutamide	Sulfonamide	Br_2/Cu^+	(*p*) $NH_2-C_6H_4-SO_2-NH-R + 2Br_2 \rightarrow NH_2-C_6Br_2H_2-SO_2-NH-R + 2HBr$	1*M* KBr, 0.05*M* $CuSO_4$ in 1 *M* HCl	Pt	Pot-1	0.05–1.5	(0.18mA)	1% E	–	119
18. Chloramine T	N-Cl	I^-/excess $S_2O_3^=/I_2$	$R-SO_2-N{<}^{Cl}_{Na} + 2I^- \rightarrow R-SO_2-NH_2 + I_2$	0.1 *M* KI in 10% acetic acid	–	Amp(15)	0.1–1	(< 12 mA)	1.3% CV	10[i]	74

19. Chlorpromazine	Phenothiazine	Ce^{4+}	$>S + 2Ce^{4+} + H_2O \rightarrow >S{\rightarrow}O + 2Ce^{3+} + 2H^+$	$0.05F\ Ce_2(SO_4)_3$ in $3F\ H_2SO_4$	Pt	Amp(500)	0.3–10	2–3	0.2% CV	6	120
20. Chlortetracycline	–	Br_2	Acid hydrolysis, then titration	4% KBr in 2% HCl	–	–	1.5–4	–	0.4% E	–	121
21. Cinchonine	Base	H^+	$R_3N + H^+ \rightleftharpoons R_3NH^+$	Same as number 6	Pt	Amp(80) (Bi/Bi)	0.9	(1.5 mA)	0.2% AE	7	112
22. Cyclobarbital	Cyclohexene	Cl_2	$C_6H_9{-}R + Cl_2 \rightarrow C_6H_9Cl_2{-}R$	10% HCl	–	Amp	1.3–3.2	(7 mA)	0.7% E	–	122
23. Cysteine	Mercaptan	Hg^{++}	$2RSH + Hg^{++} \rightarrow (RS)_2Hg\downarrow + 2H^+$	–	Hg	–	1	–	–	–	123
24. Cysteine hydrochloride	Mercaptan hydrochloride	Ag^+	$RSH{\cdot}HCl + 2Ag^+ \rightarrow RSAg\downarrow + AgCl\downarrow + 2H^+$	3.5% $HClO_4$	Ag	Pot-3	2–8	0.35–0.7	2.8% CV	6	107
25. Cystine	Disulfide	red/Hg^{++}	$RSSR + 2Hg \rightarrow (RS)_2Hg_2$ (electrode) $Hg^{++} + (RS)_2Hg_2 \rightarrow (RS)_2Hg\downarrow + Hg_2^{++}$	–	Hg	–	1	–	–	–	123
26. Cytosine	Pyrimidine	Br_2	$C_4H_5N_3O + 3Br_2 \rightarrow$ products	pH 7.5 Phosphate buffer	Pt	Amp(250)	0.01–2	(3 mA)	0.5% CV	5[f]	124
27. Cytidine	Pyrimidine	Br_2	$C_9H_{13}N_3O_5 + 2Br_2 \rightarrow$ products	Same	Pt	Same	0.02–5	Same	1.8% CV	7	124
28. Cytidine monophosphate	Pyrimidine	Br_2	$C_9H_{14}N_3O_8P + 2Br_2 \rightarrow$ products	Same	Pt	Same	0.03–6	Same	0.8% CV	6[f]	124
29. Cytidine triphosphate	Pyrimidine	Br_2	$C_9H_{16}N_3O_{14}P_3 + 2Br_2 \rightarrow$ products	Same	Pt	Same	0.05–10	Same	1.0% CV	5	124
30. Diethazine	Phenothiazine	Ce^{4+}	Same as number 19	Same as number 19	Pt	Amp(500)	1–20	2–3	0.3% CV	6	120
31. Disulfiram	Disulfide	red/Hg^{++}	Same as number 25	–	Hg	–	1	–	–	–	123
32. Ephedrine hydrochloride	Hydrochloride	Ag^+	$R_2NH{\cdot}HCl + Ag^+ \rightarrow R_2NH_2^+ + AgCl\downarrow$	3.5% $HClO_4$	Ag	Pot-3	2–10	0.35–0.7	1.8% CV	6	107
33. Ethacridine	Aromatic amine	Br_2/Cu^+	Same as number 17	Same as number 17	Pt	Pot-1	0.04–0.1	(0.18 mA)	+1.0% E	–	119
34. Ethionamide	Thioamide	Ag^+	$R{-}CS{-}NH_2 + 2Ag^+ \rightarrow R{-}CN + Ag_2S\downarrow + 2H^+$	1.3 M NH_4OH, 0.1 M NaOH	Ag	Pot(Ag/SCE)	0.25	(1 mA)	0.2% CV	–	104
35. Ethopropazine	Phenothiazine	Ce^{4+}	Same as number 19	Same as number 19	Pt	Amp(500)	1–20	2–3	1.6% CV	5	120
36. *p*-Ethylaminobenzoate	Aromatic amine	Br_2/Cu^+	Same as number 17	Same as number 17	Pt	Pot-1	0.04–0.1	(0.18 mA)	+1.0% E	–	119
37. Ethylparaben	Phenol	Br_2/Cu^+	–	30% KBr, 25% $CuSO_4$ in 10% HCl	–	–	0.01% Solution	–	0.5% AE	–	126
38. Folic acid	Aromatic amine	Cl_2	(*p*) $RNH{-}C_6H_4{-}CO{-}NH{-}R' + 2Cl_2 \rightarrow R{-}NH{-}C_6H_2Cl_2{-}CO{-}NH{-}R' + 2HCl$	10% HCl	–	Amp	–	7.7	0.5% AE	–	127
39. Glutathione	Mercaptan	Ag^+	$RSH + Ag^+ \rightarrow RSAg\downarrow + H^+$	0.05 M NH_4OH, 0.025 M NH_4NO_3, 0.01% EDTA	Ag	Amp(200)	0.01	(0.016 mA)	2–3% CV	–	128
40. Glutathione	Mercaptan	Hg^{++}	$2RSH + Hg^{++} \rightarrow (RS)_2Hg\downarrow$	–	Hg	–	1	–	–	–	123
41. Hexobarbital	Cyclohexene	Cl_2	$C_6H_9{-}R + Cl_2 \rightarrow C_6H_9Cl_2{-}R$	10% HCl	–	Amp	1.2–2.7	(7 mA)	0.7% E	–	122
42. Hydroquinone	Hydroquinone	Ce^{4+}	$C_6H_6O_2 + 2Ce^{4+} \rightarrow C_6H_4O_2 + 2Ce^{3+} + 2H^+$	0.1 M $Ce_2(SO_4)_3$ in 1 M H_3PO_4, 1.8 M H_2SO_4	–	Pot-1	0.1–0.3	–	0.5% AE	–	68
43. 8-Hydroxyquinoline (oxine)	Phenol	Br_2	$C_9H_7NO + 2Br_2 \rightarrow C_9H_5Br_2NO + 2HBr$	0.2 M NaBr in 1 mM HCl	Pt	CCP(25)	0.4–1.1	(10 mA)	0.3% CV	42[f]	129
44. 8-Hydroxyquinoline	Phenol	Br_2	Same	0.4 M NaBr in 55:45:5 acetic acid: water: pyridine	Pt	Amp(630)	0.7	1.5	1.5% CV	–	75
45. 8-Hydroxyquinoline	Phenol	Br_2/Cu^+	Same	0.42 M KBr, 0.2 M NH_4Cl in 2% HCl	–	Amp(15)	0.1–1	(0.5–5 mA)	1.8% CV	22[f]	74
46. Hydroxytetracaine	Aromatic amine, Phenol	Br_2/Cu^+	–	Same as number 17	Pt	Pot-1	0.07–1.6	(0.18 mA)	0.5% CV	11[f]	119

Table 6 (continued)
PHARMACEUTICAL APPLICATIONS OF COULOMETRIC TITRATIONS[a]

Name[b]	Electroactive group[c]	Titrants[d]	Reactions[e]	Supporting electrolyte	Working electrode	Detection[f]	Sample size (mg)	Current density[g] (mA/cm^2)	Precision[h]	Repli-cates	Ref.
47. Isoniazid	Hydrazide	Ag^{++}	$RCONHNH_2 + 4Ag^{++} + H_2O \rightarrow RCOOH + N_2\uparrow + 4H^+ + 4Ag^+$	0.01 *M* $AgNO_3$ in 5 *M* HNO_3 at −15°C	Au	Pot-1	0.5–3	2.5–5	~1.6% CV	12[i]	70
48. Isoniazid	Hydrazide	Br_2	$RCONH{-}NH_2 + 2Br_2 + H_2O \rightarrow RCOOH + N_2\uparrow + 4HBr$	0.2 *M* KBr in 1 *M* H_2SO_4	Vitreous Carbon	CCP(1)	3–15 μg	12	2.5% CV	70[i]	130
49. Isoniazid	Hydrazide	Br_2	Same	Same as number 1	–	Amp(15)	0.03–0.5	(1.5 mA)	1.6% CV	16[i]	74
50. Isoniazid	Hydrazide	I_2/excess $S_2O_3^=/I_2$	$RCONHNH_2 + 2I_2 + H_2O \rightarrow RCOOH + N_2\uparrow + 2HI$	0.1 *M* KI, pH 8	–	Amp(10)	0.3–0.5	(< 12 mA)	0.8% CV	4[i]	74
51. Isoniazid	Hydrazide	Cl_2	–	10% HCl	Pt	MO	–	10	–	–	131
52. Meprobamate	Carbamate	OBr^-	$R{-}O{-}CO{-}NH_2 \rightarrow CNO^-$(alk. alcohol) $CNO^- \rightarrow NH_4^+$ (aqueous acid) $2NH_4^+ + 3OBr^- + 2OH^- \rightarrow N_2\uparrow + 3Br^- + 5H_2O$	–	–	Amp(200)	–	(3 mA)	–	–	132
53. Methionine	Dialkyl sulfide	Br_2	$R{-}S{-}R' + 2Br_2 + H_2O \rightarrow R{-}SO_2{-}R' + 4HBr$	Same as number 1	–	Amp(15)	0.05–1	(< 12 mA)	1.5% CV	17[i]	74
54. Methionine	Amine	Cu^{++}	$2RH + Cu^{++} \rightarrow CuR_2\downarrow + 2H^+$	0.3 m*M* borax, 0.1 *M* KNO_3, pH 9.2	Cu	Pot(Cu/SCE)	2.2	(10 mA)	2% E	14	104
55. Methotrimeprazine	Phenothiazine	Ce^{4+}	Same as number 19	Same as number 19	Pt	Amp(500)	1–10	2–3	1.0% CV	5	120
56. Methylthiouracil	–	Cl_2	–	10% HCl	–	–	0.5–1	5.5	1% AE	–	133
57. Nicotine	Base	H^+	$R_3N + H^+ \rightleftharpoons R_3NH^+$	0.1 *M* $NaClO_4$ in 6:1 acetic anhydride: acetic acid	Pt	MG	1.1	64	1.8% CV	12[i]	73
58. Nitrogen	N	OBr^-	Kjeldahl digestion, then $2NH_3 + 3OBr^- \rightarrow N_2\uparrow + 3Br^- + 3H_2O$	1.5 *M* KBr, 0.075 *M* $Na_2B_4O_7$, pH 8.6	Pt	Pot-1	3–15	20	0.1–1% CV	–	134
59. Nitrogen	N	OBr^-	Digestion 1:10 H_2SO_4:HCl, then $2NH_3 + 3OBr^- \rightarrow N_2\uparrow + 3Br^- + 3H_2O$	pH 8.6 buffer	–	–	–	–	–	–	135
60. Noramidopyrine methanesulfonate	–	I_2	–	6% KI, 0.6% HCl in 60% methanol	–	Amp	3–5	(10 mA)	0.2% AE	–	136
61. Oxytetracycline	–	Br_2	Acid hydrolysis, then titration	4% KBr in 2% HCl	–	–	1.5–4	–	0.6% E	–	121
62. Papaverine	R_3N	$(C_6H_5)_4BH/Ag^+$	Same as number 14	Same as number 14	Ag	Pot-2	2–10	–	0.2% CV	3[i]	117
63. Papaverine hydrochloride	Hydrochloride	Ag^+	$R_3N{\cdot}HCl + Ag^+ \rightarrow R_3NH^+ + AgCl$	3.5% $HClO_4$	Ag	Pot-3	4–20	0.35–0.7	1.1% CV	6	107
64. Penicillins	–	Br_2	Beta lactam cleavage (alkaline hydrolysis) then titration with Br_2	–	–	–	–	–	5% E	–	137
64a. Penicillins	–	Hg^{++}	Beta lactam cleavage (alkaline hydrolysis, then titration with Hg^{++}	0.8 *M* acetate, pH 4.6	Au(Hg)[j]	Pot-1	3–4	0.5	0.4% CV	30[i]	137a
65. 6-Aminopenicillanic acid	–	Br_2 or Cl_2	Beta lactam cleavage (hydrolysis 1N NaOH), penicillamine air oxidized to disulfide, then titrate penicillamine disulfide, 7 electrons	Acidic KBr or 0.5 *M* HCl	–	–	–	–	–	–	138

66. Penicillin K	–	Cl_2 (also Br_2, I_2)	Beta lactam cleavage (hydrolysis 2.5% KOH), titration with 4 moles Cl_2 (8 electrons)	3.75% HCl	–	–	0.3–1	(5.2 mA)	–	–	139
67. Penicillin V	–	I_2/excess $S_2O_3^=/I_2$	Beta lactam cleavage (hydrolysis NaOH), add excess I_2, then excess $S_2O_3^=$, titrate with I_2	3.75% KI in dilute acetic acid	–	–	0.3–1	–	–	–	140
68. Phenylbutazone	–	Cl_2 or Br_2	–	5% HCl in 50% ethanol or 1.6% KBr, 3.6% HCl in 50% ethanol	–	Amp	3–5	–	0.2–0.3% AE	–	141
69. Phenylephrine hydrochloride	Phenol	Br_2/excess As^{3+}/Br_2	(*m*) $HO{-}C_6H_4{-}R + 3Br_2 \rightarrow HO{-}CHBr_3{-}R + 3HBr$	Same as number 1	Pt	Amp(15)	0.3–0.5	(3 mA)	0.7% CV	5^i	74
70. Phenylsalicylate	Phenol	Br_2/Cu^+	–	Same as number 35	–	–	0.01% solution	–	0.5% AE	–	126
71. Physostigmine sulfate	R_3N	$(C_6H_5)_4BH/Ag^+$	Same as number 14	Same as number 14	Ag	Pot-2	10	–	1.0% CV	3	117
72. Pilocarpine nitrate	R_3N	$(C_6H_5)_4BH/Ag^+$	Same as number 14	Same as number 14	Ag	Pot-2	2.5–10	–	0.2% CV	2^i	117
73. Procaine hydrochloride	Aromatic amine	Br_2/Cu^+	(*p*) $HO{-}C_6H_4{-}R + 2Br_2 \rightarrow HO{-}C_6H_2Br_2{-}R + 2HBr$	Same as number 17	Pt	Pot-1	0.06–0.15	(0.18 mA)	+2% E	–	119
74. Procaine hydrochloride	R_3N	$(C_6H_5)_4BH/Ag$	Same as number 14	Same as number 14	Ag	Pot-2	3–10	–	0.1% CV	3^i	117
75. Procarbazine hydrochloride	Hydrazine	I_2	$R{-}NH{-}NH{-}CH_3 + 2I_2 \rightarrow R{-}N{=}N{-}CH_3 + 2HI$	0.5 *M* KI in 0.15 *M* $NaHCO_3$, pH 8.4	Pt	CCP (10)	40 50	(96.5 mA)	0.2% CV(drug substance) 0.3% CV(capsules)	5 8	142
76. Promethazine	Phenothiazine	Ce^{4+}	Same as number 19	Same as number 19	Pt	Amp(500)	1–20	2–3	0.6% CV	5	120
77. Quinine	R_3N	$(C_6H_5)_4BH/Ag^+$	$R_3N + 2(C_6H_5)_4BH \rightarrow [(C_6H_5)_4BH]_2{\cdot}R_3N\downarrow$ $(C_6H_5)_4B^- + Ag^+ \rightarrow (C_6H_5)_4BAg\downarrow$	Same as number 14	Ag	Pot-2	2–10	–	–	–	117
78. Quinine	Base	H^+	$R_3N + H^+ \rightleftharpoons R_3NH^+$	Same as number 6	Pt	Amp(Bi/Bi)	1	(1.5 mA)	0.4% AE	7	112
79. Quinine hydrochloride	Hydrochloride	Ag^+	$R_3N{\cdot}HCl + Ag^+ \rightarrow R_3NH^+ + AgCl\downarrow$	3.5% $HClO_4$	Ag	Pot-3	4–20	0.35–0.7	2.5% CV	6	107
80. Sparteine sulfate	R_3N	$(C_6H_5)_4BH/Ag^+$	Same as number 14	Same as number 14	Ag	Pot-2	10–20	–	–	–	117
81. Strychnine sulfate	R_3N	$(C_6H_5)_4BH/Ag^+$	Same as number 14	Same as number 14	Ag	Pot-2	5–10	–	0.1% CV	4^i	117
82. Sulfadiazine	Sulfonamide	Br_2/excess As^{3+}/Br_2	(*p*) $NH_2{-}C_6H_4{-}SO_2NHR + 3Br_2 \rightarrow NH_2{-}C_6H_2Br_2{-}SO_2NHR{-}Br + 3HBr$	Same as number 1	Pt	Amp(15)	0.2–0.5	(3 mA)	1.5% CV	6^i	74
83. Sulfaguanidine	Sulfonamide	Br_2/Cu^+	Same as number 17	Same as number 17	Pt	Amp(500)	0.05–1.2	(0.18 mA)	1.6% CV	9	119
84. Sulfaguanidine	Sulfonamide	Br_2	Same as number 17	0.1 *M* KBr in 0.3 *M* HCl	Pt	Pot-1	0.9–2.8	(10 mA)	0.4% CV	12^i	143
85. Sulfamerazine	Sulfonamide	Br_2/Cu^+	Same as number 82	Same as number 17	Pt	Amp(500)	0.05–1	(0.18 mA)	1% E	–	119
86. Sulfamerazine	Sulfonamide	Br_2	Same as number 82	Same as number 84	Pt	Pot-1	0.6	(1.2 mA)	0.6% CV	5	143
87. Sulfamerazine	Sulfonamide	Br_2/excess As^{3+}/Br_2	Same as number 82	Same as number 1	Pt	Amp(15)	0.3–0.5	(3 mA)	1.4% CV	4^i	74
88. Sulfamethizole	Sulfonamide	Br_2	Same as number 17	Same as number 84	Pt	Pot-1	–	–	–	–	144
89. Sulfamethoxypyrazine	Sulfonamide	Br_2	Same as number 17	Same as number 84	Pt	Pot-1	1–2	(10 mA)	0.3% CV	12^i	145
90. Sulfamethoxypyridazine	Sulfonamide	Br_2	Same as number 17	Same as number 84	Pt	Pot-1	–	–	–	–	144
91. Sulfanilamide	Sulfonamide	Br_2/Cu^+	Same as number 17	Same as number 17	Pt	Amp(500)	0.04–1.2	(0.18 mA)	2% E	–	119
92. Sulfanilamide	Sulfonamide	Br_2	Same as number 17	Same as number 84	Pt	Pot-1	0.65–2	(10 mA)	0.2% CV	12^i	143

Table 6 (continued)
PHARMACEUTICAL APPLICATIONS OF COULOMETRIC TITRATIONS[a]

Name[b]	Electroactive group[c]	Titrants[d]	Reactions[e]	Supporting electrolyte	Working electrode	Detection[f]	Sample size (mg)	Current density[g] (mA/cm^2)	Precision[h]	Replicates	Ref.
93. Sulfanilylurea	Sulfonamide	Br_2/Cu^+	Same as number 17	Same as number 17	Pt	Amp(500)	0.05–0.13	(0.18 mA)	4% E	–	119
94. N-Sulfanilyl-3,4-xylamide	Sulfonamide	Br_2	$R{-}SO_2{-}NH{-}CO{-}R' \rightarrow RSO_2NH_2$ (3 *M* HCl) then same as number 17	Same as number 84	Pt	Pot-1	1.3–1.7	(10 mA)	0.2% CV	8[i]	143
95. Sulfaphenazole	Sulfonamide	Br_2	Same as number 82	Same as number 84	Pt	Pot-1	–	–	–	–	144
96. Sulfathiazole	Sulfonamide	Br_2/Cu^+	Same as number 82	Same as number 17	Pt	Amp(500)	0.06–0.14	(0.18 mA)	+1.5% E	–	119
97. Sulfathiazole	Sulfonamide	Br_2	Same as number 82	Same as number 84	Pt	Pot-1	0.8	(10 mA)	0.6% CV	10	143
98. Sulfathiazole	Sulfonamide	Br_2/excess As^{3+}/Br_2	Same as number 82	Same as number 82	Pt	Amp(15)	0.2–0.5	(3 mA)	0.8% CV	7[i]	74
99. Sulfisomidine	Sulfonamide	Br_2/Cu^+	Same as number 17	Same as number 17	Pt	Amp(500)	0.06–1.1	(0.18 mA)	2% E	–	119
100. Sulfisomidine	Sulfonamide	Br_2	Same as number 17	Same as number 84	Pt	Pot-1	0.8–1.5	(10 mA)	0.4% CV	12[i]	143
101. Sulfisoxazole	Sulfonamide	Br_2	Same as number 82	Same as number 84	Pt	Pot-1	0.9–1.8	(10 mA)	0.1% CV	8[i]	143
102. Tetracaine	Sulfonamide	Br_2/Cu^+	Same as number 17	Same as number 17	Pt	Amp(500)	0.15	(0.18 mA)	–5% E	–	119
103. Tetracaine hydrochloride	R_3N	$(C_6H_5)_4BH/Ag^+$	Same as number 14	Same as number 14	Ag	Pot-2	10–20	–	0.2% CV	2	117
104. Theobromine	–	excess Cl_2/ ?	$C_7H_8N_4O_2 + 2Cl_2 \rightarrow$ products	5% HCl	–	–	0.1–1.5	–	0.5% AE	–	146
105. Theophylline	–	excess Cl_2/ ?	$C_7H_8N_4O_2 + 2Cl_2 \rightarrow$ products	5% HCl	–	–	0.2–1	–	0.4% AE	–	146
106. Thiamine hydrochloride	R_3N	$(C_6H_5)_4BH/Ag^+$	Same as number 14	Same as number 14	Ag	Pot-2	5–10	–	0.1% CV	3[i]	117
107. Thiazinamium	Phenothiazine	Ce^{4+}	Same as number 19	Same as number 19	Pt	Amp(500)	0.25–5	2–3	2.0% CV	6	120
108. Thioglycolic acid	Mercaptan	Br_2	$RSH + 3Br_2 + 2H_2O \rightarrow R{-}SO_2{-}Br + 5HBr$	0.1 *M* KBr in 50% acetic acid	–	Amp(15)	0.05–0.25	(< 12 mA)	1.0% CV	8[i]	74
109. Thiopental sodium	Mercaptan	Br_2	–	30% KBr in 10% HCl	–	Amp	0.3–1.5	–	0.7% AE	–	125
110. Thiosemicarbazide	Mercaptan (tautomer)	Ag^+	$NH_2{-}N{=}C(NH_2)(SH) + Ag^+ \rightarrow RSAg\downarrow + H^+$	2 *M* HNO_3	Ag	Pot(Ag/SCE)	0.45	(3 mA)	1% CV	–	104
111. Thymine	Pyrimidine	Br_2	$C_5H_6N_2O_2 + Br_2 \rightarrow$ products	pH 7.5 phosphate buffer	Pt	Amp(250)	0.5–1	(3 mA)	0.4% CV	8[i]	124
112. Thymidine	Pyrimidine	Br_2	$C_{10}H_{14}N_2O_6 + Br_2 \rightarrow$ products	Same	Pt	Amp(250)	1–2	(3 mA)	0.6% CV	4[i]	124
113. Thymidine monophosphate	Pyrimidine	Br_2	$C_{10}H_{15}N_2O_9P + Br_2 \rightarrow$ products	Same	Pt	Amp(250)	1	(3 mA)	0.9% CV	6	124
114. Thymol	Phenol	Br_2	(1,2,4) $C_6H_3(R)(OH)(CH_3) + 2Br_2 \rightarrow C_6HBr_2(R)(OH)(CH_3) + 2HBr$	Same as number 1	Pt	Amp(15)	0.2–0.5	(< 12 mA)	0.1% CV	4[i]	74
115. Uracil	Pyrimidine	Br_2	$C_4H_4N_2O_2 + 2Br_2 \rightarrow$ products	pH 7.5 phosphate buffer	Pt	Amp(250)	0.2–0.4	(3 mA)	0.3% CV	8[i]	124
116. Uridine	Pyrimidine	Br_2	$C_9H_{12}N_2O_6 + Br_2 \rightarrow$ products	Same	Pt	Amp(250)	1	(3 mA)	1.5% CV	5	124
117. Uridine monophosphate	Pyrimidine	Br_2	$C_9H_{13}N_2O_9P + Br_2 \rightarrow$ products	Same	Pt	Amp(250)	0.75–1.5	(3 mA)	0.9% CV	11[i]	124
118. Vitamin E (alpha tocopherol)	Tocopherol	V(V)	$C_{29}H_{50}O_2 + V(V) \rightarrow$ 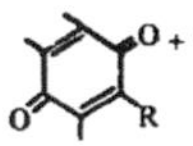 O, O, R + V(III)	0.1 *M* $VOSO_4$ in 80% acetic acid	Pt	Pot-1	0.07–1.2	0.1–2	1.4% CV	19[i]	147

119. Vitamin K_3 (menadione)	Quinone/ hydroquinone	Red/Ce^{4+}	$C_{11}H_8O_2 \rightarrow C_{11}H_{10}O_2$ (hydroquinone by Zn or electrochemical reduction) $C_{11}H_{10}O_2 + 2Ce^{4+} \rightarrow C_{11}H_8O_2 + 2Ce^{3+} + 2H^+$	Same as number 19	Pt	Pot-2	0.3–3.5	2–3	0.9% CV	15[i]	148
120. Yohimbine hydrochloride	R_3N	$(C_6H_5)_4BH/Ag^+$	Same as number 14	Same as number 14	Ag	Pot-2	5–10	–	0.7% CV	3[i]	117

[a] Data not given (or not available) indicated by a line.
[b] Names are titles assigned in the *Merck Index*, 9th ed., where applicable.
[c] Functional group reacting with electrogenerated titrant.
[d] Primary electrogenerated titrant listed first. Back titrants listed sequentially. For example, Br_2/Cu^+ indicates excess bromine is generated and then back titrated with Cu^+. All species except As(III) and $S_2O_3^=$ generated electrochemically.
[e] Abbreviated reaction where stated or known. Common back titration reactions are:

$2Br^- \rightarrow Br_2 + 2e$ (allowed to react with compound)
$2\,S_2O_3^=$ (or As^{3+}) $+ Br_2 \rightarrow S_4O_6^=$ (or As^{5+}) $+ 2Br^-$ (excess standard solution added)
$2Br^- \rightarrow Br_2 + 2e$ (at electrode)

and

$2Br^- \rightarrow Br_2 + 2e$ (allowed to react with compound)
$Cu^{++} + e \rightarrow Cu^+$ (at electrode)

$2Cu^+ + Br_2 \rightarrow Cu^{++} + 2Br^-$

[f] Abbreviations are: AMP = amperometry with two indicating electrodes (Pt/Pt unless indicated otherwise); applied voltage, mv, given in ( ); CCP = controlled current potentiometry with two indicating electrodes (Pt/Pt); applied current, microamps, given in ( ); INT = integration of charge under controlled potential electrolysis for the back titration; MG = malachite green plus hydroquinone (colorimetric); MO = methyl orange decolorization; POT-1 = potentiometric, Pt indicator vs. SCE reference; POT-2 = potentiometric, Pt indicator vs. Ag/AgCl reference; POT-3 = potentiometric, Orion chloride specific electrode vs. mercury/mercurous sulfate/sulfate reference; and POT = other potentiometric as indicated.
[g] Currents in milliamperes are given in ( ) when current densities were not available.
[h] Precision or accuracy given as: AE = average deviation or average error (from known value); E = error; σ = standard deviation in mg; SE = standard error; and CV = relative standard deviation in %.
[i] Indicates data from more than one sample or from different weight ranges were pooled to obtain the standard deviation.

$$\sigma_p = \left[\frac{\sum_i (n_i - 1)\,\sigma_i^2}{(\sum_i n_i) - i} \right]^{1/2}$$

where σ_p is the pooled standard deviation and the σ_i are individual standard deviations calculated from n_i measurements within the group.
[j] Amalgamated gold.

tical applications. The list of oxidants is unusually varied. In particular, many applications have been developed for bromocoulometry including addition reactions in alkenes, substitution in activated aromatic systems such as phenols, aromatic amines, or sulfonamides, and oxidation of hydroquinones, hydrazides, mercaptans, disulfides, α, β-unsaturated lactones, etc. Iodocoulometry and chlorocoulometry have been used for many applications similar to those described for bromine. One specific application of iodocoulometry is the determination of peroxides. Fiedler[77,78] reported data from analysis of 1 mg samples at a precision level of 0.1 to 0.25% for seven different types of peroxides.

Patriarche[74] has listed extensive applications for the determination of metals and metallic salts of pharmaceutical compounds using precipitation with 8-hydroxyquinoline followed by filtration, acid dissolution of the precipitate, and a coulometric titration of the 8-hydroxyquinoline. These include the determination of Cu, Co (II), Mg, Mn, Zn, and Bi using bromocoulometry and As, Sb, and Se using iodocoulometry. The determination of ammonia with hypobromite can be performed after digestion of complex proteins or other nitrogen containing compounds.

5. Coulometric Karl Fischer Titration Reactions

The Karl Fischer titration is extensively used to determine water in samples. Meyer and Boyd[101] apparently were the first to use a coulometrically generated reagent. The kinetic rates and mechanism of the electrogeneration reaction have been investigated by Barendrecht and Verhoef[80,80a] using voltammetry at the rotating ring disk electrode. Cedergren[81,82] investigated rates and compared the precision and accuracy of the potentiometric and amperometric endpoints. Karlsson[109] proposed a technique for heating the sample in an oven and transporting the released water vapor to the titration cell using a stream of dry gas. Since the sample is not added to the cell, chemical interferences observed with some classes of compounds (e.g., carbonates, methyl ketones, etc.) can be avoided. The amounts of water determined in various substances and observed precision levels include the following: 5 to 1000 μg water per mℓ in solvents, σ 2 μg;[101] 10 to 100 μg water per mℓ in solvents, σ 0.2 to 0.3 μg;[100] 5 to 5000 μg water in samples, CV 0.3 to 0.4% or 0.5 to 5 μg water in samples, CV 10%;[109] and 0.05 to 200 μg water in samples, σ 0.015 μg.[81]

REFERENCES

1. Baizer, M. M., Ed., *Organic Electrochemistry,* Marcel Dekker, New York, 1973.
2. Weinberg, N. L., Ed., *Techniques of Electroorganic Synthesis,* Vol. 5, Parts I and II, John Wiley & Sons, New York, 1975.
3. Fry, A. J., *Synthetic Organic Electrochemistry,* Harper & Row, New York, 1972.
4. Bard, A. J. and Santhanam, K. S. V., Application of controlled potential coulometry to the study of electrode reactions, in *Electroanalytical Chemistry,* Vol. 4, Bard, A. J., Ed., Marcel Dekker, New York, 1970, 215.
5. Meites, L., Controlled potential electrolysis, in *Physical Methods of Chemistry,* Vol. 1, Part IIA, Weissberger, A. and Rossiter, B., Eds., Wiley-Interscience, New York, 1971, chap. 9.
6. Mark, H. B. and Janata, J., Application of controlled current coulometry to reaction kinetics, in *Electroanalytical Chemistry,* Vol. 3, Bard, A. J., Ed., Marcel Dekker, New York, 1969, 1.
7. Harrar, J. E., Techniques, apparatus and analytical applications of controlled potential coulometry, in *Electroanalytical Chemistry,* Vol. 8, Bard, A. J., Ed., Marcel Dekker, New York, 1975, 1.
8. Tackett, S. L., *Chem. Tech.,* 1972, 734.
9. Skoog, D. A. and West, D. M., *Principles of Instrumental Analysis,* Holt, Rinehart & Winston, New York, 1971, chap. 21.
10. Clem, R. G., *Ind. Res.,* 15(9), 50, 1973.
11. Siegerman, H., Chang, J., and Thompson, J., *Am. Lab.,* 7, 61, 1975.
12. Zuman, P., Relation between micro and macro phenomena, in *Organic Electrochemistry,* Baizer, M. M., Ed., Marcel Dekker, New York, 1973, chap. 3.
13. Meites, L., *Polarographic Techniques,* 2nd ed., Interscience, New York, 1965, chap. 10.
14. Stephens, F. B., Jakob, F., Rigdon, L. P., and Harrar, J. E., *Anal. Chem.,* 42, 764, 1970.
15. Adams, R. N., *Electrochemistry at Solid Electrodes,* Marcel Dekker, New York, 1969, chaps. 2 and 9.
16. Kolthoff, I. M. and Lingane, J. J., *Polarography,* Vol. 2, Interscience, New York, 1952, chap. 42.
17. Meites, L. and Moros, S. A., *Anal. Chem.,* 31, 23, 1959.
18. Bates, R. G., Chm., *Analytical Chemistry Division, Committee on Electroanalytical Chemistry, Pure Appl. Chem.,* 45, 125, 1976.
19. Clem, R. G. and Goldsworthy, W. W., *Anal. Chem.,* 43, 918, 1971.
20. Woodward, W. S., Ridgway, T. H., and Reilley, C. N., *Anal. Chem.,* 46, 1151, 1974.
21. Mann, C. K., Nonaqueous solvents for electrochemical use, in *Electroanalytical Chemistry,* Vol. 3, Bard, A. J., Ed., Marcel Dekker, New York, 1969, 57.
22. Mann, C. K. and Barnes, K. K., *Electrochemical Reactions in Nonaqueous Systems,* Marcel Dekker, New York, 1970.
23. Koch, W. F., Hoyle, W. C., and Diehl, H., *Talanta,* 22, 717, 1975.
23a. Koch, W. F. and Diehl, H., *Talanta,* 23, 509, 1976.
24. Marinenko, G. and Foley, R. T., *J. Electrochem. Soc.,* 119, 232C, Abstract 198, 1972. Presented at the 142nd meeting of The Electrochemical Society, Inc., Miami Beach, Oct., 1972.
25. Marinenko, G. and Foley, R. T., *J. Res. Natl. Bur. Stand. A,* 79A, 737, 1975.
26. Zynger, J., *Anal. Chem.,* 47, 1380, 1975.
27. Rickard, E. C. and Dinner, A., Electrochemistry of Cinoxacin, presented at the 3rd annual meeting of the Federation of Analytical Chemistry and Spectroscopy Societies, Philadelphia, November, 1976, Abstract 371.
28. Richards, J. A. and Evans, D. H., *Anal. Chem.,* 47, 964, 1975.
29. Clark, B. R. and Evans, D. H., *J. Electroanal. Chem.,* 69, 181, 1976.
30. McKinney, T. M., Electron spin resonance in electrochemistry, in *Electroanalytical Chemistry,* Vol. 10, Bard, A. J., Ed., Marcel Dekker, New York, 1977, chap. 2.
31. Kuwana, T. and Winograd, N., Spectroelectrochemistry at optically transparent electrodes, I. Electrodes under semi-infinite diffusion conditions, in *Electroanalytical Chemistry,* Vol. 7, Bard, A. J., Ed., Marcel Dekker, New York, 1974, 1.
31a. Heineman, W. R., *Anal. Chem.,* 50, 390A, 1978.
31b. Kastening, B., Joint application of electrochemical and ESR techniques, in *Electroanalytical Chemistry,* Nürnberg, H. W., Ed., John Wiley & Sons, New York, chap. 6.
32. Hubbard, A. T. and Anson, F. C., The theory and practice of electrochemistry with thin layer cells, in *Electroanalytical Chemistry,* Vol. 4, Bard, A. J., Ed., Marcel Dekker, New York, 1970, 129.
33. Bard, A. J., *Anal. Chem.,* 35, 1125, 1963.
34. Clem, R. G., Jakob, F., Anderberg, D. H., and Ornelas, L. D., *Anal. Chem.,* 43, 1398, 1971.
35. Clem, R. G., *Anal. Chem.,* 43, 1853, 1971.
36. Sioda, R. E. and Kambara, T., *J. Electroanal. Chem.,* 38, 51, 1972.
37. Kenkel, J. V. and Bard, A. J., *J. Electroanal. Chem.,* 54, 47, 1974.
38. Schmulbach, C. D. and Oommen, T. V., *Anal. Chem.,* 45, 820, 1973.

39. Kihara, S., *J. Electroanal. Chem.*, 45, 31, 1973.
40. Davis, D. G., *Anal. Chem.*, 46, 21R, 1974 and previous biennial reviews of Electroanalysis and Coulometric Analysis.
41. Booman, G. L. and Holbrook, W. B., *Anal. Chem.*, 35, 1793, 1963.
42. Harrar, J. E. and Shain, I., *Anal. Chem.*, 38, 1148, 1966.
43. Claesson, J. and Lindberg, J., *J. Electroanal. Chem.*, 40, 255, 1972.
44. Lindberg, J., *J. Electroanal. Chem.*, 40, 265, 1972.
45. Goldsworthy, W. W. and Clem, R. G., *Anal. Chem.*, 44, 1360, 1972.
46. Phillips, G., Newton, D. A., and Wilson, J. D., *J. Electroanal. Chem.*, 75, 77, 1977.
47. Myers, R. L., Application of Cyclic Voltammetry to a Kinetic Study of the Fe(III) Oxalate System in Acid Solution Using a Digital Data Acquisition System, Ph.D. thesis, University of Wisconsin, Madison, 1971.
48. Gordon, B. M., *EDN*, Aug. 15, 1972, 55.
48a. Kelly, P. C. and Horlick, G., *Anal. Chem.*, 45, 518, 1973.
49. Zynger, J. and Rickard, E. C., presented at 28th Pittsburgh Conference on Analytical Chemistry and Applied Spectroscopy, Cleveland, Feb., 1977, Abstract 479.
50. Janicki, C. A., Gilpin, R. K., Moyer, E. S., Almond, H. R., Jr., and Erlich, R. H., *Anal. Chem.*, 49, 110R, 1976, and previous biennial reviews of Pharmaceuticals and Related Drugs.
51. Zuman, P., *Organic Polarographic Analysis*, Macmillan, New York, 1964.
52. Rickard, E. C. and Cooke, G. G., *J. Pharm. Sci.*, 66, 379, 1977.
53. Rickard, E. C., Eli Lilly & Co., unpublished data.
54. Dinner, A. and Rickard, E. C., *J. Heterocycl. Chem.*, 15, 333, 1978.
55. Beyer, E. B., *J. Assoc. Off. Anal. Chem.*, 52, 844, 1969.
56. Kolthoff, I. M., Lingane, J. J., and Wawzonek, S., *Polarography*, Vol. 2, Interscience, New York, 1952, chap. 43.
57. Mittal, M. L. and Pandey, A. B., *J. Electroanal. Chem.*, 36, 249, 1972.
58. Krivis, A. F., Gazda, E. S., Supp, G. R., and Robinson, M. A., *Anal. Chem.*, 35, 966, 1963.
59. Csejka, D. A., Nakos, S. T., and DuBord, E. W., *Anal. Chem.*, 47, 322, 1975.
60. Furlani, C., *Gazz. Chim. Ital.*, 85, 1646, 1955.
61. Santhanam, K. S. V. and Krishman, V. R., *Anal. Chem.*, 33, 1493, 1961.
62. Janssen, R. W. and Discher, C. A., *J. Pharm. Sci.*, 60, 798, 1971.
63. Merkel, F. H. and Discher, C. A., *Anal. Chem.*, 36, 1639, 1964.
64. Willard, H. H., Merritt, L. L., and Dean, J. A., *Instrumental Methods of Analysis*, 5th ed., D. van Nostrand, New York, 1974, chap. 24.
65. Purdy, W. C., *Electroanalytical Methods in Biochemistry*, McGraw-Hill, New York, 1965, chap. 8.
66. Purdy, W. C., Coloumetry in clinical chemistry, *CRC Crit. Rev. Clin. Lab. Sci.*, 7, 227, 1977.
67. Lingane, J. J., *Electroanalytical Chemistry*, 2nd ed., Interscience, New York, 1958, chap. 20.
68. Lingane, J. J., *Electroanalytical Chemistry*, 2nd ed., Interscience, New York, 1958, chap. 21.
69. Davis, D., *J. Electroanal. Chem.*, 1, 73, 1959.
70. Chateau-Gosselin, M., Patriarche, G. J., and Christian, G. D., *Fresenius' Z. Anal. Chem.*, 285, 373, 1977.
71. Bos, M., Iupma, S. T., and Dahmen, E., *Anal. Chim. Acta*, 83, 39, 1976.
72. Siggia, S., *Instrumental Methods of Organic Functional Group Analysis*, Wiley-Interscience, New York, 1972.
73. Stock, J. T., *J. Chem. Educ.*, 50, 268, 1973.
74. Patriarche, G., *Contribution a l'Analyse Coulometrique; Applications aux Sciences Pharmaceutiques*, Editions Arscia S. A., Brussels, 1964.
75. Kinberger, B., Edholm, L. E., Nilsson, O., and Smith, B. E. F., *Talanta*, 22, 979, 1975.
76. Sement, E. G., Rousselet, F., Girard, M. L., and Chemla, M., *Analusis*, 3, 456, 1975.
77. Fiedler, U., *Talanta*, 20, 1097, 1973.
78. Fiedler, U., *Talanta*, 20, 1309, 1973.
79. Karlsson, R., *Talanta*, 22, 989, 1975.
80. Barendrecht, E. and Verhoef, J. C., *J. Electroanal. Chem.*, 59, 221, 1975.
80a. Verhoef, J. C. and Barendrecht, E., *J. Electroanal. Chem.*, 75, 705, 1977.
81. Cedergren, A., *Talanta*, 21, 367, 1974.
82. Cedergren, A., *Talanta*, 21, 553, 1974.
82a. Koch, W. F., Poe, D. P., and Diehl, H., *Talanta*, 22, 609, 1975.
83. Tackett, S. L., *J. Chem. Educ.*, 49, 52, 1972.
84. Bates, R. G., Chm., Analytical Chemistry Division, Committee on Electroanalytical Chemistry, *Pure Appl. Chem.*, 45, 81, 1976.
85. Yoshimori, T. and Tanaka, T., *Anal. Chim. Acta*, 66, 85, 1973.
86. Yoshimori, T., Tanaka, T., Ogawa, M., and Horikoshi, T., *Anal. Chim. Acta*, 63, 351, 1973.

86a. Yoshimori, T., *Talanta*, 22, 827, 1975.
87. Johansson, G., *Talanta*, 12, 111, 1965.
88. Jaycox, L. B. and Curran, D. J., *Anal. Chem.*, 48, 1061, 1976.
89. Earle, W. E. and Fletcher, K. S., III, *Chem. Instrum.*, 7, 101, 1976.
90. Muha, G. M., *J. Chem. Educ.*, 53, 465, 1976.
91. Miyake, S. and Suto, T., *Bunseki Kagaku*, 23, 482, 1974.
92. McCracken, J. E., Guyon, J. C., Shults, W. D., and Jones, H. C., *Chem. Instrum.*, 3, 311, 1972.
93. Stock, J. T., *J. Chem. Educ.*, 46, 858, 1969.
94. Moros, S. A., Hamilton, C. M., Heveran, J. E., and Oliveri-Vigh, S., *J. Pharm. Sci.*, 64, 1229, 1975.
95. Adams, R. E., Betso, S. R., and Carr, P. W., *Anal. Chem.*, 48, 1989, 1976.
95a. Harrar, J. E. and Pomernacki, C. L., *Chem. Instrum.*, 7, 229, 1976.
95b. Hendler, R. W., Songco, D., and Clem, T. R., *Anal. Chem.*, 49, 1908, 1977.
95c. Lindberg, A. O., *Anal. Chim. Acta*, 96, 319, 1978.
96. Coulson, D. M. and Cavanagh, L., *Anal. Chem.*, 32, 1245, 1960.
97. Linnet, N., *Am. Lab.*, 5(10), 79, 1973.
98. Nebesar, B., *J. Chem. Educ.*, 49, A9, 1972.
98a. Dougherty, J. A., *Am. Lab.*, 10(6), 83, 1978.
99. Atkinson, G. F., *Microchem. J.*, 19, 52, 1974.
100. Lindbeck, M. and Freund, H., *Anal. Chem.*, 37, 1647, 1965.
101. Meyer, A. S., Jr. and Boyd, C. M., *Anal. Chem.*, 31, 215, 1959.
102. Feher, Z. and Pungor, E., *Anal. Chim. Acta*, 71, 425, 1974.
103. Stock, J. T., *Anal. Chem.*, 48, 1R, 1976, and previous biennial reviews of Amperometric, Bipotentiometric and Coulometric Titrations.
104. Sement, E., Rousselet, F., Girard, M. L., and Chemla, M., *Ann. Pharm. Fr.*, 30, 691, 1972.
105. Girard, M. L., Rousselet, F., Fouye, H., and Levillain, P., *Ann. Pharm. Fr.*, 27, 173, 1969.
106. Unterman, H. W. and Weissbuch, S., *Pharmazie*, 29, 752, 1974.
107. Lemahieu, G., Lemahieu, C., Resibois, B., and Bertrand, J., *Ann. Pharm. Fr.*, 31, 709, 1973.
108. van Oort, W. J., Veenendaal, G., Buijsman, E., and Griepink, B., *Frensenius' Z. Anal. Chem.*, 284, 125, 1977.
109. Karlsson, R., *Talanta*, 19, 1639, 1972.
110. Kalinowska, Z. E., and Jagielska, B., *Acta Polon. Pharm.*, 21, 149, 1964, from *Chem. Abstr.*, 62, 10294g, 1965.
111. Sykut, W. B., *Acta Polon. Pharm.*, 16, 21, 1959, from *Chem. Abstr.*, 53, 12587a, 1959.
112. Gaal, F. F., Siriski, J. S., Jovanovic, M. S., and Branovacki, B. D. J., *Fresenius' Z. Anal. Chem.*, 260, 361, 1972.
113. Marsh, D. G., Jacobs, D. L., and Veening, H., *J. Chem. Educ.*, 50, 626, 1973.
114. Kalinowski, K., *Rocz. Chem.*, 30, 269, 1956, from *Chem. Abstr.*, 50, 12166h, 1956.
115. Edholm, L. E., *Talanta*, 23, 709, 1976.
116. Cordova-Orellana, R. and Lucena-Conde, F., *Talanta*, 24, 124, 1977.
117. Patriarche, G. J. and Lingane, J. J., *Anal. Chim. Acta*, 37, 455, 1967.
118. Kalinowska, Z. E., *Acta Polon. Pharm.*, 20, 69, 1964, from *Chem. Abstr.*, 61, 9355h, 1964.
119. Ebel, S. and Kalb, S., *Arch. Pharm. (Weinheim)*, 307, 2, 1974.
120. Patriarche, G. J. and Lingane, J. J., *Anal. Chim. Acta*, 49, 25, 1970.
121. Kalinowski, K. and Piotrowska, A., *Acta Polon. Pharm.*, 20, 199, 1963, from *Chem. Abstr.*, 62, 1515f, 1965.
122. Kalinowski, K. and Baran, H., *Acta Polon. Pharm.*, 16, 231, 1959, from *Chem. Abstr.*, 54, 1805e, 1960.
123. Mairesse-Ducarmois, C. A., Vanderbalck, J. L., and Patriarche, G. J., *J. Pharm. Belg.*, 28, 300, 1973, from *Chem. Abstr.*, 79, 70262s, 1973.
124. O'Reilly, J. E., *Anal. Chem.*, 47, 1077, 1975.
125. Kalinowski, K. and Piotrowska, A., *Acta Polon. Pharm.*, 16, 107, 1959, from *Chem. Abstr.*, 53, 18387e, 1959.
126. Kalinowski, K. and Piotrowska, A., *Acta Polon. Pharm.*, 15, 321, 1958, from *Chem. Abstr.*, 53, 7506f, 1959.
127. Kalinowski, K. and Sykulska, Z., *Acta Polon. Pharm.*, 15, 179, 1958, from *Chem. Abstr.*, 52, 17617d, 1958.
128. Ladenson, H. and Purdy, W. C., *Clin. Chem.*, 17, 908, 1971.
129. Carsen, W. N., *Anal. Chem.*, 22, 1565, 1950.
130. Jennings, V. J., Dodson, A., and Harrison, A., *Analyst*, 99, 145, 1974.
131. Kalinowski, K., *Acta Polon. Pharm.*, 11, 113, 1954, from *Chem. Abstr.*, 48, 14123b, 1954.
132. van Vlasselaer, S. and Crucke, F., *J. Pharm. Belg.*, 28, 36, 1973, from *Chem. Abstr.*, 79, 9962d, 1973.

133. Kalinowski, K., Bersztel, J., Fecko, J., and Zwierzchowski, Z., *Acta Polon. Pharm.*, 14, 77, 1957, from *Chem. Abstr.*, 52, 12322d, 1958.
134. Bostrom, C. A., Cedergren, A., Johansson, G., and Pettersson, I., *Talanta*, 21, 1123, 1974.
135. Mndzhoyan, A. L., Kropivnitskaya, R. A., and Sarkisyan, A. A., *Arm. Khim. Zh.*, 25, 991, 1972, from *Chem. Abstr.*, 80, 19641h, 1974.
136. Kalinowski, K. and Fecko, J., *Acta Polon. Pharm.*, 20, 53, 1963, from *Chem. Abstr.*, 61, 9358c, 1964.
137. Mndzhoyan, A. L., Kropivnitskaya, R. A., Ter-Zakharyan, Y. Z., Sarkisyan, A. A., and Lusararyan, K. S., *Khim. Farm. Zh.*, 6, 5, 1972, from *Chem. Abstr.*, 76, 158393j, 1972.
137a. Forsman, U., *Anal. Chim. Acta*, 93, 153, 1977.
138. Kharlamov, V. T., Inkin, A. A., and Ermolina, G. E., *Antibiotiki (Moscow)*, 20, 99, 1975.
139. Kalinowski, K. and Czlonkowski, F., *Acta Polon. Pharm.*, 24, 31, 1967, from *Chem. Abstr.*, 67, 67632j, 1967.
140. Kalinowski, K. and Czlonkowski, F., *Acta Polon. Pharm.*, 25, 29, 1968, from *Chem. Abstr.*, 68, 117162z, 1968.
141. Kalinowski, K. and Fecko, J., *Acta Polon. Pharm.*, 21, 247, 1964, from *Chem. Abstr.*, 62, 13849a, 1965.
142. Oliveri-Vigh, S., Donahue, J. J., and Heveran, J. E., *J. Pharm. Sci.*, 60, 1851, 1971.
143. Ejima, A., Tokusawa, J., and Ishibashi, M., *J. Pharm. Soc. Japan*, 87, 769, 1967.
144. Ishibashi, M., Tokusawa, J., Kojima, S., Ejima, A., and Inoue, T., *Yakugaku Zasshi*, 93, 711, 1973, from *Chem. Abstr.*, 79, 57718a, 1973.
145. Tokusawa, J. and Ishibashi, M., *J. Pharm. Soc. Japan*, 89, 450, 1969.
146. Kalinowska, Z. E., *Acta Polon. Pharm.*, 20, 193, 1963, from *Chem. Abstr.*, 62, 1516e, 1965.
147. Cospito, M., Zanello, P., and Raspi, G., *Fresenius' Z. Anal. Chem.*, 271, 200, 1974.
148. Patriarche, G. J., and Lingane, J. J., *Anal. Chim. Acta*, 49, 241, 1970.

Chapter 3

GAS-LIQUID CHROMATOGRAPHY

R. W. Souter

TABLE OF CONTENTS

I. INTRODUCTION

A. Gas Chromatography

Gas chromatography is a method used to separate components of mixtures of volatile compounds. Normally the separations are made to identify or quantitate the constituents in a sample mixture, but in some applications the separations are made for preparative purposes.

Chromatography was first applied by Ramsey[1] in 1905 to separate gaseous mixtures. During the following year Tswett[2] obtained colored bands of plant pigments on a chromatographic column, and he coined the term "chromatography" meaning "color writing", which is a misnomer when applied to current techniques. In 1952, James and Martin introduced gas-liquid chromatography[3,4] based on the suggestion of Martin and Synge.[5] As is obvious to all those familiar with its literature, the sensitivity, selectivity, speed, accuracy, and precision of this technique for the separation, identification, and determination of volatile compounds has resulted in almost explosive growth.

Gas chromatography separates volatile substances by percolating a gas stream over a stationary phase. If the stationary phase is a solid, the technique is known as gas-solid chromatography (GSC), and the separation is based on the adsorptive properties of the column packing. When the stationary phase is a liquid, the technique is gas-liquid chromatography (GLC). The basis for separation in GLC is the partitioning of the sample in and out of a thin film of liquid spread over an inert solid. GLC is probably the most selective and versatile form of gas chromatography since there exists a wide range of liquid phases usable up to about 450°C. This chapter will deal primarily with GLC.

B. Applications of Gas Chromatography to Pharmaceutical Analysis

GLC has found increasing application in the analysis of drugs and their metabolites. GLC methods often are more sensitive and specific than colorimetric, spectrofluorometric, and spectrometric assays, and thus may be used for assay purposes per se or to validate the simpler methods. The speed, resolution, and sensitivity of GLC separation and detection methods, especially with compounds amenable to some of the new, specific detectors, make these very attractive for drug analysis problems dealing with bioavailability, raw materials, and pharmaceutical formulations.

C. Basic Gas Chromatography Apparatus

The relatively simple instrumentation for gas chromatography is shown in the block diagram in Figure 1. Inert carrier gas from a compressed gas cylinder passes to a controller whose purpose is to maintain a constant gas flow through the column. The gas passes next to the head of the column, at the inlet to which is an injection port through which the sample is introduced. The carrier gas then moves the sample components through the column to the far end where the detector is located. The function of the latter is to detect or sense the components as they emerge, and this is done by use of some chemical or physical property of the vapors. From the detector the signal is sent to a recorder and/or to a computer for calculation of peak areas.

II. GAS CHROMATOGRAPHY THEORY

A. Introduction

The separation of two substances depends on the quality of the performance by the column (its efficiency) and on the relative retention or stationary phase efficiency (selectivity).

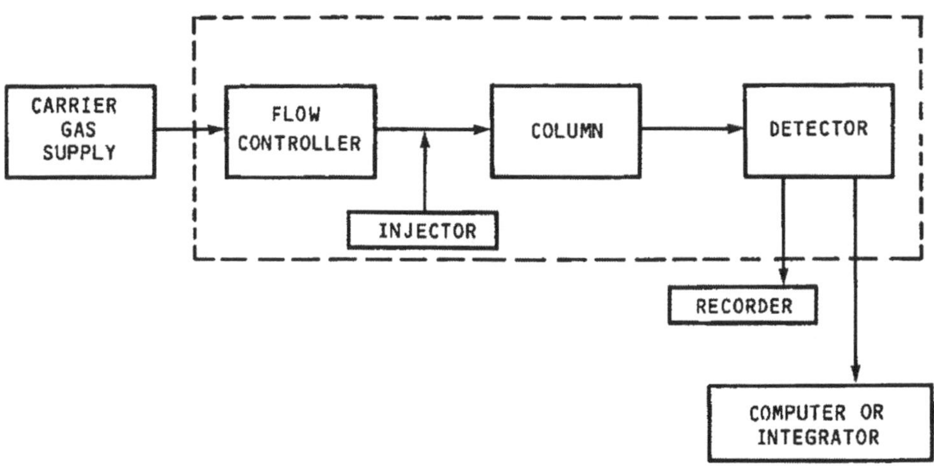

FIGURE 1. Block diagram of the apparatus for a basic gas chromatograph. The dashed line outlines the components under temperature control.

Column performance determines the width of peaks relative to the length of time they spend in the column. A zone of vapor is generally introduced into the column as quickly as possible so that it takes the form of a very narrow band. The band will broaden while being eluted through the column, and the degree of broadening may be defined and related to the factors upon which it depends. Historically, an artificial model of the chromatographic column, the "plate" model, was developed to treat the behavior of sample vapor molecules moving in and out of a stationary phase.[5,6] More useful yet is a consideration of what happens to the vapor molecules whose motion is controlled by diffusion and by column geometry.[7]

Stationary phase or solvent efficiency results from the interaction of the sample vapor (solute) and the stationary phase (solvent), and it determines the relative positions of the solute bands on the chromatogram. In a particular chromatographic run, column variables are the same for each vapor, so the separation is determined by the values of the partition coefficients; if they are sufficiently different, an adequate separation will result. The partition coefficient is the ratio of solute concentration in the liquid phase to solute concentration in the gas phase. Relative retention is the ratio of the partition coefficients for two substances, and may be predicted from a knowledge of the factors determining partition coefficients. This information may subsequently be applied to predict the best stationary phase for a separation.

B. Column Efficiency

The concept of a "theoretical plate" is carried over from distillation processes, and is used to evaluate column performance. Plates are useful for comparison of columns or to set standards for column-packing techniques. From an experimental chromatogram it is easy to calculate the number of theoretical plates.

In a GC experiment the carrier gas flows in a tube of uniform cross-sectional area, and causes the sample zone or zones to migrate. The flow rate f is the volume swept out by a cross section of carrier gas per unit time. The interval which elapses between injecting the sample and the elution of the center of mass of the zone or peak is called retention time, t_r. The product of flow rate and retention time is retention volume, V_r. With reference to Figure 2, the number of theoretical plates, N, is defined as

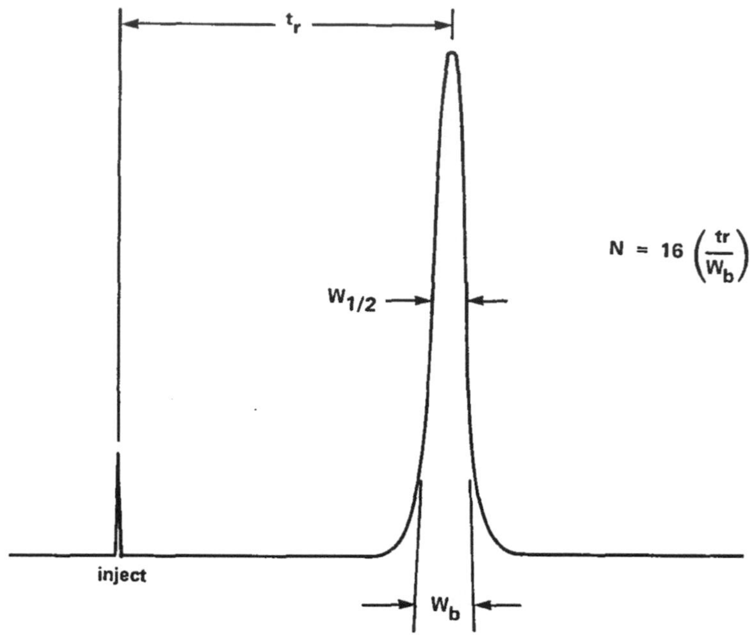

FIGURE 2. Calculation of the number of theoretical plates for a gas chromatographic column.

$$N = 16\left(\frac{t_r}{W_b}\right) \quad (1)$$

or as

$$N = 5.54\left(\frac{t_r}{W_{1/2}}\right)^2 \quad (2)$$

where W_b is the base width of the peak and $W_{1/2}$ is the full width of the peak at half-height. The higher the value of N for a column, the more efficient it will be. Columns with high efficiency allow smaller samples to be injected onto shorter columns at lower temperatures, and consequently give greater separations in less time with reduced risk of thermal decomposition.

To compare the performance of a column under different conditions or to compare different columns, the height equivalent to a theoretical plate (HETP) may be compared:

$$\text{HETP} = \frac{L}{N} \quad (3)$$

where L is the column length, usually in centimeters.

A major development in chromatographic column theory was the rate theory of van Deemter et al.[8] which was extended by Glueckauf.[9] In its simplest form under normal conditions, the van Deemter equation relates the physical factors determining the theoretical plate height.[10]

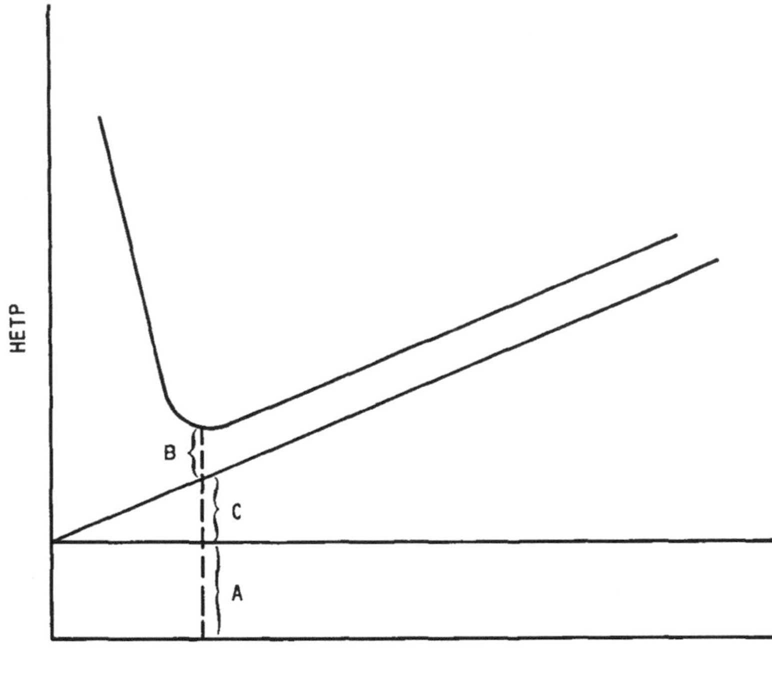

$$\text{HETP} = A + \frac{B}{\mu} + C_{\mu}$$

FIGURE 3. Plot of HETP vs. flow velocity.

$$\text{HETP} = 2\lambda d_p + \frac{2\gamma D_g}{\mu} + \frac{8k'd_f^2}{\pi^2(1+k')^2 D_l}\mu \tag{4}$$

where λ is a constant related to the geometry of the column packing particles, d_p is the average diameter of the solid support particles, γ is a factor to correct for the "tortuosity" of the column's gas channels, D_g and D_l are solute diffusion coefficients in the gas and liquid phases, respectively, d_f is the liquid film (stationary phase) thickness, k' is the partition coefficient of the solute, and μ is the linear gas velocity. For a particular column, all the parameters in Equation 4 are fixed except μ, and the equation may be rewritten in simpler form:

$$\text{HETP} = A + \frac{B}{\mu} + C\mu \tag{5}$$

where A is the eddy diffusion term, B is the molecular diffusion term, and C reflects resistance to mass transfer in the liquid phase. The linear gas velocity, μ, is calculated from

$$\mu = \frac{\text{length of column, cm}}{\text{retention time of air, sec}} \tag{6}$$

An isothermal plot of observed HETP vs. μ for a single solute yields a chracteristic hyperbola with a minimum HETP. The minimum is the flow rate at which the column is operating most efficiently. The plot is shown in Figure 3 and shows how the A, B, and C terms contribute to the HETP.

Table 1
EFFECTS OF PARAMETERS ON COLUMN PERFORMANCE, ACCORDING TO VAN DEEMTER RATE THEORY

Variable	Method to optimize column performance
Solid support	Use uniform small particle sizes: a narrow size range and as small a particle size as will allow an acceptable pressure drop. The most common compromise is 80/100 or 100/120 mesh.
Liquid phase	Thin films of liquid phase (1 to 5%) give fast analyses and lower operating temperature, but reduce the sample capacity. The liquid phase should normally be used with inactive solid supports. The liquid phase should be low viscosity and should have a low vapor pressure.
Carrier gas	For best efficiency, use a high molecular weight gas (will be affected by the detector choice), but for rapid analysis with reduced efficiency, use hydrogen or helium. For best efficiency, use the optimum flow rate from an experimental plot of HETP vs. flow rate. Efficiency may be sacrificed for time savings by using flows slightly higher than optimum without materially affecting HETP.
Temperature	Separations can often be slightly improved by lowering the temperature of the column, but analysis times will eventually become impractically long.

The eddy diffusion term describes peak broadening due to travel of solute along paths of different lengths. To minimize the A term and increase efficiency, small, uniform particles and small diameter columns should be used. The molecular diffusion term is proportional to the diffusivity in the carrier gas. If diffusivity increases, solute bands broaden. Diffusivity in the liquid phase is negligible. The "tortuosity" correction, γ, adjusts linear gas velocity to actual velocity in a packed column. The resistance to mass transfer term represents transfer of solute vapor along several paths, including from liquid to gas and from gas to liquid in the flow stream. Several conclusions which may be drawn from van Deemter's theory and which are of practical value in improving column performance are summarized in Table 1.

Column performance may also be described in terms of the factors that contribute to broadening a zone as it elutes. Broadening factors each introduce a component of variance σ_i^2 into the width of the peak eluted after instantaneous injection of the sample. The components of variance are combined by the usual rule. For addition of the effects of independent sources of random error on a distribution,

$$\sigma^2_{\text{total}} = \sum_i \sigma_i^2 \tag{7}$$

where σ is the standard deviation of a Gaussian peak. The observed variance of the chromatographic peak — σ^2_{total} — may be envisioned as being composed of variances due to the column partition process and to the band spreading outside the column. The former was discussed already in terms of column efficiency while the latter are summarized in Table 2.

Column efficiency and variance has been treated in detail by Grubner[11] and by Littlewood.[7] In addition, the latter author presents an excellent discussion of the relationships of plate height (HETP) and the broadening of a peak expressed as a variance.[12] HETP is proportional to peak variance; therefore, each broadening factor may be regarded as adding a term to HETP,[7] thus decreasing the efficiency of the chromatographic system.

C. Stationary Phase Efficiency

The stationary phase (solvent) efficiency results from the solute-solvent interaction,

Table 2
VARIANCE CONTRIBUTIONS DUE TO BAND SPREADING OUTSIDE THE CHROMATOGRAPHIC COLUMN

Excessive bandwith contribution	Means to reduce variance
Sample injection volume is too great	Inject a smaller volume
Sample injection too slow	Improve injection technique
Too much "dead" volume between point of injection and column	Reduce space by adding column packing or modifying injection system
Detector volume	Keep volume of detector small to minimize diffusion
Detector-column connecting line	Minimize length or eliminate completely so gas flow is not slowed down; keep volume of line small to minimize diffusion

and it determines the relative position of solute bands on the chromatogram. A striking advantage of gas chromatography is that appropriate selection of the liquid phase will separate substances which have the same vapor pressure. The combined effects of several types of interaction forces determine the solubilities of the solutes in the stationary phase (hydrogen bonding, complex formation, etc.) and consequently the observed separation. The combined effects of these forces are expressed by the partition coefficient, k, where

$$k = \frac{\text{amount of solute per unit volume of liquid phase}}{\text{amount of solute per unit volume of gas phase}} \qquad (8)$$

The greater the difference in partition coefficients, the larger the separation of two compounds and the shorter the column length needed to effect the separation.

The efficiency of a stationary phase for a particular separation is measured by α, the relative retention, which is the ratio of the partition coefficients for the substances being separated:

$$\alpha = \frac{k_2}{k_1} \qquad (9)$$

Calculation of the solvent efficiency is shown in Figure 4. Note that adjusted retention times are used to calculate relative retention.

Although k and therefore α are temperature dependent, α is essentially constant over a limited temperature range. With an increase in temperature, the solute spends less time in the stationary phase and more time in the carrier gas, so elution time and separation will both decrease.

To properly evaluate the separation of two consecutive peaks, one must consider resolution, R, which is demonstrated in Figure 5. It is a measure of both column and solvent efficiencies and relates peak width and maxima separation. R must be 1.5 or greater to achieve complete (baseline) separation of two peaks.

For two peaks which are spaced closely enough that they have nearly equal widths, Purnell[13] proposed an equation to relate the relative retention and the resolution to the number of theoretical plates for a separation:

$$N_{req} = 16\left(\frac{\alpha}{\alpha - 1}\right)\left(\frac{k_2 + 1}{k_2}\right)^2 R \qquad (10)$$

where k_2 is the partition coefficient of the longer retained component. This equation is used to determine the number of plates required for a specified separation. Figure 6

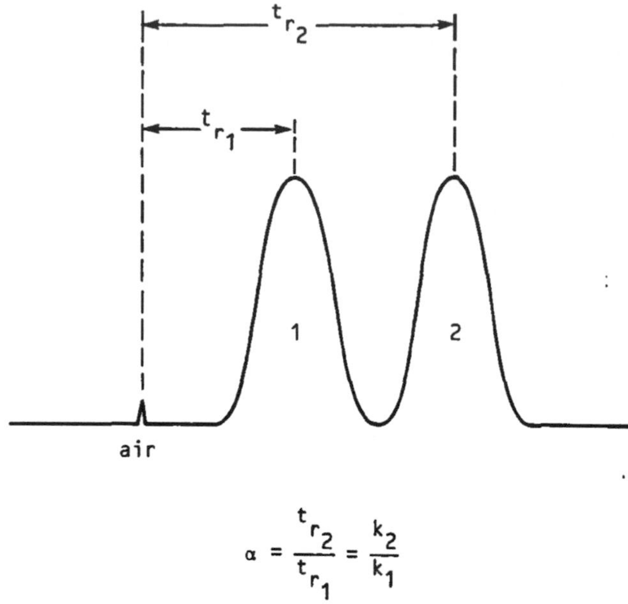

FIGURE 4. Calculation of the stationary phase efficiency.

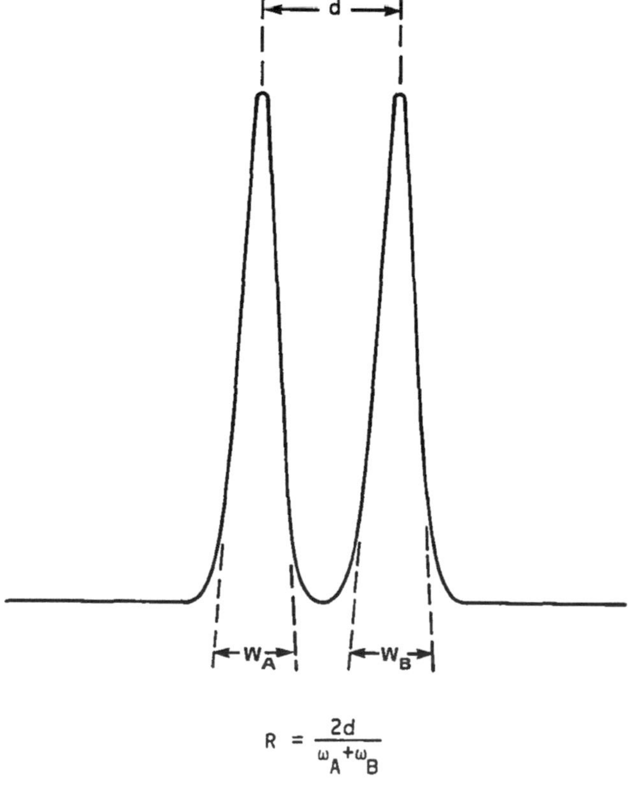

FIGURE 5. Calculation of the resolution, R, of two consecutive peaks.

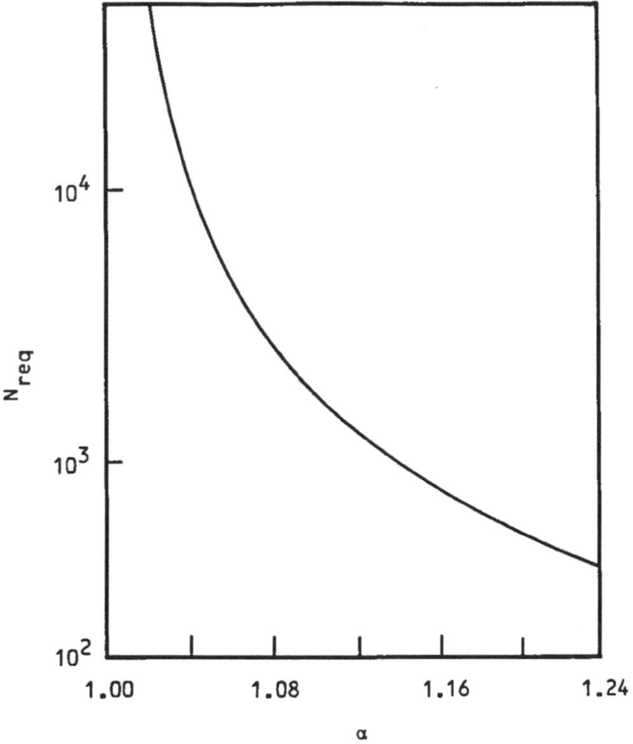

FIGURE 6. Plot of log N vs. α to illustrate the importance of relative retention on the number of theoretical plates required for a separation (k_2 large). (From Karger, B. L., *Anal. Chem.*, 39, 24A, 1967. With permission.)

illustrates that N_{req} is very dependent on relative retention, but at large α values, N_{req} decreases at a slower rate.[14] For hard-to-resolve components (low α values), large Ns are needed, and for easier separations low N values are sufficient.

D. Gas Chromatography Theory and Practice

It should now be apparent that gas chromatographic separation of materials depends on a variety of related factors. GC can be used for qualitative identification of sample components or for quantitative analysis with a precision approaching ± 0.2%.[15] The technique is widely applicable because of the wide range of operating conditions, but it is this very flexibility that creates difficulty in determining the optimum conditions needed for a specific separation to be done in a minimum time. It is hoped that this chapter, especially the practical sections to follow, will be useful to analysts wishing to apply GC to specific pharmaceutical analysis problems.

III. THE GAS CHROMATOGRAPHIC SYSTEM

A. Introduction

The basic apparatus for gas chromatography was illustrated in Figure 1. In addition to the components shown, a host of related features are available for wider applicability and easier operation. This section describes the components, the considerations in selecting features for specific applications, and the relative advantages and disadvantages of components.

Table 3
PROPERTIES OF SOME CARRIER GASES USED FOR GAS CHROMATOGRAPHY

Gas	MW	C_p^a	Viscosity[b] (cP × 10²)	Thermal conductivity[c] 0°C	100°C
Air	28.98	7.0	1.71	5.8	7.5
Hydrogen	2.016	6.8	0.84	41.6	53.4
Helium	4.003	5.1	1.87	34.8	41.6
Nitrogen	28.02	6.9	1.66	5.8	7.5
Oxygen	32.00	7.0	2.04(23°C)	5.9	7.6
Argon	39.95	5.0	2.10	4.0	5.2
Carbon dioxide	44.01	8.8	1.37	3.5	5.3
Carbon monoxide	28.01	6.9	1.66	5.6	7.2
Methane	16.04	8.5	1.02	7.2	10.9
Ethane	30.07	11.6	0.85	4.3	7.3
Propane	44.10			3.6	6.3

[a] Heat capacity in cal/deg-mol at 15°C, 1 at.
[b] At 0°C, 1 at in centipoise (cP).
[c] In 10^{-5} cal/sec-cm²/(°C/cm).

From *A User's Guide to Chromatography*, Regis Chemical Company, Morton Grove, Ill., 1976, 22. With permission.

B. Carrier Gas Supply

Generally the source of carrier gas for laboratory gas chromatographs is a high-pressure cylinder. In principle, any gas might be used as the GC mobile phase. The principal limiting factor, however, is the requirement that the gas be inert to the sample and to the stationary phase at the operating temperature. In practice, the choice of carrier will determine the efficiency of the GC system since, as was shown in Equation 4, the HETP depends on solute diffusivity in the carrier. Normally a compromise among inertness, efficiency, and operating cost make nitrogen or helium the most common GC carrier gases. Properties of several possible carrier gases are shown in Table 3.[16]

The carrier gas affects the separation and the analysis time, and it provides a suitable matrix for the detector to measure each sample component. Since the optimum flow velocity in a column is proportional to the diffusivity of the solute in the carrier gas[17] (van Deemter equation), it will be higher for a low-density gas such as hydrogen or helium than for carbon dioxide or nitrogen. This means that nitrogen will yield the highest efficiency (lowest HETP) at a lower flow rate than helium. If both gases were used at their minimum HETPs, nitrogen would yield 15% more plates, but the analysis time would be roughly twice as long. With helium, small losses in column efficiency can be offset by large gains in analysis speed. Despite its higher efficiency, nitrogen is a much poorer choice than helium for carrier when using a thermal conductivity detector. Maximum sensitivity requires the use of helium or hydrogen, and these will also ensure that all the peaks move in the same direction from the baseline. Note that hydrogen cannot be used with thermistor bead detectors, however, due to reaction with the rare earth oxides. There are significant advantages in using nitrogen instead of helium as the carrier with a flame detector.[18] Typical flow rates are not discussed here since they vary from preparative to analytical to capillary columns, and are usually determined either experimentally or from manufacturer's literature.

Table 4
COMMON CARRIER GASES IN GAS CHROMATOGRAPHY

Carrier	Method of manufacture	Available purity	Detector	Impurities
He	Separation from natural gases	99.995—99.9999%	TC (HeD)	0.1 ppm (Ar, N_2, H_2)-5 ppm (Ne)
N_2	Air separation	99.9995%	FID, ECD (TC)	0.5 ppm (O_2, CO_2, THC)-1 to 5 ppm (H_2, Ar, Ne, He)
Ar	Air liquefaction	99.9997—99.9999%	ArID (TC)	0.1 ppm (CO_2, THC)-3 ppm (N_2, H_2)
H_2	Electrolysis	99.99—99.9999%	TC	0.1 ppm (CO_2, THC)-1 to 5 ppm (N_2, O_2)

From Perretta, A. T., *Am. Lab.*, 5, 35, 1976. With permission.

Use of high-purity carrier gas is extremely important for good GC results.[19] Traces of hydrocarbons can lower the sensitivity of a flame detector, water traces can desorb contaminants leading to high detector background or "ghost peaks", and oxygen and water traces can chemically change the liquid phase and thus change the retention times. Oxygen can be removed with a special filter[20] and water with a 5Å molecular sieve filter[20] which can be replaced or regenerated (heat the sieves for 3 hr at 300°C). Carrier gas purity in gas chromatography has been extensively discussed by Perretta,[16] and some common carrier gases with suitable detectors, sources, and typical impurities are listed in Table 4.

C. Flow Indicators and Controllers

The carrier gas cylinder is usually equipped with a pressure regulator to reduce the gas pressure to a workable level. Two-stage regulators reduce the cylinder pressure to an intermediate level, which is fed to the inlet of the second stage. Because the inlet pressure to the second stage is regulated, its delivery pressure is unaffected by changes in the cylinder pressure. However, regulators control only pressure and do not control flow rate.

Rotometers are available for most commercial gas chromatographs, and are used for measuring mass gas flows. A rotometer is, however, only a convenient means of reestablishing a given flow rate and does not control flow rate.

For isothermal operation, constant pressure delivery (i.e., from a regulator) will suffice to establish a constant flow provided that the column has a constant pressure drop. In temperature programming, even with constant inlet pressure, flow rate will decrease as the column temperature increases. This is due primarily to increasing viscosity of the carrier gas with increasing temperature and to thermal expansion of the liquid phase.[19] For this reason, gas chromatographs equipped for temperature programming should have a differential flow controller to assure a constant mass flow rate independent of the resistance of the column.

A typical flow controller is set with its needle valve for a given flow rate. Its spring-loaded diaphragm is positioned by the inlet pressure of the incoming gas on one side, and is held in position by a balance with a combination of a spring and the outlet pressure. With a pressure disruption beyond the controller outlet, the diaphragm repositions itself causing the flow controller valve to open or close, ensuring a constant mass flow delivery. Normally 10 to 20 psig difference (minimum) is needed between the controller inlet and outlet for proper sensing of pressure fluctuations.

The operating principle of a Brooks® flowmeter is illustrated in Figure 7. A precision power supply produces a constant amount of heat at the midpoint of a sensor

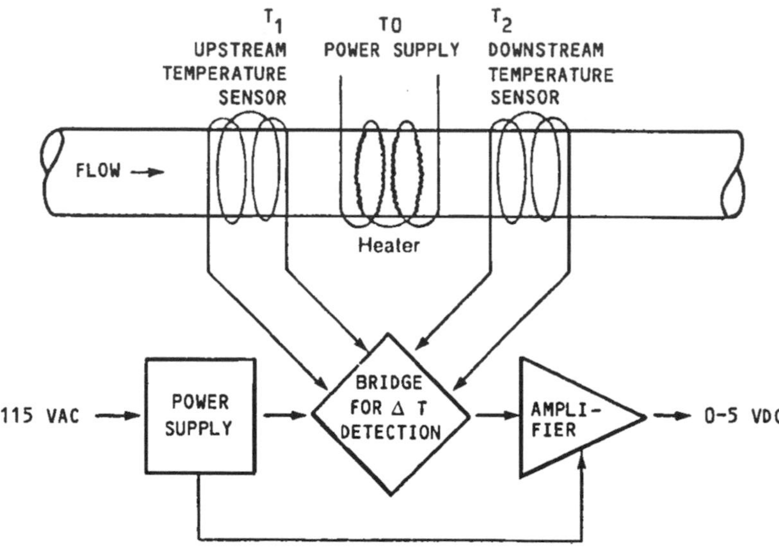

FIGURE 7. Operating principle of a Brooks® mass flow meter. (From Karasek, F. W., *Res. Dev.*, 41, July 1976. With permission.)

tube carrying the gas flow. Equidistant from the heat input are resistance temperature-measuring elements which are positioned upstream and downstream. With no flow, heat reaching each measuring element is equal. As the flow increases, an increasing amount of heat is carried toward downstream element T_2, and an increasing temperature difference develops between the two elements. The difference is proportional to the amount of gas flowing or to the mass flow rate.

A dual mass flow controller is also available from Brooks[21] which combines two flow sensors with control valves and electronics for use especially in gas chromatography. The flow rate may be set with a calibrated potentiometer and is indicated by a digital display.

D. Sample Introduction

A well-designed sample inlet system will receive the sample, vaporize it instantly, and deliver the vaporized material to the head of the column in a sharp, concentrated band with a minimum of "fronting" or "tailing". Since peak widths are directly related to injection band widths, an inferior injection system design might destroy resolution of close-boiling materials.

Typically, liquids or solutions are injected with a microsyringe through a self-sealing septum into a narrow metal or glass injection port liner maintained about 50°C above column temperature or directly onto the head of the column. In some instruments, the first few inches of the glass column may serve as the injection port liner. Stainless steel or other metal vaporizers may be used only with thermally stable samples that will not react with the hot metal surfaces.

A very wide variety (cost and performancewise) of GC septa is available, but unfortunately their colors or other obvious characteristics are unreliable guides to their properties. They may have widely different lifetimes depending on temperature of operation, number of injections, and type of syringe needle. The septum type used in an assay should be given careful consideration since some may bleed and give rise to "ghost peaks", particularly if temperature programming and high injection port tem-

Table 5
PROPERTIES OF SOME GAS CHROMATOGRAPHY SEPTA

Source	Construction	Operating characteristics
Alltech	Metasep™: heat-treated silicone rubber and metal foil barrier combined	Creates positive barrier between column and septum-minimal bleed
	Blue septa	For nonspecialized applications; heat-treated, high injection life
	Microsep F-138®	Special silicone rubber with Teflon® backing; have had high temperature (200°C) vacuum treatment to minimize bleed
	Hamilton "sandwich"	Three layers of silicone rubber-soft inside layer quickly reshapes to minimize "blow back" while hard outer layers protect and hold shape
Regis[b]	Puresep® polyimide faced silicone septa	Stable to 400°C and no halogenated bleed to interfere with electron capture detection
Applied Science[c]	W-type	Long life at 300°C; satisfactory for all isothermal GC work, but not for high-sensitivity GC or for GC/MS
	HT-type	Long life and low bleed in temperature programming at high sensitivity and for GC/MS
Supelco[d]	Gray	Inlet ≤225°C; isothermal column to 225°C
	Green	Inlet ≤300°C; isothermal or programmed up to 300°C
	White teflon-coated	Inlet ≤350°C; isothermal or programmed up to 400°C

[a] Alltech Assoc., Inc., 202 Campus Drive, Arlington Heights, Ill. 60004.
[b] Regis Chemical Co., 8210 Austin Ave., Morton Grove, Ill. 60053.
[c] Applied Science Laboratories, Inc., P. O. Box 440, State College, Pa. 16801.
[d] Supelco, Inc., Supelco Park, Bellefonte, Pa. 16823.

peratures are employed in the analysis. Such bleed can easily render submicrogram analysis impractical. Table 5 summarizes the characteristics of a number of commercially available septa for GC.

For introducing gases and liquids, syringes are most often used. Gas-tight syringes are commonly available at relatively low cost and are quite flexible. Sampling valves are more expensive, but have the advantages of better reproducibility, less skill required for operation, and ease in automation. For liquid samples, the desired sizes usually range from 1 to 25 $\mu\ell$, since liquids expand considerably when vaporized. A wide variety of syringes and sampling loops is available for analytical and preparative purposes and the appropriate manufacturer's literature should be consulted for these.

To introduce very small sample volumes for open tubular columns, an inlet splitter device is used. Several designs are available, and one is shown in Figure 8. The sample-gas stream is split by passing the vapor past a needlelike restrictor that connects to the column. Early splitter designs often suffered poor quantitative precision due to sample factionation associated with poor evaporation rates and turbulence problems.[21]

Solid sample injection presents the greatest introduction problem in GC, but has recently been conveniently handled.[22] Perkin-Elmer Corp. markets a manual or automatic solid sampling system in which a sealed aluminum capsule containing the sample is inserted into a pressure-tight heated injector where it is pierced and the sample escapes. Such a system is useful for solids and viscous liquids, and the small metal capsule may also serve as a minireactor for derivatization.

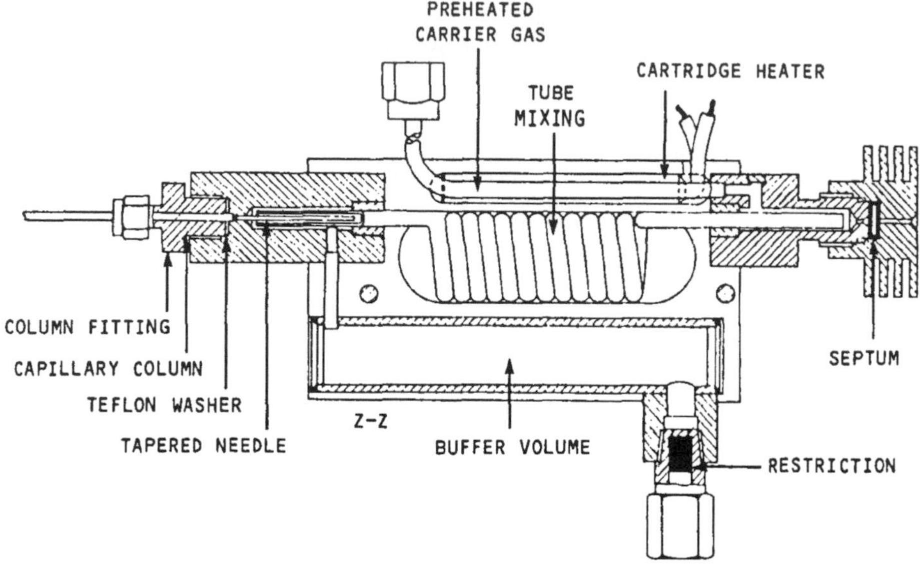

FIGURE 8. A typical inlet splitter designed for gas chromatography. (From Catalog No. H-75, Hamilton Company, Reno, Nevada, 26. With permission.)

E. The Column

1. General Considerations

The stationary phase, solid support, column tubing type, column shape and inside diameter, percent liquid loading, temperature, and other factors all determine the suitability of a column for a particular use. Columns are generally prepared in three size ranges depending on the objective: (1) preparative (packed) to effect the separation and collection of quantities of individual components, (2) analytical (packed) for qualitative and quantitative assays, and (3) capillary (not packed) with very narrow inside diameters for high resolution applications such as separations of isomers or close-boiling components. The considerations needed for choosing a useful column in any of the three categories will be evaluated here.

Column tubing material may be any of a number of substances, but glass and stainless steel are by far the most used. Table 6 lists several common tubing materials and some characteristics useful in choosing among them. Major requirements for column tubing are that it must be physically strong, should not expand or contract significantly with applied temperature or pressure changes, should be able to be coiled reasonably easily, and should be inert to avoid peak "tailing".

The large majority of gas chromatography columns are "packed columns", meaning the tubing is packed with the material used to effect the separation. The packing may consist of porous polymers, of liquid phases coated on inert solid supports, or of materials such as charcoal, molecular sieves, or silica gel. The latter materials and the porous polymeric packings are used for gas-solid chromatography (GSC) where samples such as solvents, permanent gases, and low molecular weight compounds are separated on an active solid rather than on a liquid phase. These packing materials for GSC will be discussed first, followed by those for GLC.

Table 7 lists some miscellaneous GSC packings with their characteristics and typical applications. There are fundamentally three types of adsorbents:[23] nonpolar, acidic (electron-accepting surface groups), and basic (electron-donating surface groups), the second and third being classified as polar adsorbents. GSC possesses a number of

Table 6
A GUIDE TO COLUMN TUBING MATERIAL

Material	Comments
Glass	Most inert and must be used in the analysis of drugs, steroids, and pesticides. To avoid breakage, use one fourth in O.D. columns, not one eighth in O.D. To render the column walls more inert, silanize with 5% dimethyldichlorosilane in toluene (and rinse with methanol) before use.
Stainless steel	Reasonably inert and most widely used. Not suitable in the trace analysis of polar compounds, drugs, biomedical compounds, water, etc. Polar compounds frequently decompose on contact with hot metal surfaces, particularly those in the injection port.
Teflon®	Used in the analysis of trace sulfur dioxide and water, and in the analysis of corrosive and reactive inorganic halides: COS, HF, H_2S, F_2 etc. However, air can diffuse through the walls of the Teflon® column and adversely effect the response of an electron capture detector, or shorten the life of thermal conductivity detector filaments.
Aluminum	The inner column walls are coated with aluminum oxide which can be reactive towards certain compounds. Aluminum columns have been found successful for the analysis of certain chlorinated pesticides. However, brass ferrules do not seal well on aluminum after several high temperature programmed runs, and continued tightening fractures tne column.
Copper	The inner column walls contain copper oxide which can be reactive. Therefore, oxygen must be excluded. Do NOT use in the analysis of unsaturated compounds. Good for trace water analysis.

From Catalog No. 18, Analabs, North Haven, Conn., 1978, 47. With permission.

Table 7
CHARACTERISTICS OF MISCELLANEOUS PACKAGINGS USED FOR GSC

GSC column packing	Characteristics
Activated charcoal	Large surface area; heterogeneous adsorption site distribution; nonpolar, for separations by molecular weight
Graphitized thermal carbon black	Homogeneous; low surface area; nonpolar support; separations are fairly dependent on molecular size
Silica	Separations depend highly on the method used to prepare the silica; used for separations of gases
Porous glass	Porous silica beads with adsorption properties of siliceous materials; exists in a broad range of specific surface areas
Zeolites (molecular sieves)	Calcium aluminum silicates with fairly constant interatomic distances between layers; uniform shape and size of pores yields sharp separations according to size; these have a strong affinity for water; used for separations of higher molecular weight *n*-alkanes from branched and cyclic derivatives and for separations of gases
Alumina	Utility depends on prior treatment; used for separations of hydrocarbon gases and the oxides of carbon and nitrogen
Carbon molecular sieves	Very inert; useful for part-per-million and part-per-billion assays of permanent gases, light hydrocarbons, and hydrocarbons in water

advantages over GLC:[24] (1) useful for low-boiling samples, (2) no bleed since there is no stationary liquid phase, (3) good selectivity, and (4) good efficiency at reasonable flow rates.

Porous polymer packings are produced by Johns-Manville Corp. (Chromosorb® Century Series — 101 through 108) and by Waters Associates (Porapak® Series) and provide some unique separations. They can be used directly without further coating with a stationary phase, and require very little conditioning. These materials are copolymers of styrene and divinyl benzene with chemically incorporated vinyl monomers, such as acrylonitrile or vinyl pyridine. The Chromosorbs® and Porapaks® differ in pore size and surface area, and have unique selectivities by functional group, size, and shape of the molecules being separated. Table 8 lists physical properties and some applications for Chromosorb® packings, and Table 9 presents physical properties and general applications for Porapak® polymers. The reader is referred to the manufacturer's literature for further details, including retentions of various compounds.[25,26]

A very useful modification of the most widely used Porapak® Q is silanization. Porapak QS® has had its active sites rendered inactive by dimethyldichlorosilane treatment.[27] Tailing is eliminated, but there is virtually no effect on retention characteristics. Consequently, aliphatic acids, amines, and other compounds are eluted as much sharper peaks without ghosting problems.[28]

Tenax-GC® is one of the newest polymer column packing materials. It is made from 2,6-diphenyl-phenylene oxide and may be operated up to 400°C.[29] Its applications include separation of nitroaniline positional isomers, aromatic diamines, ketones, and alkyl halides.[30] The physical properties, conditioning procedure, and separation characteristics are available.[30,31]

Due to the large number of available solid supports and the much larger variety of liquid stationary phases, the selection of a column for GLC is much more complicated than for GSC.

The purpose of a solid support is to hold a thin film of liquid phase. The optimum support should (1) have a large specific surface area, (2) not be extremely friable, (3) have a uniform small pore diameter, (4) be inert, and (5) have regularly shaped, uniform-sized particles. A number of materials are available which meet most of these requirements. Table 10 lists typical physical properties for Chromosorb® diatomite supports.

It is known that solid support surfaces are covered with active sites which may lead to tailing.[32] Therefore, a number of deactivating treatments may be applied and are described by Ottenstein[33] in an excellent article on the role of the support in GC. Acid washing and dimethylchlorosilane treatment of Chromosorbs® gives the most inert surface. A number of suppliers offer diatomaceous earth supports for GLC, and a conversion chart based on treatment is available.[34]

Column efficiency improves with the use of small and narrow-range mesh solid support particles. However, for roughly 1/8 in. I.D. columns, 100/120 mesh is the practical lower limit due to pressure drop problems. The amount of liquid phase should be enough to coat the particles with a thin, uniform layer. Too low a load can leave active sites exposed, creating reproducibility problems. Too much liquid phase collects in pools between particles and decreases column efficiency. The current trend is toward lightly loaded columns for rapid analyses. Equivalent stationary phase loadings are about 5, 9, and 4% by weight on Chromosorbs® P, W and G, respectively.

A chromatographic system must show some selectivity toward the compounds of interest, and this is obtained by proper choice of the stationary phase. This selectivity depends on differences in solubility of the sample components in the liquid phase. GLC stationary phases have been traditionally arranged according to "polarity", a vague and sometimes misleading term. "Polarity" really depends on both the sample type and the stationary phase, and can be expressed by a set of parameters describing interactions between them. These parameters, the McReynold's constants,[35] are directly related to the retention times for selected test compounds, and can be used to arrange and compare the separating abilities of GC phases for various compounds. The larger the value of the McReynold's constant, the greater the retention time for that particular compound. McReynold's constants are presented in the form of tabu-

Table 8
PHYSICAL PROPERTIES OF CHROMOSORB® POLYMERS

Chromosorb®	Type of polymer	Surface area (m^2/g)	Free fall density (g/cc)	Isothermal temperature Limit (°C)	Programmed temperature Limit (°C)	Average pore diameter (μ)	Separation applications
101	Styrene-divinylbenzene	<50	0.30	275	325	0.3—0.4	Acids, glycols, alkanols, ketones, aldehydes, other oxygenated compounds
102	Styrene-divinylbenzene	300—400	0.29	250	300	0.0085	Light permanent gases, low molecular weight compounds; some oxygenated compounds
103	Cross-linked polystyrene	15—25	0.32	275	300	0.3—0.4	Amines, amides, alcohols, hydrazines
104	Acrylonitrile-divinylbenzene	100—200	0.32	250	275	0.06—0.08	Nitriles, nitroparaffins, nitrogen oxides, xylenols
105	Polyaromatic	600—700	0.34	250	275	0.04—0.06	Gases, low-boiling organics, acetylene and formaldehyde from lower hydrocarbons
106	Cross-linked polystyrene	700—800	—	250	—	—	C_2 to C_5 acids from C_2-C_5 alcohols
107	Cross-linked acrylic ester	400—500	—	250	—	—	Formalin from other organics, miscellaneous compounds
108	Cross-linked acrylic ester	100—200	—	250	—	—	Gases, oxygenated compounds

From product literature, Johns-Manville, Celite Division, Denver, Colo. With permission.

Table 9
USES OF PORAPAK® POLYMERS

Porapak®	Maximum operating temperature (°C)	Applications
P	250	Least polar Porapak®; separation of alcohols, glycols, and carbonyl compounds
P-S	250	Surface-silanized Porapak® P (eliminates tailing and ghosting); aldehydes, glycols
Q	250	Nitrogen oxides, organic compounds in water, hydrocarbons, esters, sulfur compounds, ketones
Q-S	250	Surface-silanized Porapak® Q—organic acids and other polar compounds separate without tailing
R	250	Resolution of esters
S	250	Normal and branched-chain alcohols
N	190	Will separate water, ammonia, carbon dioxide, acetylene, and air from C_2 hydrocarbons
T	190	Greatest retention for water of all Porapaks®

From *Porapak®-A Selected Bibliography of Applications,* Waters Associates, Milford, Mass., 1970. With permission.

Table 10
PHYSICAL PROPERTIES OF CHROMOSORB® DIATOMITE SUPPORTS

Properties	Chromosorb A	G	P	W
Color	Pink	Oyster white	Pink	White
Type	Flux-calcined	Flux-calcined	Calcined	Flux-calcined
Density, gm/cc				
Loose weight	0.40	0.47	0.38	0.18
Packed	0.48	0.58	0.47	0.24
Surface area, m^2/gm	2.7	0.5	4.0	1.0
Surface area, m^2/cc	1.3	0.29	1.88	0.29
Maximum liquid phase loading	25%	5%	30%	15%
pH	7.1	8.5	6.5	8.5
Handling characteristics	Good	Good	Good	Slightly friable

From product literature, Johns-Manville, Celite Division, Denver, Colo., 1969. With permission.

lations of Δ I values for seven test compounds on a large number of popular stationary phases in Table 11.

The manner in which McReynold's constants are determined has been discussed.[35,36] The ways in which they are used makes them quite valuable for choosing a stationary phase for a particular application. The phases are arranged in order of their increasing polarity, as measured by the sum of the Δ I values. To selectively retain compounds with a particular functional group, one must choose a stationary phase with a high Δ I value for the test compound representing that functional group. For instance, to elute an alcohol after a ketone, one would need a phase with a high $\Delta I_{1\text{-butanol}}$ with respect to the $\Delta I_{2\text{-pentanone}}$. Besides predicting selectivities, the McReynold's constants can be used to choose substitute phases. Differences of 50 to 100 in McReynold's constants

Table 11
SELECTED McREYNOLD'S CONSTANTS

Phase name	Benzene 1	1-Butanol 2	2-Pentanone 3	Nitropropane 4	Pyridine 5	2-Methyl-2-pentanol 6	2-Octyne 7	$\sum_1^5$
	ΔI Values[a]							
Squalane	0	0	0	0	0	0	0	0
Nujol®	9	5	2	6	11	2	2	33
Apiezon M®	31	22	15	30	40	12	10	138
Apiezon L®	32	22	15	32	42	13	11	143
SF 96®	12	53	42	61	37	31	21	205
Apiezon J®	38	36	27	49	57	23	15	207
Apiezon N®	38	40	28	52	58	25	15	216
SE 30®	15	53	44	64	41	31	22	217
OV-1®	16	55	44	65	42	32	23	222
M and B silicone oil	14	57	46	67	43	33	22	227
DC 200 (12,500 centistokes)®	16	57	45	66	43	33	23	227
AST 100 methyl®	17	57	45	67	43			229
OV-101®	17	57	45	67	43	33	23	229
DC-410®	18	57	47	68	44	34	24	234
Versilube F-50®	19	57	48	69	47	36	23	240
DC 11®	17	86	48	69	56	36	23	276
SE 52®	32	72	65	98	67	44	36	334
SE 54®	33	72	66	99	67	46	36	337
OV-3®	44	86	81	124	88	55	35	423
Dexsil 300®	47	80	103	148	96	55	46	474
Fluorolube HG 1200®	51	68	114	144	118	68	53	495
Kel F wax	55	67	114	143	116	73	57	495
Apiezon H®	59	86	81	151	129	46	23	506
OV-7®	69	113	111	171	128	77	66	592
DC 550®	74	116	117	178	135	81	72	620
Di (2-ethylhexyl)sebacate	72	168	108	180	125	132	49	653
Diisodecyl adipate	71	171	113	185	128	134	52	668

Table 11 (continued)
SELECTED McREYNOLD'S CONSTANTS

Phase name	Benzene 1	1-Butanol 2	2-Pentanone 3	Nitropropane 4	Pyridine 5	2-Methyl-2-pentanol 6	2-Octyne 7	$\sum_{1}^{5}$
	ΔI Values[a]							
Octyl decyl adipate	79	179	119	193	134	141	57	704
Dilauryl phthalate	79	158	120	192	158	120	52	707
Bis(2-Ethylhexyl) tetrachlorophthalate	112	150	123	168	181	110		734
Diisodecyl phthalate	84	173	137	218	155	133	59	767
Dinonyl phthalate	83	183	147	231	159	141	65	803
DC 710®	107	149	153	228	190	107	98	827
Dioctyl phthalate	92	186	150	236	167	143	66	831
POLY-1 110®	115	194	122	204	202	152	55	837
Hallcomid M-18®	79	268	130	222	146	202	48	845
ASI 50-methyl-50-phenyl®	119	158	162	243	202			884
OV-17®	119	158	162	243	202	112	105	884
UCON LB-550-X®	118	271	158	243	206	177	91	996
Span 80®	97	266	170	216	268	207	66	1017
Castorwax	108	265	175	229	246	202	73	1023
POLY-A 103®	115	331	149	263	214	221	62	1072
OV-22®	160	188	191	283	253	133	132	1075
Polypropylene glycol	128	294	173	264	226	196	98	1085
Trimer acid	94	271	163	182	378	234	57	1088
POLY-A 101A®	115	357	151	262	214	233	64	1099
UCON LB-1715®	132	297	180	275	235	201	100	1119
Acetyl tributyl citrate	135	268	202	314	233	214	102	1152
Didecyl phthalate	136	255	213	320	235	201	101	1159
OV-25®	178	204	208	305	280	144	147	1175
Polyphenyl ether OS-124 (5 rings)	176	227	224	306	283	177	135	1216
Tributyl citrate	135	286	213	324	262	226	102	1220

Polyphenyl ether OS-138 (6 rings)	182	233	228	313	293	181	136	1249
POLY-A 135®	163	389	168	340	269	282	—	1329
Neopentyl glycol sebacate (HI-EFF-3CP®)	172	327	225	344	326	257	109	1394
Squalene	152	341	238	329	344	248	101	1404
UCON 50-HB-280X®	177	362	227	351	302	252	130	1419
Tricresyl phosphate	176	321	250	374	299	242	131	1420
Sucrose acetate isobutyrate	172	330	251	378	295	264	128	1426
QF-1®	144	233	355	463	305	203	53	1500
ASI 50-phenyl-50-cyanopropyl®	319	495	446	637	531			2428
OV-210®	146	238	358	468	310	206	56	1520
Ethofat 60/25®	191	382	244	380	333	277	131	1530
Igepal CO-630®	192	381	253	382	344	277	136	1552
DC LSX-3-0295®	152	241	366	479	319	208	55	1557
UCON 50-HB-2000®	202	394	253	392	341	277	147	1582
Emulphor ON-870®	202	395	251	395	344	282	140	1587
Triton X-100®	203	399	268	402	362	290	145	1634
UCON 50-HB-5100®	214	418	278	421	375	301	155	1706
Siponate DS-10®	99	569	320	344	388	466	61	1720
Tween 80®	277	430	283	438	396	310		1747
XE 60®	204	381	340	493	367	289	120	1785
ASI 50-methyl-25-phenyl-25-cyanopropyl	228	369	338	492	386			1813
OV-225®	228	369	338	492	386	282	150	1813
Neopentyl glycol adipate (HI-EFF-3AP®)	232	421	311	461	424	335	156	1849
UCON 75-H-90000®	255	452	299	470	406	321	180	1882
Igepal CO 880®	259	461	311	482	426	334	180	1939
Triton X-305®	262	467	314	488	430	336	183	1961
HI-EFF-8BP®	271	444	330	498	463	346	175	2006
Quadrol®	214	571	357	472	489	431	142	2103
Neopentyl glycol succinate (HI-EFF-3BP)	272	469	366	539	474	371	184	2120
Igepal CO 990®	298	508	345	540	475	366	205	2166
EGSP-Z	308	474	399	548	549	373	220	2278

Table 11 (continued)
SELECTED McREYNOLD'S CONSTANTS

Phase name	Δ I Values[a] Benzene 1	1-Butanol 2	2-Pentanone 3	Nitropropane 4	Pyridine 5	2-Methyl-2-pentanol 6	2-Octyne 7	$\sum_1^5$
Carbowax® 20M	322	536	368	572	510	387	221	2308
Carbowax® 20M (TPA)	321	537	367	573	520	387	220	2318
EPON 1001®	284	489	406	539	601	378	207	2319
Carbowax® 6000	322	540	369	577	512	390	222	2320
Carbowax® 4000	325	551	375	582	520	399	224	2353
Silicone ASI 50-phenyl-50-cyanopropyl	319	495	446	637	531	379	216	2428
SILAR-5 CP®	319	495	446	637	531	379	216	2428
Ethylene glycol isophthalate (HI-EFF-2EP®)	326	508	425	607	561	400	213	2427
XF-1150®	308	520	470	669	528	401	174	2495
Carbowax® 1000	347	607	418	626	589	449	240	2587
Ethylene glycol adipate (HI-EFF-2AP®)	372	576	453	655	617	462	250	2673
Butane-1,4-diol succinate (HI-EFF-4BP®)	369	591	457	661	629	476	243	2707
Phenyldiethanolamine succinate (HI-EFF-10BP®)	386	555	472	674	654	437	242	2741
Reoplex 400®	364	619	449	647	671	482	245	2750
LAC-1-R-296®	377	601	458	663	655	477	253	2754
Diethylene glycol adipate (HI-EFF-1AP)®	378	603	460	665	658	479	254	2764
Carbowax® 1540	371	639	453	666	641	479	255	2770
LAC-2-R-466®	387	616	471	679	667	489	257	2820
EGSS-Y	391	597	493	693	661	469	261	2835
Hyprose SP-80®	336	742	492	639	727	565	227	2936
EGSP-A	397	629	519	727	700	496	278	2972

SILAR-7CP	440	638	605	844	673	492	268	3200
ECNSS-M	421	690	581	803	732	548	259	3227
ECNSS-S	438	659	566	820	722	530	286.	3205
EGSS-X	484	710	585	831	778	566	316	3388
Ethylene glycol phthalate (HI-EFF-2GP®)	453	697	602	816	872	560	306	3410
SILAR-9CP	489	725	631	913	778	566	292	3536
Diethylene glycol succinate (HI-EFF-1BP)	499	751	593	840		595	323	3543
LAC-3-R-728®	502	755	597	849	852	599	329	3555
SILAR-10C	523	757	659	942	801	584	298	3682
THEED	463	942	626	801	893	746	269	3725
Tetracyanoethylated pentaerythritol	526	782	677	920	837	621	333	3742
Ethylene glycol succinate (HI-EFF-2BP®)	537	787	643	903	889	633	348	3759
1,2,3,4,5,6-Hexakis (CYCLO-N)	567	825	713	978	901	620		3984
1,2,3-Tris (2-cyanoethoxy)-propane	593	857	752	1028	915	672	375	4145
1,2,3,4-Tetrakis (CYANO-B)	617	860	773	1048	941	685		4239
Cyanoethylsucrose	647	919	797	1043	976	713	388	4382
N,N,Bis(2-cyanoethyl)-formamide	690	991	853	1110	1000	773	371	4644
OV-275®	781	1006	885	1177	1089			4938
Absolute ΔI values for squalene	653	590	627	652	699	690	841	

[a] ΔI is the retention index difference of a solute between a polar liquid phase and the nonpolar phase, squalane. From Catalog No. 19, Applied Science Laboratories, State College, Pa., 1976, 15. With permission.

are necessary to show significant separating ability differences among columns. For instance, similar constants and separations are observed for OV-1, OV-101, and SE-30.

A wide range of separation capability is represented in the OV-series of stationary phases. These materials are easily available with little lot-to-lot variation and have quite high maximum operating temperatures. Structures and data for OV-phases are given in Table 12 and elsewhere.[37]

Certain types of specialized liquid phases have found application in GLC. Liquid crystals are used for separations of various isomers, including positional isomers.[38-41] Some enantiomeric species may be resolved on optically active liquid phases.[42] Recently available chemically bonded packings have offered high resolution separations.[43,44]

2. Capillary Columns

When very high efficiencies are required for resolving closely spaced peaks, capillary or open tubular columns are recommended. Very good separations of components of complex mixtures or of closely related isomers may be obtained in reasonable times. The theoretical plates per foot of column length are comparable with packed columns, but capillaries may be of much greater length due to their relatively small resistance to gas flow. However, their small internal volumes and thin stationary phase coatings require very small samples and low flow rates. Injection of very small samples may be accomplished by use of an inlet splitter (see Figure 8). Comparison of the performance of a packed column with open tubular columns in general shows that the open tubular columns always give better resolution in the same or shorter analysis time.[45]

Two major types of capillary columns are generally available.[46,47] The first type is the wall-coated open tubular (WCOT) columns where the liquid phase is coated as a thin film on the inside surface of the tube. The other type is the support-coated open tubular (SCOT) columns where a layer of fine porous particles is coated on the column walls followed by a thin film of liquid phase. Open tubular columns have been discussed in detail by Ettre.[45]

A comparison of the WCOT and SCOT columns yields some interesting observations. While early-eluting peaks with small partition coefficients may not be separable on a given WCOT column, their separation may be possible on a shorter SCOT column. The use of SCOT columns reduces the bleeding of the liquid phase, so the resultant baseline drift during temperature programming is much less than in the case of WCOT columns. SCOT columns are more generally applicable to trace analysis since they provide a better detectability (sample reaching the detector per unit time) than WCOT columns. Open tubular columns are used in most cases with flame ionization detectors because of the small gas volumes of the peaks, but small volume hot wire detectors may be used with SCOT columns. These and other problems are discussed in detail by Ettre.[45]

While the use of open tubular columns has been subject to a number of problems,[48-50] several papers relating successful capillary GLC analyses have recently been reported.[51-53]

3. Preparative Columns

To separate large quantities of individual components and collect them for subsequent use, gas chromatography on a preparative scale may be used. The amount of material to be handled, the resolution of the compounds, and the time needed to accomplish the separation are related factors which all need to be considered in developing preparative GLC conditions.

In the scale-up from analytical conditions to preparative separation conditions, a loss of resolution is normally encountered. This efficiency loss in large-diameter col-

Table 12
STRUCTURES AND DATA FOR OV-SERIES STATIONARY PHASES

Name	Type		Structure	Solvent	Temperature limit (°C)
OV-1	Dimethylsilicone gum		$[-Si(CH_3)_2-O-]_n$	Toluene	325–375
OV-101	Dimethylsilicone fluid		$[-Si(CH_3)_2-O-]_n$	Toluene	325–375
OV-3	Phenylmethyldimethylsilicone	10% phenyl	$[-Si(CH_3)(C_6H_5)-O-]_n\,[Si(CH_3)_2-O-]_m$	Acetone	325–375
OV-7	Phenylmethyldimethylsilicone	20% phenyl	$[-Si(CH_3)(C_6H_5)-O-]_n\,[Si(CH_3)_2-O-]_m$	Acetone	325–375
OV-11	Phenylmethyldimethylsilicone	35% phenyl	$[-Si(CH_3)(C_6H_5)-O-]_n\,[Si(CH_3)_2-O-]_m$	Acetone	325–375

Table 12 (continued)
STRUCTURES AND DATA FOR OV-SERIES STATIONARY PHASES

Name	Type		Structure	Solvent	Temperature limit (°C)
OV-17	Phenylmethylsilicone	50% phenyl	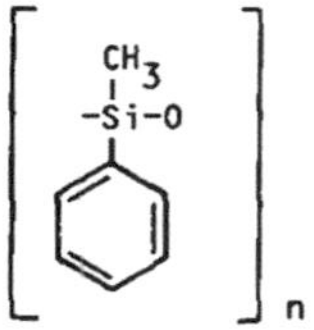 $[-Si(CH_3)(C_6H_5)-O]_n$	Acetone	350—375
OV-61	Diphenyldimethylsilicone	33% phenyl	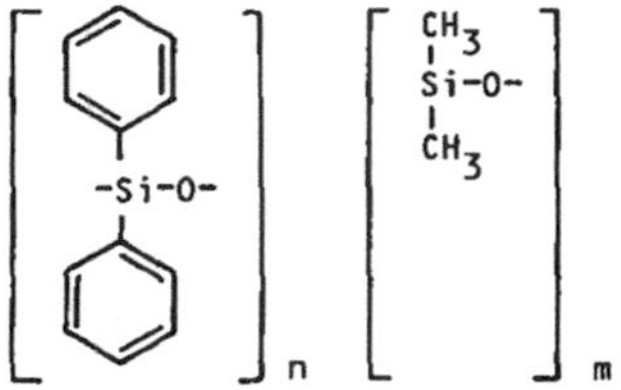 $[-Si(C_6H_5)_2-O-]_n[Si(CH_3)_2-O-]_m$	Acetone	325—375
OV-22	Phenylmethyldiphenylsilicone	65% phenyl	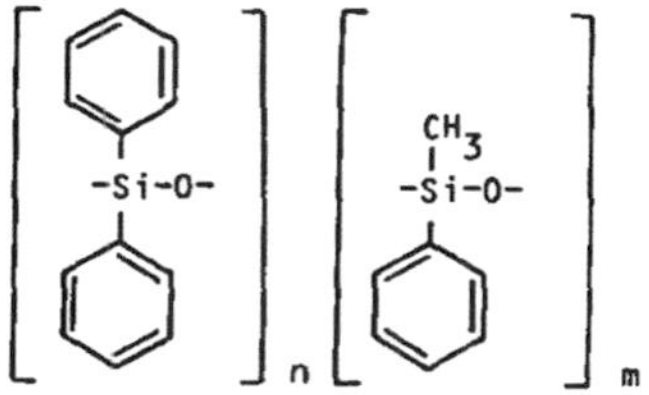 $[-Si(C_6H_5)_2-O-]_n[-Si(CH_3)(C_6H_5)-O-]_m$	Acetone	325—375
OV-25	Phenylmethyldiphenylsilicone	75% phenyl	$[-Si(C_6H_5)_2-O-]_n[-Si(CH_3)(C_6H_5)-O-]_m$	Acetone	325—375

OV-210	Trifluoropropylmethylsilicone	$[-Si(CH_3)(C_2H_4CF_3)-O-]_n$	Acetone	250—325
OV-225	Cyanopropylmethyl-phenylmethylsilicone	$[-Si(CH_3)(C_3H_6C{\equiv}N)-O-Si(CH_3)(C_6H_5)-O-]_n$	Acetone	250—300
OV-105	Cyanopropylmethyl-dimethylsilicone	$[-Si(CH_3)(C_3H_6C{\equiv}N)-O-]_n[-Si(CH_3)_2-O-]_m$	Acetone	250—275
OV-275	A cyano silicone		Acetone	250—275

From Catalog No. 25, Ohio Valley Specialty Chemicals, Marietta, Ohio, 1975-1976, 2. With permission.

umns is associated with a lack of uniformity of the packed particles, which contributes added resistance to mass transfer in the gas phase.[54] Other problems are involved in using large-scale columns. First, one must consider economy of operation: very large volumes of carrier gas may be required, and, since preparative-scale columns are often packed with heavily coated packings,[55] large quantities of stationary liquid phase may be called for. The temperature stability of the column must also be considered, as heavily loaded packings may yield extreme bleed of liquid phase which can contaminate the collected samples and shorten column life. Preparative-scale gas chromatography has been discussed in detail by Karasek,[54] Roz et al.,[56] and Rijnders.[57] The latter authors treat column efficiency preparation, apparatus, trapping, methods to increase yield, and automation.

4. Preparation and Preservation of Columns

Probably the best means of obtaining column packings is to purchase them from a reputable supplier. Many are available as pretested materials. Purchase of packings avoids the problem of analyst-to-analyst variation in their preparation.

Columns should be thoroughly cleaned before use. They may be rinsed with methanol, acetone, and then chloroform and finally air-dried. Glass columns should be treated with a dilute solution of dimethylchlorosilane in toluene to deactivate the surface, except for use with packing that might react with the silylated glass.

Any glassware or equipment to be used in preparing packings from silicone polymer phases should be treated to remove any traces of base; washing with acid and then distilled water is sufficient. In addition, base-washed supports should not be used with silicone stationary phases.

One of three techniques may be used to coat the solid support with the liquid phase. The "rotary evaporator", "pan coating", and "funnel coating" methods have been described in detail.[58,59] Care should be exercised so that the support particles are not crushed in the coating process, since the "fines" produced may create excess pressure drops in the column. Preparation of high-efficiency packed GC columns has been described by Leibrand and Dunham.[60]

The maximum operating temperatures for liquid phases are generally available in the distributor's literature and are not listed here except for the OV phases in Table 12. Near or above the maximum temperature the phase will begin to bleed off the column and give a drifting baseline or excessive spiking. Under these conditions the liquid phase is being lost, resulting in decreased retention times, poorer resolution of peaks, and tailing.

The older, less pure siloxane phases required a "no-flow" conditioning which is not normally necessary or beneficial with newer liquid phases.[59] In the column conditioning process generally used, the freshly prepared column is heated with carrier flow to remove traces of solvent and to help distribute the liquid phase evenly over the solid support surface. The column should not be connected to the detector until after it is conditioned. The conditioning temperature should be 20 to 30°C above the expected operating temperature and still 10 to 20°C lower than the maximum operating temperature. It is best to condition the column overnight, but often a usable baseline may be obtained in a few hours.

Hot columns should never be disconnected from carrier flow since organic liquid phases will decompose at high temperatures in the presence of oxygen. Columns should be heated for 1 to 2 hr above the operating temperature before each use to insure removal of materials from previous assays.

The packing should be carefully added to the tube to fit the dimensions specified for inlet space, etc. No technique should be used which might crush the packing; therefore, only very gentle electric vibrators should be used if they are used at all. In most cases the tubing can be filled by adding the packing via a funnel and concurrently tapping with a pencil on the sides. Filling of coiled columns may be aided by applying a vacuum at the outlet. Metal columns may be packed and then coiled.[58]

F. The Detector

In a gas chromatograph the detector is a device to indicate and measure the amount of each separated component which elutes. Littlewood[61] classifies detectors as "integrating" and "differentiating". Differentiating detectors give responses proportional to concentrations or to mass flow rates of the eluted components. They are used in the majority of applications because of their convenience and accuracy. In integrating detectors, the response is proportional to the total quantity of substance which elutes. When only carrier gas is flowing through an integrating detector, a straight line is observed on the recorder. As a zone of material elutes from the column, a deflection from the baseline is observed on the recorder whose distance is proportional to the total mass of the component in the zone. Chromatograms from integrating detectors consist of series of steps.

A variety of detectors is currently in use for gas chromatography. They vary widely in principle of operation but may be compared using several characteristics which indicate their utility. These have been discussed in detail by McNair and Bonelli[62] and by Littlewood[63] and will be only briefly outlined here.

The linear range of a detector is expressed as the ratio of the largest to smallest concentration within which the detector output is linear. Few detectors attain perfect linearity, but several cover a very wide range and have an excellent practical utility. Accurate quantitative analysis depends on having a linear relationship between the sample concentration and the detector response. The minimum detectable quantity is the amount of sample component which gives a detector response equal to twice the noise level. Electrical noise from the electronics and the detector is amplified with the signal and limits the concentration of component which may be detected. Sensitivity is normally expressed as the detector response in millivolts per unit concentration of sample component. Detector drift is a very low-frequency type of noise, and is typically regarded as a baseline change which occurs during a time period much longer than that needed to run a chromatogram. Drift may be expressed as average voltage drift per unit time. Many extraneous variables may affect detectors: flow rate changes, pressure, temperature, and electrical supply variations are but a few. It is desirable that a detector be as free of these effects as possible, but in practice much of the noise and drift are due to them.

One very widely used detector is the katharometer or thermal conductivity (TC) detector.[64] In its simplest form the TC unit consists of a spiral filament wire supported in a cavity inside a metal block. The resistance of the filament varies widely with temperature. With pure carrier gas flowing past the filament, the heat loss of the wire (heated by passing a constant current through it) is constant and so is the filament temperature. When a sample component passes through the thermal conductivity cell, it represents a different gas composition and therefore a different rate of heat loss from the filament. The filament temperature changes, creating an electrical resistance change which is measured by a Wheatstone bridge circuit. Filament metals must have a high-temperature coefficient of resistance and be highly resistant to chemical corrosion. Some TC cells use thermistor beads instead of filaments: these are very sensitive, but lack good stability and have a limited temperature range. Carrier gases with a high thermal conductivity, such as hydrogen or helium, should be used (rather than nitrogen) for best sensitivity. Increasing the filament current will increase the output signal, but too high a current will produce baseline instability and may burn out the filament. Excessive noise and baseline drift are typically due to corroded filaments (due to samples like halogens or other reactive materials) or condensation of high-boiling components on them. Finally, TC detectors are sensitive to flow changes; therefore, carrier flow should be carefully controlled, especially during temperature-programmed operation. Thermal conductivity detectors have been extensively described by Aue[65] and by Johns and Stapp.[66]

The basis for the widely used, highly sensitive flame ionization detector is as follows. A sample component elutes from the column and enters a hydrogen flame whose conductivity is very small. The sample vapor burns in the flame and provides an appreciable concentration of free electrons resulting from ionization, thus giving the flame an electrical conductivity. Using a stable voltage supply a potential is applied between the jet where the sample is burned and the collector, which is usually a metal ring. The presence of charged particles within the electrode gap allows a current to flow through a measuring resistor, and the resulting voltage drop is amplified by an electrometer and fed into a recorder.

The flame ionization detector (FID) will respond to almost all organic compounds except the inert gases, air gases, carbon disulfide, ammonia, some oxides of carbon, nitrogen, and sulfur, and a few others.[67] Good performance of the FID depends on proper choice of gas flows for carrier and for the flame, and on the geometry of the collector electrode and burner jet. The response of the FID depends on the number of molecules per unit time entering the detector. Consequently, greater sensitivity is obtained at high flow rates; thus increasing the linear velocity by using a smaller diameter jet improves sensitivity. The effect of carrier gas on the sensitivity of the FID has been recently described by Blades.[68] Blades has also discussed the operating characteristics of the FID.[69]

The electron capture detector (ECD) is widely used, especially for low-level analysis of halogenated compounds. Three operational modes are available and should be evaluated carefully in light of the capability needed. As the carrier gas flows through the ECD, β radiation from a ^{63}Ni or ^{3}H source ionizes the carrier molecules to form slow electrons which migrate to the anode under a fixed voltage termed the "cell voltage". When an electron-capturing substance elutes from the column into the ECD, some of the electrons are captured. Since the ions thus formed move slowly, they cannot carry an appreciable current, and the resulting current drop provides the response of the ECD.[70]

Molecules such as alkyl halides, conjugated carbonyls, and nitriles will give high ECD responses, while very little response is noted for hydrocarbons or alcohols.

Properties of the FID, TCD, and ECD are summarized in Table 13, which also lists characteristics of some other detector types.

The areas of the peaks obtained in a chromatogram are not directly proportional to the percent compositon of the sample. In most cases, different compounds give different detector responses, so it is necessary to use correction factors or response factors unless samples and standards of the compound are directly compared. Once the factors are known, they may be used to calculate percent composition. Dietz[71] has determined a variety of response factors for the thermal conductivity and flame ionization detectors, and some of these are presented in Table 14.

G. Related Equipment

1. Temperature Programmer

Controlled change of the column temperature during an analysis is called temperature programming. Temperature programming is used to elute components which boil over a wide range, and is easily, reproducibly accomplished with the temperature programmer available with most commercial instruments. The most popular is the linear

Table13
SUMMARY OF DETECTOR PARAMETERS

Detector	Principle of operation	Selectivity	Sensitivity (g/sec)	Response (C/g)	Linear range	Minimum detectable quantity (g)	Stability	Temperature limit(°C)	Carrier gas	Remarks
Thermal conductivity	Measures thermal conductivity of gases	Universal—responds to all compounds	6×10^{-10}	—	10^4	10^{-5} of CH_4 per mℓ of detector effluent	Good	450	He, H_2, Ni_2	Nondestructive — requires good temperature and flow control; simple, inexpensive, and rugged
Gas density balance	Difference of molecular weight of gases	Universal—responds to all compounds whose MW differs from carrier gas	Variable in range of T.C.D.	—	10^3		Good	Better sensitivity, 150	N_2, CO_2, Ar	Good for analysis of corrosive compounds, nondestructive
Flame ionization	H_2-O_2 flame 2000°C plasma	Responds to organic compounds, not to fixed gases or water	9×10^{-13} for alkanes	0.01	10^7	2×10^{-11} for alkanes	Excellent	400	He, N_2	$H_2O + CS_2$ good solvents because no response, destructive
Electron capture H^3	$N_2 + \beta \rightarrow e^-$; e^- + sample → loss of 1	Response to electron adsorbing compounds, esp. halogens, nitrates, and conjugated carbonyls.	2×10^{-14} for CCl_4	—	5×10^2	10^{-13} for lindane,	Fair	225	N_2 or Ar + 10% CH_4	Detector is easily contaminated and easy to clean; sensitive to water, carrier gas must be dry; can be operated in pulsed or DC mode; nondestructive
Ni^{63}			5×10^{-14} for CCl_4	—	50	4×10^{-12} for lindane	Fair	350		
Alkali flame P compound	Alkali modified H_2-O_2 flame 1600°C plasma	Enhanced response to phosphorus compounds	4×10^{-14}	—	10^3	2×10^{-12} parathion	Fair	300	N_2	Destructive — requires flow controller for hydrogen and air
N compound		Enhanced response to nitrogen compounds	7×10^{-12}	—	10^3	2×10^{-10} azobenzene	Fair	300	He	Destructive — requires flow controller for hydrogen and air; high sensitivity operating in starved O_2 mode
Helium	He + $\beta \rightarrow$ He*; Sample He* 1.	Universal — responds to all compounds	2×10^{-14} for methane	28	5×10^3	10^{-12} for fixed gases	Poor	100	He	High sensitivity to bleed precludes its use with columns other than active solids; nondestructive
Cross section	β + sample → 1.	Universal — responds to all compounds	10^{-9}	—	10^4	10^{-5}	Good	225	H_2 or He + 3% CH_4	

From McNair, H. M. and Bonelli, E. J., *Basic Gas Chromatography*, Consolidated printers, Berkeley, Calif., 1968, 118. With permission.

programmer, which increases the temperature at a set rate from a given start to a set finish temperature.

Temperature programming allows proper selection of a convenient analysis temperature for isothermal operation or provides a set, reproducible heating cycle for elution and quantitation of components having a wide range of boiling points.

General requirements for programmed-temperature GC (PTGC) include (1) an oven with a low mass for fast heating and cooling, (2) a flow controller to hold the carrier gas flow constant with changing temperature of the column, (3) good insulation of the oven from the injection ports and detectors, and (4) a temperature programming unit capable of reproducing a range of programming rates. In addition, consideration must be given to the thermal stability of the liquid phase over the intended temperature range, since vaporization of that phase leads to noise and baseline shifts from the column bleed. Provided that a precisely reproducible programmer is employed, quantitative analysis by PTGC can yield the same accuracy and precision as isothermal operation.

To compensate for the column bleed often found in PTGC, a dual column arrangement may be used. Here the signal from one column is opposed to the signal from the other, so the bleed from the compensating column is "subtracted" from the chromatogram from the test column. Dual columns ought to contain the same liquid phase but need not be of the same length.

2. Automatic Sampler

The automatic sampler is a device which periodically, automatically injects a sample onto the column. Typically, a syringe is pneumatically controlled to inject liquid samples, but solid samples may also be handled.[22] Large numbers of samples can be assayed unattended with a precision surpassing that achieved by any skilled operator. Automatic samplers can be used for isothermal assays or in conjunction with temperature programs. In assays where relatively poor precision is acceptable, the volume delivered by the syringe is reproducible enough that the samples may be run without an internal standard. Bidirectional operation of an autoinjector allows reinjection of selected samples after data evaluation by a remote processing device.

In the Hewlett-Packard® Model 7671A autosampler design, samples from up to 35 vials may be injected up to three times each.[72] Microvials may be used with low-volume samples. The Varian Series 8000® autosampler offers a larger vial capacity (60 vials) and can be operated bidirectionally via a remote control device.[73]

The Perkin-Elmer Autosampler AS 41 is in the encapsulation category.[74] Liquid or solid samples including highly viscous materials may be measured into small aluminum or gold capsules which are cold-weld sealed.[75] Up to 100 capsules may be loaded onto coded racks for automatic sampling, which is accomplished by piercing each capsule on a zirconium thorn after it is driven into the injection port. Each capsule is automatically ejected at the end of the analysis.

3. Ancillary Apparatus

A gas chromatograph can be coupled with an IR spectrometer to obtain IR spectra of components as they elute from the column. Problems associated with combining IR spectroscopy with gas chromatography have been detailed by Leathard.[76] Commercial GC/IR instruments are available.

A gas chromatograph can also be coupled to a mass spectrometer (MS) to allow both identification and quantitation of components eluting from the column. The mass spectrometer is a specific and extremely sensitive GC detector, and its use in trace analysis has grown rapidly over the last few years. Applications of this technique in organic analysis have been reviewed in a recent book by McFadden.[77] McFadden dis-

Table 14
RESPONSE FACTORS FOR GAS CHROMATOGRAPHIC ANALYSES

Relative Sensitivity Data for Hydrogen Flame Detector

Compound	Relative sensitivity
Normal paraffins	
Methane	0.97
Ethane	0.97
Propane	0.98
Butane	1.09 (1.03)
Pentane	1.04
Hexane	1.03
Heptane	1.00
Octane	0.97
Nonane	0.98
Branched paraffins	
Isopentane	1.05
2,2-Dimethylbutane	1.04
2,3-Dimethylbutane	1.03
2-Methylpentane	1.05
3-Methylpentane	1.04
2-Methylhexane	1.02
3-Methylhexane	1.02
2,2-Dimethylpentane	1.02
2,3-Dimethylpentane	0.99
2,4-Dimethylpentane	1.02
3,3-Dimethylpentane	1.03
3-Ethylpentane	1.02
2,2,3-Trimethylbutane	1.02
2-Methylheptane	0.97
3-Methylheptane	1.01
4-Methylheptane	1.02
2,2-Dimethylhexane	1.01
2,3-Dimethylhexane	0.99
2,4-Dimethylhexane	0.99
2,5-Dimethylhexane	1.01
3,4-Dimethylhexane	0.99
3-Ethylhexane	1.00
2-Methyl-3-ethylpentane	0.98
2,2,3-Trimethylpentane	1.02
2,2,4-Trimethylpentane	1.00
2,3,3-Trimethylpentane	1.01
2,3,4-Trimethylpentane	0.99
2,2-Dimethylheptane	0.97
3,3-Dimethylheptane	1.00
2,4-Dimethyl-3-ethylpentane	0.99
2,2,3-Trimethylhexane	1.01
2,2,4-Trimethylhexane	0.99
2,2,5-Trimethylhexane	0.99
2,3,3-Trimethylhexane	1.00
2,3,5-Trimethylhexane	0.96
2,4,4-Trimethylhexane	1.01
2,2,3,3-Tetramethylpentane	1.00
2,2,3,4-Tetramethylpentane	0.99
2,3,3,4-Tetramethylpentane	0.99
3,3,5-Trimethylheptane	0.99
2,2,3,4-Tetramethylhexane	1.01
2,2,4,5-Tetramethylhexane	1.00
Cyclopentanes	
Cyclopentane	1.04
Methylcyclopentane	1.01
Ethylcyclopentane	1.00
1,1-Dimethylcyclopentane	1.03
T-1,2-Dimethylcyclopentane	1.01
C-1,2-Dimethylcyclopentane	1.00
T-1,3-Dimethylcyclopentane	1.00
C-1,3-Dimethylcyclopentane	1.00
1MT2-Ethylcyclopentane	1.01
1MC2-Ethylcyclopentane	1.00
1MT3-Ethylcyclopentane	0.97
1MC3-Ethylcyclopentane	1.00
1,1,2-Trimethylcyclopentane	1.03
1,1,3-Trimethylcyclopentane	1.04
T-1,2C3-Trimethylcyclopentane	1.00
T-1,2C4-Trimethylcyclopentane	0.98
C-1,2T3-Trimethylcyclopentane	0.98
C-1,2T4-Trimethylcyclopentane	0.99
Isopropylcyclopentane	0.98
n-Propylcyclopentane	0.97
Cyclohexanes	
Cyclohexane	1.01
Methylcyclohexane	1.01
Ethylcyclohexane	1.01
1,1-Dimethylcyclohexane	1.03
T-1,2-Dimethylcyclohexane	1.01
C-1,2-Dimethylcyclohexane	0.99
T-1,4-Dimethylcyclohexane	0.99
1MT4-Ethylcyclohexane	0.98
1MC4-Ethylcyclohexane	0.96
1,1,2-Trimethylcyclohexane	1.01
Isopropylcyclohexane	0.98
Cycloheptane	1.01
Aromatics	
Benzene	1.12
Toluene	1.07
Ethylbenzene	1.03
p-Xylene	1.00
m-Xylene	1.04

Table 14 (continued)
RESPONSE FACTORS FOR GAS CHROMATOGRAPHIC ANALYSES

Relative Sensitivity Data for Hydrogen Flame Detector

Compound	Relative sensitivity	Compound	Relative sensitivity
Aromatics		**Acids**	
o-Xylene	1.02	Formic	0.01
1M2-Ethylbenzene	1.02	Acetic	0.24
1M3-Ethylbenzene	1.01	Propionic	0.40
1M4-Ethylbenzene	1.00	Butyric	0.48
1,2,3-Trimethylbenzene	0.98	Hexanoic	0.63
1,2,4-Trimethylbenzene	0.97	Heptanoic	0.61
1,3,5-Trimethylbenzene	0.98	Octanoic	0.65
Isopropylbenzene	0.97		
n-Propylbenzene	1.01	**Esters**	
1M2-Isopropylbenzene	0.99		
1M3-Isopropylbenzene	1.01	Methylacetate	0.20
1M4-Isopropylbenzene	0.99	Ethylacetate	0.38
sec-Butylbenzene	1.00	Isopropylacetate	0.49
tert-Butylbenzene	1.02	*sec*-Butylacetate	0.52
n-Butylbenzene	0.98	Isobutylacetate	0.54
		n-Butylacetate	0.55
Unsaturates		Isoamylacetate	0.62
		Methylamylacetate	0.63
Acetylene	1.07	Ethyl-(2)-ethylhexanoate	0.72
Ethylene	1.02	Hexylcaproate	0.78
Hexene-1	0.99	Cellosolve acetate	0.50
Octene-1	1.03		
Decene-1	1.01	**Nitrogen compounds**	
Alcohols		Acetonitrile	0.39
		Trimethylamine	0.46
Methanol	0.23	*tert*-Butylamine	0.54
Ethanol	0.46	Diethylamine	0.61
n-Propanol	0.60	Aniline	0.75
Isopropanol	0.53	di-*n*-Butaylamine	0.75
n-Butanol	0.66		
Isobutanol	0.68	**Ketones**	
sec-Butanol	0.63		
tert-Butanol	0.74	Acetone	0.49
Amyl alcohol	0.71	Methylethylketone	0.61
Methylisobutylcarbinol	0.74	Methylisobutylketone	0.71
Methylamyl alcohol	0.65	Ethylbutylketone	0.71
Hexyl alcohol	0.74	Diisobutylketone	0.72
Octyl alcohol	0.85	Ethylamylketone	0.80
Decyl alcohol	0.84	Cyclohexanone	0.72
Pentoxol	0.60	Pentoxone	0.56
Aldehydes		**Others (solvents)**	
Butyraldehyde	0.62	Cellosolve	0.45
Heptanoic aldehyde	0.77	Butyl cellosolve	0.62
Octaldehyde	0.78	Isophorone	0.85
Capric aldehyde	0.80	Thiophane	0.57

Note: Divide peak area by relative sensitivity, then normalize values to get weight percent.

Table 14 (continued)
RESPONSE FACTORS FOR GAS CHROMATOGRAPHIC ANALYSES

Response Factors for Thermal Conductivity Detectors

bp (°C)	Compound	MW	Thermal response	Weight factor
	Normal paraffins			
−161	Methane	16	35.7	0.45
−89	Ethane	30	51.2	0.59
−42	Propane	44	64.5	0.68
−0.5	Butane	58	85	0.68
+36	Pentane	72	105	0.69
68	Hexane	86	123	0.70
98	Heptane	100	143	0.70
126	Octane	114	160	0.71
151	Nonane	128	177	0.72
174	Decane	142	199	0.71
196	Undecane	156	198	0.79
254	Tetradecane	198	234	0.85
	C_{20}—C_{36}	—	—	0.72
	Branched paraffins			
−12	Isobutane	58	82	0.710
+28	Isopentane	72	102	0.707
10	Neopentane	72	99	0.727
50	2,2-Dimethylbutane	86	116	0.741
58	2,3-Dimethylbutane	86	116	0.741
60	2-Methylpentane	86	120	0.714
63	3-Methylpentane	86	119	0.725
79	2,2-Dimethylketone	100	133	0.752
81	2,4-Dimethylpentane	100	129	0.775
90	2,3-Dimethylpentane	100	135	0.741
86	3,5-Dimethylpentane	100	133	0.750
81	2,2,3-Trimethylbutane	100	129	0.775
90	2-Methylhexane	100	136	0.735
92	3-Methylhexane	100	133	0.752
93	3-Ethylpentane	100	131	0.763
99	2,2,4-Trimethylpentane	114	147	0.775
	Unsaturates			
−104	Ethylene	28	48	0.585
−48	Propylene	42	64.5	0.652
− 7	Isobutylene	56	82	0.683
− 6	Butene-1	56	81	0.697
+1	*trans*-Butene-2	56	85	0.658
+4	*cis*-Butene-2	56	87	0.643
20	3-Methylbutene-1	70	99	0.707
31	2-Methylbutene-1	70	99	0.707
30	Pentene-1	70	98.5	0.710
36	*trans*-Pentene-2	70	104	0.673
37	*cis*-Pentene-2	70	98.5	0.710
39	2-Methylpentene-2	70	96	0.750
101	2,4,4-Trimethylpentene-1	112	158	0.71
−35	Propadiene	40	53	0.76
− 4	1,3-Butadiene	54	80	0.674
43	Cyclopentadiene	66	68	0.97
34	Isoprene	68	92	0.738
110	1-Methylcyclohexene	96	115	0.837
−23	Methylacetylene	40	58	0.69

Table 14 (continued)
RESPONSE FACTORS FOR GAS CHROMATOGRAPHIC ANALYSES

Response Factors for Thermal Conductivity Detectors

bp (°C)	Compound	MW	Thermal response	Weight factor
170	Dicyclopentadiene	132	76	1.00
130	4-Vinylcyclohexene	108	130	0.83
44	Cyclopentene	68	80.5	0.844
	Other Unsaturates			
	Norbornene	94	113	0.830
	Norbornadiene	92	111	0.828
	Cycloheptatriene	92	104	0.885
	1,3-Cyclooctadiene	108	127	0.852
	1,5-Cyclooctadiene	108	131	0.826
	1,3,5,7-Cyclooctatetraene	104	114	0.910
	Cyclododecatriene (TTT)	162	168	0.966
	Cyclododecatriene	162	153	1.06
	Aromatics			
80	Benzene	78	100	0.780
110	Toluene	92	116	0.794
136	Ethylbenzene	106	129	0.818
139	*m*-Xylene	106	131	0.812
138	*p*-Xylene	106	131	0.812
144	*o*-Xylene	106	127	0.840
152	Isopropylbenzene	120	142	0.847
159	*n*-Propylbenzene	120	145	0.826
169	1,2,4-Trimethylbenzene	120	150	0.800
176	1,2,3-Trimethylbenzene	120	149	0.806
163	*p*-Ethyltoluene	120	150	0.800
165	1,3,5-Trimethylbenzene	120	149	0.805
171	*sec*-Butylbenzene	120	158	0.847
254	Biphenyl	154	169	0.912
332	*o*-Terphenyl(1,2-diphenylbenzene)	230	217	1.060
363	*m*-Terphenyl(1,3-diphenylbenzene)	230	230	1.00
383	*p*-Terphenyl(1,4-diphenylbenzene)	230	224	1.03
359	Triphenylmethane	244	232	1.05
218	Naphthalene	128	139	0.923
207	Tetralin	132	145	0.910
234	1-Methyltetralin	146	158	0.927
242	1-Ethyltetralin	160	170	0.944
186	*trans*-Decalin	138	150	0.920
195	*cis*-Decalin	138	151	0.913
	Cycloparaffins			
49	Cyclopentane	70	97	0.720
72	Methylcyclopentane	84	115	0.730
88	1,1-Dimethylcyclopentane	98	124	0.787
103	Ethylcyclopentane	98	126	0.775
100	*cis*-1,2-Dimethylcyclopentane	98	125	0.780
91	*cis*+ *trans*-1,3-Dimethylcyclopentane	98	125	0.780
116	1,2,4-Trimethylcyclopentane (CTC)	112	136	0.825
109	1,2,4-Trimethylcyclopentane (CCT)	112	143	0.783
81	Cyclohexane	84	114	0.735
101	Methylcyclohexane	98	120	0.820

Table 14 (continued)
RESPONSE FACTORS FOR GAS CHROMATOGRAPHIC ANALYSES

Response Factors for Thermal Conductivity Detectors

bp (°C)	Compound	MW	Thermal response	Weight factor
	Cycloparafins			
120	1,1-Dimethylcyclohexane	112	141	0.794
119—124	1,4-Dimethylcyclohexane	112	146	0.769
132	Ethylcyclohexane	112	145	0.775
155	*n*-Propylcyclohexane	126	158	0.800
139	1,1,3-Trimethylcyclohexane	126	139	0.907
	Inorganic compounds			
	Argon	40	42	0.95
	Nitrogen	28	42	0.67
	Oxygen	32	40	0.80
	Carbon dioxide	44	48	0.915
	Carbon monoxide	28	42	0.67
	Carbon tetrachloride	154	108	1.43
	Iron carbonyl ($Fe(CO_5)$)	195	150	1.30
	Hydrogen sulfide	34	38	0.89
	Water	18	33	0.55
	Hetero compounds			
131	Pyrrole	67	86	0.780
132	Hexylamine	101	104	0.970
11	Ethyleneoxide	44	58	0.758
35	Propyleneoxide	58	80	0.730
−62	Hydrogen sulfide	34	38	0.890
7	Methyl mercaptan	48	59	0.810
35	Ethyl mercaptan	62	87	0.720
68	1-Propanethiol	76	101	0.750
66	Tetrahydrofuran	72	83	0.870
119	Thiophane (cyclic sulfide)	88	103	0.855
165	Ethyl silicate	208	208	0.995
21	Acetaldehyde	44	65	0.680
135	Cellosolve	90	107	0.840
	Nitrogen compounds			
77	*n*-Butylamine	73	114	0.64
104	*n*-Pentylamine	87	152	0.57
131	Pyrrole	67	86	0.78
90	Pyrroline	69	83	0.83
89	Pyrrolidine	71	91	0.78
115	Pyridine	79	100	0.79
	1,2,5,6-Tetrahydropyridine	83	103	0.81
106	Piperidine	85	102	0.83
79	Acrylonitrile	53	78	0.68
97	Propionitrile	55	84	0.65
118	*n*-Butyronitrile	69	105	0.66
184	Aniline	93	114	0.82
238	Quinoline	129	194	0.67
203	*trans*-Decahydroquinoline	139	117	1.19
206	*cis*-Decahydroquinoline	139	117	1.19
33	Ammonia	17	40	0.42

Table 14 (continued)
RESPONSE FACTORS FOR GAS CHROMATOGRAPHIC ANALYSES

Response Factors for Thermal Conductivity Detectors

bp (°C)	Compound	MW	Thermal response	Weight factor
	Oxygenated compounds			
	Ketones			
56	Acetone	58	86	0.68
80	Methylethylketone	72	98	0.74
102	Diethylketone (3-pentanone)	86	110	0.78
124	3-Hexanone	100	123	0.81
127	2-Hexanone	100	130	0.77
106	3,3-Dimethylbutanone-2	114	118	0.97
155	Methyl-*n*-amylketone	114	133	0.86
173	Methyl-*n*-hexylketone	128	147	0.87
131	Cyclopentanone	84	106	0.79
157	Cyclohexanone	98	125	0.785
195	2-Nonanone	142	161	0.84
118	Methylisobutylketone	100	118	0.86
127	Methylisoamylketone	114	138	0.83
	Alcohols			
65	Methanol	32	55	0.58
78	Ethanol	46	72	0.64
97	*n*-Propanol	60	83	0.72
82	Isopropanol	60	85	0.71
117	*n*-Butanol	74	95	0.78
108	Isobutanol	74	96	0.77
100	*sec*-Butanol	74	97	0.76
83	*tert*-Butanol	74	96	0.77
153	3-Methylpentanol-1	86	107	0.80
119	2-Pentanol	88	110	0.80
116	3-Pentanol	88	109	0.81
102	2-Methyl-2-butanol	88	106	0.83
157	*n*-Hexanol	102	118	0.87
135	3-Hexanol	102	125	0.80
140	2-Hexanol	102	130	0.77
176	*n*-Heptanol	116	128	0.91
210	Decanol-5	158	184	0.86
255	Dodecanol-2	186	198	0.93
139	Cyclopentanol	86	109	0.79
161.5	Cyclohexanol	100	112	0.89
	Acetates			
77	Ethylacetate	88	111	0.79
89	Isopropylacetate	102	121	0.84
126	*n*-Butylacetate	116	135	0.86
148	*n*-Amylacetate	130	146	0.89
142	Iso-amylacetate	130	145	0.90
192	*n*-Heptylacetate	158	170	0.93
	Ethers			
35	Diethyl ether	74	110	0.67
67	Diisopropyl ether	102	130	0.79
91	Di-*n*-propyl ether	102	131	0.78
91	Ethyl-*n*-butyl ether	102	130	0.79
142	Di-*n*-butyl ether	130	160	0.81
190	Di-*n*-amyl ether	158	183	0.86

Table 14 (continued)
RESPONSE FACTORS FOR GAS CHROMATOGRAPHIC ANALYSES

Response Factors for Thermal Conductivity Detectors

bp (°C)	Compound	MW	Thermal response	Weight factor
	Diols			
	2,5-Hexanediol	118	127	0.93
250	1,6-Hexanediol	118	121	0.98
179	1,10-Decanediol	174	108	1.62
	1,12-Decanediol	174	110	1.58
	C_{14} Diol (from *sec* C_{14} alcohol)	230	128	1.80

From Dietz, W. A., *J. Gas Chromatogr.*, 5, 68, 1967. With permission.

cusses components, operation, the GC/MS interface, computer applications, and some special techniques. GC/MS is also discussed by Leathard.[76]

IV. APPLICATIONS OF GAS CHROMATOGRAPHY TO PHARMACEUTICAL ANALYSIS

A. General

In pharmaceutical analysis, GLC is typically applied to three types of problems: (1) assay of the raw material drug substance or of a synthetic precursor as the primary component, (2) quantitation of the drug in formulations, and (3) assays of impurities and/or solvents (i.e., minor components) in the raw material drug substance. This discussion will apply specifically to (1) and (3). The separation of drugs from excipients is covered in a separate chapter, and the quantitation of the drug can then be handled as in (1).

A major parameter for consideration before attempting gas chromatography is the choice of solvent. The drug or related substance must be soluble enough to provide workable concentration levels and, if an internal standard is to be used, the solvent must also dissolve the internal standard. For low-level impurity assays, the main component must be very soluble to provide a reasonable level of the impurity. In many cases of relatively insoluble materials, solution may be effected by derivatization methods. The choice of solvent is especially important in assays involving porous polymer column packings, since the solvent may not elute first as the typical "solvent front", and may create interference problems. Also, with the latter packings, solvent impurities can become critical problems — especially in trace level assays, since they are often retained by the column. Finally, derivatization reactions often cannot be done in certain solvents, or the product may be destroyed by dilution with a reactive solvent.

The vast majority of gas chromatographic assays reported in the literature use injection port-column configurations which are, as nearly as possible, all glass. Often the sample is injected into a space in the end of the column where it is vaporized by a surrounding heated injection block and is carried to the detector in an all-glass column. In other cases, the sample is injected into a glass liner where it vaporizes and is carried into the column. Use of a glass column is not generally as critical for heat-labile samples as use of a glass-lined injection port, but both help avoid potential problems.

In beginning the development of a gas chromatography assay for a previously uncharacterized compound, or for reevaluation to find improved assay conditions, one is faced with the selection of a column which will yield the best combination of peak

Table 15
GAS CHROMATOGRAPHY SCREENING CONDITIONS[a]

Liquid phase	McReynold's[b] Σ $\frac{5}{1}$	Maximum operating temperature (°C)
OV-101	229	350
OV-17	884	350
OV-25	1175	350
OV-225	1813	275
Carbowax® 20M	2308	225
Butanediol succinate	2707	225

[a] All columns are 3 ft × 1/8 in. I.D. glass, and packings are 3% liquid phase on 100/120 mesh AW DMCS Chromasorb G.
[b] See reference 35.

From Bishara, R. H. and Souter, R. W., *J. Chromatogr. Sci.*, 13, 593, 1975. With permission.

shape, reasonable retention time with stable operating conditions, and separations of impurities and related materials. To accomplish this objective in a simple, straightforward manner, a screening process for gas chromatography has been suggested.[78] The term "screening" refers to a standardized process of examining the behavior of the sample over a wide range of gas chromatographic conditions. A wide "polarity" range of column materials may be covered by use of a small variety of stationary phases. The summation of McReynold's constants for a particular phase is used as an evaluation of its polarity; the greater the sum, the more polar the liquid phase toward a particular compound. The screening columns listed in Table 15 vary from the nonpolar OV-101 to very polar butanediol succinate. The column loading, mesh size, lengths, and solid support are all identical so that in each case the initial chromatographic results vary only because of the liquid phase.

In the screening process, the sample is temperature-programmed off each column (generally 8 to 10°C/min). Based on the elution temperatures and peak shapes, isothermal experiments are then run to obtain corresponding peak profiles and to check retention times of potential interfering compounds. Such a screening program can also yield valuable data about impurities in the sample since, by varying polarity and temperature, one may elute materials which might not have been apparent from less systematic or less rigorous chromatography experiments.

Analysts typically want GC assays to be completed in a minimum amount of time, since replicate determinations on several samples must often be completed. The highest column temperature which maintains required resolution of peaks will thus be chosen. Extremes of temperature operation are ultimately set by the choice of stationary phase, but the liquid phase, column length, percent liquid loading, carrier flow rate, and choice of solid support all partially account for the separation speed. Beginning with a good choice of column materials, this screening procedure[78] is a straightforward means of finding good assay conditions or will help suggest further chromatographic experiments for specific problems.

Subsequent to the choice of column and operating conditions, it is normally necessary to choose an internal standard for quantitative analyses. The internal standard must meet several requirements, the first of which is that it should elute reasonably close to the compound being measured. It should also be stable in the solution of which it becomes part. The internal standard must be completely resolved from other sample components and should be present at roughly the same level as the other components being measured. Finally, it should ideally be of the same homologous series (or at least the same class of compounds) as the component being measured.

To insure freedom from interference in assays, known potential interferences such as synthetic precursors, degradation products, etc. should be chromatographed under identical conditions chosen for the sample and internal standard. If no overlap problems exist, the final step in developing a rigorously quantitative GC assay is statistical testing. A procedure for this has been described by Rickard and Zynger.[79] It includes testing of (1) linearity of detector response to concentration over a wide working range, (2) precision of response with enough replicates to insure that the reported result will be within ± 2% of the true value at the 95% confidence level, and (3) stability of the samples as judged by their detector responses after a specified elapsed time period. The analyst should also, if possible, check the accuracy of the method by comparison of the GC results to results obtained from some absolute method where applicable.

B. Quantitative Analysis

The simplest method of quantitation of the components of a sample is by area normalization. The area of each peak in the chromatogram is measured, and the percent composition is calculated by dividing the individual areas by the total area. If the components are close boiling members of a homologous series, this method yields weight percent. If the sample components are quite varied in volatility and in functional group type, two problems may arise which can lead to gross errors in the result. First, if not all components are eluted, the results will be too high. For instance, if the sample contained a large proportion of material which did not elute mixed with a component which did, the sample could be judged pure while actually being quite impure. The other problem relates to detector response. Components of a sample may represent several organic classes, and these (see Table 14) may have very different detector responses. Normalization assumes the same response for all which could yield large errors in results.

The relative weight percentages of the components in a gas chromatogram can be calculated if correction factors are applied to overcome the problem of different detector responses. The procedure to calculate correction factors has been discussed,[80] but their use is of more importance: peak areas are divided by relative sensitivity (Table 14), and then the values are normalized to get weight percent.[80] To use weight factors (Table 14) for TC detectors, the peak areas are measured and multiplied by the weight factors. These values are then normalized to give weight percentages of each compound.[80]

Much safer but more time-consuming is the use of internal standards. Accurately weighed samples and standards of a compound are diluted to the same volume with a solution of the internal standard. (Note: at this point solvent evaporation ceases to be a problem since the ratio of compound weight to internal standard weight is set.) The replicate values of

$$\text{Factor} = \frac{\left(\dfrac{\text{compound area}}{\text{internal standard area}}\right)}{(\text{compound weight})} \tag{11}$$

are calculated for the standards and samples and these are averaged in each case. The purity of the sample is then calculated using Equation 12:

$$\frac{\text{average sample factor}}{\text{average standard factor}} \times \text{percent potency standard} = \text{percent purity of sample} \tag{12}$$

In contrast to the area normalization method, use of an internal standard circumvents the problems caused by noneluting materials. Any nonvolatile materials — or even inorganic salts which are impurities in the sample — would be reflected by low sample purity if an internal standard was used in the assay. Other advantages of this technique are that quantities injected onto the column need not be accurately measured or even be the same, and detector response need not be known or remain constant since changes will not alter the area ratio. A disadvantage is that a reference standard of the analyte of known purity must be available.

There are a number of means of measuring peak areas or peak heights for use in quantitation, and the choice among these for any particular laboratory is often a compromise among cost, precision and accuracy required, degree of difficulty of measurement, and physical space limitations. These techniques vary from simple pencil-and-ruler peak triangulation to sophisticated computer systems capable of handling data output from a large number of instruments. These have been described by Littlewood[81] and by McNair and Bonelli.[82] Highest precision assays are obtained by use of an internal standard with data handling via a digital computer. The recent review of gas chromatography by Cram and Juvet[83] discusses advances in computerization for gas chromatography users. Many new GC instruments may be purchased with their own digital processors to control the instrument and process the data. Grant and Clarke[84] have reported on the contribution of the peak measurement method to accuracy and precision in gas chromatography.

C. Quantitation of Impurities in the Raw Material Drug Substance

The determination of low-level substances in a raw material is handled in essentially the same manner as quantitation of the main component, but several special problems may be encountered. It may be difficult to find a solvent in which the substance is sufficiently soluble to allow either simple dissolution of the sample and injection onto the chromatograph or efficient extraction of the impurity prior to chromatography. Slurry extractions, Soxhlet extractions, distillation, and other procedures must then be used to separate the impurity from the bulk sample. Careful recovery studies must be carried out for any of these extraction procedures. Trace impurities in the solvents used often become problems in trace level assays forcing one to use highly purified, distilled-in-glass solvents. By-products from derivatization reactions may also lead to interference problems, so derivatization reagents and reactions ought to be carefully considered before use.

Gas chromatography provides an especially convenient and specific means for determination of solvent residues in pharmaceuticals. The solvents elute from columns containing porous polymeric packings while less volatile components of the sample elute later or not at all. The latter sometimes causes increased background in the chromatogram, changes in column behavior (like peak broadening), or plugging of the column. Volatile impurities may be removed from the column in many cases by temperature programming before the next injection.

D. Qualitative Analysis

A variety of techniques have been developed for identification of peaks eluting from GC columns. These include retention behavior, sample modification, microchemical tests on eluted materials, selective detection, and identification by auxiliary instrumentation. The latter was discussed previously under ancillary apparatus, so this section will cover the former techniques.

Retention time at constant temperature is the most widely used method for identification of gas chromatographic peaks. The retention time is useful to confirm an identity strongly suggested by other considerations, but is dangerous for analysis of an uncharacterized sample, since the probability that other similar materials would have identical retention times is reasonably large. Comparing the retention times of the known and unknown on columns of two different polarities substantially reduces the chance of error but still does not eliminate it. The use of this method for peak identification requires an instrument capable of excellent retention time reproducibility. The demands on the instrument are reduced if relative retention is used instead of retention time for peak identification. Relative retention, α, is retention time relative to that of a standard component which is added into the sample and is much more reliable than simple retention time, especially for measurements made over an extended period of time where factors such as aging of the stationary phase and reduced liquid phase loading due to bleed become important. Retention indices, which are calculated from retention times, can also be of great value in identifying molecular structure.[36]

Modification of the sample has been discussed in a recent review by Leathard[76] who describes three approaches: one or more of the sample components may be completely or partially removed by means of chemical agents, converted to other materials by chemical agents, or degraded by pyrolytic or photolytic techniques.

Removal is normally accomplished by placing "on stream" with the column a physically or chemically active material distributed on a solid support. Depending on the agent chosen, members of various organic classes may be abstracted from the chromatogram. These include sulfur, olefinic, carboxyl, hydroxyl, and carbonyl compounds, and suitable abstraction reagents have been described.[85-87]

Just as selective removal of a GC peak provides information about identity of the peak, chemical modification can assist identification in two ways: the retention time of the new peak and the fact that a change occurred both provide information about identity and for functionality. Conversion may be effected by normal derivatization (to be discussed later) or by other types of reactions including hydrogenation for classification as to saturation and ozonolysis to locate double bonds.

Hoff and Feit[88] have described a general method for microanalysis using gas chromatography and prior chemical reaction effected in a hypodermic syringe. A list of several suitable reagents and reactions used are given in Table 16.

Some of the classical microchemical tests may also be used to identify chromatographic peaks. Sensitive color reactions may be used to help identify the compound in the effluent, which is bubbled into the solutions. The effluent may even be divided with a stream splitter to test for several functional groups simultaneously.[89] Table 17 gives a number of functional group classification tests and the minimum detectable amounts.[89]

Sample modification can also include degradation techniques, the most popular of which are pyrolysis- and photolysis-GC. In pyrolysis GC, the sample for analysis is heated to a very high temperature in an extremely short time (up to 600°C in 8 msec[90]), and the fragments of the thermally destroyed sample are swept into the gas chromatograph. Many materials give characteristic profiles of peaks called "pyrograms" which aid in identifying a compound just as IR or NMR spectra do. The rapid growth of pyrolysis GC has been largely due to its use for plastic and polymer identification, but it has been applied to pharmaceutical compounds such as barbituric acids,[91,92] indole alkaloids,[93] and drugs and poisons.[94] Photolysis GC has been applied to a lesser extent, but research has included alcohols, aldehydes and ketones, esters, and ethers.[95,96]

Detectors which are selective for certain elements are very useful for qualitative analysis. Alkali metal flame detectors are frequently used for detection of nitrogen, sulfur, and phosphorus-containing compounds. The sensitivity of these detectors is generally similar to that of the flame ionization detector, but their great value lies in the fact that they are far less sensitive to compounds not containing nitrogen, sulfur, or phos-

Table 16
REAGENTS FOR SYRINGE REACTIONS FOR GAS CHROMATOGRAPHY

Reagent	Preparation of reagents	Amount of reagent $\mu\ell$	Syringe		Exposure time (min)	Additional treatment necessary	Purpose of reagent
			Type	Size ($m\ell$)			
Sodium metal	Slice on tip of plunger	Thin slice	Cornwall	10	3	None	Cleanup, leaving ethers and hydrocarbons
Sulfuric acid	Concentrated H_2SO_4	5	Yale	2	3	None	Cleanup, leaving paraffins and aromatic hydrocarbons
7:3 sulfuric acid	7 $m\ell$ H_2SO_4 (conc) and 3 $m\ell$ water, cooled to room temperature	5	Yale	2	3	None	Leaves olefins, paraffins, and aromatic hydrocarbons
Hydrogen gas	H_2 and a few mg PtO_2		Yale	5	3	None	Saturates unsaturated compounds
Hydrogen iodide	2 $m\ell$ 90 to 95% H_3PO_4 warmed and stirred with a few milligrams KI	Wetting of plunger tip	Cornwall	10	3	$NaHCO_3$	Cleaves ethers
Bromine water	Saturated solution of Br_2 in water, freshly prepared	5	Yale	2	5	None	Brominates (removes) unsaturated compounds
Hydroxylamine	4 g NH_2OH, HCl in 50 $m\ell$ water	5	Yale	2	3	None	Removes carbonyl compounds
Sodium borohydride	1 g $NaBH_4$ in 2 $m\ell$ water	5	Yale	2	3	None	Removes carbonyl compounds producing the corresponding alcohols
Potassium permanganate	Saturated solution in water	5	Yale	2	3	None	Removes (oxidizes) aldehydes leaving ketones; produces ketones from secondary alcohols
Sodium nitrite	Freshly prepared cold mixture of equal amounts of 2.5 g $NaNO_2$ in 50 $m\ell$ water and 1 N H_2SO_4	5	Yale	2	3	one	Produces nitrites from alcohols

Acetic anhydride	5 mℓ acetic anhydride and two drops H_2SO_4 (conc)	5	Yale	2	3	$NaHCO_3$	Produces esters from alcohols
Sodium hydroxide	2.5 g NaOH in 50 mℓ water	10	Yale	2	5	None	Hydrolyzes esters, producing alcohols
Ozone	Ozone in oxygen		Yale	5	3	$NaAsO_2$	Removes unsaturated compounds, producing carbonyl compounds
Hydrogen chloride	2.5 mℓ HCl (conc) in 50 mℓ water	5	Yale	2	3	None	Removes amines
Water		5	Yale	2	3	None	Decreases water-soluble compounds
Sodium arsenite	5 g $NaAsO_2$ in 50 mℓ water	5	Yale	2	3	None	Eliminates excess ozone, reduces ozonides
Sodium bicarbonate	2.5 g $NaHCO_3$ in 50 mℓ water	5	Yale	2	3	None	Eliminates acidic compounds

From Hoff, J. E. and Feit, E. D., *Anal. Chem.*, 36, 1002, 1964. With permission.

Table 17
FUNCTIONAL GROUP CLASSIFICATION TESTS

Compound Type	Reagent	Type of positive test	Minimum detectable: Amount (μg)	Compounds tested
Alcohols	$K_2Cr_2O_7$-HNO_3	Blue color	20	C_1—C_8
	Ceric nitrate	Amber color	100	C_1—C_8
Aldehydes	2,4-DNP	Yellow ppt	20	C_1—C_6
	Schiff's	Pink color	50	C_1—C_6
Ketones	2,4-DNP	Yellow ppt	20	C_3—C_8 (methyl ketones)
Esters	Ferric hydroxamate	Red color	40	C_1—C_6 acetates
Mercaptans	Sodium nitroprusside	Red color	50	C_1—C_9
	Isatin	Green color	100	C_1—C_9
	$Pb(OAc)_2$	Yellow ppt	100	C_1—C_9
Sulfides	Sodium nitroprusside	Red color	50	C_2—C_{12}
Disulfides	Sodium nitroprusside	Red color	50	C_2—C_6
	Isatin	Green color	100	C_2—C_6
Amines	Hinsberg	Orange color	100	C_1—C_4
	Sodium nitroprusside	Red color, 1°	50	C_1—C_4
		Blue color, 2°		Diethyl and diamyl
Nitriles	Ferric hydroxamate-propylene glycol	Red color	40	C_2—C_5
Aromatics	$HCHO$-H_2SO_4	Red-wine color	20	ØH—ØC_4
Aliphatic unsaturation	$HCHO$-H_2SO_4	Red-wine color	40	C_2H_8
Alkyl halide	Alcoholic $AgNO_3$	White ppt	20	C_1—C_5

From Walsh, J. T. and Merritt, C., Jr., *Anal. Chem.*, 32, 1378, 1960. With permission.

phorus. This specificity gives such detectors great potential in drug analysis[97] since background interference can be greatly reduced. Sensitivity, specificity, selectivity, response, and sample preparation have been discussed by Riedmann.[98,99] Operating parameters of alkali flame ionization detectors (thermionic detectors) must be carefully optimized, since selectivity, sensitivity, and stability depend critically on parameters like flow rates and jet geometry. Commercial thermionic detectors are available.

In a flame photometric detector, the intensity of atomic or molecular emission from a hydrogen-air flame is measured at a particular wavelength. Operation in sulfur-selective and phosphorus-selective modes are most common, and simultaneous operation in these modes is also possible. Specificity for fluorine is also possible by mixing the column effluent with an argon stream containing calcium vapor. The intensity of emission from CaF at 529.9 nm is monitored after the sample-containing stream is fed to an oxyacetylene microburner.[100]

Coulometric detectors may be used to determine chlorine, sulfur, and nitrogen absolutely — and sometimes specifically. The gas chromatograph effluent is made to undergo oxidation or reduction, and is then passed through an electrolyte. If an eluate-electrolyte reaction changes the cell potential, it is offset by a coulometric titration system to maintain constant electrolyte concentration. The coulombs passed during peak elution is an absolute measure of the amount of eluate.

Much information is to be gained about the composition of a sample by comparison of results obtained with two or more different detectors. Comparison, for instance, of chromatograms from a flame ionization detector and a nitrogen thermionic detector can indicate those peaks for nitrogen-containing components. Selective detection systems have been described by Leathard[76] and by Vanden Heuvel and Zacchei.[97]

E. Derivatization

Modification of the sample structure may be accomplished by a number of relatively simple reactions and is done for a variety of reasons, including improvement of peak shape, increase (or decrease) of volatility, relocation of an interfering peak, and improvement of sensitivity. Derivatization may also aid in cases of thermally labile compounds and can be used to improve separations of closely related compounds. The types of functional groups most widely treated by derivatization for the reasons named above include carboxyls, hydroxyls, amines, thiols, imines, aldehydes, and ketones.

In planning a derivatization for use in a quantitative analysis, a number of points should be considered. If the sample will not dissolve in a suitable solvent, it may well dissolve by treatment with the chemical agent, and this can be followed by dilution to the appropriate volume. Use of impure derivatization agents or ones that lead to side reactions can complicate an analysis or even make it impossible due to extra peaks. Stability of the derivatives must be considered since some may decompose over time and/or may be affected by light, heat, or moisture. The degree of conversion to the desired derivative should also be determined. The ratio of sample to derivatizing agent and the reaction time and temperature conditions should be carefully tested to show that complete or at least constant conversion is obtained. In some reactions, the by-products may adversely affect the reaction yield.

One of the most widely applied general derivatizations is silylation, which is routinely used for carboxylic acids, amines, alcohols, phenols, imines, and compounds with an enolized carbonyl group. The addition of the trimethylsilyl (TMS) group to polar compounds is represented by the following set of reactions:

$$R{-}COOH \qquad\qquad R{-}COOSi(CH_3)_3$$

$$R{-}OH \qquad\qquad R{-}O{-}Si(CH_3)_3$$

$$R{-}SH \xrightarrow{\text{TMS DONOR}} R{-}S{-}Si(CH_3)_3$$

$$R{=}NH \longrightarrow R{=}N{-}Si(CH_3)_3$$

$$R{-}NH_2 \longrightarrow R{-}N(H){-}Si(CH_3)_3 \quad \text{and} \quad R{-}N[Si(CH_3)_3]_2$$

The methods of preparation of TMS and other silyl derivatives have been described in detail.[101,102] Often the sample or a solution of it may be treated with a slight excess of silylating agent and heated. Solvents must be free of active hydrogen, which excludes water, alcohols, and enolizable ketones. Most nonpolar solvents like hexane, ether, and benzene will dissolve the reagents, but are not conducive to rapid reaction. Solvents such as dimethylsulfoxide, dimethylformamide, pyridine, acetonitrile, and tetrahydrofuran provide excellent reaction media as well as excellent solvent characteristics. The reactions can normally be performed in a small vial sealed with an inert closure.

For silyl derivatives the choice of stationary phases is generally not critical; care should be taken, however, not to inject silylated materials onto a phase with which they may react (e.g., Carbowax®). In some instances the column must be conditioned or "primed" with some of the silylated solution in order to obtain reproducible results, but this must be experimentally determined. Some useful data for widely used silylating agents is given in Table 18, but the reader should refer to the manufacturer's literature or to References 101 and 102 for specific applications and procedures.

Unlike the hydroxyl and carboxyl groups, the amine group is not very reactive towards silylation reactions, so it is fairly difficult to silanize. Whereas *N*-trimethylsilylimidazole will only react with carboxyl and hydroxyl groups, *N,O*-bis-(trimethylsilyl)-trifluoroacetamide (BSTFA) can be used to derivatize amines. In more difficult cases, a catalyst may be added to the BSTFA, or other derivatizing agents may be used.[101]

Acyl derivatives of amines, alcohols, and thiols are common and are easy to prepare. Typically, acylated products are prepared by reaction with an excess of reagent, usually the anhydride of the corresponding acid in a solvent such as pyridine, which is able to bind the acid produced. Anhydrous conditions are essential since derivatives are hydrolyzed easily. Acid chlorides or acylimidazoles may also be used as the acylating agents. *N*-methyl-bis-trifluoroacetylamide[103] will react with $-NH_2$, $-OH$, and $-SH$ to trifluoroacetylate the compound. Perfluorinated anhydrides are often used to increase electron capture detector responses of compounds.

N-acyl imidazoles[104] provide several advantages in many cases over the use of acid chlorides and anhydrides. The by-product is relatively inert imidazole, and the reaction is smooth. Fluorinated acyl derivatives are surprisingly volatile despite the bulk of the perfluoro group. Hydroxyl groups and primary and secondary amines may be easily acylated. Selective acylation of amines in the presence of hydroxyl or carboxyl is possible if these are protected first by silylation.[104]

Pierce Chemical Company offers a series of alkylating reagents, the Alkyl-8® reagents, which undergo reactions with a large number of compounds.[105] Many of the uses of these dimethylformamide dialkylacetals are represented in Figure 9. A variety of techniques for chemical derivatization were recently described by Drozd.[106] Greeley has discussed approaches to derivatization of barbiturates.[107] Gejvall has shown that

Table 18
PROPERTIES OF SOME REAGENTS FOR SILYLATION

Abbreviation	Chemical name	Formula	By-product of silylation	MW	bp at 760 mm
BCM-TMDS	1,3-Bis(Chloromethyl)-tetramethyldisilazane	$((CH_2Cl)(CH_3)_2Si)_2NH$	NH_3	230.2	103 /10 mm
BDSA	Bis(Dimethylsilyl)acetamide	$OSiH(CH_3)_2$ $CH_3C = NSiH(CH_3)_2$	CH_3CONH_2[a]	175.3	55 /35 mm
BSA	*N,O*-Bis(trimethylsilyl)-acetamide	$OSi(CH_3)_3$ $CH_3C = NSi(CH_3)_3$	CH_3CONH_2[a]	203.4	71 /35 mm
BSEDA	Bis(trimethylsilyl)-ethylenediamine	$((CH_3)_3SiNHCH_2)_2$	$H_2NCH_2CH_2NH_2$	202.4	46/5 mm
BSTFA	*N,O*-Bis(trimethylsilyl)-trifluoroacetamide	$OSi(CH_3)_3$ $CF_3C = NSi(CH_3)_3$	CF_3CONH_2[a]	257.4	40 /12 mm
CM-DMCS	Chloromethyldimethyl-chlorosilane	$(CH_2Cl)(CH_3)_2SiCl$	HCl	143.1	115
DMCS	Dimethylmonochlorosilane	$H(CH_3)_2SiCl$	HCl	94.1	36
DMDCS	Dimethyldichlorosilane	$(CH_3)_2Si(Cl)_2$	HCl	129.1	70.5
HMDS	Hexamethyldisilazane	$(CH_3)_3SiNHSi(CH_3)_3$	NH_3	161.4	126
MSTFA	*N*-Methyl-*N*-Trimethylsilyl-trifluoroacetamide	$CF_3CONCH_3(Si(CH_3)_3)$	$CF_3CONHCH_3$	199.3	132
TMCS	Trimethylchlorosilane	$(CH_3)_3SiCl$	HCl	108.7	57
TMDS	1,1,3,3-Tetramethyldisilazane	$H(CH_3)_2SiHNSi(CH_3)_2H$	NH_3	133.4	99
TMSA	*N*-Trimethylsilylacetamide	$CH_3CONHSi(CH_3)_3$	CH_3CONH_2	131.3	M.P. 52—54
TMS-DMA	Trimethylsilyldimethylamine	$(CH_3)_3SiN(CH_3)_2$	$(CH_3)_2NH$	117.3	86
TMS-DEA	Trimethylsilyldiethylamine	$(CH_3)_3SiN(C_2H_5)_2$	$(C_2H_5)_2NH$	145.3	130
TMSI	*N*-Trimethylsilylimidazole	$(CH_3)_3SiNCH = NCH = CH$	HNCH = NCH CH	140.3	100 /14 mm

[a] Also the mono substituted silyl acetamide.

From Catalog No. 25, Ohio Valley Specialty Chemical Co., Marietta, Ohio, 1975-1976, 16. With permission.

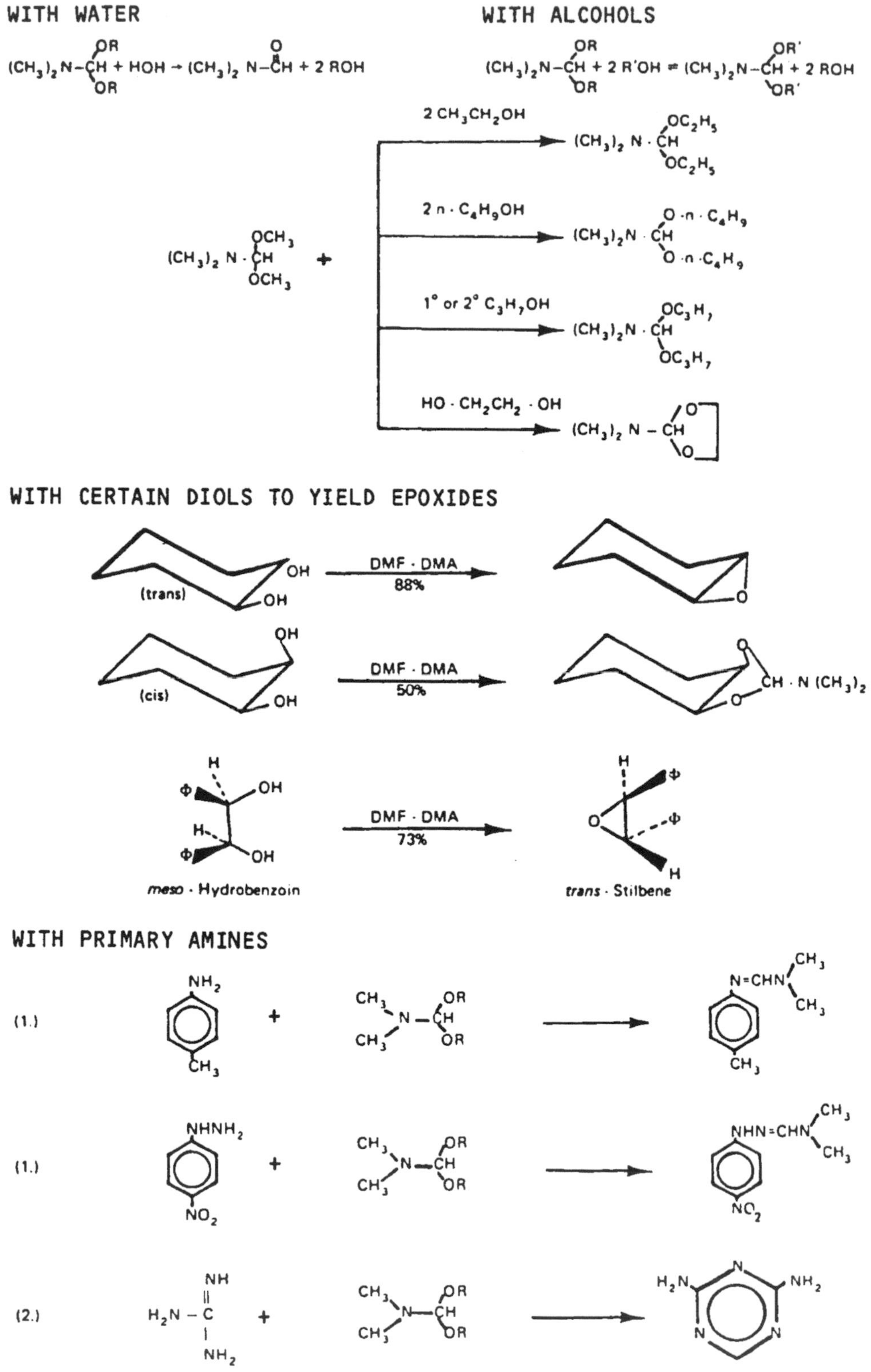

WITH WATER

$$(CH_3)_2N{-}CH(OR)_2 + HOH \rightarrow (CH_3)_2N{-}CHO + 2\,ROH$$

WITH ALCOHOLS

$$(CH_3)_2N{-}CH(OR)_2 + 2\,R'OH \rightleftharpoons (CH_3)_2N{-}CH(OR')_2 + 2\,ROH$$

$(CH_3)_2N \cdot CH(OCH_3)_2$ +

- $2\,CH_3CH_2OH$ → $(CH_3)_2N \cdot CH(OC_2H_5)_2$
- $2\,n\text{-}C_4H_9OH$ → $(CH_3)_2N \cdot CH(O\text{-}n\text{-}C_4H_9)_2$
- 1° or 2° C_3H_7OH → $(CH_3)_2N \cdot CH(OC_3H_7)_2$
- $HO \cdot CH_2CH_2 \cdot OH$ → $(CH_3)_2N - CH$ (cyclic, $-O-CH_2CH_2-O-$)

WITH CERTAIN DIOLS TO YIELD EPOXIDES

(trans) cyclohexanediol (OH, OH) —DMF · DMA, 88%→ epoxide (O)

(cis) cyclohexanediol (OH, OH) —DMF · DMA, 50%→ cyclic acetal (O, O, $CH \cdot N(CH_3)_2$)

meso · Hydrobenzoin (Φ, H, OH; Φ, H, OH) —DMF · DMA, 73%→ *trans* · Stilbene (oxide: O, Φ, H, Φ, H)

WITH PRIMARY AMINES

(1.) p-toluidine (NH_2, CH_3) + $(CH_3)_2N{-}CH(OR)_2$ → $N{=}CHN(CH_3)_2$ (aryl, CH_3)

(1.) p-nitrophenylhydrazine ($NHNH_2$, NO_2) + $(CH_3)_2N{-}CH(OR)_2$ → $NHN{=}CHN(CH_3)_2$ (aryl, NO_2)

(2.) $H_2N-C(=NH)-NH_2$ + $(CH_3)_2N{-}CH(OR)_2$ → triazine (H_2N, NH_2; N, N, N)

FIGURE 9. Chemical reactions of dimethylformamide dialkyl acetals. (From Handbook and General Catalog, Pierce Chemical Company, Rockford, Ill., 1977—1978, 254. With permission.)

WITH PHENOLS

$$ArOH + (CH_3)_2N\cdot CH(OR)_2 \longrightarrow ArOR + ROH + DMF$$

WITH CARBOXYLIC ACIDS

$$R\cdot COOH + (CH_3)_2NCH(OR')_2 \longrightarrow RCOOR' + R'OH + DMF$$

Methyl, ethyl, propyl and n-butyl esters form with equal ease. The mechanism is believed to involve the carboxylate ion and an alkoxycarbonium ion from the reagent.

WITH UNHINDERED CARBONYLS (TRANS-KETALISATION)

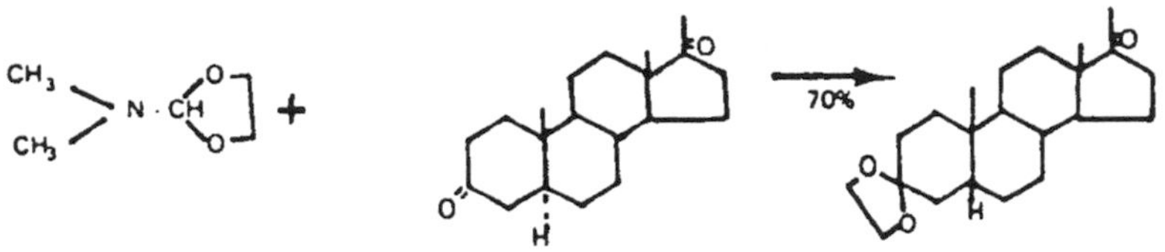

CONDENSATION WITH ACTIVE METHYLENE COMPOUNDS

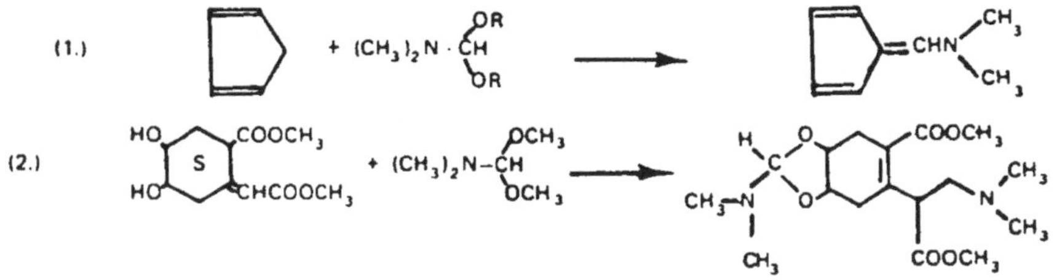

WITH SECONDARY ALIPHATIC AMINES

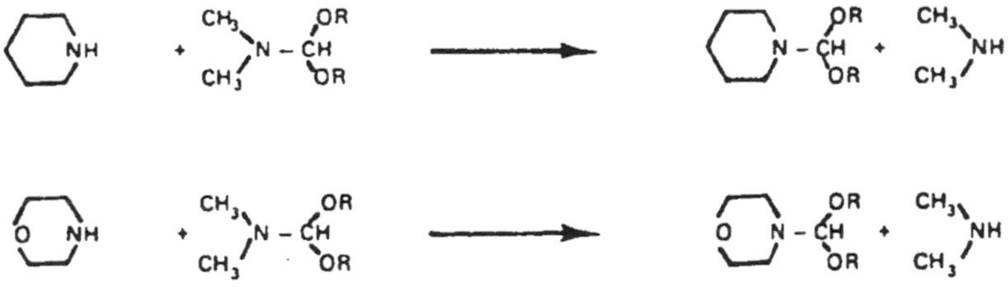

WITH AMINO ACIDS

Amino acids form volatile derivatives with good G. C. and mass spec. characteristics

FIGURE 9. (continued)

low molecular weight amines may be separated as the urethanes.[108] Cimbura and Kofoed reviewed derivatizations useful in forensic toxicology.[109] Applied Science Labortories offers a trimethylsilylating reagent, (isopropenyloxy) trimethylsilane (IPOTMS), which is insensitive to small amounts of water and yields relatively volatile acetone as the by-product.[110] Gehrike, Nakamoto, and Zumwalt have discussed in detail the GC of trimethylsilylated protein amino acids.[111] Regis Chemical Company provides an excellent discussion of various derivatization techniques in its *User's Guide.*[112] For optical purity analysis, *N*-TFA-L-prolylchloride and (−)-menthylchloroformate are available commercially from Regis Chemical Company.[112]

Many naturally occurring organic compounds such as catecholamines, steroids, amino acids, etc. contain two or more polar functional groups. One approach to derivatization of such compounds when the polar groups are present on adjacent carbon atoms is to react them with a bifunctional reagent. The most generally applicable derivatives are cyclic alkylboronates which are heterocycles formed from the reaction of an alkylboronic acid and the bifunctional sample.

Preparation of the cyclic alkylboronates is generally a very simple, straightforward, rapid process. The reactants are mixed in a 1:1 molar ratio or with a slight excess of the boronic acid reagent in an appropriate solvent for several minutes at room temperature, and the reaction mixture is then analyzed directly. Cyclic alkylboronates are stable towards other derivatization reactions such as trimethylsilylation and acylation. Methylboronic acid and *n*-butylboronic acid are the most widely used reagents since they usually lend excellent volatility. Brooks and Maclean have studied the derivatization of diols, hydroxy acids, and hydroxyamines with *n*-butylboronic acid.[113] Corticosteroid 17α,20,21-triols, diastereomeric triols, and certain catecholamines form stable cyclic boronates.[114] Ahuja has discussed the use of derivatization in pharmaceutical analysis and covered a wide variety of compounds.[115]

V. SPECIFIC APPLICATIONS OF GLC IN PHARMACEUTICAL ANALYSIS

GLC is a valuable technique for analysis of compounds of pharmaceutical interest. Usual objectives for GLC of pharmaceuticals are determination of: (1) drugs in tissues and fluids for forensic applications, (2) pharmacokinetics of drugs, (3) metabolic pathways and toxicology of drugs, and (4) purity and stability of drugs.

Both the *U.S.P.*[116] and the *N.F.*[117] describe applications of gas chromatography to the assay of raw materials, impurities, and solvents in pharmaceutical products. In some cases GLC serves for identity only. Most of the uses of gas chromatography in the *U.S.P.* and *N.F.* are collected in Tables 19 and 20, respectively. The *Official Methods of Analysis* of the Association of Official Analytical Chemists[118] includes assays for alcohol, acetone, amphetamine, phenothiazine, and water determinations. Moffat et. al.[119] have studied the gas chromatographic behavior of 62 basic drugs in eight GLC systems, and the reader may find the retention indices useful.

A novel approach in derivatization was used recently by Lauback, Balitz, and Mays[120] in the determination of carboxylic acid chlorides and residual carboxylic acid precursors used in the production of some penicillins. The acid chlorides were reacted with diethylamine (DEA) to form volatile amides which were not attacked upon further treatment of the samples with a silylating reagent to form trimethylsilylethers of the carboxylic acids. The derivatives were both of relatively low molecular weight and could be separated and quantitated in the same run. DEA had an advantage over hydroxylamine or alcohols in that there was little likelihood of side reactions, and the HCl released was neutralized by excess DEA.

Table 19
APPLICATIONS OF GAS CHROMATOGRAPHY IN THE *U.S. PHARMACOPEIA*

Monograph	Application
Amobarbital	Assay
Amobarbital Tablets	Assay
Amyl Nitrite	Total nitrites
Amyl Nitrite Inhalant	Total nitrites
Atropine Sulfate Injection	Assay
Atropine Sulfate Ophthalmic Ointment	Assay
Atropine Sulfate Ophthalmic Solution	Assay
Atropine Sulfate Tablets	Assay
Belladonna Extract	Assay
Belladonna Extract Tablets	Assay
Belladonna Leaf	Assay
Belladonna Tincture	Assay
Benzethonium Chloride Tincture	Alcohol and acetone content
Compound Benzoin Tincture	Alcohol content
Bupivacaine Hydrochloride	Residual solvents
Butabarbital Sodium Capsules	Assay
Butabarbital Sodium Elixir	Assay and alcohol content
Butabarbital Sodium Tablets	Assay
Camphor Spirit	Alcohol content
Castor Oil Emulsion	Assay
Chlorotrianisene	Volatile related compounds
Ciclopirox Olamine Cream	Benzyl Alcohol content
Cinoxate Lotion	Alcohol content
Clindamycin Hydrochloride	Assay
Clindamycin Hydrochloride Capsules	Assay
Clindamycin Palmitate Hydrochloride	Assay
Clindamycin Palmitate Hydrochloride for Oral Solution	Assay
Clindamycin Phosphate	Assay
Clindamycin Phosphate Injection	Assay
Clindamycin Phosphate Topical Solution	Assay
Sterile Clindamycin Phosphate	Assay
Clioquinol	Assay
Clioquinol Cream	Assay
Clioquinol Ointment	Assay
Colchicine	Chloroform and Ethyl Acetate
Collodion	Alcohol content
Flexible Collodion	Alcohol content
Desoximetasone Gel	Alcohol content
Dexamethasone Sodium Phosphate	Alcohol
Dexamethasone Sodium Phosphate Inhalation Aerosol	Alcohol content
Dexpanthenol Preparation	Pantolactone
Dimethyl Sulfoxide	Dimethyl Sulfone
Dimethyl Sulfoxide Irrigation	Assay
Diphenoxylate Hydrochloride and Atropine Sulfate Tablets	Assay for Atropine Sulfate
Dobutamine Hydrochloride	Assay
Dobutamine Hydrochloride for Injection	Assay
Doxapram Hydrochloride Injection	Assay
Doxepin Hydrochloride	Assay
Doxylamine Succinate	Volatile related substances
Echothiophate Iodide for Ophthalmic Solution	Water
Enflurane	Assay
Estradiol	Assay
Estradiol Tablets	Assay
Conjugated Estrogens	Assay
Conjugated Estrogens Tablets	Assay

Table 19 (continued)
APPLICATIONS OF GAS CHROMATOGRAPHY IN THE *U.S. PHARMACOPEIA*

Monograph	Application
Esterified Estrogens	Assay
Esterified Estrogens Tablets	Assay
Sterile Estrone Suspension	Assay
Ether	Low-boiling hydrocarbons
Ethosuximide	2-Ethyl-2-methylsuccinic Anhydride and other impurities
Fenoprofen Calcium	Assay
Fenoprofen Calcium Capsules	Assay
Fenoprofen Calcium Tablets	Assay
Flurothyl	2,2,2-Trifluoroethanol and assay
Guiafenesin Capsules	Assay
Guiafenesin Tablets	Assay
Halothane	Chromatographic Purity
Hexachlorophene	2,3,7,8-Tetrachloro-dibenzo-*p*-dioxin
Hydroxypropyl Methylcellulose 2208	Assay
Hydroxypropyl Methylcellulose 2906	Assay
Hydroxypropyl Methylcellulose 2910	Assay
Hyoscyamine Tablets	Assay
Hyoscyamine Sulfate Elixir	Assay
Hyoscyamine Sulfate Injection	Assay
Hyoscyamine Sulfate Oral Solution	Assay
Hyoscyamine Sulfate Tablets	Assay
Ibuprofen	Assay
Ibuprofen Tablets	Assay
Isoflurane	Assay
Isofluorophate	Assay
Isofluorophate Ophthalmic Ointment	Assay
Isopropyl Alcohol	Assay
Azeotropic Isopropyl Alcohol	Volatile impurities
Isoproterenol Hydrochloride Inhalation Aerosol	Alcohol content
Isosorbide Concentrate	Methyl Ethyl Ketone and assay
Isosorbide Oral Solution	Assay
Isoxsuprine Hydrochloride	Related compounds
Levopropoxyphene Napsylate	Related compounds
Levopropoxyphene Napsylate Capsules	Related compounds and assay
Levopropoxyphene Napsylate Oral Suspension	Alcohol content, related compounds, and assay
Lincomycin Hydrochloride	Assay
Lincomycin Hydrochloride Capsules	Assay
Lincomycin Hydrochloride Injection	Assay
Lincomycin Hydrochloride Syrup	Assay
Lindane Cream	Assay
Lindane Lotion	Assay
Lindane Shampoo	Assay
Meperidine Hydrochloride	Assay
Meperidine Hydrochloride Injection	Assay
Meperidine Hydrochloride Tablets	Assay
Mepivacaine Hydrochloride	Chromatographic purity
Methadone Hydrochloride Injection	Assay
Methadone Hydrochloride Tablets	Assay
Methdilazine Hydrochloride Syrup	Alcohol content
Methoxyflurane	Assay
Methylcellulose	Assay
Naloxone Hydrochloride Injection	Assay
Nitromersol Tincture	Alcohol content
Nitrous Oxide	Assay
Nortriptyline Hydrochloride Oral Solution	Alcohol content

Table 19 (continued)
APPLICATIONS OF GAS CHROMATOGRAPHY IN THE *U.S. PHARMACOPEIA*

Monograph	Application
Opium tincture	Alcohol content
Oxandrolone Tablets	Assay
Paramethadione	Assay
Paramethadione Capsules	Assay
Paramethadione Oral Solution	Assay
Paregoric	Alcohol content
Pentobarbital Elixir	Assay
Pentobarbital Sodium	Assay
Pentobarbital Sodium Injection	Assay
Peppermint Spirit	Alcohol content
Phendimetrazine Tartrate	L-*erythro* isomer
Phendimetrazine Tartrate Capsules	Assay
Phendimetrazine Tartrate Tablets	Assay
Phenobarbital Elixir	Alcohol content
Phenytoin Sodium Injection	Alcohol and Propylene Glycol content
Potassium Chloride Elixir	Alcohol content
Potassium Gluconate Elixir	Alcohol content
Primidone Oral Suspension	Assay
Primidone Tablets	Assay
Procyclidine Hydrochloride	Related compounds
Propoxyphene Hydrochloride	Related compounds
Propoxyphene Hydrochloride Capsules	Related compounds and assay
Propoxyphene Hydrochloride and Acetaminophen Tablets	Related compounds and assay
Propoxyphene Hydrochloride, Aspirin and Caffeine Capsules	Related compounds, assay for Propoxyphene Hydrochloride and Caffeine, dissolution
Propoxyphene Napsylate	Related compounds
Propoxyphene Napsylate Oral Suspension	Alcohol content, related compounds, and assay
Propoxyphene Napsylate Tablets	Related compounds and assay
Propoxyphene Napsylate and Acetaminophen Tablets	Related compounds, assays for Propoxyphene Napsylate and Acetaminophen
Propoxyphene Napsylate and Aspirin Tablets	Related compounds, assay for Propoxyphene Napsylate
Propylene Glycol	Assay
Saccharin Calcium	Toluenesulfonamides
Saccharin Sodium	Toluenesulfonamides
Scopolamine Hydrobromide Injection	Assay
Scopolamine Hydrobromide Ophthalmic Ointment	Assay
Scopolamine Hydrobromide Ophthalmic Solution	Assay
Scopolamine Hydrobromide Tablets	Assay
Secobarbital Elixir	Assay
Secobarbital Sodium Capsules	Assay
Secobarbital Sodium and Amobarbital Sodium Capsules	Assay
Sterile Spectinomycin Hydrochloride	Assay
Tamoxifen Citrate	Related impurities
Terpin Hydrate	Assay
Terpin Hydrate Elixir	Alcohol content and assay
Terpin Hydrate and Codeine Elixir	Alcohol content, assays for Terpin Hydrate and Codeine
Tetracaine and Menthol Ointment	Assay for Menthol
Thiamine Hydrochloride Elixir	Alcohol content
Thimerosal Topical Aerosol	Alcohol content
Thimerosal Tincture	Alcohol content
Trihexyphenidyl Hydrochloride	Assay
Trihexyphenidyl Hydrochloride Elixir	Assay

Table 19 (continued)
APPLICATIONS OF GAS CHROMATOGRAPHY IN THE *U.S. PHARMACOPEIA*

Monograph	Application
Trihexyphenidyl Hydrochloride Tablets	Assay
Trimethadione	Assay
Trimethadione Capsules	Dissolution and assay
Trimethadione Oral Solution	Assay
Trimethadione Tablets	Assay
Triprolidine Hydrochloride Syrup	Alcohol content
Tuaminoheptane Inhalant	Identification
Compound Undecylenic Acid Ointment	Assay for Undecylenic Acid
Vitamin E	Assay for Alpha Tocopherol, Alpha Tocopherol Acetate and Alpha Tocopherol Acid Succinate
Vitamin E Preparation	Assay
Vitamin E Capsules	Assay
Xylose	Identification

Derivatization using the versatile boronic acids was applied to determination of sphingosines.[121] The reaction of the sphingosines with methane-, *n*-butane-, or benzeneboronic acids was very fast and yielded cyclic derivatives which gave well-defined molecular ions for gc/ms where silyl derivatives did not.

Optically active derivatizing reagents have been used to form diastereomers which were easily separable by GLC for optical purity analysis. Gal[122] reacted amphetamine and eight related amines from biological fluids with (−)-α-methoxy-α-(trifluoromethyl)phenylacetyl chloride. The enantiomers of the anti-inflammatory Ibuprofen® in biological specimens were determined by GLC after reaction with *l*-(−)-α-methylbenzylamine to form diastereomeric amides.[123]

The value of a careful choice of internal standard was noted in a recent published assay for methadone.[124] A new GLC assay was developed which incorporated a basic internal standard rather than using a hydrocarbon. The internal standard was added at the start of the extraction and was quantitatively extracted, eliminating an evaporation step. The technique was found broadly applicable to a variety of concentrations in body fluids.

Burgett, Smith, and Bente[125] demonstrated high sensitivity of detection for underivatized barbiturates using a nitrogen-phosphorus detector which they developed. Other alkali flame detectors were found to have little or no response. Hucker and Stauffer[126] found that a nitrogen detector offered very distinct advantages in rapidity, specificity, and sensitivity for the measurement of lidocaine in plasma.

For nonvolatile pharmaceuticals not easily or conveniently derivatized for GLC, pyrolysis GLC offers a means of quantitation. Roy and Szinai[127] used pyrolysis techniques for qualitative and quantitative analysis of penicillins and cephalosporins. Direct characterization and differentiation was possible in most of the 14 cases studied. Even where no reference chromatograms are available, pyrolysis gas chromatograms may be analyzed using pattern recognition techniques.[128] The problem of deviations of retention times of corresponding pyrolysis peaks produced from different polymers can also be handled by this technique.

Often direct attempts at gas chromatography of salts of compounds (such as amine hydrochlorides or sulfates) fail because of irreproducible breakdown to the free amine or because of elongated peak tails which make the peaks difficult to quantitate. This can be avoided by extraction of the free base after neutralization or by direct derivatization of the salt. The latter approach was adopted by Fisher and Gillard[129] for the GLC determination of opium alkaloids in Papaveretum® (a synthetic mixture of the

Table 20
APPLICATIONS OF GAS CHROMATOGRAPHY IN THE *NATIONAL FORMULARY*

Monograph	Application
Acetone	Water
Amylene Hydrate	Assay
Aromatic Elixir	Alcohol content
Aspartame	5-Benzyl-3,6-dioxo-2- piperazine acetic acid
Benzalkonium Chloride	Ratio of alkyl components
Butane	Assay
Butyl Alcohol	Butyl Ether
Butylated Hydroxyanisole	Assay
Cetostearyl Alcohol	Assay
Cetyl Alcohol	Assay
Diacetylated Monoglycerides	Free Glycerin
Glyceryl Monostearate	Free Glycerin and assay for Mono-Glycerides
Isobutane	Assay
Isopropyl Myristate	Assay
Isopropyl Palmitate	Assay
Methyl Alcohol	Assay
Methylene Chloride	Assay
Mono- and Di-Glycerides	Free Glycerin and assay for Mono-glycerides
Mono- and Di-Acetylated Monoglycerides	Free Glycerin
Nitrogen	Assay
Polyethylene Glycol	Limit of Ethylene Glycol and Diethylene Glycol
Polyoxyl 10 Oleyl Ether	Free Ethylene Oxide
Polyoxyl 20 Cetostearyl Ether	Free Ethylene Oxide
Propane	Assay
Saccharin	Toluenesulfonamides
Sorbitol	Assay
Squalane	Chromatographic purity
Stearic Acid	Assay
Purified Stearic Acid	Assay
Stearyl Alcohol	Assay
Tolu Balsam Syrup	Alcohol content
Xanthan Gum	Isopropyl Alcohol

hydrochlorides of the opium alkaloids). These were derivatized directly and then chromatographed, thus avoiding an extremely lengthy extraction-titration-gravimetric analysis procedure.

GLC has been successfully applied to pharmaceutical kinetic studies, such as that reported by Souter and Dinner.[130] The stabilities of the two related amine uptake inhibitors DL-3-(*p*-trifluoromethylphenoxy)-*N*-methyl-3-phenylpropylamine and DL-3-(*O*-methoxyphenoxy)-*N*-methyl-3-phenylpropylamine in acidic and basic solutions were conveniently monitored at time intervals after neutralization and extraction. The drugs (hydrochloride salts) were refluxed in acid and base, samples were withdrawn at set times, and these were neutralized and then extracted with a chloroform solution of the internal standard. A portion of the chloroform solution was then chromatographed.

Although successful capillary column GC separations have been observed for years, the versatility of such columns and their application for quantitative analyses is a controversial topic. Because of insufficient accuracy and precision often related to unreliable sampling and because of instrument-installation problems, relatively few quantitative analyses have been reported using capillary GC techniques. Only very recently has the determination of the antidepressant psychotropic drug Nomifensine® in human plasma[131] been reported. The authors showed that improvements in specificity as well as method simplification were possible through the use of capillaries with high separation efficiencies. Extensive purification of biological extracts or use of specialized detectors is not always necessary in problem analyses of drugs and metabolites if capillary column technology is properly applied.

The low volatility, high polarity, and low chemical stability of some steroids, alkaloids, and other drugs makes their gas chromatographic analysis more difficult than usual. The gas chromatography of free steroids and alkaloids containing amino groups is often a problem due to tailing which decreases resolution and reduces the sensitivity. Baydrovtseva et al.[132] have shown that decreased retention times and peak shape improvements can often be effected by use of steam as the mobile phase. Cholesterol, sitosterol, alkaloids of the atropine, morphine and pyrrolizidine series, some steroidal terpenoids, and other compounds were studied. No baseline disturbance after the sample injections was observed, and either the free bases or salts gave satisfactory chromatograms.

Often adsorption problems occur in the GLC of basic nitrogenous compounds. To overcome these problems, supports incorporating basic material such as KOH[133] or highly silanized supports[133-135] have been used along with alkaline precolumns.[133,136] If the nitrogenous compounds are relatively nonvolatile, their GLC analysis is even more complicated. Knight[137] suggested the addition of *n*-hexylamine to the carrier gas to facilitate the analysis of nitrogenous compounds. More recently the effects of basic additives to the carrier gas for analysis of some local anaesthetics has been investigated. Greenwood and Nursten[138] evaluated this technique using lignocaine hydrochloride, butacaine sulfate, and cinchocaine hydrochloride.The important functions of the additives were to liberate the bases from the salts on-column and to overcome the column adsorption sites by being preferentially adsorbed themselves. These relatively nonvolatile basic nitrogen drugs were analyzed successfully as the salts.

For an excellent review including gas chromatographic pharmaceutical applications, the reader is referred to Janicki et al.[139] In addition, Table 21 lists a variety of examples of GC applications of pharmaceutical interest from recent literature.

VI. TROUBLESHOOTING GAS CHROMATOGRAMS

Many problems which occur in gas chromatography may be recognized from the chromatograms. Most may be solved relatively simply after they are diagnosed, and Table 22 presents common problems, causes, and suggested corrective actions. Analabs,[243] McNair and Bonelli,[244] and Walker, Jackson, and Maynard[245] all discuss troubleshooting in gas chromatography and these are reasonably detailed for the average user. Supelco Bulletin 739A describes purification of carrier gas and column lifetime.[246] Purcell et al. have discussed ghost peaks and their causes and elimination.[247]

Table 21
SELECTED GC APPLICATIONS OF PHARMACEUTICAL INTEREST

Application description	Ref.
Parabens in pharmaceutical dosage forms	140
Diphenhydramine in human plasma with nitrogen/phosphorus detection	141
Hydrazine in pharmaceuticals (Hydralazine and Isoniazid)	142
Cyproheptadine in plasma and urine with nitrogen-sensitive detection	143
Serum salicylates, after silylation	144
Busulfan in plasma, using GC/MS with selected-ion monitoring	145
Carbinoxamine and Hydrocodone in human serum by capillary GLC with nitrogen-sensitive detection	146
Simultaneous alkyl 4-hydroxybenzoates and 4-hydroxybenzoic acid determinations in liquid antacid formulations	147
Diltiazem and Deacetyldiltiazem in human plasma	148
Diuretic-antihypertensive agent, (±)-[[6,7-dichloro-2-(4-fluoro-phenyl)-2-methyl-1-oxo-5-indanyl]-oxy]acetic acid, in biological fluids	149
N-(*trans*-2-dimethylaminocyclopentyl)-N-(3′,4′-dichlorophenyl) propanamide and N-demethyl metabolite in dog serum	150
Captopril in blood, and Captopril and its disulfide metabolites in plasma	151
Quazepam and two major metabolites in human plasma	152
17β-Hydroxy-1α-methyl-17α-propyl-5α-androstran-3-one in plasma by GC/MS with single-ion detection	153
4-Methylpyrazole in plasma and urine using nitrogen-sensitive detection	154
Amphetamine, Norephedrine and their phenolic metabolites in rat brain	155
Nitroglycerin in plasma at picogram concentrations by capillary GC with on-column injection	156
Captopril (total) in human plasma by GC/MS with selected-ion monitoring after reduction of disulfides	157
Phenylpropanolamine in human plasma using trifluoroacetic anhydride derivatization and electron-capture detection	158
Flecainide acetate, a new antiarrythmic agent in biological fluids, using electron-capture detection	159
Captopril S-methyl metabolite in human plasma by selected-ion monitoring GC/MS	160
Naldol and deuterated analog in human serum and urine by GC with selected-ion monitoring mass spectrometry	161
Solvent residues in drug raw materials	162
Cetiedil (candidate antisickling agent) in human plasma with nitrogen-sensitive detection	163
Hydrazine and benzylhydrazine in isocarboxazid	164
17α-Methyltestosterone bioavailability study with stable-isotope methodology using gas chromatography-mass spectrometry	165
Turinabol (oral) and its metabolites by GC and GC/MS	166
Aldicarb, Butocarboxime, and their metabolites by GC/CI/MS analysis	167
Nifedipine in serum using electron capture detection	168
Urinary organic acids by capillary GC	169
Urinary succinylacetone by capillary GC	170
Residual volatiles in pharmaceuticals by dynamic headspace analysis	171
Diethylstilbestrol and its phosphorylated precursors in plasma and tissues by capillary GC-selection-ion monitoring mass spectrometry	172
Nomifensine by GC/MS	173
Triazolam (benzodiazepine) in plasma	174
Nitrendipine in plasma by capillary GC/MS	175
Metapramine and its dimethylated metabolites in plasma by GC/MS	176
Cantharidin in biological materials	177
3-Methylclonazepam in plasma by capillary GC	178
Cyclobenzaprine in plasma and urine by capillary GC with nitrogen-sensitive detection	179
Mefloquine in human and dog plasma with electron-capture detection	180
Phenteramine in human plasma	181
Nicotine and cotinine in urine	182
Viloxazine and its acetyl derivative in human plasma	183
Alprenolol and its 4-hydroxy metabolite in urine	184
Cianopramine in human plasma	185
Cetiedil in plasma using automated capillary GC with nitrogen-phosphorus detection	186
S-Benzoyl captopril in human urine by capillary GC with electron capture detection	187
Nitroglycerin in plasma	188

Table 21 (continued)
SELECTED GC APPLICATIONS OF PHARMACEUTICAL INTEREST

Application description	Ref.
Nitrazepam (underivatized) in plasma	189
Pentacaine in rat serum	190
Isosorbide-5-mononitrate	191
Bencyclane in biological samples	192
Cibenzoline in human plasma by GC-negative-ion chemical ionization MS	193
Propisomide (new antiarrythmic) in biological samples with a thermionic detector	194
D-Penicillamine using nitrogen-phosphorus detection	195
Tamoxifen citrate in pharmaceuticals	196
Nitroglycerin in human plasma by capillary GC with electron capture detection	197
Tertatolol in biological fluids by GC/MS	198
Veralipride in plasma and urine	199
Doxapram by nitrogen-phosphorus capillary GC	200
l-α-Acetylmethadol and metabolites in human serum using nitrogen detection	201
Oxcarbazepine and main metabolites in plasma	202
Azintamide (Oragallin) in pharmaceutical formulations	203
Haloperidol in serum and cerebrospinal fluid using nitrogen-phosphorus detection	204
Stirocainide (antiarrythmic) in plasma using nitrogen-sensitive detection	205
Dicyclomine (nanogram amounts) using nitrogen-sensitive detection	206
Chloroquine and two metabolites using capillary columns	207
5-Nitroimidazole class of antimicrobials in blood using electron-capture detection	208
Amitriptyline and its basic metabolites in plasma	209
Barbiturates (underivatized) by capillary column GC	210
Morphine in serum and cerebrospinal fluid	211
Mequitazin in human plasma and urine by capillary column GC/MS	212
Mexiletine routine serum monitoring	213
Metoprolol and metabolites in urine	214
Residual solvents in bulk pharmaceuticals by an automated method	215
Mefloquine (antimalarial) in human plasma using electron-capture detection	216
Diclofensine in human plasma using electron-capture detection following derivatization	217
Cocaine in plasma by automated method	218
Diclofenac in human plasma by GC/MS	219
Disopyramide and its mono-dealkylated metabolite in human plasma by capillary GC using nitrogen-phosphorus specific detection	220
Baclofen in plasma and urine	221
Fenfluramine and norfenfluramine in plasma using a nitrogen-sensitive detector	222
Diphenhydramine by capillary GC	223
Diethylcarbamazine in blood with alkali flame ionization detection	224
Opium and crude morphine by capillary GC (impurity profile comparison)	225
Gemfibrozil and its metabolites in plasma and urine	226
Bevantolol in plasma	227
Methoxyphenamine and three of its metabolites in plasma	228
Isosorbide dinitrate and its mononitrate metabolites in human plasma	229
Tocainide enantiomers in blood plasma using electron-capture detection	230
Sobrerol in biological fluids using electron-capture detection	231
Methamphetamine and metabolites in rat tissue using nitrogen-phosphorus detection	232
Lidocaine and deethylated metabolites with nitrogen-phosphorus detection	233
Heroin impurities using capillary columns and electron-capture detection	234
Tricyclic antidepressants in plasma	235
Methohexital in plasma or whole blood with electron-capture detection of the pentafluorobenzyl derivative	236
Fenfluramine and norfenfluramine in heparinized plasma	237
Indomethacin and metabolite in human plasma and urine	238
Narcotics, adulterants, and diluents using capillary columns	239
Metoclopramide in biological fluids using capillary columns and electron-capture detection	240
Amitriptyline, nortriptyline, and 10-hydroxy metabolite isomers in plasma using capillary GC with nitrogen-sensitive detection	241
Methylparaben in pharmaceutical dosage forms	242

Table 22
COMMON GAS CHROMATOGRAPHIC PROBLEMS, CAUSES, AND SUGGESTED CORRECTIONS

Problem	Cause	Suggested correction
Excessive random noise	Excessive column bleed	Lower the column temperature
Excessive peak tailing	1. Wrong choice of column	1. Use column of different polarity
	2. Injector is dirty	2. Clean the injector
	3. Oven temperature too low	3. Increase oven temperature
Leading side of peak tails badly	1. Sample injection poor	1. Review injection techniques
	2. Sample overload of column	2. Use smaller sample
	3. Sample condensing in the gas chromatograph	3. Check oven, injector, and detector temperatures
Flame will not light (FID)	1. Wrong hydrogen or air flow	1. Readjust flows to instrument manual settings
	2. Igniter wire burned out	2. Replace wire
	3. Detector jet dirty	3. Replace or clean the jet
	4. Column is loose	4. Tighten column connections
Flame extinguishes upon injection	1. Sample volume too large	1. Reduce volume
	2. Detector jet dirty	2. Clean or replace the jet
	3. Carrier gas flow too fast	3. Reduce flow rate
Peak tail above original baseline	1. Sample interacting with TC detector filaments	1. Raise detector temperature
	2. Water or other solvent washing out contamination; zeroing requirement decreased	2. Remove water from sample and recondition the column.
Negative dip after peak	1. If for all peaks, dirty ECD	1. Clean the detector
	2. If only for solvent peak, solvent is overloading column	2. Reduce sample volume
Peak tail below original baseline; flame goes out (FID)	1. Sample volume too large	1. Reduce sample volume
	2. Air or hydrogen cut off	2. Reestablish flows
	3. Carrier flow too high	3. Establish proper flow rate
	4. Sample contains more O_2 than the air for the flame does, causing flashback	4. Use O_2 for the flame gas
Extraneous peaks	1. Heavy residual material from previous injection(s) eluting	1. Allow more time between injections; always "bake off" column before use (heat above maximum use temperature for 1—2 hr)
	2. Moisture or impurities from carrier gas condensed on cool column and eluted during temperature program	2. Regenerate or replace the carrier gas filter

Table 22 (continued)
COMMON GAS CHROMATOGRAPHIC PROBLEMS, CAUSES, AND SUGGESTED CORRECTIONS

Problem	Cause	Suggested correction
	3. Ghost peaks	3. Make several solvent injections and recondition column
	4. Sample is decomposing	4. Lower injector temperature; use glass injector liner; use a different column packing
	5. Dirty sample or contaminated glassware	5. Clean up as appropriate
	6. Contaminants from septum	6. Condition septum before use
Square top peaks	1. Electrometer is being saturated (FID)	1. Reduce sample size or change the range setting to next highest
	2. Mechanical binding in the recorder drive, or defective recorder slidewire	2. Repair recorder or relieve binding problem
No peaks	1. Detector power "off"	1. Turn on detector
	2. No carrier flow	2. Start flow
	3. Loose column connections	3. Tighten column
	4. Injector cold	4. Increase injector temperature
	5. Syringe plugged	5. Clean or replace syringe
	6. Column too cold	6. Increase column temperature
Flame lights but soon extinguishes	Column is loose	Tighten column connections
Retention time too long and sensitivity poor	1. Septum is leaking continuously	1. Replace septum
	2. Carrier leak downstream from injector	2. Find and correct leak
	3. Carrier flow too low	3. Increase flow
Retention time ok but low sensitivity	1. Poor FID response	1. Readjust collector position; adjust flame gases ratio
	2. Sample too small	2. Concentrate sample or inject larger volume
	3. Attenuation too high	3. Adjust attenuation
Baseline drifts irregularly during isothermal operation	1. Detector, injector, or column temperatures not stabilized	1. Allow time to stabilize
	2. Column bleeding	2. Recondition and allow to stabilize — column may never stabilize completely under desired conditions
	3. Leak in carrier gas line or flow controller not operating properly	3. Fix leak or flow controller; flow controller inlet pressure may be too low
	4. Poor flame gases flow regulation (FID)	4. Check regulator operation
	5. Defective electrometer (FID)	5. Fix/replace electrometer
	6. TCD power supply defective	6. Replace power supply

Cannot zero baseline	1. Dirty detector 2. Excess column bleed 3. Bucking voltage battery voltage low (older instrument); baseline may be noisy	1. Clean detector 2. Reduce temperature or change column 3. Replace batteries
Resolution keeps changing	1. In packed columns, is usually due to channeling 2. In capillaries, due to deterioration of coating	Replace the column
Poor sensitivity and peaks tail	Interaction between sample and metal connector between column end and detector	Replace connector, using glass connection if possible; use another instrument

REFERENCES

1. **Ramsey, W.,** *Proc. R. Soc. (London)*, A76, 111, 1905.
2. **Tswett, M.,** *Ber Dtsch. Bot. Ges.*, 24, 316, 384, 1906.
3. **James, A. T. and Martin, A. J. P.,** *Biochem. J. Proc.*, 48, vii, 1951.
4. **James, A. T. and Martin, A. J. P.,** *Analyst (London)*, 77, 915, 1952.
5. **Martin, A. J. P. and Synge, R. L. M.,** *Biochem. J.*, 35, 1358, 1941.
6. **Mayer, S. W. and Tompkins, E. R.,** *J. Am. Chem. Soc.*, 69, 2866, 1947.
7. **Littlewood, A. B.,** *Gas Chromatography, Principles, Techniques, and Applications*, 2nd ed., Academic Press, New York, 1970, 164.
8. **van Deemter, J. J., Zuiderweg, F. J., and Klinkenberg, A.,** *Chem. Eng. Sci.*, 5, 271, 1956.
9. **Glueckauf, E.,** *Soc. Chem. Ind.*, London, 34, 1955.
10. **McNair, H. M. and Bonelli, E. J.,** *Basic Gas Chromatography*, Consolidated Printers, Berkeley, Calif., 1968, 25.
11. **Grubner, C.,** *Advances in Chromatography*, Vol. 6, Giddings, J. and Keller, R., Eds., Marcel Dekker, New York, 1968, 173.
12. **Littlewood, A. B.,** *Gas Chromatography, Principles, Techniques, and Applications*, 2nd ed., Academic Press, New York, 1970, 144.
13. **Purnell, J. H.,** *J. Chem. Soc.*, 1268, 1960.
14. **Karger, B. L.,** *Anal. Chem.*, 39, 24A, 1967.
15. **Scott, R. P. W.,** *Advances in Chromatography*, Vol. 9, Giddings, J. C. and Keller, R. A., Eds., Marcel Dekker, New York, 1970, 193.
16. **Perretta, A. T.,** *Am. Lab.*, May 1976, 35, 6.
17. **Dal Nogare, S. and Juvet, R. S., Jr.,** *Gas-Liquid Chromatography*, John Wiley & Sons, New York, 1962.
18. **Sternberg, J. C., Gallerway, W. S. and Jones, D. T. L.,** *Gas Chromatography*, Littlewood, A. B., Ed., Institute of Petroleum, London, 1961.
19. **McNair, H. M.,** *Chromatography*, 3rd ed., Heftmann, E., Ed., Van Nostrand Reinhold, New York, 1975, 189.
20. *A User's Guide to Chromatography*, Regis Chemical Co., Morton Grove, Ill., 1976, 23.
21. **Lochmüller, C. H.,** *Chemical Instrumentation: A Systematic Approach*, Strobel, H. A., Ed., Addison-Wesley, Reading, Mass., 1973, 859.
22. **Hartigan, M. J., Hoberecht, H. D., and Purcell, J. E.,** *Am. Lab.*, 10, 1973.
23. **Karger, B. L., Snyder, L. R., and Horvath, C.,** *An Introduction to Separation Science*, John Wiley & Sons, New York, 1973, 412.
24. **Karger, B. L., Snyder, L. R., and Horvath, C.,** *An Introduction to Separation Science*, John Wiley & Sons, New York, 1973, 415.
25. Johns-Manville, Celite Division, Greenwood Plaza, Denver, Colo., 1969.
26. **Porapak, A.,** *Selected Bibliography of Applications*, Waters Associates, Milford, Mass., 1971.
27. **Bombaugh, K. J., Dark, W. A., Horgan, D. F., Jr., and Farlinger, P. W.,** *Surface Modified Porapak for Gas Chromatography*, Waters Associates, Milford, Mass., 1971.
28. **Ackman, R. G. and Burgher, R. D.,** *Anal. Chem.*, 35, 647, 1963.
29. **Sakodynskii, K., Panina, L., and Klinskaya, N.,** *Chromatographia*, 7, 339, 1974.
30. **Daemen, J. M. H., Dankelman, W., and Hendriks, M. E.,** *J. Chromatogr. Sci.*, 13, 79, 1975.
31. Technical Bulletin No. 24, Applied Science Laboratories, State College, Pa., 1975.
32. **McNair, H. M. and Bonelli, E. J.,** *Basic Gas Chromatography*, Consolidated Printers, Berkeley, Calif., 1968, 53.
33. **Ottenstein, D. M.,** *J. Chromatogr. Sci.*, 11, 137, 1973.
34. Catalog No. 18, Analabs, North Haven, Conn., 1976, 10.
35. **McReynolds, W. O.,** *J. Chromatogr. Sci.*, 8, 685, 1970.
36. **Ettre, L. S.,** Pamphlet No. GCD-32, Perkin-Elmer Corp., Norwalk, Conn., 1972.
37. Catalog No 25, Ohio Valley Specialty Chemical Co., Marietta, Ohio, 1976, 2.
38. **Barrall, E. M., II, Porter, R. S., and Johnson, J. F.,** *J. Chromatogr.*, 21, 392, 1966.
39. **Richmond, A. B.,** *J. Chromatogr. Sci.*, 9, 571, 1971.
40. **Andrews, M. A., Schroeder, D. C., and Schroeder, J. P.,** *J. Chromatogr.*, 71, 233, 1972.
41. **Kelker, H. and von Schivizhoffen,** *Advances in Chromatography*, Vol. 6, Giddings, J. C. and Keller, R. A., Eds., Marcel Dekker, New York, 1968, 294.
42. **Lochmüller, C. H. and Souter, R. W.,** *J. Chromatogr.*, 113, 283, 1975.
43. **Karasek, F. W. and Hill, H. H., Jr.,** *Res. Dev.*, Dec., 1975, p. 30.
44. **Karasek, F. W.,** *Res. Dev.*, June, 1971, p. 34.
45. **Ettre, L. S.,** Pamphlet No. GCD-35, Perkin-Elmer Corp., Norwalk, Conn., 1973.

46. Catalog No. 26, *Chromatography Supplies*, Altech Associates, Arlington Heights, Ill., 1977, 32.
47. Brochure PL-125K, *Gas Chromatography Columns and Accessories*, Perkin-Elmer Instrument Division, Norwalk, Conn., 1976.
48. **Grob, K. and Grob, G.,** *Chromatographia*, 5, 3, 1972.
49. **Schomburg, G., Dielmann, R., Husmann, H., and Weeke, F.,** *J. Chromatogr.*, 122, 55, 1976.
50. **Verzele, M., Verstappe, M., Sandra, P., Van Luchere, E., and Vuye, A.,** *J. Chromatogr. Sci.*, 10, 668, 1972.
51. **Bailey, E., Fenoughty, M., and Richardson, L.,** *J. Chromatogr.*, 131, 347, 1977.
52. **Krijgsman, W. and van De Camp, C.,** *J. Chromatogr.*, 131, 412, 1977.
53. **Mattsson, Per E. and Nygren, S.,** *J. Chromatogr.*, 124, 265, 1976.
54. **Karasek, F. W.,** *Res. Dev.*, Sept., 1976, p. 31.
55. **McNair, H. M. and Bonelli, E. J.,** *Basic Gas Chromatography*, Consolidated Printers, Berkeley, Calif., 1968, 209.
56. **Roz, B., Bonmati, R., Hagenbach, G., Valentin, P., and Guiochon, G.,** *J. Chromatogr. Sci.*, 14, 367, 1976.
57. **Rijnders, G. W. A.,** *Advances in Chromatography*, Vol. 3, Giddings, J. C. and Keller, R. A., Eds., Marcel Dekker, New York, 1966, 215.
58. **McNair, H. M. and Bonelli, E. J.,** *Basic Gas Chromatography*, Consolidated Printers, Berkeley, Calif., 1968, 64.
59. *A User's Guide to Chromatography*, Regis Chemical Co., Morton Grove, Ill., 1976, 64.
60. **Liebrand, R. and Dunham, L.,** *Res. Dev.*, Sept., 1973, p. 32.
61. **Littlewood, A. B.,** *Gas Chromatography, Principles, Techniques, and Applications*, 2nd ed., Academic Press, New York, 1970, 6.
62. **McNair, H. M. and Bonelli, E. J.,** *Basic Gas Chromatography*, Consolidated Printers, Berkeley, Calif., 1968, 85.
63. **Littlewood, A. B.,** *Gas Chromatography, Principles, Techniques, and Applications*, 2nd ed., Academic Press, New York, 1970, 291.
64. **McNair, H. M. and Bonelli, E. J.,** *Basic Gas Chromatography*, Consolidated Printers, Berkeley, Calif., 1968, 91.
65. **Aue, W. A. J.,** *Chromatogr. Sci.*, 13, 329, 1975.
66. **Johns, T. and Stapp, A.,** *J. Chromatogr. Sci.*, 11, 234, 1973.
67. **Condon, R. D., Scholly, P. R., and Averill, W.,** *Gas Chromatography 1960*, Scott, R. P. W., Ed., Academic Press, New York, 1960, 30.
68. **Blades, A. T.,** *J. Chromatogr. Sci.*, 14, 145, 1976.
69. **Blades, A. T.,** *J. Chromatogr. Sci.*, 11, 251, 1973.
70. **Littlewood, A. B.,** *Gas Chromatography, Principles, Techniques, and Applications*, 2nd ed., Academic Press, New York, 1970, 315.
71. **Dietz, W. A.,** *J. Gas Chromatogr.*, 5, 68, 1967.
72. Model 7671A, Hewlett-Packard Co., Avondale, Pa., 1977.
73. Series 8000, Varian Aerograph; Varian Associates, Palo Alto, Calif., 1977.
74. Perkin-Elmer Corporation, Norwalk, Conn., 1973.
75. **Rhodes, C. T. and Hone, R. E.,** *Automated Analysis of Drugs*, Butterworths, London, 1973, 36.
76. **Leathard, D. A.,** *Advances in Chromatography*, Vol. 13, Giddings, J. C., Grushka, E. A., Keller, R. A., and Cazes, J., Eds., Marcel Dekker, New York, 1975, 265.
77. **McFadden, W.,** *Techniques of Combined Gas Chromatography/Mass Spectrometry: Applications in Organic Analysis*, John Wiley & Sons, New York, 1973.
78. **Bishara, R. H. and Souter, R. W.,** *J. Chromatogr. Sci.*, 13, 593, 1975.
79. **Rickard, E. C. and Zynger, J.,** *Anal. Chem.*, submitted.
80. **McNair, H. M. and Bonelli, E. J.,** *Basic Gas Chromatography*, Consolidated Printers, Berkeley, Calif., 1968, 140.
81. **Littlewood, A. B.,** *Gas Chromatography. Principles, Techniques, and Applications*, 2nd ed., Academic Press, New York, 1970, 409.
82. **McNair, H. M. and Bonelli, E. J.,** *Basic Gas Chromatography*, Consolidated Press, Berkeley, Calif., 1968, 151.
83. **Cram, S. P. and Juvet, R. S., Jr.,** *Anal. Chem.*, 48, 411R, 1976.
84. **Grant, D. W. and Clarke, A.,** *J. Chromatogr.*, 92, 257, 1974.
85. **Leathard, D. A. and Shurlock, B. C.,** *Identification Techniques in Gas Chromatography*, John Wiley & Sons, New York, 1970, 67.
86. **Ettre, L. S. and McFadden, W. H., Eds.,** *Ancillary Techniques of Gas Chromatography*, John Wiley & Sons, New York, 1969, 128.
87. **Littlewood, A. B.,** *Chromatographia*, 1, 133, 1968.

88. **Hoff, J. E. and Feit, E. D.,** *Anal. Chem.*, 36, 1002, 1964.
89. **Walsh, J. T. and Merritt, C., Jr.,** *Anal. Chem.*, 32, 1378, 1960.
90. Pyroprobe 100 from Chemical Data Systems, Oxford, Pa., 1977.
91. **Nelson, D. F. and Kirk, P. L.,** *Anal. Chem.*, 34, 899, 1962.
92. **Nelson, D. F. and Kirk, P. L.,** *Anal. Chem.*, 36, 875, 1964.
93. **Van Binst, G., Denolin-Deuaersegger, L., and Martin, R. H.,** *J. Chromatogr.*, 16, 34, 1964.
94. **Kirk, P. L.,** Pyrolysis-gas-liquid chromatography in the identification of drug and poisons, in 17th Pittsburgh Conference on Analytical Chemistry and Applied Spectroscopy, Abstr., Vol. 71, 73, 1966.
95. **Juvet, R. S. and Turner, L. P.,** *Anal. Chem.*, 37, 1464, 1965.
96. **Juvet, R. S., Turner, R. L., and Tsao, J. C. Y.,** *J. Gas Chromatogr.*, 5, 15, 1967.
97. **Vanden Heuvel, W. J. A. and Zacchei, A. G.,** *Advances in Chromatography,* Vol. 14, Giddings, J. C., Grushka, E., Cuzes, J., and Brown, P. R., Eds., Marcel Dekker, New York, 1976, 217.
98. **Riedmann, M.,** *J. Chromatogr.*, 88, 376, 1974.
99. **Riedmann, M.,** *J. Chromatogr.*, 92, 55, 1974.
100. **Gutsche, B., Herrmann, R., and Ruediger, K.,** *Z. Anal. Chem.*, 258, 273, 1972.
101. **Pierce, A. E.,** *Silylation of Organic Compounds,* Pierce Chemical, Rockford, Ill., 1968.
102. Handbook and General Catalog, Pierce Chemical, Rockford, Ill., 1977—1978, 212.
103. **Donike, M.,** *J. Chromatogr.*, 78, 273, 1973.
104. Handbook and General Catalog, Pierce Chemical, Rockford, Ill., 1977—1978, 245.
105. Handbook and General Catalog, Pierce Chemical, Rockford, Ill., 1977—1978, 253.
106. **Drozd, J.,** *J. Chromatogr.*, 113, 303, 1975.
107. **Greeley, R. H.,** *Clin. Chem. (New York)*, 20, 192, 1974.
108. **Gejvall, T.,** *J. Chromatogr.*, 90, 157, 1974.
109. **Cimbura, G. and Kofoed, J.,** *J. Chromatogr. Sci.*, 12, 261, 1974.
110. Catalog No. 20, Applied Science Laboratories, State College, Pa., 1977, 119.
111. **Gehrke, C. W., Nakamoto, H., and Zumwalt, R. W.,** *J. Chromatogr.*, 45, 24, 1969.
112. *A User's Guide to Chromatography,* Regis Chemical Co., Morton Grove, Ill., 1976, 71.
113. **Brooks, C. J. W. and Maclean, I.,** *J. Chromatogr. Sci.*, 9, 18, 1971.
114. **Anthony, G. M., Brooks, C. J. W., Maclean, I., and Sangster, I.,** *J. Chromatogr. Sci.*, 7, 623, 1969.
115. **Ahuja, S.,** *J. Pharm. Sci.*, 65, 163, 1976.
116. *The United States Pharmacopeia XXI,* United States Pharmacopeial Convention, Rockville, Md., 1985.
117. *National Formulary XVI,* United States Pharmacopeial Convention, Rockville, Md., 1985.
118. **Horwitz, W., Ed.,** *Official Methods of Analysis of the Association of Official Analytical Chemists,* 12th ed., Association of Official Analytical Chemists, Washington, D.C., 1975.
119. **Moffat, A. C., Stead, A. H., and Smalldon, K. W.,** *J. Chromatogr.*, 90, 19, 1974.
120. **Lauback, R. G., Balitz, D. F., and Mays, D. L.,** *J. Chromatogr. Sci.*, 14, 240, 1976.
121. **Gaskell, S. J. and Brooks, C. J. W.,** *J. Chromatogr.*, 122, 415, 1976.
122. **Gal, J.,** *J. Pharm. Sci.*, 66, 169, 1977.
123. **Vangiessen, G. J. and Kaiser, D. G.,** *J. Pharm. Sci.*, 64, 799, 1975.
124. **Lynn, R. K., Keger, R. M., Gordon, W. P., Olsen, G. D., and Gerber, N.,** *J. Chromatogr.*, 131, 329, 1977.
125. **Burgett, C. A., Smith, D. H., and Bente, H. B.,** *J. Chromatogr.*, 134, 57, 1977.
126. **Hucker, H. B. and Stauffer, S. C.,** *J. Pharm. Sci.*, 65, 926, 1976.
127. **Roy, T. and Szinai, S. S.,** *J. Chromatogr. Sci.*, 14, 580, 1976.
128. **Kullik, E., Kaljurand, M., and Koel, M.,** *J. Chromatogr.*, 126, 249, 1976.
129. **Fisher, G. and Gillard, R.,** *J. Pharm. Sci.*, 66, 421, 1977.
130. **Souter, R. W. and Dinner, A.,** *J. Pharm. Sci.*, 65, 457, 1976.
131. **Bailey, E., Fenoughty, M., and Richardson, L.,** *J. Chromatogr.*, 131, 347, 1977.
132. **Baydrovtseva, M. A., Rudenko, B. A., Kucherov, V. F., and Kuleshova, M. I.,** *J. Chromatogr.*, 104, 277, 1975.
133. **Umbreit, G. R., Nygren, R. E., and Testa, A. J.,** *J. Chromatogr.*, 43, 25, 1969.
134. **Reiser, R. W.,** *Anal. Chem.*, 36, 96, 1964.
135. **Albert, L. L., J. Assoc. Off.,** *Anal. Chem.*, 52, 1995, 1969.
136. **Thompson, G. F. and Smith, K.,** *Anal. Chem.*, 37, 1591, 1965.
137. **Knight, H. S.,** *Anal. Chem.*, 30, 2030, 1958.
138. **Greenwood, N. D. and Nursten, H. E.,** *J. Chromatogr.*, 92, 323, 1974.
139. **Janicki, C. A., Gilpin, R. K., Moyer, E. S., Almond, R. H., Jr., and Erlich, R. H.,** *Anal. Chem. Rev.*, 49, 110R, 1977.
140. **De Croo, F., De Schutter, J., Van den Bossche, W., and De Moerloose, P.,** *Chromatographia,* 18, 260, 1984.
141. **Abernathy, D. R. and Greenblatt, D. J.,** *J. Pharm. Sci.*, 72, 941, 1983.

142. **Matsui, F., Robertson, D. L., and Lovering, E. G.,** *J. Pharm. Sci.*, 72, 948, 1983.
143. **Hucker, H. B. and Hutt, J. E.,** *J. Pharm. Sci.*, 72, 1069, 1983.
144. **Bélanger, P. M., Lalande, M., Doré, F., and Labrecque, G.,** *J. Pharm. Sci.*, 72, 1092, 1983.
145. **Ehrsson, H. and Hassan, M.,** *J. Pharm. Sci.*, 72, 1203, 1983.
146. **Hoffman, D. J., Leveque, M. J., and Thompson, T.,** *J. Pharm. Sci.*, 72, 1342, 1983.
147. **Schieffer, G. W., Palermo, P. J., and Pollard-Walker, S.,** *J. Pharm. Sci.*, 73, 128, 1984.
148. **Clozel, J. P., Caillé, G., Théroux, P., Biron, P., and Besner, J. G.,** *J. Pharm. Sci.*, 73, 207, 1983.
149. **Weidner, L. L. and Zacchei, A. G.,** *J. Pharm. Sci.*, 73, 268, 1984.
150. **Lakings, D. B., Stryd, R. P., and Gilbertson, T. J.,** *J. Pharm. Sci.*, 73, 317, 1984.
151. **Bathala, M. S., Weinstein, S. H., Meeker, F. S., Jr., Singhvi, S. M., and Migdalof, B. H.,** *J. Pharm. Sci.*, 73, 340, 1984.
152. **Hilbert, J. M., Ning, J. M., Murphy, G., Jimenez, A., and Zampaglione, N.,** *J. Pharm. Sci.*, 73, 516, 1984.
153. **Krause, W. and Jakobs, U.,** *J. Pharm. Sci.*, 73, 563, 1984.
154. **Achari, R. and Mayersohn, M.,** *J. Pharm. Sci.*, 73, 690, 1984.
155. **Coutts, R. T., Prelusky, D. B., and Baker, G. B.,** *J. Pharm. Sci.*, 73, 808, 1984.
156. **Noonan, P. K., Kanfer, I., Riejelman, S., and Benet, L.Z.,** *J. Pharm. Sci.*, 73, 923, 1984.
157. **Ivashkiv, E., McKinstry, D. N., and Cohen, A. I.,** *J. Pharm. Sci.*, 73, 1113, 1984.
158. **Crisologo, N., Dye, D., and Bayne, W. F.,** *J. Pharm. Sci.*, 73, 1313, 1984.
159. **Johnson, J. D., Carlson, G. L., Fox, J. M., Miller, A. M., Chang, S. F., and Conard, G. J.,** *J. Pharm. Sci.*, 73, 1469, 1984.
160. **Cohen, A. I., Ivashkiv, E., McCormick, T., and McKinstry, D. N.,** *J. Pharm. Sci.*, 73, 1493, 1984.
161. **Cohen, A. I., Delvin, R. G., Ivashkiv, E., Funke, P. T., and McCormick, T.,** *J. Pharm. Sci.*, 73, 1571, 1984.
162. **Matsui, F., Lovering, E. G., Watson, J. R., Black D. B., and Sears, R. W.,** *J. Pharm. Sci.*, 73, 1664, 1984.
163. **Henderson, J. D., Mankad, V. N., Glenn, T. M., and Cho, Y. W.,** *J. Pharm. Sci.*, 73, 1748, 1984.
164. **Lovering, E. G., Matsui, F., Robertson, D., and Curran, N. M.,** *J. Pharm. Sci.*, 74, 105, 1985.
165. **Shinohara, Y., Baba, S., Kasuya, Y., Knapp, G., Pelsor, F. R., Shah, V. P., and Honigberg, I. L.,** *J. Pharm. Sci.*, 75, 161, 1986.
166. **Dürbeck, H. W., Büker, I., Scheulen, B., and Telin, B.,** *J. Chromatogr. Sci.*, 21, 405, 1983.
167. **Muszkat, L. and Aharonson, N.,** *J. Chromatogr. Sci.*, 21, 411, 1983.
168. **Lesko, L. J., Miller, A. K., Yeager, R. L., and Chatterji, D. C.,** *J. Chromatogr. Sci.*, 21, 415, 1983.
169. **Tuchman, M., Bowers, L. D., Fregien, K. D., Crippin, P. J., and Krivit, W.,** *J. Chromatogr. Sci.*, 22, 198, 1984.
170. **Tuchman, M., Whitley, C. B., Ramnaraine, M. L., and Bowers, L. D.,** *J. Chromatogr. Sci.*, 22, 211, 1984.
171. **Wampler, T. P., Bowe, W. A., and Levy, E. J.,** *J. Chromatogr. Sci.*, 23, 64, 1985.
172. **Abramson, F. P. and Lutz, M. P.,** *J. Chromatogr.*, 339, 87, 1985.
173. **Bagchi, S. P., Lutz, T., and Jindal, S. P.,** *J. Chromatogr.*, 344, 362, 1985.
174. **Baktir, G., Bircher, J., Fisch, H.-U., and Karlaganis, G.,** *J. Chromatogr.*, 339, 192, 1985.
175. **Beck, O. and Ryman, T.,** *J. Chromatogr.*, 337, 402, 1985.
176. **Bougerolle, A. M., Chabard, J. L., Bargnoux, H., Petit, J., Berger, J. A., and Dordain, G.,** *J. Chromatogr.*, 345, 59, 1985.
177. **Carrel, J. E., Doom, J. P., and McCormick, J. P.,** *J. Chromatogr.*, 342, 411, 1985.
178. **Coassolo, P., Aubert, C., and Cano, J.-P.,** *J. Chromatogr.*, 338, 347, 1985.
179. **Constanzeo, M. L., Vincek, W. C., and Bayne, W. F.,** *J. Chromatogr.*, 339, 414, 1985.
180. **Dadgar, D., Climax, J., Lambe, R., and Darragh, A.,** *J. Chromatogr.*, 337, 47, 1985.
181. **Dadgar, D., Climax, J., Lambe, R., and Darragh, A.,** *J. Chromatogr.*, 337, 136, 1985.
182. **Godin, J., Girard, F., and Hellier, G.,** *J. Chromatogr.*, 343, 424, 1985.
183. **Groppi, A. and Papa, P.,** *J. Chromatogr.*, 337, 142, 1985.
184. **Gyllenhaal, O.,** *J. Chromatogr.*, 349, 447, 1985.
185. **Hojabri, H. and Glennon, J. D.,** *J. Chromatogr.*, 342, 97, 1985.
186. **Holland, M. L. and Ng, K. T.,** *J. Chromatogr.*, 345, 178, 1985.
187. **Jemal, M. and Cohen, A. I.,** *J. Chromatogr.*, 342, 186, 1985.
188. **Langseth-Maurique, K., Bredesen, J. E., and Greibrokk, T.,** *J. Chromatogr.*, 349, 421, 1985.
189. **Locniskar, A., Greenblatt, D. J., and Ochs, H. R.,** *J. Chromatogr.*, 337, 131, 1985.
190. **Marko, V., Stefek, M., and Soltés, L.,** *J. Chromatogr.*, 339, 410, 1985.
191. **Marzo, A. and Treffner, E.,** *J. Chromatogr.*, 345, 390, 1985.
192. **Marzo, A., Treffner, E., Quadro, G., and Rapelli, S.,** *J. Chromatogr.*, 343, 77, 1985.
193. **Min, B. H. and Garland, W. A.,** *J. Chromatogr.*, 336, 403, 1985.

194. **Necciari, J., Mery, D., Sales, Y., Berthet, D., and Cautreels, W.,** *J. Chromatogr.*, 341, 202, 1985.
195. **Rushing, L. G., Hansen, E. B., Jr., and Thompson, H.C., Jr.,** *J. Chromatogr.*, 337, 37, 1985.
196. **Sane, R. T., Desai, S. V., Sonawne, K. K., and Nayak, V. G.,** *J. Chromatogr.*, 331, 432, 1985.
197. **Sioufi, A. and Pommier, F.,** *J. Chromatogr.*, 339, 117, 1985.
198. **Staveris, S., Blaise, P., Efthymiopoulos, C., Schneider, M., Jamet, G., Jung, L., and Koffel, J. C.,** *J. Chromatogr.*, 339, 97, 1985.
199. **Staveris, S., Jung, L., Jamet, G., and Koffel, J. C.,** *J. Chromatogr.*, 338, 79, 1985.
200. **Torok-Both, G. A., Coutts, R. T., Jamali, F., Patsulto, F. M., and Barrington, K. J.,** *J. Chromatogr.*, 344, 372, 1985.
201. **Verebey, K., De Pace, A., and Mulé, S. J.,** *J. Chromatogr.*, 343, 339, 1985.
202. **Von Unruh, G. E. and Parr, W. D.,** *J. Chromatogr.*, 345, 67, 1985.
203. **Abdel-Moety, E. M.,** *J. Chromatogr.*, 324, 475, 1985.
204. **Abernethy, D. R., Greenblatt, D. J., Ochs, H. R., Willis, C. R., Miller, D. D., and Shader, R. I.,** *J. Chromatogr.*, 307, 194, 1984.
205. **Backhaus, J., Dingler, E., and Weyhenmeyer, R.,** *J. Chromatogr.*, 307, 190, 1984.
206. **Beretta, E. and Vanazzi, G.,** *J. Chromatogr.*, 308, 341, 1984.
207. **Berggvist, Y. and Eckerbom, S.,** *J. Chromatogr.*, 306, 147, 1984.
208. **Bhatia, S. C. and Schanbhag, V. D.,** *J. Chromatogr.*, 305, 325, 1984.
209. **Burch, J. E., Roberts, S. G., and Raddats, M. A.,** *J. Chromatogr.*, 308, 165, 1984.
210. **Chow, W. M. L. and Caddy, B.,** *J. Chromatogr.*, 318, 255, 1985.
211. **Drost, R. H., Van Ooijen, R. D., Ionescu, T., and Maes, R. A. A.,** *J. Chromatogr.*, 310, 193, 1984.
212. **Fourtillan, J.-B., Girault, J., Bouquet, S., and Lefèbvre, M.-A.,** *J. Chromatogr.*, 309, 391, 1984.
213. **Grech-Bélanger, O.,** *J. Chromatogr.*, 309, 165, 1984.
214. **Gyllenhaal, O. and Hoffmann, K.-J.,** *J. Chromatogr.*, 309, 317, 1984.
215. **Haky, J. E. and Stickney, T. M.,** *J. Chromatogr.*, 321, 137, 1985.
216. **Heizmann, P. and Geschke, R.,** *J. Chromatogr.*, 311, 411, 1984.
217. **Hojabri, H., Dadgar, D., and Glennon, J. D.,** *J. Chromatogr.*, 311, 189, 1984.
218. **Jacob, P., III, Elias-Baker, B. A., Jones, R. T., and Benowitz, N. L.,** *J. Chromatogr.*, 306, 173, 1984.
219. **Kadowaki, H., Shiino, M., Uemura, I., and Kobayashi, K.,** *J. Chromatogr.*, 308, 329, 1984.
220. **Kapil, R. P., Abbott, F. S., Kerr, C. R., Edwards, D. J., Lalka, D., and Axelson, J. E.,** *J. Chromatogr.*, 307, 305, 1984.
221. **Kochak, G. and Honc, F.,** *J. Chromatogr.*, 310, 319, 1984.
222. **Krebs, H. A., Cheng, L. K., and Wright, G. J.,** *J. Chromatogr.*, 310, 412, 1984.
223. **Meatherall, R. C. and Guay, D. R. P.,** *J. Chromatogr.*, 307, 295, 1984.
224. **Nene, S., Anjaneyulu, B., and Rajagopalan, T. G.,** *J. Chromatogr.*, 308, 334, 1984.
225. **Neumann, H.,** *J. Chromatogr.*, 315, 404, 1984.
226. **Randinitis, E. J., Kinkel, A. W., Nelson, C., and Parker, T. D., III,** *J. Chromatogr.*, 307, 210, 1984.
227. **Randinitis, E. J., Nelson, C., and Kinkel, A. W.,** *J. Chromatogr.*, 308, 345, 1984.
228. **Roy, S. D., McKay, G., Hawes, E. M., and Midha, K. K.,** *J. Chromatogr.*, 310, 307, 1984.
229. **Santoni, Y., Rolland, P. H., and Cano, J.-P.,** *J. Chromatogr.*, 306, 165, 1984.
230. **Sedman, A. J. and Gal, J.,** *J. Chromatogr.*, 306, 155, 1984.
231. **Shukla, U. A., Stetson, P. L., and Wagner, J. G.,** *J. Chromatogr.*, 308, 189, 1984.
232. **Terada, M.,** *J. Chromatogr.*, 318, 307, 1985.
233. **Willis, C. R., Greenblatt, D. J., Benjamin, D. M., and Abernethy, D. R.,** *J. Chromatogr.*, 307, 200, 1984.
234. **Moore, J. M.,** *J. Chromatogr.*, 281, 355, 1983.
235. **Corona, G. L., Bonferoni, B., Frattini, P., Cucchi, M. L., Santagostino, G.,** *J. Chromatogr.*, 277, 347, 1983.
236. **Björkman, S., Idvall, J., and Stenberg, P.,** *J. Chromatogr.*, 278, 424, 1983.
237. **Morris, R. G. and Reece, P. A.,** *J. Chromatogr.*, 278, 434, 1983.
238. **Guissou, P., Cuisinaud, G., and Sassard, J.,** *J. Chromatogr.*, 277, 368, 1983.
239. **Comparini, I. B., Centini, F., and Pariali, A.,** *J. Chromatogr.*, 279, 609, 1983.
240. **Riggs, K. W., Axelson, J. E., Rurak, D. W., Hasman, D. A., McErlane, B., and Bylsma-Howell, M., McMorland, G. H., Ongley, R., and Price, J. D. E.,** *J. Chromatogr.*, 276, 319, 1983.
241. **Jones, D. R., Lukey, B. J., and Hurst, H. E.,** *J. Chromatogr.*, 278, 291, 1983.
242. **Majlát, P. and Barthos, E.,** *J. Chromatogr.*, 294, 431, 1984.
243. **Lynn, T. R., Walker, J. Q., and Kondrath, P. F.,** *Guide to Stationary Phases for Gas Chromatography*, Analabs, North Haven, CT, 1983, 160.
244. **McNair, H. M. and Bonelli, E. J.,** *Basic Gas Chromatography*, Consolidated Printers, Berkeley, CA, 1968, 259.

245. **Walker, J. Q., Jackson, M. T., Jr., and Maynard, J. B.,** *Chromatographic Systems, Maintenance and Troubleshooting,* Academic Press, New York, 1972.
246. Bulletin 739A, Supelco, Inc., Supelco Park, Bellefonte, PA.
247. **Purcell, J. E., Downs, H. D., and Ettre, L. S.,** *Chromatographia,* 8, 605, 1975.

Chapter 4

HIGH-PERFORMANCE LIQUID CHROMATOGRAPHY

L. J. Lorenz

TABLE OF CONTENTS

I. INTRODUCTION

High-performance liquid chromatography (HPLC) is a chromatographic separation technique in which separation is accomplished by partitioning between a mobile solvent and a stationary column packing material. HPLC differs from other types of liquid chromatography only in that packing materials of small, uniform particle size are used. The small size of the particles gives very high column efficiencies, but can also result in a high resistance to flow through the column and therefore a high pressure drop across the column. For this reason, HPLC has often been called high-pressure liquid chromatography.

Although HPLC is basically a separation technique, detectors are usually incorporated directly in the column effluent line so that qualitative and quantitative information can be obtained for the sample as the separated components exit from the column. HPLC can also be used in a preparative mode to provide pure materials.

HPLC should be considered complementary to gas chromatography (GC). Since HPLC depends upon the solubility of the substance rather than upon its vapor pressure, separations can usually be carried out under much milder conditions than would be required if GC were used. HPLC is typically performed at or slightly above room temperature. Although derivatization is used for some very specialized cases, the majority of compounds including ionic and polar compounds can be handled without such treatment. HPLC is also applicable for compounds with higher molecular weights than can be accommodated by GC. Figure 1 displays the approximate molecular weight ranges applicable to each technique.

The combined advantages of HPLC have led to very rapid growth in its use since it was introduced as a unique discipline in the late 1960s. The developments in HPLC now make it the preferred technique for many, if not most, pharmaceutical analyses. In most pharmaceutical applications, a simple dissolution and filtration suffice to prepare the sample for HPLC analysis.

The following sections attempt to provide a practical discussion of the components involved in HPLC. This discussion will serve as a guide for use of HPLC techniques in solving one's own problems. Many sections cover only basic knowledge essential for developing applications of this technique. As the practitioner becomes more experienced, he will certainly want to delve into the theory and extended discussions of various aspects of HPLC. The first eight references are books which provide good discussions of HPLC, in general, as well as some specific HPLC subjects.[1-8] Snyder and Kirkland,[3] Deyl et al.,[4] and Parris[5] provide very good discussions of basic chromatographic theory and its application to HPLC.

In addition to books, several journals routinely provide a great deal of guidance in the advancement of HPLC. The most popular journals for HPLC are *Journal of Chromatography*, *Journal of Liquid Chromatography*, and *LC-GC*. Other journals which have good HPLC articles include *Chromatographia*, *Journal of High Resolution Chromatography*, and *Ana-*

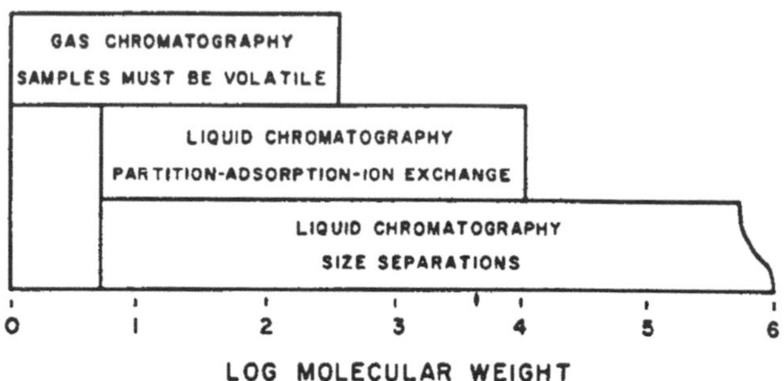

FIGURE 1. Molecular weight ranges for separation by several chromatographic techniques.

lytical Chemistry. The pharmaceutical and biomedical journals provide many articles on specific pharmaceutical applications using HPLC.

The last section of this chapter tabulates many different drugs for which HPLC procedures have been published. Many of these examples are from the official compendia for control of the drugs, while other examples are from recent publications pertaining to the analysis of drug materials.

II. THEORY AND NOMENCLATURE

Each analytical technique has its own theory and nomenclature. To understand the literature of a particular technique as well as to be able to communicate intelligently with others regarding the technique, a knowledge of some of its terms is mandatory. This section is therefore presented to familiarize the reader with some of the common terminology used in HPLC.

One of the fundamental quantities in any HPLC system is the *elution volume* of the compound. The elution volume is that volume of solvent which elutes from the column over the period from the time of sample injection to the time when the maximum concentration per unit time is eluting from the column. Retention time, which is the time between sample injection and elution of the peak maximum, is frequently used in place of retention volume. As long as the flow rate is constant, retention time and retention volume can be used interchangeably.

Another important term in HPLC is the *column volume, void volume,* or *dead volume.* All of these terms have the same meaning. The column void volume is the total volume of solvent within the column at any one time. The void volume is the total volume in the spaces between the particles, the spaces between the column walls and the packing material, and the spaces within the porous network of the particle. The column void volume is sometimes approximated as the elution volume of the sample solvent. Since most organic solvents are retained to some extent on common column packings, this procedure does not always provide a realistic value for the void volume.

The *interstitial column volume* is another related term that arises in exclusion chromatography. The interstitial volume is only the physical volume within the column between the particles and/or wall. The interstitial volume does not include the volume within the physical confines of the particulate packing material.

In most chromatographic procedures it is desirable to measure parameters relative to the nonretained peak or to the column void volume. The common term used for this parameter is the *capacity factor.* The capacity factor is usually indicated by k′ and is defined by the expression

$$k' = (V_R - V_0)/V_0 \tag{1}$$

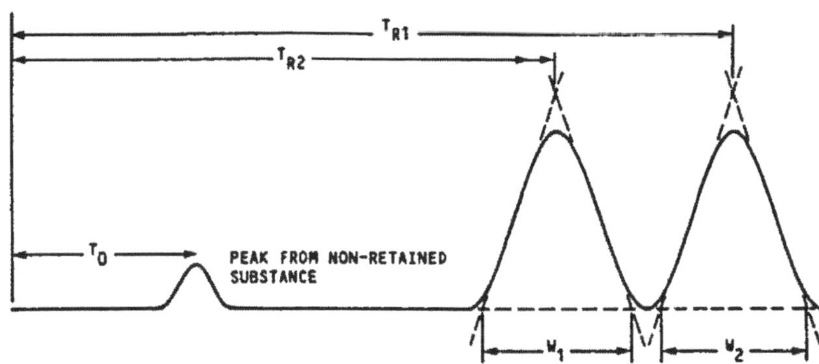

FIGURE 2. Definitions of retention times and peak widths commonly used in liquid chromatography.

where V_R is the retention volume of the sample and V_0 is the column void volume. An equivalent expression, applicable when the column flow rate is constant, is

$$k' = (T_R - T_0)/T_0 \tag{2}$$

where T_R is the retention time of the sample component and T_0 is the retention time for a nonretained substance on the column. These measurements are demonstrated in Figure 2.

The degree of separation of two components under a given set of conditions is described by the *separation factor.* The separation factor is defined as the ratio of the capacity factors.

$$\alpha = k'_B/k'_A \tag{3}$$

From this expression it can be seen that α must be greater than one to have separation of the two components. The larger the value of α, the greater the separation.

A quantity commonly used to describe the quality of chromatographic columns is *column efficiency.* This quantity is measured in *theoretical plates,* a carry-over from the terminology of distillation theory. The number of theoretical plates in a column is defined as

$$N = 16\ T_R^2/w^2 \tag{4}$$

where N is the number of theoretical plates, T_R is the retention time of the peak, and w is the width of the peak at the base measured in units of time. Since time, retention volume, and chart distance are all related by a constant conversion factor, any of them may be used in calculating the number of theoretical plates for a column.

Another term which is commonly encountered in discussions of HPLC columns is the *height of a theoretical plate.* This is usually referred to as H or HETP and is defined by

$$H = \text{column length}/N \tag{5}$$

The height of a theoretical plate is a measure of column efficiency and is a convenient method for judging the condition of a column. It provides a way to determine if one column is equal to or better than another column with the same packing material. As a column becomes more efficient, the peaks elute in a smaller volume (are narrower) while having the same retention volume. Thus, two components are more easily resolved on a column with a high efficiency than on a column of similar packing but possessing fewer theoretical plates. However, a high efficiency is not a sufficient condition for separation.

In addition to a high plate count, the column must also show different selectivities for the substances being separated. Assuming that there are differences in the behavior of the different sample components on the column, then as the column becomes more efficient, the separation of the two components will improve.

In practice, most work can be performed without paying a great deal of attention to column efficiency. Many of the compounds of interest in pharmaceutical analysis are not the best compounds for providing high efficiencies. In fact, most compounds will provide their own value for a column efficiency measurement, and the value may vary considerably from that reported by the manufacturer for a particular test compound. However, column efficiencies can be used to determine if a column is similar to a previous column and to detect the deterioration of columns with use.

The resolution of components from one another is an important parameter in HPLC applications. In analyses of raw material or formulated products, the drug substance should be resolved from its intermediates, degradation products, excipients, or other potential impurities. To assure that a column will perform adequately, a system suitability test is often included in an assay procedure. The usual measurement called for in such tests is the resolution of two critical components. The resolution is given by the following expression:

$$R = (V_2 - V_1)/[{}^1/_2(W_1 + W_2)] \tag{6}$$

where V_1 and V_2 are the retention volumes for the two components and W_1 and W_2 are the baseline widths of the two peaks expressed in terms of volume. For equal size peaks with a resolution of one, there is about 2% of the peak area which overlaps. For an R of about 1.5, the overlap is on the order of 0.1%.

The resolution of two components on a column can change as the column loses efficiency or as the characteristics of the column change. The column characteristics can change as materials remain on the column and are not eluted by the column solvent. In some cases, samples placed on the column may react with it, and thus change its characteristics.

Two other important terms are commonly found in the description of solvent systems used in liquid chromatographic separations. The most common mode of operation is the isocratic mode. The term isocratic describes a solvent system which remains constant in composition throughout the analysis. The second mode of operation commonly encountered in HPLC is gradient elution. In this mode of operation, the composition of the solvent changes during the chromatographic run.

III. INSTRUMENTATION

A high-performance liquid chromatograph contains several distinct components. As a result, much of the HPLC instrumentation that is or has been commercially available has been of modular design. This design feature has been very beneficial since instruments can be updated by adding components or by replacing antiquated components.

A block diagram of a high-performance liquid chromatograph is given in Figure 3.

An HPLC instrument consists of a solvent supply system feeding a pumping device. The pump is coupled to a column through some sort of sample introduction system. The eluate from the column then passes through a detector, and the response of the detector is measured by an analogue or digital output device.

A. Pumps

The pump is one of the most important parts of any HPLC system. Many different types of pumps have been employed in the practice of HPLC. The different pumping techniques are given in Table 1. Pumps may be operated as constant pressure delivery systems or constant displacement systems.

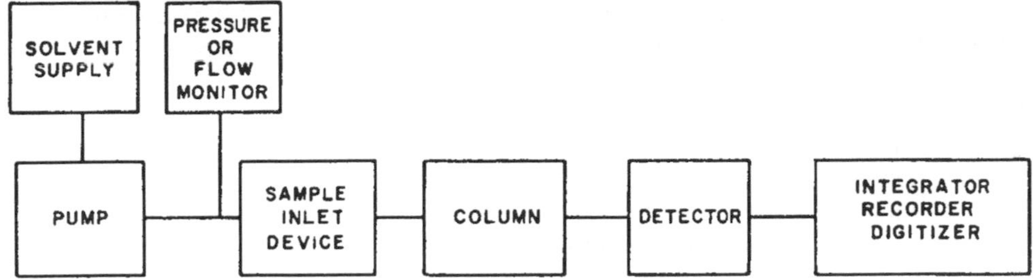

FIGURE 3. Block diagram of an HPLC instrument.

Table 1
PUMPING DEVICES

Constant pressure devices
- Pneumatic
- Pneumatic amplifier

Constant displacement devices
- Syringe pumps
- Reciprocating pumps

The pneumatic pump is the simplest of the pumps. This device is of the constant-pressure type and consists of a high-pressure reservoir which contains the HPLC solvent. Gas pressure is applied to the head of the fluid to provide the driving force to push the liquid through the column and detector. The major problem with this device is that it often saturates the solvent with the driving gas. This will eventually cause problems, for the solvent will degas or form bubbles in the detector. More sophisticated versions of the pump have placed a flexible wall container holding the solvent inside a high-pressure reservoir.[9] In this configuration, the gas pressure is applied to the high-pressure reservoir, but the gas does not come in direct contact with the chromatographic solvent. The pressure limitations of these devices are typically 1000 to 2000 psi or the pressure limit for the cylinder of the driving gas.

Pneumatic amplifier pumps are also constant pressure devices and were used in some early HPLC instruments. In these devices, a low-pressure gas stream is used to drive one piston with a large surface. A standard hydraulic design is used to amplify the force and apply it to a smaller piston which then drives the solvent through the chromatograph. These systems generally suffer from the limited volume of solvent which can be delivered in one cycle and the difficulty of using them in gradient systems. This type of pump has almost totally disappeared from use in HPLC except for one application. That one remaining application is column packing. The dominant pump for column packing is a pneumatic amplifier pump manufactured by Haskell.

The second class of pumping devices is the constant displacement pumps. These are often called ''constant-volume pumps'', but few of them make any allowance for the compressibility of the solvent, which may result in variations of a few percent in the flow rate. There are two basic types of constant-displacement pumps.

The syringe pump is one form of the constant-displacement pump. These pumps have a delivery volume of 200 to 500 mℓ of solvent. When this volume of solvent has been expended, the HPLC system must be shut down and the pump refilled. The syringe pumps are reasonably pulse-free devices. The major drawback with this type of pump is the limited volume of solvent that is available before a slow refilling process must be done. Syringe pumps have therefore fallen from favor for use in general HPLC. Today, they are used only in very specialized applications such as microbore HPLC. In this application the volume limitation is of small concern because the pump volume is still large with respect to the volume of solvent required for any run. The pulse-free flow characteristic of these pumps is also very desirable for microbore operations.

The reciprocating pump is the last of the major types of pumps which have been used in HPLC. The favorable characteristics of the reciprocating pump, such as fast cycle times, constant volume delivery, and low solvent retention in the pump, have made this the most popular type of pump. In fact, reciprocating pumps are used almost exclusively in HPLC today. The major disadvantage of this type of pump is the pulsation in its output. Many different principles have been applied to overcome this undesirable characteristic of the pump.

The first reciprocating pumps used in HPLC were single-piston devices. These suffered from the fact that the solvent is only being delivered by the pump during one half of the pump cycle. During the remainder of the cycle the pump is refilling with solvent. The flow and pressure across the column both decay during the fill stroke of the pump. This leads to a flow rate and pressure pulse in the system. While this is averaged over time to give a constant volume delivery over a long period, the instantaneous flow rates vary considerably. Various pressure- and/or flow-regulating devices which are described later in this section can be added to minimize the magnitude of the pulse seen by the total HPLC system. The Milton-Roy minipump is the major pump of this design used in HPLC applications.

One of the early methods used to alleviate the pulsing problem with the single-piston pump is one which allowed the filling cycle of the pump to be very fast with respect to the total pump cycle. With this approach, the period of no solvent delivery is small compared to the time of one complete pump cycle. Some of the more popular designs which fit into this classification are the Beckman 110 series pumps and the Kontron and Shimadzu pumps.

The single-piston pump with fast refill can be taken one step further by monitoring pressure and allowing the first part of the delivery cycle to run fast to bring the pressure in the pump head up to the pressure in the remainder of the system. The major pumps in this category are the Beckman 114 series and the Varian 5000 series pumps.

Another principle that has been used to reduce the pulsing nature of the piston pump is to add multiple pistons. By adding a second piston the pulsing of the pumps can be reduced because the only time a pulse can occur is during the depressurization of one pump head, while the second pump head is being pressurized at the start of its delivery cycle. This has been a very popular approach in modern HPLC instruments. Some of the pumps that use this configuration are the Waters 6000 series, Varian (Jasco), Micromeritics, and Perkin Elmer.

The concept of adding pistons to reduce pulsations has also been tried. The addition of a third piston eliminates yet more of the pulsing from the pump system. The major commercial pumps in this category are the DuPont and IBM systems. The DuPont system is now manufactured and marketed by Anspec.

All of the reciprocating pumps described above use an inlet and outlet check valve to control the direction of flow in and out of each piston. For these pumps to function properly, the check valves must work perfectly. In reality, the check valves are a major point of failure for any of these pumps. To some extent, the reliability of these systems is inversely proportional to the number of check valves in the system. To partially address this problem as well as reduce manufacturing costs, a set of tandem piston pumps has been marketed. These pumps use two pistons, but in a different manner than normal two-piston pumps. In these systems, the size and delivery volume of each piston and chamber are different. The first piston in these pumps delivers solvent to fill the second piston while also delivering solvent to the column. With this type of system, check valves are needed only for the first piston, since the second piston always operates on the high-pressure side of the pump. The Waters M45 and Kratos pumps belong in this category.

In all of the pumps discussed to this point, the driving piston comes in direct contact with the HPLC solvent. A potential problem with this arrangement is that the solvent may also contact the piston seal. The salt concentrations and lubricating nature of the solvent can affect the seals, leading to early failure of the seal. A diaphragm pump has been used in some HPLC applications as a means for avoiding the interaction of the solvent with the pump seal. In a diaphragm system, an oil or other hydraulic fluid fills a portion of the pump head. A second part of the head is exposed to the HPLC solvent with a diaphragm separating the two chambers. The driving piston and seal for this system work on the side with hydraulic fluid and are not exposed to the potentially harsh effects of the HPLC solvent. Hewlett-Packard pumps operate on this principle.

Various devices can be incorporated into these pumps to further reduce pulsation. One of the best devices is a chamber containing solvent that is incorporated within the pump system. Pulse dampening increases as the chamber volume increases. The primary disadvantage of this pulse-dampening system is that the relatively large volume of the chamber severely restricts the rate at which solvent composition can be changed for gradient operation. Alternates to the solvent chamber that are often used in pump designs are the coiled compression spring and Bourdon tube. These devices can be made to contain rather low volumes of solvent. As the pulse occurs, the device flexes and part of the energy of the pulse is taken up in this mechanical operation. More recently, the trapped solvent approach has been revived by incorporating a diaphragm in the chamber between the HPLC solvent and the solvent trapped within the damper. These devices work very well and are now being incorporated into many gradient HPLC systems. The SSI "LO-PULSE" is a good example of this kind of device.

Reciprocating pumps with various types of pulse dampers are the pumps of choice for HPLC. Essentially all commercial HPLC instruments today incorporate some form of reciprocating pump. These pumps are convenient in that they can be operated at both low and high flow rates. Therefore they can be used in both very slow flow rate analytical work or in very high flow rate preparative HPLC applications. These pumps can have very good flow rate precision. For most of these devices, the flow rate precision has a standard deviation of 1% or less. For the better HPLC pumps, flow rate precision can be better than 0.1%.

B. Gradient Pump Systems

The power of HPLC separations can often be increased by use of gradient techniques. In the gradient mode, the composition of the column eluting solvent is changed during the run. Depending upon the mode of separation, the solvent change may consist of a change in pH, ionic strength, or solvent strength by changing the percentages of the component solvents comprising the eluting solvent. Gradient pump systems are used to achieve these changes in solvent composition. Gradient elution is often used during method development to select appropriate solvent compositions for a given separation. Gradients are also used to achieve separation when analyzing samples containing many different types of components, or when evaluating impurities in a sample.

Two different instrument configurations are commonly used for the formation of gradients. The first uses multiple pumps and mixes the solvents on the high-pressure side of the pumps before reaching the HPLC column. The second configuration uses a single pump with other auxiliary hardware to mix the solvents on the low-pressure side of the pump. Frequently this is accomplished as the solvent is pulled into the pump.

Most of the early HPLC systems used the multiple-pump approach with formation of the gradient on the high-pressure side of the HPLC system. In this type of system, the outputs of two or more pumps are brought together in a tee or other appropriate mixing device. The composition of the solvent is then changed by varying the speed of the pumps to achieve the proper proportion of each of the "pure" solvents to the system. The "pure" solvent in

Table 2
HPLC INJECTION DEVICES

Septum injectors
- Septumless injectors
- Stop flow injectors
- Valve injectors

Autoinjector systems
- Syringe systems
- Loop valve system

each pump maintains the same composition throughout the run, although each may in fact be a mixture of several components. This system for forming gradients has several disadvantages associated with its use. First, the gradient formed is dependent upon the ability of the pumps to deliver the set solvent composition at any time. In reality, many of the pumps do not perform well at very low speeds. Thus the accuracy of the gradient is often inadequate when the delivery of one of the pumps falls below 5 to 10% of the total volume pumped by the system. A second and major disadvantage of this approach lies in the fact that an expensive pump is needed for each solvent. This becomes even more of a disadvantage as more than two solvents are used in a gradient run. The high equipment cost associated with this type of gradient HPLC system has caused it to fall from favor for general use. More recently, however, these systems are seeing a resurgence in work with proteins. This type of system seems to give better results in many protein separations which require very good control over a very narrow range of solvent compositions.

The single-pump gradient HPLC systems are by far the most popular today. These systems control the composition of the solvent at the point of entry of the solvents into the pump. A set of proportioning valves are placed in the pump inlet manifold. The proportioning valves are controlled by the system controller and their delivery times are adjusted during the filling cycle of the pump so that the correct composition of solvent is supplied to the pump. The solvents mix as they enter the pump. In addition, a mixing chamber is incorporated into the system between the pump and the sample inlet valve.

Various types of controllers are available for gradient HPLC units. Normally the pump and controller system are purchased as an integrated unit from one manufacturer. Many of the earlier programmer designs for gradient instruments allow one to select the shape of the gradient from a series of linear or exponential curves with the duration of the curve also selectable. Once a run is initiated, the programmer controls the pumps and/or the supply of solvent to the pumps to generate the solvent profile over the specified time. A second approach to gradient control utilizes a matrix programmer which breaks the curve into a series of segments. The operator is then allowed to control the curve by varying the slope or rate of change for each segment along with the time period for execution of the section.

More recently, computer-based programmers have become the norm in HPLC instruments. Frequently these systems are capable of controlling mixing of three or four different solvents in the HPLC system. The computer-based programmers are generally fashioned after the matrix-type gradient programmers. They can also control other functions related to the HPLC such as control of sampling valves, data systems, and column switching valves.

C. Sample Inlet System

The sample introduction system is an important part of any HPLC system. Table 2 lists the primary types of injection systems that have been used. Early sample introduction systems made use of septum and stop-flow injectors in which syringes were used to introduce the sample into the HPLC system near the head of the column. With septum injectors, the syringe penetrated the septum and the sample was injected into the flowing stream. In stop-flow systems, the column flow was stopped and the system was actually opened near the

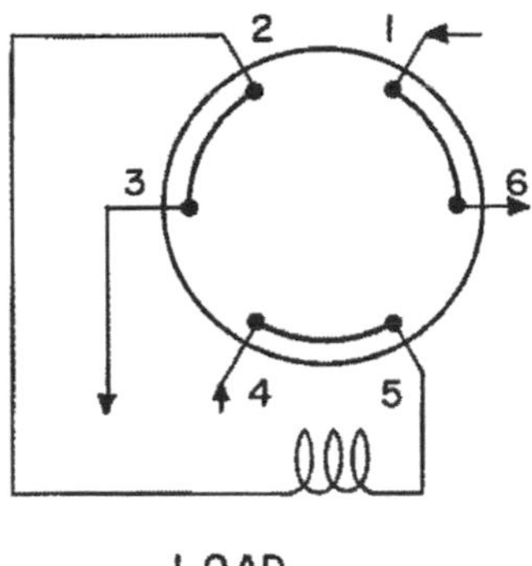

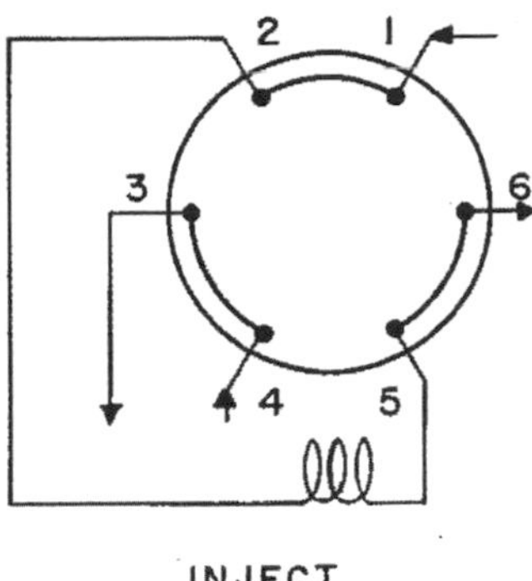

FIGURE 4. Typical configuration of a six-port, two-position sampling valve for HPLC.

head of the column. The sample was carefully placed near the head of the column using a syringe. The system was then resealed and the flow restarted. These devices were very cumbersome and subject to leaks. They have been almost universally replaced by various forms of sampling valves.

Most sampling valve systems in use today are designed around a six-port sampling valve which has two operating positions. The basic valve configuration is shown in Figure 4. In this type of valve, a sample loop is connected between two of the valve ports. For volumes of 10 $\mu\ell$ and larger, these loops are external to the body of the valve. For sample loops smaller than 10 $\mu\ell$, the loop will normally be internal to the body of the valve. One port of the valve serves as the connection to the pump and serves as the solvent inlet to the remainder of the HPLC system. An adjacent port serves as the outlet of the valve and is connected to the HPLC column. The remaining two adjacent ports serve as the sample inlet and outlet ports.

In operation, a sample is first loaded into the sample loop of the valve. This is accomplished while the valve is in the "load" position. The sample port is plumbed so that the sample stream will flow through the loop and to the outlet port for waste sample. In the load mode, the pump and column are connected directly through the valve. When a sample has been loaded, the valve is switched to the second position. Normally this involves turning the valve stator by 60°. The valve switch can be done manually or by an electric or pneumatic driver. The sample loop is now placed in the flow between the pump and the column. The sample is then swept from the sample loop directly onto the HPLC column.

Sampling valves are operated in one of two modes. One mode of operation is the filled-loop mode. The second mode uses only partial filling of the sample loop with the sample.

The filled-loop mode of operation requires that the entire volume of the loop be filled with sample in the loading operation. This mode of operation requires that the loop be flushed well so that all of the chromatographic solvent in the loop is fully replaced with sample. This normally requires flushing the loop with a volume of sample which is eight to ten times the volume of the loop. A syringe or an automatic sampling device can be used to push the sample through the loop of the sampling valve.

The filled-loop technique generally provides the best precision in sample volume. The good precision characteristic of this mode of operation is due to the fact that the volume of the loop is fixed and invariant. To assure that this advantage is obtained, it is crucial that adequate sample be passed through the valve to insure that the sample has replaced all solvent which was in the valve loop. Care must also be taken that bubbles are not formed in the sample flow system. Bubbles can lodge in the loop and cause a variation in the volume which is eventually injected onto the column.

The filled-loop mode of operation for HPLC sampling valves is generally the preferred method of operation. This type of operation does, however, have its disadvantages and limitations. The major disadvantage is the size of the injection. The size of the injection is determined by the loop. To change the size of the injection volume, the loop must be physically replaced with a loop of a different size. Loop volumes of 10, 20, and 25 $\mu\ell$ are the common sizes in routine use. For special applications, larger loops such as 50 or 100 $\mu\ell$ may be used. Smaller sample loops may be used in some specialized micro-HPLC techniques. Valves using smaller sample loops usually have the loop machined directly in the rotor. Thus the valve must be disassembled to change the injection volumes when 0.1- to 5-$\mu\ell$ loops are used. A second limitation of the filled-loop technique lies in the volume of sample required to give reproducible results. Normally a sample volume of eight to ten times the volume of the loop should be flushed through the loop before a sample volume is trapped for injection. These valves are therefore very wasteful of sample. This mode of operation may not be desirable when the amount of sample is limited.

Most HPLC injection valves can also be operated in a partially filled mode. With this technique, a syringe is used to force a known volume of sample into the loop of the sampling valve. Sample loops with volumes of 50 to 200 $\mu\ell$ are typical in this mode of operation. The valve is plumbed so that the sample is placed in the loop on the side adjacent to the column. This configuration avoids much of the band spreading which would be associated with the dead volume in the system. Care must be exercised in selecting the amount of sample to be placed into the loop. Normally one should limit the volume of sample to 60 to 70% of the volume of the loop. If the sample volume is allowed to exceed this limit, part of the sample may be lost out of the exit of the sample loop. In the filling operation, the nature of the flow is not plug-like, but will probably spread. Therefore sample will be lost from the loop long before one replacement of the volume is completed. The advantage of this mode of operation is that the actual sample volume can easily be varied. A second advantage is that it requires only the volume of sample actually injected on the column. The negative side is that the precision of the sample volume is now determined manually by measurement in a syringe.

Sampling valves can usually be operated manually or used as part of an automated sampling system. Most automated HPLC sampling systems transfer a solution from a turntable to the loop of an HPLC sampling valve. Thus the sample must be in an appropriate solvent before these devices are functional. Automated sampling systems generally consist of circuitry to switch the sample valve as well as to drive the liquid from a sample vial into the sample loop. Many sample systems pressurize the vials with a gas to force the solution through the valve. Some sample systems use a piston-like cap which is forced down the vial to push the solution through the sampling valve. These devices usually require significant volumes of sample. One half milliliter to several milliliters of solution is generally required. Today many automated samplers can be refrigerated and operated in a cold mode. This is often advantageous when working with biological samples or protein materials.

There are some automatic samplers that operate in the partially filled mode. They usually incorporate a syringe which pulls the sample from a vial and then transfers it to the loop of the sampling valve. The Water's WISP works on this principle. Hewlett-Packard offers a sampler which uses a high-pressure syringe that can function with a small sample volume. Both of these devices are capable of handling samples with limited sample volume.

Robotics are also having an impact in automated sample handling for HPLC. Chromatographic applications are the major users of laboratory robotics today.[10] Laboratory robotics offer significant advantages where special sampling handling is needed. Robotics are advantageous whenever it is desirable to keep the sample dry in order to maintian its stability until the time of injection. This is often important when one is looking for impurities or degradation products which may change with time. Robotics can also be used when the sample must be processed or derivatized in some manner. Robotic systems are, however, fairly complex and require a great amount of effort in their implementation. Ideally chromatographic applications should be developed using a genetic sampling scheme so that the same robotic sample preparation will work for many different chromatographic procedures. Although this field is developing rapidly, the complexity and reliability of the robotic systems limit them to specialized applications at present.

D. Columns

The HPLC column is the heart of all HPLC systems. HPLC columns take many different forms today, and can be purchased from a variety of sources. Only a few of the primary instrument manufacturers still sell columns packed with proprietary stationary phases. Most packings are available in bulk as well as in prepacked columns. When bulk packings are available, columns can be packed in the users laboratory. The means by which these columns are packed will be discussed in a later section.

For many years, the standard HPLC columns were 15 or 25 cm in length and were packed with 5- to 10-μm materials. More recently a greater emphasis has been placed on the speed of analysis, resulting in the introduction of columns packed with 2- and 3-μm particles. With these small materials, large numbers of theoretical plates can be generated in a short period of time. With these small particle diameters, short columns give results equivalent to longer columns packed with coarser particles. Thus 3- 5- and 10-cm columns have become available packed with the smaller particles. These columns often give 1- to 2-min chromatographic runs. The short columns do, however, place great demands upon the HPLC system. Column inlets and detectors must be designed with extremely small dead volumes. In reality, very few HPLC systems have been designed for effective use of these very short columns.

In spite of the rising use of small particles and short columns, the standard HPLC column is still the 15- to 25-cm tube packed with 5- to 10-μm particulate packing. The analytical versions of these columns normally have a 4- to 4.6-mm bed diameter. These columns are available with a host of incompatible end fittings. Almost every manufacturer has developed his own column end fitting. Most fittings use an inverted nut system for plumbing into and out of the column. Figure 5 shows a typical inverted nut fitting. These fittings are much less susceptible to damage than the unions that used to be used as terminating devices.

HPLC columns generally have a limited life. Typical problems that develop with the use of a column are discussed in Section V.E. A major cost in column replacement has been the replacement cost of the column hardware. If terminating fittings could be reused or recycled in a convenient manner, the procurement cost of HPLC columns might be reduced. The advent of cartridge HPLC columns has made a great advance along this line. Today, there are many different cartridge column systems available in which the columns are purchased prepacked and the end fittings are attached when the column is used. Most major HPLC column manufacturers now offer some type of cartridge system. Most of the cartridge columns consist of a stainless steel tube with an imbedded frit to retain the packing within the tube.

In addition to the steel cartridge systems, Millipore (Waters) offers a cartridge system using a plastic casing that is marketed under the Radial Pak trademark. These columns are used in a special device which compresses the outer walls of the column to eliminate column void areas. Thus the flexible plastic wall allows for the compression of the column bed. The columns used with this system are of slightly larger diameter than those normally used.

E. Detectors

An integral part of any HPLC system is the detector at the end of the column. Ideally, a detector should have good sensitivity to all eluting components. It should be reasonably linear so that it can be used in quantitative analysis. It should not significantly degrade the separation obtained in the column. Finally, the detector should be reliable and easy to operate.

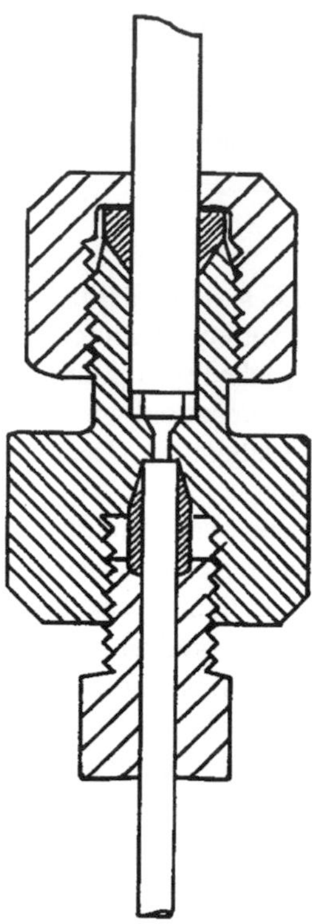

FIGURE 5. Column end fitting using an inverted nut configuration. When properly installed, the tubing ends butt against the body of the nut leaving no dead space.

Table 3
HPLC DETECTORS

Photometric detectors
- Single wavelength UV
- Variable wavelength UV
- Diode array detectors
- Fluorescence detectors
- IR detectors

Miscellaneous detectors
- Refractive index detector
- Flame ionization
- Electron capture
- Conductivity
- Electrochemical
- Radioactivity detector
- Chemical reaction detectors
- Mass spectrometers
- Photoionization detectors

Many detectors have been proposed for HPLC, and a number of these devices are commercially available. Table 3 gives a partial listing of the types of detectors which have been employed in HPLC.

Unfortunately, no practical universal detector has been developed for HPLC. Therefore, a detection device must be chosen which provides adequate sensitivity for a particular HPLC problem.

1. Photometric Detectors

Photometric detectors are by far the most important detectors in pharmaceutical analysis, and the most important of the photometric detectors is the UV detector. Both single wavelength and variable wavelength detectors are available. The single wave-length detectors generally use a low-pressure mercury arc lamp which has a high intensity output at about 254 nm. Some of these devices use phototubes or photomultipliers to measure the light intensity passing through the cell, but most detectors utilize photoresistors. The photoresistors are most sensitive in the visible region around 450 nm, so the light must be changed from 254 nm to 450 nm prior to detection. This is accomplished by inserting a phosphor that absorbs at 254 nm and emits near 450 nm into the optical system. The detectors may also be operated at longer wavelengths by inserting a suitable phosphor into the light path ahead of the sample cell. The phosphor absorbs light at 254 nm and emits light at some higher wavelength. The major disadvantage of these devices is that the phosphors emit in a rather broad band so that the usable source intensity is very low. Therefore, the response of these detectors is only linear over a very narrow range. Several phosphor combinations are available, but only the 280 nm option is in routine use.

Operation of HPLC detectors at lower wavelengths offers the potential for better sensitivity and detection of a wider range of compound classes. To obtain this goal, a stable light source is required with a high intensity line at some lower wavelength. A zinc source has been developed which meets many of the criteria for a single wavelength source. The zinc source has a 214 nm emission and is being used routinely in many instruments. Cadmium sources for use at 229 nm and magnesium sources for use as 206 nm are also now available.

HPLC detectors with micro flow cells present a very good compromise for optimization of HPLC instrumentation and detection sensitivity. However, spurious peaks caused by light scattering and refractive index changes of substances coming through the cell are observed with these detectors. The scattered light causes a decrease in the light intensity coming through the cell and of course shows as an absorption peak. Later versions of these detectors from some manufacturers have incorporated angled cell walls which tend to minimize the scattering effects. These single wavelength detectors can provide very good sensitivity at each of their operating wavelengths. The signal-to-noise ratios of these devices is generally very high, and these devices probably cannot be surpassed for sensitivity and noise when one of the three available wavelengths coincide with a major absorption band on the analyte. A major problem with these devices is the rather short life of the phosphors used in them.

There are many compounds which are not ideally suited for monitoring at the available fixed wavelengths. Therefore, in an effort to increase detector sensitivity, people started using variable spectrophotometers to allow monitoring at the adsorption maxima of the compound. The first variable wavelength detectors were regular spectrophotometers fitted with micro flow cell assemblies. Many of these early devices suffered from stray light problems, low light levels, low sensitivity and narrow linearity. Today, many devices have been designed and built for HPLC applications. Many devices are now available which permit monitoring at two or more wavelengths simultaneously. Such devices are advantageous when working with complex sample mixtures or when trying to differentiate between two or more poorly resolved components.

In choosing a spectrophotometric detector, consideration should be given to the intended use of the device and to the sensitivity which will be required. The single wavelength 254 nm detector usually has lower noise levels and can be operated at more sensitive settings than most variable wavelength detectors. However, the sensitivity for a given compound can often be improved by monitoring the column eluate at the absorption maxima for the compound rather than at 254 nm. In pharmaceutical analysis, there are many important classes of compounds that have major absorbance bands below 254 nm.

Other considerations must also be given to the construction and design of the HPLC

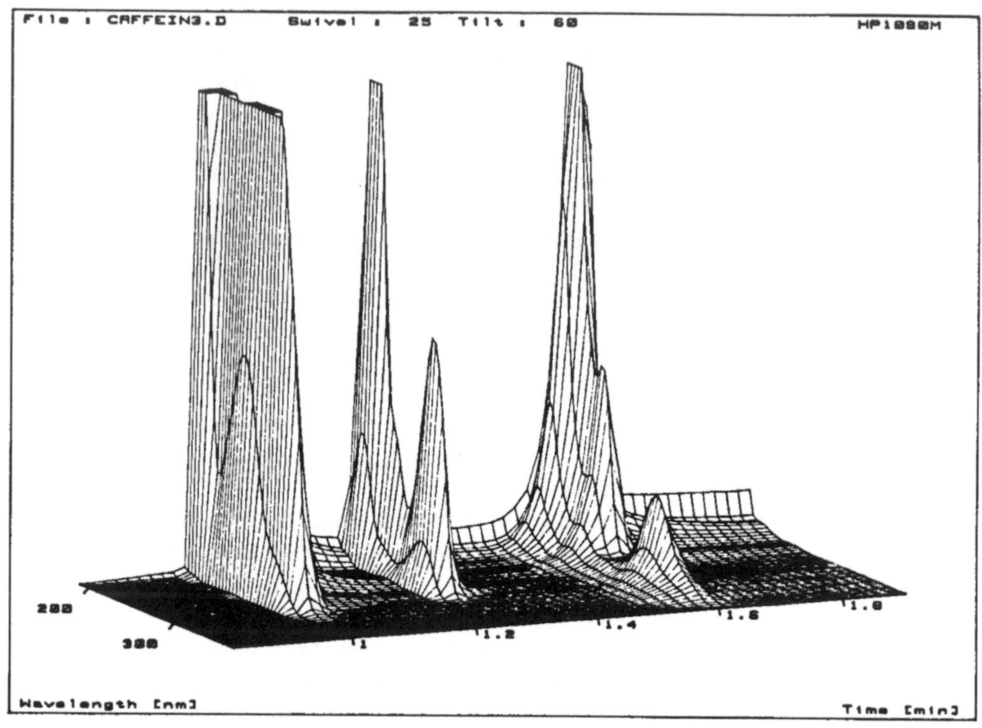

FIGURE 6. Three-dimensional plot of the absorbance of the column eluent as a function of both time and wavelength during an HPLC separation of caffeine.

detector. There are many commercially available HPLC spectrophotometric detectors which severely affect the separation obtained from the HPLC column. Most spectrophotometric HPLC detectors and in particular the optical cells used in these devices have been designed to give good sensitivity while offering little resistance to flow. As a result, many of these devices are compromise systems where the separating efficiency (plates) of the column are sacrificed for versatility of the device. In selecting a detector, care should be exercised to select a design which does not significantly degrade the separation while offering good sensitivity with wide linear range of operation. Unfortunately, there are many commercial detectors which have severe drawbacks in one or more of the basic goals for a good UV detector.

Photometric detectors for liquid chromatography can be used to observe changes of about 5×10^{-5} absorbance units in the column eluate. For a sample with a molar absorptivity of 10,000 and a molecular weight of 200, this is equivalent to about 10^{-9} g/mℓ in the column eluate.

Most variable wavelength detectors also operate into the visible region of the electromagentic spectrum, but this capability is seldom used. The few exceptions are generally instances where a colored derivative of a compound is formed and monitored, or where a dye or pigment is being analyzed.

The photodiode array detectors are a valuable advancement in specialized HPLC photometric detectors. These devices allow continuous monitoring of the entire ultraviolet spectrum throughout the chromatographic run. This device is ideal for monitoring the chromatogram for multiple components. The spectrum of each component in the sample can be obtained. Several different plotting techniques can then be used to display the data collected from the detector.

Figure 6 is a three-dimensional plot of the wavelength, time, and absorbance of the

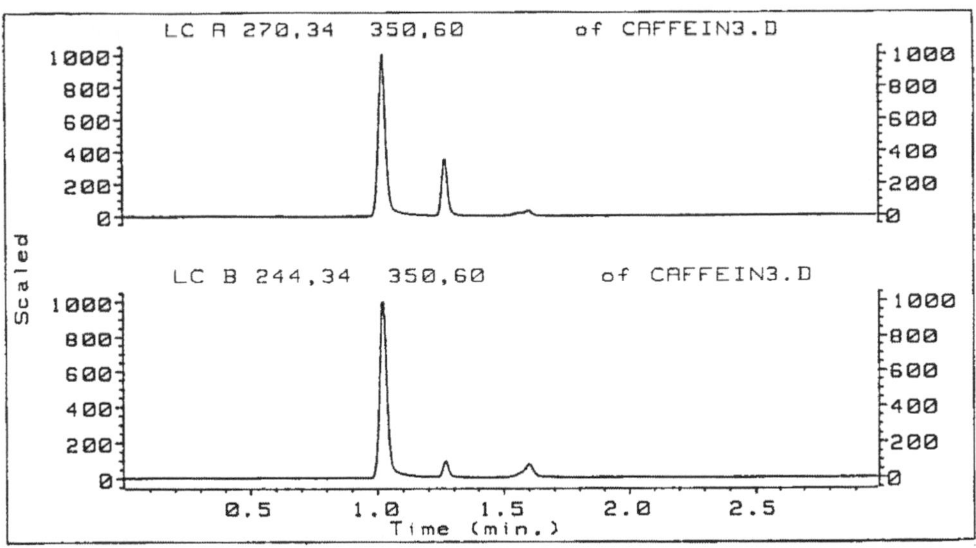

FIGURE 7. Plots of the absorbance of the column eluent vs. time at 350 and 360 nm during the same separation of caffeine shown in Figure 6.

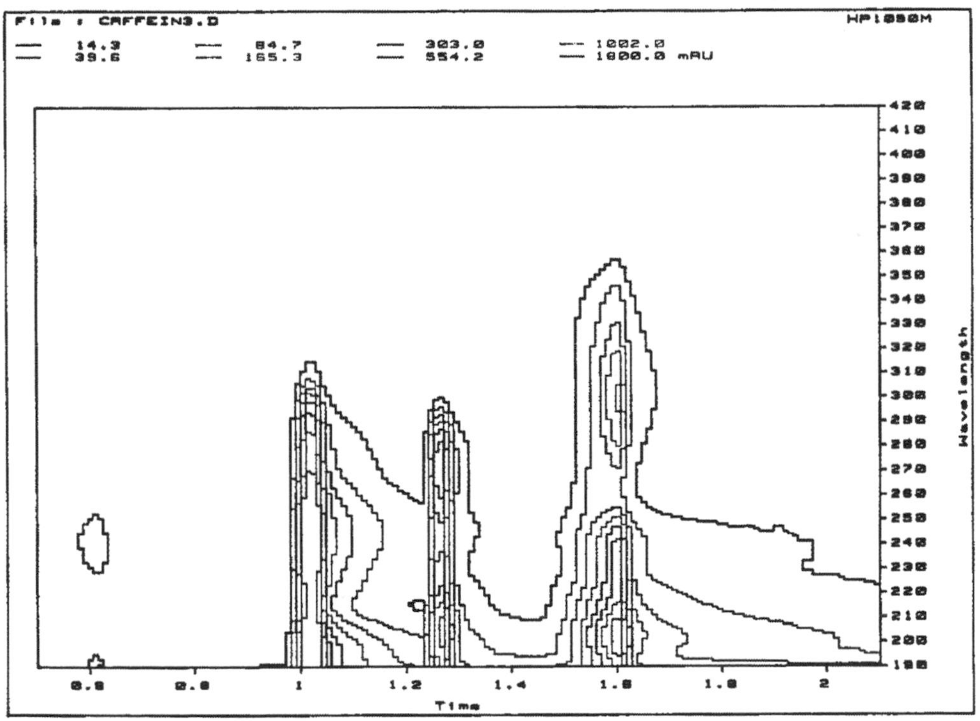

FIGURE 8. Contour plot of the absorbance vs. time and wavelength during the same separation of caffeine that is shown in Figure 6.

chromatographic eluent through a single run. This type of plot can also be rotated so that one might see hidden components lying behind a major peak. Figure 7 is a cut across the chromatogram at one wavelength. This would be the typical output from a single wavelength detector. Figure 8 is a contour plot of the components in the chromatogram. This provides an indication of where the absorption maxima are located in the time-wavelength plane.

The diode array detector provides a wealth of information from a single chromatographic run. It is ideal for monitoring purity of a component eluting from the column. It also offers a means for detecting all of the components which have a chromophore absorbing in the ultraviolet.

IR photometers for monitoring column eluates are also commercially available, but the solvents that can be used with these devices are very limited and they have not been used to any appreciable extent in pharmaceutical analysis. The principal applications of IR detectors are in polymer analysis.

Fluorimetric detectors are generally available and offer the advantages of higher sensitivity and greater selectivity than UV detectors. The high sensitivity of this detector, which is typically on the order of 10^{-11} g,[10] has resulted in its extensive use in environmental and biomedical analysis. However, fluorimetric detection of most substances requires that a fluorescent derivative be prepared prior to chromatography, and therefore it is the detector of choice only in those cases where sensitivity is a serious problem and where a means can be devised for preparation of a suitable derivative.

2. Electrochemical Detectors

Electrochemical detectors are now widely used for easily oxidized and reduced organic compounds. Detection limits of about 0.1 pmol can frequently be achieved. Electrochemical detectors operate on either a direct-current or a pulsed-current amperometric principle. Carbon paste electrodes may be used, although vitreous ("glassy") carbon electrodes are preferred for oxidation reactions. Mercury pools or amalgamated metal electrodes are preferred for electrochemical reductions.

Electrochemical detectors are often used for determination of phenols and aromatic amines. Thiols are another good electroactive group which can be easily oxidized to a disulfide. Quinones and nitro compounds are good candidates for reduction with an electrochemical detector.

3. Chemical Reaction Detectors

The popularity of postcolumn reactors has increased considerably since the introduction of HPLC. The reactors are used to convert the analyte into products that can be detected more easily or with greater selectivity than the analyte itself. Postcolumn reactors have been used with ultraviolet, visible, fluorescence, chemiluminescence, and electrochemical detection.

Postcolumn reaction detectors minimize the formation of undesirable by-products because sample components are separated before reaction. However, the need for mixing reagents and effecting a reaction following separation on the column results in detectors with greater effective dead volume than more conventional detectors. The increased dead volume leads to band broadening with loss of apparent efficiency in the separation. To overcome this disadvantage, many reaction detectors utilize an air segmented flow stream or use reaction chambers which are packed with small particles to reduce dead volume and minimize band broadening. The devices work well for fast reactions, but lose their effectiveness if the chemical reaction takes longer than about 5 min. Restrictions on reaction rates are partially alleviated because the reaction does not have to be driven to completion, but only to a reproducible point. Another limitation in applying these detectors results from the need to use solvents in the chromatographic separation that are compatible with the proposed reaction.

4. Refractive Index Detectors

The refractive index (RI) detector was one of the first devices used routinely as an HPLC detector. This detector responds to most substances, and its sensitivity is related to the difference in refractive index between the analyte and mobile phase. The newer models offer sensitivities on the order of 10^{-8} g/mℓ under favorable conditions. The high sensitivity of these detectors is due in part to use of temperature control in the cell area as well as in the solvent transport area ahead of the cell. The temperature coefficient of a typical RI detector is 1.1×10^{-4} RIU/°C.[11] The cell temperature must be controlled to a few thousandths of a degree to keep noise to tolerable levels. The RI detector is also very sensitive to pressure variations. Therefore pump pulsation must be kept to an absolute minimum when using the RI detector at high sensitivity. The one big disadvantage of the RI detector is the inability of the device to perform in the gradient elution mode of HPLC. It responds to changes in the solvent composition and is therefore only useful in isocratic separations.

The RI detector is often the device of choice in preparative HPLC. The universal nature of the RI detector is a very desirable characteristic in preparative work. The RI detector is also better suited than most other detectors to measurements at the high concentrations typically found in preparative HPLC.

5. Radioactivity Detectors

Several commercial devices are available for monitoring beta radiation in HPLC eluents. Some of these devices mix a scintillator fluid with the column eluent before passing it through the cell. In others, the eluent is allowed to flow directly to a cell which is packed with a solid scintillator such as anthracene or Ce-activated glass.

Radioactive detectors are relatively insensitive. The counting efficiency for tritium may be 1% or less. To compensate for this deficiency, compromises are often made. Large cells are often used which considerably degrade the separation obtained on the column. Low flow rates must be used to increase the residence time of the sample in the counting chamber. Reeve and Crozier[12] present a good guide for the use and application of these devices. Overall, radioactivity detectors find very limited use with their main utility lying in biological studies of drug metabolism and distribution.

6. Electron Capture Detectors

Electron capture detectors are becoming available. The device could find some interesting applications in the analysis of special types of compounds. Unfortunately, there is a solvent compatibility problem which may limit its overall effectiveness as an HPLC detector

7. Photoionization Detectors

The photoionization detector is a relatively new type of HPLC detector and may prove to be one of the more universal detectors for HPLC use. In this device, the sample and solvent are vaporized as they elute from the column. The sample stream is then passed by a UV lamp emitting 10.2 eV photons, which ionize many materials. A series of acceleration electrodes and collection electrodes are then used to measure the current in this system. As more ions are present, the current levels of the detector increase. Fortunately, most of the common HPLC solvents such as water, methanol, methylene chloride, and acetonitrile all have ionization potentials greater than 10.2 eV and are not ionized by the source. The sensitivity for this device is reported to be about 4.5×10^{-12} g of material. The major disadvantage would appear to be in the fact that the sample must be vaporized for this device to function. Currently, the vaporization causes a large dilution of the sample which affects the sensitivity limit of the detector.

8. Conductivity Detectors

The conductivity detector is a very specialized detector which is used in aqueous mobile phases for the detection of ionic species. This detector is widely used in ion chromatography (IC). The sensitivity of these devices is about 10^{-8} g/mℓ. These detectors require good temperature control because they are sensitive to temperature fluctuations.

Two types of conductivity detectors intended for use in IC are available commercially. One type of IC system employs a conductivity suppression system to enhance the sensitivity of the conductivity detector. In this approach, the conductivity of the mobile phase is reduced by use of membrane technology or ion exchange suppressor columns. The second type of IC system employs buffers which have relatively low conductivity and are not removed from the column eluent prior to passage through the detector. These devices are designed to be sensitive to small changes in the conductivity on a large background signal. This type of detector uses electronic suppression of the background current.

9. Mass Spectrometric Detectors

The use of mass spectrometers as detectors for HPLC systems is growing rapidly. Several systems for interfacing the mass spectrometer to the chromatograph are now being used.[13,14] Most LC-MS systems require removal of the eluting solvent from the eluted sample before introduction into the mass spectrometer. A wire or belt transport device is frequently used for this purpose. In this type of system, the column eluent is deposited upon a moving belt or wire. The belt is then passed through a heating device to remove the solvent. After the solvent is evaporated, the belt is passed through an air lock into the mass spectrometer. Here the sample is ionized with one of the many mass spectrometric ionization techniques.

Another means for sample introduction into the mass spectrometer is the thermospray technique. The column eluent is sprayed into the mass spectrometer where most of the solvent evaporates and is pumped away before reaching the ionization chamber. Neither of these systems work well when the column eluent contains a buffering salt.

Interfaces for LC-MS are still evolving. The use of microcolumn or microbore HPLC techniques offers many advantages in the interfacing of LC-MS, since these systems use much smaller volumes of solvent than do the typical packed column HPLC techniques. With reduced solvent volumes, the entire column eluent can often be pumped directly into the mass spectrometer.

The field of LC-MS is still a very active field. The structural information that can be gained by this technique is very valuable, since great effort is expended in trying to identify all impurities that might be present in a drug substance above 0.1%. The use of LC-MS systems can save days and weeks that would normally be required to carry out chemical isolation procedures to obtain samples for structural characterization.

10. Other Less Popular Detectors

Many devices have appeared in the HPLC literature for use as HPLC detectors. Many of these have very limited utility, although a few are commercially available. The flame ionization detector is one of these. The flame ionization detector is combined with a transport device for HPLC use. The column eluent or a fraction of the column eluent is deposited onto a moving wire or belt, and passed through a heater to remove the solvent. The residue is then usually oxidized to carbon dioxide. The carbon dioxide is reduced to methane, and the methane is finally burned in the flame detector to produce a signal. The sensitivity of the device is about 2×10^{-8} g/mℓ. Flame detectors often do not work well with reverse phase systems when buffering salts are present.

The same type of transport system has also been used with other devices such as a thermal ionization detector[15] and an alkali flame ionization detector.[16] The alkali flame ionization detector is ideal for organophosphorous compounds.

The thermal energy analyzer is useful for the determination of nitrogen-containing compounds and especially for nitroso-containing substances. The device is sensitive to about 1×10^{-8} g/mℓ.

The photoionization detector is a very sensitive device which can have detection limits of about 1×10^{-11} g/s. In this device, the entire sample and solvent must be vaporized as they elute from the column. The gaseous stream is then passed in front of an ultraviolet lamp which emits 10.2-eV photons which ionize many materials. The ions are then passed through a series of accelerating electrodes until the ions reach the sensing or collector electrodes. The advantage of this detector is that sample components can often be ionized, whereas the common HPLC solvents such as water, methanol, acetonitrile, and methylene chloride all have ionization potentials above 10.2 eV, and are therefore not seen with this device.

F. HPLC Connections

The liquid chromatograph is composed of many critical components. The fittings used to connect these components can be very detrimental to the HPLC system if not properly selected and installed. The volume associated with fittings located beyond the sample injector should be kept to a minimum because any volume outside of the column causes broadening of the eluted peaks.

HPLC connections are usually made of stainless steel tubing with small internal diameter. Capillary tubing 0.007 in. (0.02 mm) I.D. to about 0.012 in. (0.04 mm) I.D. is used. In addition to the use of small diameter tubing, the length of the connecting pieces should be kept to a minimum.

Various types of compression tube fittings such as Swagelok, Parker Hanifen, Gyrolok, Valco, Rheodyne, Upchurch, Waters, and Beckman as well as others are used to fasten together the various parts in the flow system of a liquid chromatograph. Although all of these fittings are similar, they also have distinct differences. The dimensions of the ferrule spacing for several of the popular fittings are shown in Figure 9. Most ferrules are interchangeable, but it is preferable and often safer to match the manufacturer of the mating fitting. After a fitting has been made with a given type of ferrule and mating fitting, the prepared tube and ferrule should not be mixed with other manufacturer's mating fittings. When fittings are mixed, a void may remain within the connector where the tube does not fill the entire cavity provided for it within the fitting. Another possibility when connecting fittings from different manufacturers is that the tubing may extend further than it should. In such cases the ferrule will not mate properly with the seat of the other fitting, as shown in Figure 10, and the connection will probably leak. Stainless steel ferrules are generally used for the system interconnections. The major exception has been the use of soft polymer ferrules on the column outlet and on connections with the detector.

Several types of soft polymer fittings have recently become available for use in HPLC applications. Many of these fittings are rated for system pressures to 5000 psi. Figure 11 shows several fittings of this type that are now being used. The advantage of these fittings lies in the fact that they can be moved when column or components are changed without preparation of a new set of leads.

There are many options for the interconnection of HPLC components. The major consideration in selection and installation of connectors is minimization of dead volume in the system.

IV. THE SEPARATION PROCESS

In liquid chromatography a dissolved substance in a mobile liquid phase interacts with an immobile phase. The nature of this interaction governs the speed at which the solute moves through the column. Many different types of physical interactions can occur, and several may be involved simultaneously in practical chromatographic experiments. For simplicity the principal types of interactions will be discussed individually. The four principal modes of liquid chromatographic separation are listed in Table 4.

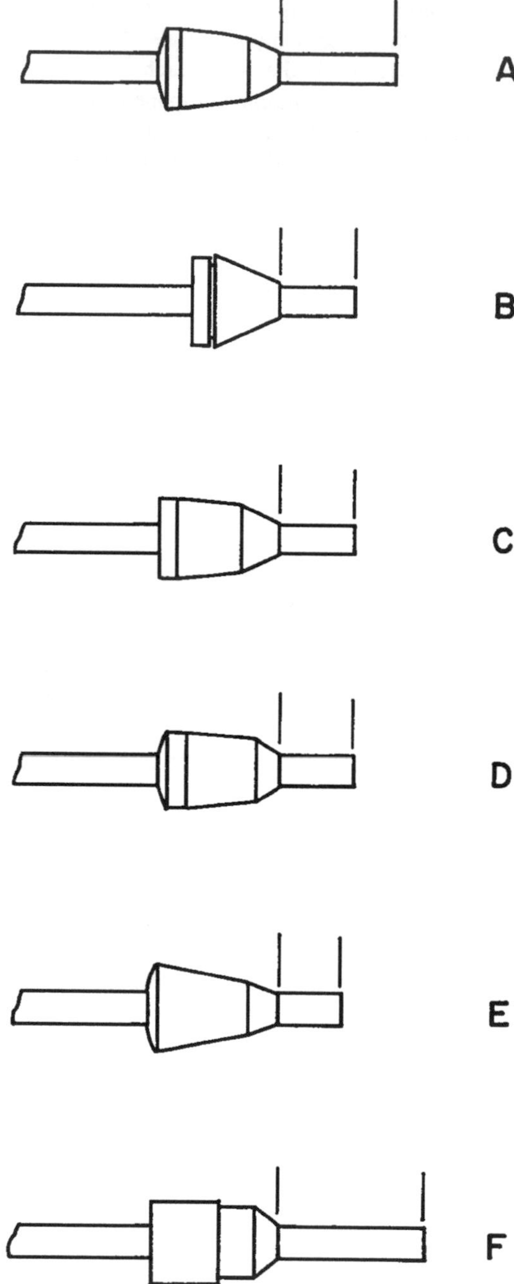

FIGURE 9. Ferrule spacing in several popular compression fittings used to connect tubing in HPLC systems. The distances from the end of the ferrule to the end of the tube are (A) Waters, 0.130"; (B) Swagelock, 0.090"; (C) Parker, 0.090"; (D) Uptight, 0.090"; (E) Valco, 0.080"; (F) Rheodyne, 0.170".

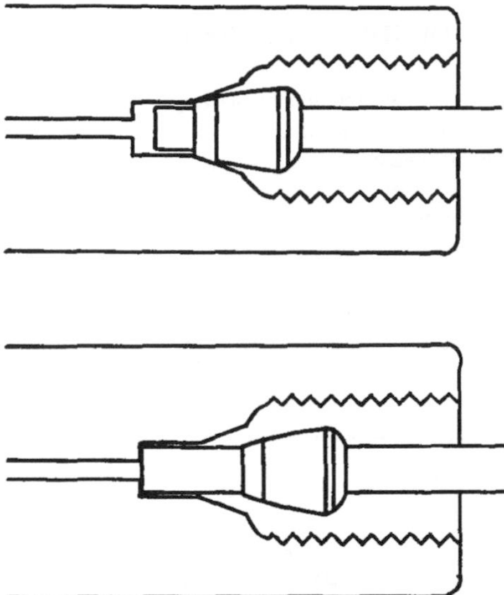

FIGURE 10. Misfit of ferrules in nuts due to mixing components from different manufacturers. Variations in distance between the ferrule and tube end lead to dead space (top) at the end of the tubing or improper seating of the ferrule (bottom) resulting in leaks.

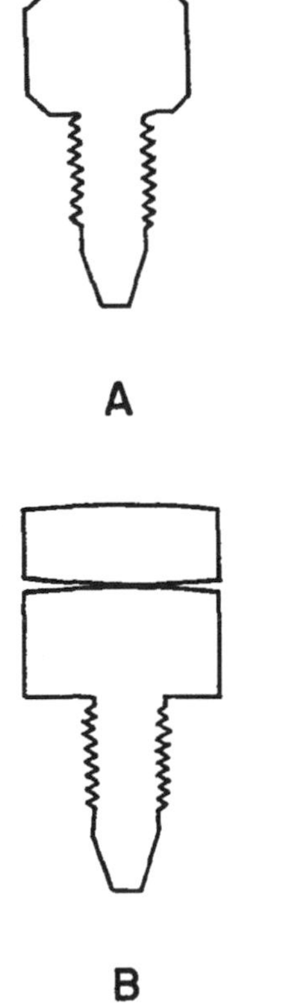

FIGURE 11. Soft polymer fittings for HPLC tubing connections.

Table 4
MODES OF HPLC SEPARATIONS

Liquid-solid adsorption chromatography
Liquid-liquid chromatography
  Normal phase
  Reverse phase
Ion exchange
Exclusion chromatography

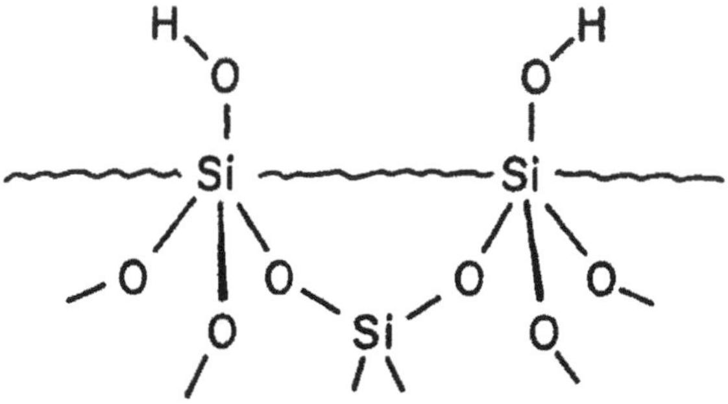

FIGURE 12. Silanol group on the surface of a silica particle.

A. Liquid-Solid Chromatography

One common mode of liquid chromatography is liquid-solid chromatography or adsorption chromatography. In this mode a substance with an active surface is packed into the chromatographic column. Common packings for this mode of separation include silica gel and alumina. In the case of silica gel, the silica structure contains polar silanol groups, as shown in Figure 12.

In adsorption chromatography, the mobile phase normally interacts with the active sites on the column. If silica gel is being used, the active sites are the silanol groups. When a solute molecule is present and in the proximity of the active column site, the equilibrium of the column is upset. There is a competition between the solute and the mobile phase for the active site on the packing surface. If the mobile phase is a strong solvent, it will form a much more stable association with the surface than does the sample. Under these conditions, the sample is washed through the column very rapidly. If the solvent is a weak solvent with respect to the sample, the sample will form the more stable association with the surface. In this situation the sample will be slow to move through the column. In some instances the solvent may be so weak that it will not significantly move the sample in the column. For best separation, a solvent should be used which allows competition between the sample and the mobile phase for the active sites of the column. Competition allows separation to occur since the interactions of different species will be slightly different.

Adsorption chromatography is based upon a surface effect of the packing material and should therefore be a good method for the separation of isomers. Indeed, this is the case with adsorption chromatography generally being good at resolving samples which contain geometrical and positional isomers. However, the lengths of the various alkyl chains in a molecule are not as important as the pattern of substitution and the polar functional groups in determining its adsorption properties; therefore, adsorption chromatography is not very effective in separating members of a homologous series.

B. Liquid-Liquid Chromatography

Liquid-liquid chromatography utilizes partitioning between two immiscible liquid phases to effect the separation of the components of a sample. One liquid is the mobile solvent, and the other is the stationary phase held on the packing of the column. The separation is the result of differences in the relative solubilities of the sample in the mobile and stationary phases. The stationary phase should be a good solvent for the sample and the mobile phase a poor solvent for the sample.

The nature of liquid-liquid chromatography dictates that polar samples be run on columns containing polar stationary phases. Likewise, nonpolar samples should be run on columns containing a nonpolar stationary phase. The terms "normal phase liquid chromatography" and "reverse phase liquid chromatography" have evolved to differentiate these two modes of operation. Normal phase liquid chromatography employs a polar stationary phase and a nonpolar mobile solvent, while reverse phase liquid chromatography employs a nonpolar stationary phase and a polar mobile solvent.

In classical liquid-liquid chromatography the column is coated with the stationary phase by simple physical adsorption. Since the stationary phase will have a small but finite solubility in the mobile phase, the eluant must be saturated with the stationary phase in order to avoid stripping the stationary phase from the column. However, small changes in temperature and batch-to-batch differences in eluate lead to slightly different equilibrium amounts of stationary phase on the column, thus causing changes in column characteristics. The presence of stationary phase in the eluate also complicates the collection and isolation of compounds eluting from the column. For these reasons, bonded stationary phases are almost universally used in HPLC today. The stationary phase in these columns is chemically bonded to the solid support. There is no problem with the removal of stationary phase from the column or contamination of the sample with the stationary phase when using this type of column. The use of these types of column materials in a chromatographic separation is sometimes called bonded phase chromatography. The actual separation mechanism in bonded phase chromatography may not be entirely a liquid partitioning effect, but the results obtained in this mode of operation are similar to those expected in liquid partitioning systems.

C. Ion Exchange Chromatography

Ion exchange chromatography is a relatively old technique in which the column packing contains polar ionizable groups such as carboxylic acids, sulfonic acids, or quaternary amines. In the appropriate pH range, these groups will be ionized and will attract substances with the opposite charge. Ionic substances can be separated in this way because their interaction with the column packing will depend on their size, charge density, and structure.

Buffers are usually employed as eluants in ion exchange chromatography. Both the pH and the ionic strength of the buffer will affect the elution of components from the column. By varying the pH around the pK_A of the sample, the degree of ionization of the compound can be altered. Decreases in the fraction of the sample that is ionized decrease the interaction of the sample with the column, and the sample component is then eluted from the column more rapidly. By increasing the ionic strength of the mobile phase the equilibrium can be shifted to disfavor interaction of the column packing with the sample. This again causes the sample to elute from the column more rapidly.

D. Exclusion Chromatography

Exclusion chromatography is a mode of liquid chromatography based on separations by molecular size. The packing materials used in this mode of chromatography have small pores with a carefully controlled pore size distribution. The smaller molecules can enter into all of the pores and thus see the entire liquid volume of the column. This total column volume is called the void volume of the column. As the size of sample molecules increases, a size is finally reached which can only penetrate some fraction of the pores present. Molecules in this size range will be separated from smaller ones because they experience a smaller effective column volume. Molecules so large that they cannot penetrate any of the

Table 5
SURFACTANTS EMPLOYED IN MICELLAR CHROMATOGRAPHY

Anionic	
Sodium dodecyl sulfate	(SDS)
Sodium polyoxyethylene 12(dodecyl ether)	(SDS 12EO)
Cationic	
Cetyl trimethylammonium bromide	(CTAB)
Hexadecyl trimethylammonium bromide	
Nonionic	
Polyoxyethylene (6) dodecanol	
Polyoxyethylene (23) dodecanol	(Brij 35)
Ethyoxylated C9 to C11 alcohols	(Neodol 91-6)
β-Octylglucoside	

porous structure of the packing elute together at the volume corresponding to the interstitial column volume, and are therefore the first substances to elute from the column. The interstitial volume is the volume of solvent present in the column between the particles of packing material. In this technique, it should be kept in mind that the large molecules are the first materials to elute from the column, and the smaller molecules are the last materials to elute from the column. Packing materials available today will resolve components differing in molecular weight by a few hundred mass units, and are available with pore sizes that will exclude molecules from about 600 to several million mol wt.

In exclusion chromatography, all available information is obtained by the time the solvent volume within the column is totally replaced once. This is in contrast to the previous techniques where no information is obtained until the column solvent volume has been replaced at least once.

Exclusion chromatography is often used to characterize polymers or other high molecular weight materials, and in fractionation of complex materials such as tissue extracts. It has not seen much application in pharmaceutical analysis.

E. Micellar Chromatography

Micellar chromatography is an extension and specialty form of reverse phase HPLC. Armstrong[17] has prepared an excellent review of micellar chromatography. Armstrong and Henry[18] were the first to demonstrate the usefulness of mobile phases which contained surfactants above the critical micelle concentration. In this technique, the micelles provide a hydrophobic site for interaction with the solute in the mobile phase. By varying the nature of the surfactant, one can tune the system as to its behavior for charged species or neutral entities. Increasing the micelle concentration increases the mobility of components that interact with the micelles. Anionic, cationic, and neutral surfactants are available for use in micellar chromatography. Table 5 lists several surfactants that have been used in micellar chromatography.[19-26]

The first attempts at micellar chromatography gave rather poor column efficiencies when compared to conventional reverse phase HPLC. Dorsey and others[22] have shown means for obtaining reasonable efficiencies with this technique. They now recommend that the mobile phase contain about 3% of a modifier such as propanol. Additionally they suggest that the systems be run at slightly elevated temperatures such as 40°C.

One advantage of micellar chromatography is the rapid recovery of the column after a sample elutes. Rapid recovery occurs because there is always a constant amount of free surfactant present to interact with the stationary phase. In gradient elution micellar chromatography, the number of micelles are increased during the separation to achieve the desired resolution. Since the stationary phase is not altered as it would be in conventional reversed phase chromatography with gradient elution, the column recovers almost instantaneously.

Micellar chromatography is therefore another means of generating different selectivities in liquid chromatographic systems. Micellar systems offer a unique means for solving certain separation problems. It is particularly attractive in situations where rapid screening of a series of samples is desirable. With micellar chromatography, one can recycle the system to the initial conditions without the long equilibrium times required in typical reverse phase HPLC.

V. HPLC COLUMNS

Many different classes and types of HPLC packings are now available. The two major classes are the pellicular or superficially porous materials, and the small, totally porous particle packing materials. In each of these classes, there are reverse phase, normal phase, and ion exchange packings as well as adsorbants. Most of the materials use silica particles. Descriptions of the bonding techniques used to attach the phase to the particles are given in the following sections, along with a list of the resulting commercial packing materials.

There are many entries in each of the classes and types of packings. The packings can vary as to the amount of material actually bonded to the silica, endcapping procedures used to eliminate residual silanols, and the physical characteristics of the silica particles including shape, pore size, and surface area. These parameters can all affect the behavior of the packing material in an actual separation.

The regulatory agencies in the pharmaceutical arena have discouraged the multiplicity of HPLC packings. In this effort, they have adopted a set of generic definitions for each class of materials. Ideally, anyone can select a packing from the class and use it in the assay described in the USP monograph for the drug substance. Since the properties of the columns in any category are quite varied, care should be taken during method development to include steps in the procedure that will minimize differences between columns from the same class that are obtained from different sources. For example, this might be accomplished by adding an amine to interact with residual silanols to block interactions of these groups with the sample. Ideally, steps could be taken to eliminate all of the differences between columns except for the amount of carbon on the packing. Slight adjustments in the aqueous-to-organic ratio of the chromatographic solvent can be made to compensate for the differences in carbon loading while maintaining the same selectivity for all of the sample components.

The USP as well as the Food and Drug Administration has adopted a standard classification for acceptable columns. Table 6 gives the definitions of these standard column packing materials. The most popular class by far is L1 class which has become synonymous with small particulate reverse phase packings.

A. Column Packing Materials

1. Pellicular or Superficially Porous Materials

The pellicular or superficially porous materials were the first packings designed specifically for HPLC. The pellicular materials shown in Figure 13 consist of a core which is surrounded by a crust or layer of materials which is chromatographically active. The core is usually a nonporous glass bead, and the outer layer usually consists of porous particles. These materials have the advantage of small or limited pores and thus their mass transfer characteristics are much better than those of a totally porous particle of similar size. The size of these particles generally range from about 30 to 45 μm. This is a particle size which is fairly easy to pack. These materials are also relatively cheap with respect to other state of the art HPLC column packing materials, so these materials are often used in expendable situations such as in the packing of protective precolumns and guard columns.

The porous layer packings suffer form several physical limitations. The thin chromatographic layer with limited pore size severely reduces the surface area available in the column.

Table 6
USP DEFINITIONS OF HPLC PACKINGS

L1	Octadecylsilane chemically bonded to porous silica or ceramic microparticles, 5 to 10 μm in diameter
L2	Octadecylsilane chemically bonded to silica gel of a controlled surface porosity that has been bonded to a solid spherical core, 30 to 50 μm in diameter
L3	Porous silica particles, 5 to 10 μm in diameter
L4	Silica gel of controlled surface porosity bonded to a solid spherical core, 30 to 50 μm in diameter
L5	Alumina of controlled surface porosity bonded to a solid spherical core, 30 to 50 μm in diameter
L6	Strong cation-exchange packing consisting of a sulfonated fluorocarbon polymer coated on a solid spherical core, 30 to 50 μm in diameter
L7	Octylsilane chemically bonded to totally porous silica particles, 5 to 10 μm in diameter
L8	An essentially monomolecular layer of aminopropylsilane chemically bonded to totally porous silica gel support, 10 μm in diameter
L9	10 μm irregular, totally porous silica having a chemically bonded, strongly acidic cation-exchange coating
L10	Nitrile groups chemically bonded to porous silica particles, 5 to 10 μm in diameter
L11	Phenyl groups chemically bonded to porous silica particles, 5 to 10 μm in diameter
L12	A strong anion-exchange packing made by chemically bonding a quaternary amine to a solid silica spherical core, 30 to 50 μm in diameter
L13	Trimethylsilane chemically bonded to porous silica particles, 5 to 10 μm in diameter
L14	Silica gel 10 μm in diameter having a chemically bonded strongly basic quaternary ammonium anion-exchange coating
L15	Hexylsilane chemically bonded to totally porous silica particles, 3 to 10 μm in diameter

FIGURE 13. Cross section of a particle of pellicular packing material.

The limited surface area reduces the adsorptive sites in adsorption chromatography, reduces the bonded phase coverage in bonded phase chromatography, and reduces the ion exchange capacity in ion exchange chromatography. The result is that the capacity of these types of packings are quite low compared to conventional porous packing materials. Pellicular packings are generally not suited for preparative separations or for trace component assays where large amounts of sample must be placed on the column in order to obtain the desired sensitivity for the trace constituent.

The limitations of the porous layer materials plus advances made in other HPLC column packings have relegated them to a position of very limited use in current HPLC practice. The characteristics of some of the pellicular materials are summarized in Table 7. Today these materials are often used in guard columns to hold highly retentive materials. The packing in the guard columns is frequently replaced to avoid transfer of the harmful components onto the analytical column.

2. Small Porous Particles

Most columns being used in HPLC are based on totally porous particles that range in size from about 3 to 10 μm in diameter. In this range there are basically three different sizes in routine use. The 10 μm nominal particles were the first that were routinely used in HPLC. As more difficult separations were approached and as the technology for sizing and packing smaller particles became more defined, the size has gone smaller and smaller with the 5 μm particles and the 3 μm particle becoming commercially available. With the 5 to 10 μm particles, plate counts of 30,000 to 60,000 plate per meter were commonplace. With the 3 μm particles, plate counts of up to 100,000 or more are being obtained.

Most HPLC applications have been based on silica particles. The first of these small particles were generally irregular shaped materials. As one looks at packing densities, it rapidly becomes apparent that spherical particles would probably be more desirable. With this realization, many small particulate silicas have become available for HPLC applications. Many of these spherical particulate silicas are made from colloidal silica which are agglomerated under carefully controlled conditions in the presence of a suitable binder. Ureas have often been used as the binder for this type of particle technology. The agglomerated particles are then sintered to give a stable particle. Table 8 gives a listing of many of the silicas that are used in HPLC as adsorbents. These same particles also serve as the silica backbone for the bonded phase packings which will be discussed later.

Almost all HPLC work has been done using siliceous materials. The chemical characteristics of silica presents some very severe limitations for its use. Silica is soluble in basic media. Thus there are some pH limitations that must be placed upon the use of the material. Generally silica is not used in pHs higher than about 8. This limitation of silica presents some very severe limitations on the bonded phase packings attached to silica used in reverse phase liquid chromatography.

Table 8 also includes a listing of some small particulate aluminas which have been used in HPLC. These are generally of very limited use and one finds almost no literature references to the use of these materials.

As one looks to the future, it seems quite possible that some other types of support or adsorbents may become available in HPLC. Various organic polymers have been used in ion exchange chromatography and may very well find increased usage in other modes of HPLC separations. It would also seem possible that other inorganic supports may become available. Supports which would be stable in basic media would be desirable.

3. Bonded Phase Packings

In surveying the literature involving HPLC pharmaceutical analysis, one finds that probably better than 90% of the applications involve reverse phase liquid chromatography on a

Table 7
PELLICULAR COLUMN PACKINGS

Packing	Functionality	Manufacturer[a]	Comments
		Adsorbents (Silica)	
Corasil I®	Silica	Waters	
Corasil II®	Silica	Waters	Large active surface
HC Pellosil®	Silica	Whatman	High-capacity packing
HS Pellosil®	Silica	Whatman	High-speed packing
Perisorb A®	Silica	E. Merck	
VYDAC SC Adsorbent®	Silica	Separations Group	
		Adsorbents (Alumina)	
Chromosorb LC3®	Alumina	Johns Manville	
HC Pellumina®	Alumina	Whatman	High-capacity packing
HS Pellumina®	Alumina	Whatman	High-speed packing—less active surface
		Adsorbents (Other)	
Pellamidon®	Polyamide	Whatman	
Perisorb PA 6®	Polyamide	E. Merck	
		Reverse Phase	
Bondapak Phenyl/Corasil®	Alkylphenyl	Waters	
Bondapak C_{18}/Corasil®	Octadecyl	Waters	
CO:Pell ODS®	Octadecyl	Whatman	
Perisorb RP 2®	Silanized silica	E. Merck	
Perisorb RP 8®	Octyl	E. Merck	
Perisorb RP 18®	Octadecyl	E. Merck	
VYDAC SC Reverse Phase®	Octadecyl	Separations Group	
Permaphase ODS®	Octadecyl	Dupont	
Zipax HCP®	Hydrocarbon	Dupont	Hydrocarbon polymer coated about a glass bead
		Normal Phase	
Durapak CW/400/Corasil®	Carbowax 400	Waters	Made with an ester linkage; the material is not compatible with acidic solvents
CO: Pell PAC®	Nitrile	Whatman	
VYDAC SC Polar Bonded Phase®	Nitrile	Separations Group	
Permaphase ETH®	Ether	Dupont	
Zipax CWT®	Carbowax 400	Dupont	The Zipax® series are not chemically bonded materials, but physically coated supports
Zipax BOP®	Oxynitrile	Dupont	
Zipax TMG®	Trimethylene glycol	Dupont	
Zipax ANH®	Cyanoethylsilicon	Dupont	
		Anion Exchangers	
Bondapak AX/Corasil®	Quaternary ammonium	Waters	pH range 2—7.5
AS Pellionex SAX®	Quaternary ammonium	Whatman	Styrene divinylbenzene copolymer-coated bead, pH range 2—12

Table 7 (continued)
PELLICULAR COLUMN PACKINGS

Packing	Functionality	Manufacturer[a]	Comments
AE Pellionex SAX®		Whatman	An aromatic polyester polymer-based coating
AL Pellionex WAX®	Amino	Whatman	An amino function in an aliphatic matrix
Perisorb AN®	Quaternary ammonium	E. Merck	pH range 1—8.5
VYDAC SC Anion Exchange®	Quaternary ammonium	Separations Group	
Permaphase AAX®		Dupont	
Permaphase ABX®		Dupont	
Zipax SAX®	Quarternary ammonium	Dupont	Polymer-coated material
Zipax WAX®		Dupont	
Cation Exchangers			
Bondapak CX/Corasil®	Sulfonic acid	Waters	pH range 2—7
HC Pellionex SCX®	Sulfonic acid	Whatman	Polystyrene matrix, pH range 2—10
HS Pellionex SCX®	Sulfonic acid	Whatman	Polystyrene matrix, pH range 2—10
Perisorb KAT®	Sulfonic acid	E. Merck	pH range 1—8.5
VYDAC SC Cation Exchange®	Sulfonic acid	Separations Group	
Zipax SCX®	Sulfonic acid	Dupont	Polymer-based material

[a] Addresses of column manufacturers are given in Table 8.

Table 8
MICROPARTICULATE PACKINGS

Packing base	Designation	Manufacturer[a]	Shape	Pore size (Å)	Surface area (m^2/g)	Comments
Adsorbosphere	Silica	Alltech	S	80	200	
Adsorbosphere HS	Silica	Alltech	S	60	350	
Econosil	Silica	Alltech	I	60	450	
Econosphere	Silica	Alltech	S	80	200	
RSIL	Silica	Alltech	I	80	550	
Versapack	Silica	Alltech	I	80	200	Similar to Waters μBondapak
Martex	Silica	Amicon		60, 100, 250		
Sepralyte	Silica	Analytichem				
Bakerbond Silica	Silica	Baker		170		
Bakerbond Prep	Silica	Baker		150		
Ultrasphere Si	Silica	Beckman	S			
Ultrasil Si	Silica	Beckman	I			
Bio Sil HP10	Silica	Bio Rad	I	100	550	
Bio Sil HP6	Silica	Bio Rad	I	60	400	
Spheri Si	Silica	Brownlee	S	80	220	
Chemcosorb	Silica	Chemco-Dychrom	S	100, 300		
Zorbax Sil	Silica	DuPont	S	80	300	

Table 8 (continued)
MICROPARTICULATE PACKINGS

Packing base	Designation	Manufacturer[a]	Shape	Pore size (Å)	Surface area (m^2/g)	Comments
LiChrosorb SI60	Silica	EM Science	I	60	500	
LiChrosorb SI100	Silica	EM Science	I	100	300	
LiChrosorb SI400	Silica	EM Science	I	400		
LiChrosorb ALOX T	Alumina	EM Science				
LiChrospher SI-100	Silica	EM Science	S	100	300	
Chromegabond	Silica	ES Industries	I	60	500	
	Silica	ES Industries	I	100	300	
Chromegasphere	Silica	ES Industries	S			
Chromegabond	Alumina	ES Industries				
Hitachi Gel 3030	Silica	Hitachi	I		500	
Techsphere	Silica	HPLC Technology	S		200	
Ultra-Techsphere	Silica	HPLC Technology	S	100	385	
Techsil	Silica	HPLC Technology	I			
	Alumina	HPLC Technology	I			
IBM	Silica	IBM	S	100	200	Same as Apex I
ICN Silica	Silica	ICN	I		500	
Finepak Sil	Silica	Jasco	I	95	380	
Apex I	Silica	Jones	S	100	170	
Apex WP	Silica	Jones	S	300	100	
Separon	Silica	Laboratory Instrument Works		130	450	
Nucleosil 50	Silica	Macherey-Nagel	S	50	500	
Nucleosil 100	Silica	Macherey-Nagel	S	100	350	
Nucleosil 120	Silica	Macherey-Nagel	S	120	200	
Nucleosil 300	Silica	Macherey-Nagel	S	300	100	
Nucleosil 500	Silica	Macherey-Nagel	S	500	35	
Nucleosil 1000	Silica	Macherey-Nagel	S	1000	25	
Polygosil	Silica	Macherey-Nagel	I	60	500	
Alox 60	Alumina	Macherey-Nagel	I	60	60	
Polygosil Si	Silica	Macherey-Nagel	I	60		
Microsil	Silica	Micromeritics	S	65	400	
Resolve	Silica	Millipore-Waters	S	90	200	
µPorasil	Silica	Millipore-Waters	I	125	300	
Nova-Pak	Silica	Millipore-Waters	S	60	120	
SIL X-1	Silica	Perkin Elmer	I		400	
Spherisorb	Silica	Phase Separation	S	80	220	
	Alumina	Phase Separation	S	130	93	
Bondex	Silica	Phenomenex				
Maxsil	Silica	Phenomenex	I	65	520	
	Alumina	Phenomenex	I	65	60	
W-Porex	Silica	Phenomenex		300		
Ultremex	Silica	Phenomenex	S			
Spherex	Silica	Phenomenex	S	100	180	
IB-SIL	Silica	Phenomenex	S			Similar to IBM Silica
Microsorb	Silica	Rainin	S	100		
Dynamax-60A	Silica	Rainin	I	60		
Dynamax-150A	Silica	Rainin	I	150		

Table 8 (continued)
MICROPARTICULATE PACKINGS

Packing base	Designation	Manufacturer[a]	Shape	Pore size (Å)	Surface area (m^2/g)	Comments
Dynamax-300A	Silica	Rainin	S	300		
Hichrom SI	Silica	Regis	S	80	220	
Spherosil	Silica	Rhone Poulenc	S	80	600	
Vydac HS	Silica	Separations Group	S	80	500	
Vydac TP	Silica	Separations Group	S	300	100	
Hypersil	Silica	Shandon	S	60	170	
Hypersil WP300	Silica	Shandon	S	300	60	Wide Pore
Supelcosil LC-Si	Silica	Supelco	S	100	170	
Chrom Sel SI	Silica	Tracor	I	60	500	Lichrosorb SI60
Micropak SI	Silica	Varian	I	60	500	Lichrosorb SI60
Partisil	Silica	Whatman	I	85	350	
Partisphere	Silica	Whatman	S	120	160	
YMC-Gel	Silica	YMC	S	100		
	Silica	YMC	I			
	Silica	YMC	S	300		

[a] Addresses of column manufacturers are given in Table 15.

bonded phase packing material. Most of the bonded phase HPLC packings consist of a silica support with an organic moiety bonded to it through a silicon oxygen silicon carbon covalent bonding system. These packings are prepared with a functionalized chlorosilane in the following manner (see Figure 14).

Many different functionalities have been used for preparation of bonded phase packings. The principal materials in use today involve the following substituent groups.

R = Methyl
= Ethyl
= Trimethyl
= Hexyl
= Octyl
= Octadecyl
= Phenyl
= Cyanopropyl
= Propylamine
= Diol

The functionalities attached to silica create two different classes of packings. These are the normal phase packings and the reverse phase packings. With the exception of the diol and the amine functionality, all of the above packings are commonly used in the reverse phase mode.

As the octadecyl reverse phase materials began to proliferate it rapidly became apparent that the materials were not all equivalent. This was shown by the fact that a solvent strength suitable for the desired elution of a sample component on one column would be much too weak or too strong for obtaining the elution of the same compound with a similar k′ value on another manufacturers packing. This phenomemon was the result of the carbon load or the amount of the modifier which was actually bonded to the silica. Generally the retention of nonpolar samples increases with increased carbon content of the packing.

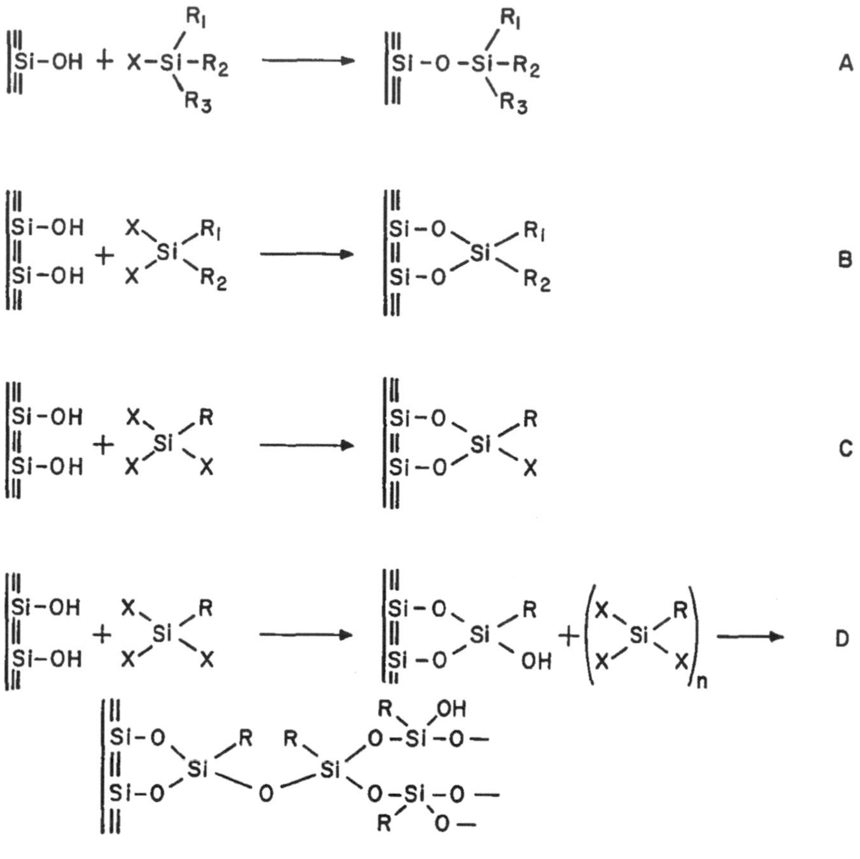

FIGURE 14. Schematic of the reaction of chloro- and alkoxysilanes with silica gel. X = chloro- or methoxyl-. The monofunctional silanes cause simple capping of surface silanols (A), whereas the bifunctional and trifunctional silanes produce bridges between surface silanol positions (B and C). Trifunctional silanols can also produce polymeric substitution on the surface (D).

Another problem with reverse phase packings became very evident as more polar sample materials were run on reverse phase columns. Many packings gave a great deal of peak tailing which was attributed to adsorption or interaction of the sample with the silica backbone of the packing rather than with the organic modifier chemically attached to the silica. Since the early reverse phase packings contained rather large bulky functional groups, steric effects might limit the number of surface hydroxyl groups on the silica surface which are available for reaction with the organic silane used to prepare the bonded phase packing. As a result, there are exposed polar groups on the silica which might interact with the polar sample.

The procedure of capping the material arose as a solution to the exposed silanol groups. After the silica is treated with the organic silane to prepare the reverse phase packing, the material is then treated with a smaller organic chlorosilane which has access to the exposed polar groups on the silica. This smaller silane can then cover the exposed polar groups remaining on the silica. Trimethylchlorosilane is commonly used in such reactions to "cap" the column. This type of chemical procedure has greatly improved the characteristics of reverse phase packings for polar sample materials.

Today there are a number of reverse phase packings on silicas with different pore sizes. In most of the early silica used for HPLC packings, the pore sizes in the silica were about 60 Å. When working with large molecules and proteins in particular, the pore size was too small to allow much interaction of the protein with the packing. Packings made from silica with larger pore sizes were found to perform much better with protein materials. When dealing with proteins by reverse phase liquid chromatography, the shorter chain alkyl packings and the nitrile packings are often preferred.

Today, manufacturers are preparing many different reverse phase packings on many different silica support materials. Table 9 provides a listing of the most of packings which are commercially available.

Although many of the manufacturers are supposedly using similar materials, each packing seems to have its own unique characteristics with respect to selectivity and retentivity of the packing. Goldberg[27] has evaluated a fairly large series of the more popular reverse phase packings and rated the packings for their retention characteristics for several different types of compounds.

Figure 15 shows the k′ for anthracene, a very nonpolar hydrocarbon on a series of reverse phase packings. In this set of data one can observe that there is quite a variation in the retentivity of these packings. To a great extent, this study shows the effect of carbon loading on the reverse phase packing. As the carbon load on the packing is increased, the retention of the material towards other hydrocarbon materials increases.

Figure 16 describes the behavior of diethyl phthalate, a material of moderate polarity on a series of reverse phase packings. In this set of data there are some strange orderings with the Zorbax® C8 having the highest retention properties of all of the tested packings. This is probably the result of end capping procedures which are used on such packings.

Figure 17 shows the behavior of caffeine, a basic compound, or a series of reverse phase packings. In this set of data some packings which had low retentivity towards hydrocarbons now show increased retentivity towards polar materials. In this type of system one might expect that exposed surface hydroxyl groups would cause increased adsorption of the basic sample.

The effects of an acidic sample on the different reverse phase packings is shown in Figure 18. The retention characteristics of these packings vary considerably for the acidic probe.

This set of data has shown that there are many differences between the various reverse phase column packings. Care must be used in substituting one reverse phase packing for another material in a published separation. Drastic changes in both the retentivity and the selectivity can be obtained by such a substitution. Goldberg's data allow one to select the most similar packings when substitutions need to be made. Care should still be exercised to establish that all of the critical substances remain separated on the substituted packing.

More recently, several polymeric reverse phase packings have become available. Figure 19 shows the structures of these materials. The newer materials, such as Interaction's ACT-1, produce efficient columns that work well for many different compounds. The advantage of these materials lies in the fact that there are no residual silanols to interact with the sample components and solvent. Tailing is therefore reduced with these packing materials. These materials are also stable to pH and can be used in a pH range of 1 to 14.

In addition to completely polymeric packings, a set of packings are now available in which an octadecyl moiety is polymerized about a silica or alumina particle with gamma radiation. This produces a polymer-coated particle. With this procedure, the silica or alumina is well-protected from the column solvent and samples. These columns can also be used in a much broader pH range than conventional silica-based packings and will tolerate a range of 1 to 13. Table 10 provides a listing of polymeric- and gamma-bonded polymeric materials.

A number of bonded phase materials are also available for normal phase chromatography. The most common of these materials are the nitrile, amino, and diol packings. A host of other functionalities are available, but they are not as widely used. Table 11 lists most of the bonded normal phase materials that are available today. The normal phase packings are not very popular and are used as an alternative when no reverse phase system has been found for the separation of a specific set of compounds, or when dealing with materials that are unstable to hydrolytic solvents.

Table 9
REVERSE PHASE PACKINGS

Support material	Designation	Functionality	Manufacturer[a]	Shape	Carbon load	Pore size (Å)	Endcapped	Notes
Adsorbosphere	C18	Octadecyl	Alltech	S	12	80	No	
	C18	Octadecyl	Alltech	S	12	80	Yes	
	C8	Octyl	Alltech	S	8	80	Yes	
	Phenyl	Phenyl	Alltech	S		80	Yes	
	TMS	Methyl	Alltech	S		80		
Adsorbosphere HS	C18	Octadecyl	Alltech	S	20	60	Yes	
Econosil	C18	Octadecyl	Alltech	I	15	60	Yes	
	C8	Octyl	Alltech	I	10	60	Yes	
Econosphere	C18	Octadecyl	Alltech	S	10	80	Yes	
	C8	Octyl	Alltech	S	5	80	Yes	
Econosphere 300	C4	Butyl	Alltech	S		300	Yes	
	C8	Octyl	Alltech	S		300	Yes	
	C18	Octadecyl	Alltech	S		300	Yes	
RSIL	C18HL	Octadecyl	Alltech	I	16	80	Yes	Similar to LiChrosorb Series
	C8	Octyl	Alltech	I	9	80	Yes	
	Phenyl	Phenyl	Alltech	I	5	80	Yes	
	C3	Propyl	Alltech	I	7	80	Yes	
Versapack	C18	Octadecyl	Alltech	I	10	80	Yes	Similar to μBondapak ODS
Reversed Phase	OD5	Octadecyl	American Burdick & Jackson	S		80		
Matrex	C8	Octyl	Amicon			60, 100, 250		
Matrex	C18	Octadecyl	Amicon			60, 100, 250		
Sepralyte	C8	Octyl	Analytichem					
	CE	Ethyl	Analytichem					
	C18	Octadecyl	Analytichem					
	Phenyl	Phenyl	Analytichem					
Astec Reversed Phase	C18	Octadecyl	Astec	I				
	C18	Octadecyl	Astec	S				
Astec Reversed Phase C8	C8	Octyl	Astec	S				
Astec Reversed Phase C1	C1	Methyl	Astec	S				
Bakerbond	WP-octadecyl	Octadecyl	Baker	S		300		
	WP-octyl	Octyl	Baker	S		300		
	WP-diphenyl	Diphenyl	Baker	S		300		
	WP-butyl	Butyl	Baker	S		300		

	WP-HI-propyl	Propyl	Baker	S		300		
	Octadecyl	Octadecyl	Baker	S	13	170	Yes	
	Octyl	Octyl	Baker	S	7	170	Yes	
	Phenylethyl	Phenylethyl	Baker	S	4.5	170	Yes	
	Octadecyl	Octadecyl	Baker	I	18	150	Yes	
	Octyl	Octyl	Baker	I	6	150	Yes	
	Cyclohexyl	Cyclohexyl	Baker	I	5	150	Yes	
	Butyl	Butyl	Baker	I	4.2	150	Yes	
	Phenyl	Phenyl	Baker	I	7	150	Yes	
Ultrasphere	ODS	Octadecyl	Beckman-Altex	S	12		Yes	Monomeric surface
	Octyl	Octyl	Beckman-Altex	S	6.5		Yes	Monolayer
	IP	Octadecyl	Beckman-Altex	S			Yes	
Ultrasil	ODS	Octadecyl	Beckman-Altex	I			No	
	Octyl	Octyl	Beckman-Altex	I			No	
Bio-Sil	ODS	Octadecyl	Bio Rad	I	15	100	Yes	
Spheri	RP-8	Octyl	Brownlee	S		80	Yes	Monofunctional
	OSS	Octyl	Brownlee	S		80		Polyfunctional
	RP-18	Octadecyl	Brownlee	S	7	80		Monofunctional
	ODS	Octadecyl	Brownlee	S		80		Polyfunctional
	Phenyl	Phenyl	Brownlee	S		80		
Aquapore	C8	Octyl	Brownlee	S		300		
	C4	Butyl	Brownlee	S		300		
	Phenyl	Phenyl	Brownlee	S		300		
Chemcosorb	ODS-H	Dimethyl-octadecyl	Chemco-Dychrom	S		120		High carbon load
	ODS-H	Dimethyl-octadecyl	Chemco-Dychrom	S	20	60		Medium carbon load
	ODS-UH	Octadecyl	Chemco-Dychrom	S	30	60		
	ODS-L	Octadecyl	Chemco-Dychrom	S		60		
	300-C18	Octadecyl	Chemco-Dychrom	S		300		
	I-C18	Methyl-octadecyl	Chemco-Dychrom	I		100		
	C8	Octyl	Chemco-Dychrom	S		100, 300		
	TMS	Trimethyl	Chemco-Dychrom	S		100, 300		
	C18	Octadecyl	Chemco-Dychrom	I				Replacement for μBondapak C18
	DPH	Diphenyl	Chemco-Dychrom	S		100, 300		
	C4	Butyl	Chemco-Dychrom	S		300		
CP-Spher	C8	Octyl	Chrompack	S			Yes	
	C18	Octadecyl	Chrompack	S			Yes	
Zorbax	ODS	Octadecyl	DuPont	S	15		No	
	C8	Octyl	DuPont	S	15		Yes	

Table 9 (continued)
REVERSE PHASE PACKINGS

Support material	Designation	Functionality	Manufacturer[a]	Shape	Carbon load	Pore size (Å)	Endcapped	Notes
	Phenyl	Phenyl	DuPont	S				
	TMS	Trimethyl	DuPont	S				
LiChrosorb	RP2	Ethyl	EM Science	I		60		
	RP8	Octyl	EM Science	I	9	60		
	RP18	Octadecyl	EM Science	I	16	60		
Lichrospher II	SI-100 RP18	Octadecyl	EM Science	S		100		
	SI-100 RP8	Octyl	EM Science	S		100		
Chromegabond	C1	Methyl	ES Industries	I	8	60, 100		
	C2	Ethyl	ES Industries	I		60, 100		
	C2	Dimethyl	ES Industries	I		60, 100		
	TMS	Trimethyl	ES Industries	I		60, 100		
	C3	*n*-Propyl	ES Industries	I		60, 100, 300		
	C4	*n*-Butyl	ES Industries	I		60, 100, 300, 500		
	NC6	*n*-Hexyl	ES Industries	I		60, 100		
	CC6	Cyclohexyl	ES Industries	I	10	60, 100		
	C8	Octyl	ES Industries	I	15	60, 100, 300, 500		
	C10	Decyl	ES Industries	I		60		
	C12	Dodecyl	ES Industries	I		60		
	C18	Octadecyl	ES Industries	I		100		Polymer
	MC-18	Octadecyl	ES Industries	I	18	60, 100, 300, 500	Yes	Monolayer
	C22	Docosyl	ES Industries	I		100		
	P	Phenyl	ES Industries	I	10	60		
	AP	Alkyl phenyl	ES Industries	I		60		
	DP	Diphenyl	ES Industries	I		100, 300, 500		
Hitachi Gel 3050	ODS	Octadecyl	Hitachi					
Techopak	C18	Octadecyl	HPLC Technology					
	Phenyl	Phenyl	HPLC Technology					
Techsphere	C1	Methyl	HPLC Technology	S				
	Hexyl	Hexyl	HPLC Technology	S				
	C8	Octyl	HPLC Technology	S	10		No	
	C8	Octyl	HPLC Technology	S	10		Yes	
	ODS	Octadecyl	HPLC Technology	S	10		No	
	ODS	Octadecyl	HPLC Technology	S	10		Yes	
	C22		HPLC Technology	S				
	Phenyl	Phenyl	HPLC Technology	S				

Techsil	C8	Octyl	HPLC Technology	I	10	100	No	
	C8	Octyl	HPLC Technoloogy	I	10	100	Yes	
	C2	Ethyl	HPLC Technology	I		100		
	C6	Hexyl	HPLC Technology	I		100		
	C22		HPLC Technology	I		100		
	Phenyl	Phenyl	HPLC Technology	I		100		
	C18	Octadecyl	HPLC Technology	I	10	100		
Ultra-Techsphere	ODS	Octadecyl	HPLC Technology	S				
	C8	Octyl	HPLC Technology	S				
IBM	Octadecyl	Octadecyl	IBM	S	9	100	Yes	IBM columns equivalent to
	Octyl	Octyl	IBM	S	6.5	100	No	Jones Apex II series
	Octyl	Octyl	IBM	S	6.5	100	Yes	
	Methyl	Methyl	IBM	S	2.5	100	Yes	
	Phenyl	Phenyl	IBM	S	2.6	100	No	
Finepak SIL	C18	Octadecyl	Jasco	I		95		Monolayer
	C8	Octyl	Jasco	I		95		Monolayer
Apex I	Octadecyl	Octadecyl	Jones	S	10	100	Yes	
	Octyl	Octyl	Jones	S	5	100	Yes	
	Ethyl	Ethyl	Jones	S	3	100		
Apex II	Octadecyl	Octadecyl	Jones	S		100		
	Octyl	Octyl	Jones	S		100		
	Phenyl	Phenyl	Jones	S	5	100		
Apex WP	Octadecyl	Octadecyl	Jones	S		300		
	Octyl	Octyl	Jones	S		300		
	Butyl	Butyl	Jones	S		300		
	Phenyl	Phenyl	Jones	S		300		
Separon SI	C1	Methyl	Laboratory Instrument Works	S				
	C18	Octadecyl	Laboratory Instrument Works	S	20	130	Yes	
Nucleosil	C18	Octadecyl	Macherey-Nagel	S	16	100		
	C8	Octyl	Macherey-Nagel	S	11	100		
	C8	Octyl	Macherey-Nagel	S		300		
Polygosil	C18	Octadecyl	Macherey-Nagel	I	18	60		
	C8	Octyl	Macherey-Nagel	I	11	60		
Nucleosil	C18	Octadecyl	Macherey-Nagel	S		300		
	Phenyl	Phenyl	Macherey-Nagel	S	10	100		
Microsil	C18	Octadecyl	Micromeritics	S	18	65	Yes	Monolayer
	C8	Octyl	Micromeritics	S	15	65	Yes	Monolayer
Nova-Pak	C18	Octadecyl	Millipore-Waters	S	7	60	Yes	
Resolve	C18	Octadecyl	Millipore-Waters	S	12	90	No	
μBondapak	C18	Octadecyl	Millipore-Waters	I	10	125	Yes	

Table 9 (continued)
REVERSE PHASE PACKINGS

Support material	Designation	Functionality	Manufacturer[a]	Shape	Carbon load	Pore size (Å)	Endcapped	Notes
	Phenyl	Phenyl	Millipore-Waters	I	8	60	Yes	
Nova-Pak	Phenyl	Phenyl	Millipore-Waters	S	4	60	Yes	
Resolve	C8	Octyl	Millipore-Waters	S	6	90	No	
SIL X-1	ODS	Octadecyl	Perkin Elmer					
Spherisorb	ODS1	Octadecyl	Phase Separation	S	7	80	No	
	ODS2	Octadecyl	Phase Separation	S	12	80	Yes	
	C1	Methyl	Phase Separation	S	2	80	Yes	
	C6	Hexyl	Phase Separation	S	6	80	Yes	
	C8	Octyl	Phase Separation	S	6	80	Yes	
	Phenyl	Phenyl	Phase Separation	S	3	80	Yes	Major silanols capped
Bondex	C18	Octadecyl	Phenomenex					Replacement for μBondapak C18
	Phenyl	Phenyl	Phenomenex					
Maxsil	C8	Octyl	Phenomenex	I	8	65		
	C18	Octadecyl	Phenomenex	I	12	65		
W-Porex	C4	Butyl	Phenomenex			300		
	C8	Octyl	Phenomenex			300		
	C18	Octadecyl	Phenomenex			300		
	Phenyl	Phenyl	Phenomenex			300		
	Diphenyl	Diphenyl	Phenomenex			300		
Ultremex	C1	Methyl	Phenomenex	S			Yes	Monomeric
	C6	Hexyl	Phenomenex	S			Yes	
	C8	Octyl	Phenomenex	S			Yes	
	C18	Octadecyl	Phenomenex	S	13		Yes	Similar to Ultrasphere and Microsorb
	Phenyl	Phenyl	Phenomenex	S				
Spherex	C1	Methyl	Phenomenex	S		100	Yes	
	C6	Hexyl	Phenomenex	S		100	Yes	
	C8	Octyl	Phenomenex	S		100	Yes	
	C18	Octadecyl	Phenomenex	S	11	100	Yes	
	C22		Phenomenex	S		100	Yes	
	Phenyl	Phenyl	Phenomenex	S		100	Yes	

IB-SIL	C1	Methyl	Phenomenex	S			Yes	Similar to IBM packings
	C8	Octyl	Phenomenex	S			Yes	Similar to IBM packings
	C18	Octadecyl	Phenomenex	S			Yes	Similar to IBM packings
	Phenyl	Phenyl	Phenomenex	S			Yes	Similar to IBM packings
Carbosphere	ODS	Octadecyl	Phenomenex	S	22		Yes	
Microsorb	C8	Octyl	Rainin	S		100		
	C18	Octadecyl	Rainin	S		100		
Dynamax	60A-C8	Octyl	Rainin	I		60		
	60A-C18	Octadecyl	Rainin	I		60		
	150A-C4	Butyl	Rainin	I		150		
	150A-C8	Octyl	Rainin	I		150		
	150A-C18	Octadecyl	Rainin	I		150		
	300A-C4	Butyl	Rainin	S		300		
	300A-C8	Octyl	Rainin	S		300		
	300A-C18	Octadecyl	Rainin	S		300		
Hichrom	ODS	Octadecyl	Regis	S		80		Spherisorb ODS
	C6	Hexyl	Regis	S	6	80	Yes	Spherisorb C6
Spherisil	C18	Octadecyl	Rhone Poulenc	S	22			
Vydac	C18 peptide	Octadecyl	Separations Group	S		330		
	C4	Butyl	Separations Group	S		330	Yes	
	Diphenyl	Diphenyl	Separations Group	S		330		
	pH stable RP	Octyl	Separations Group	S		300		
	HS-C18	Octadecyl	Separations Group	S	13.5	80	Yes	
	TB-C18 (201)	Octadecyl	Separations Group	S	8	300	No	
	TP-C4	Butyl	Separations Group	S	3	300	Yes	
	TP-C18 (218)	Octadecyl	Separations Group	S	8	300	Yes	
	TP-phenyl	Phenyl	Separations Group	S	5	300		
	TB-C8 (228)	Octyl	Separations Group	S		300	Yes	
	TP-diphenyl	Diphenyl	Separations Group	S		300	Yes	
Hypersil	ODS	Octadecyl	Shandon	S	10	120	Yes	Monolayer C18
	MOS1	Dimethyloctyl	Shandon	S	7	120	No	
	MOS2	Dimethyloctyl	Shandon	S	7	120	Yes	
	SAS	Trimethyl	Shandon	S	2.6	120		
	Phenyl	Phenyl	Shandon	S	5	120		
Hypersil WP	Octyl	Octyl	Shandon	S	2.7	300		
	Butyl	Butyl	Shandon	S	2	300		
Supelcosil	LC-1	Methyl	Supelco	S	2.5	100	Yes	
	LC-8	Octyl	Supelco	S	6.6	100	Yes	

Table 9 (continued)
REVERSE PHASE PACKINGS

Support material	Designation	Functionality	Manufacturer[a]	Shape	Carbon load	Pore size (Å)	Endcapped	Notes
	LC-18	Octadecyl	Supelco	S	11.3	100	Yes	
	LC-8-08	Octyl	Supelco	S		100	Yes	Deactivated for basic compounds
	LC-18-08	Octadecyl	Supelco	S		100	Yes	Deactivated for basic compounds
	LC-308	Octyl	Supelco	S		300	Yes	
	LC-318	Octadecyl	Supelco	S		300	Yes	
Synchropak	RP-PC18	Octadecyl	Synchrom	S		300		
	RP-PC18	Octyl	Synchrom	S		300		
	RP-PC4	Butyl	Synchrom	S		300		
	RP-PC1	Methyl	Synchrom	S		300		
Micropak	MCH-C18	Octadecyl	Varian	I	12	60	Yes	Monolayer of LiChrosorb SI60
	MCH-C18	Octadecyl	Varian	I	12	60	No	Monolayer of LiChrosorb SI60
	CH	Octadecyl	Varian	I	22	60	Yes	Polymeric on Li-Chrosorb SI60
Partisil	ODS-3	Octadecyl	Whatman	I	10.5	85	Yes	Polymeric
	C8	Octyl	Whatman	I	9	85	Yes	Monomeric
	ODS-2	Octadecyl	Whatman	I	15	85	No	Polymeric
	ODS	Octadecyl	Whatman	I	5	85	No	Polymeric
Partisphere	C18	Octadecyl	Whatman	S	11	120	Yes	Monomeric
	C8	Octyl	Whatman	S	6	120	Yes	Monomeric
Protesil	Octyl	Octyl	Whatman	S	7.5	300	Yes	
	Diphenyl	Diphenyl	Whatman	S	8	300	Yes	
YMC-GEL	ODS	Octadecyl	YMC	S		100		
	Octyl	Octyl	YMC	S		100		
	Phenyl	Phenyl	YMC	S		100		
	TMS	Trimethyl	YMC	S		100		
	ODS	Octadecyl	YMC	S		300		
	Butyl	Butyl	YMC	S		300		

[a] Addresses of column manufacturers are given in Table 15.

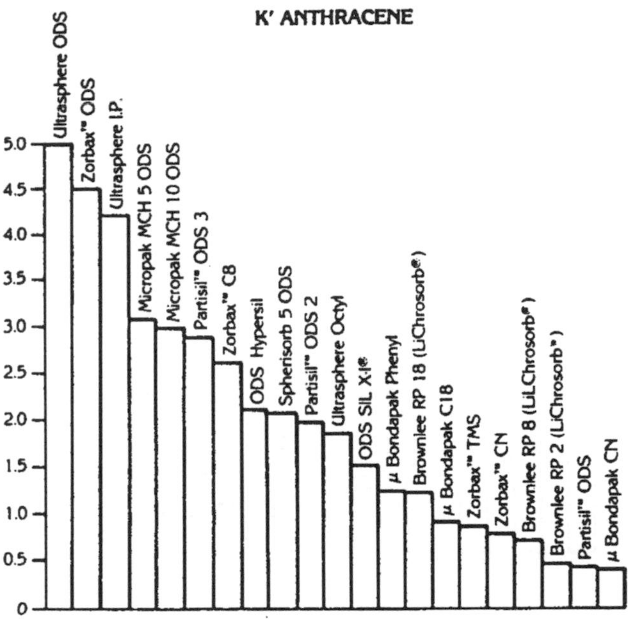

FIGURE 15. K′ for anthracene on reverse phase packings. The test solvent was methanol/water (85:15).

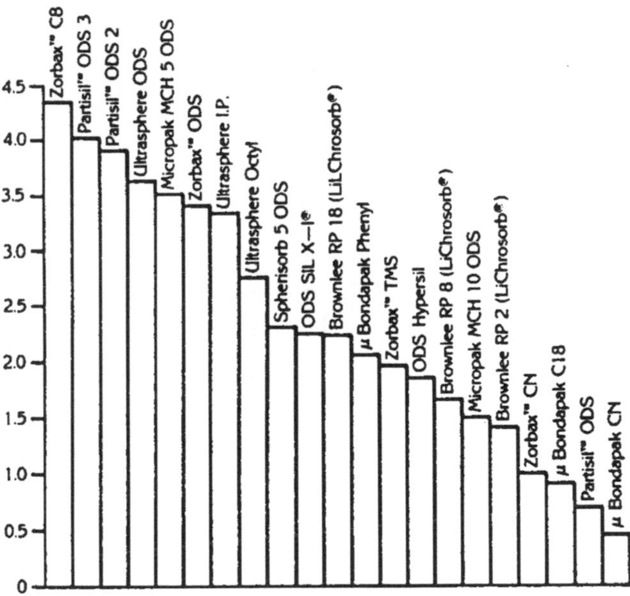

FIGURE 16. K′ for diethyl phthalate on reverse phase packings. The test solvent was methanol/water (65:35).

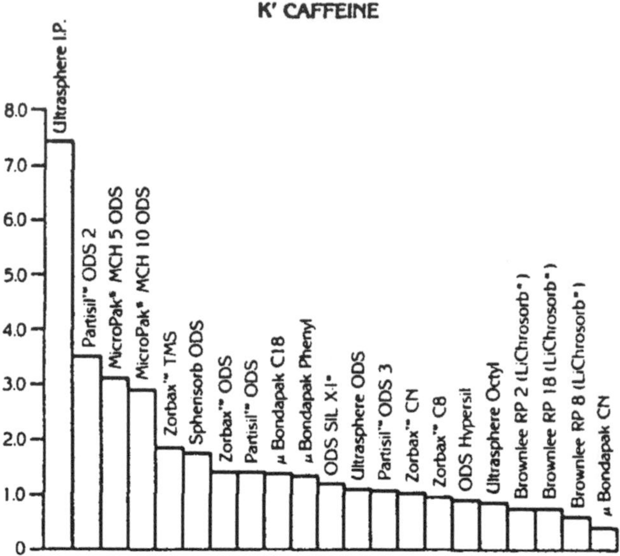

FIGURE 17. K′ for caffeine on reverse phase packings. The test solvent was acetonitrile/10 m*M* sodium acetate pH 4.5 (20:80).

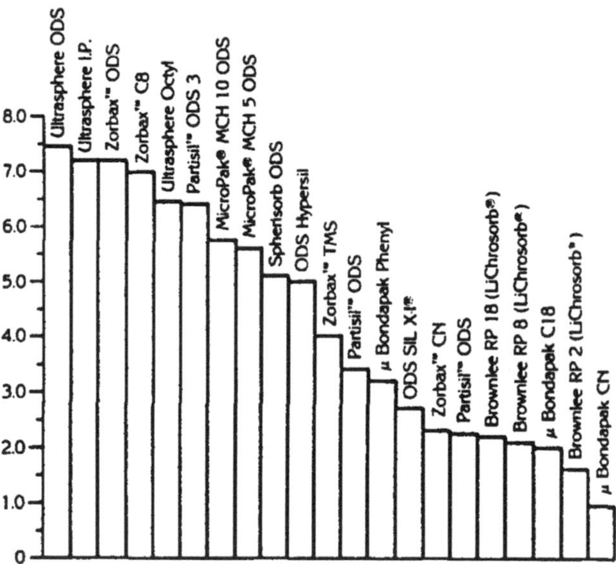

FIGURE 18. K′ for toluic acid on reverse phase packings. The test solvent was acetonitrile/10 m*M* sodium acetate pH 4.5 (20:80).

Many columns have been proposed for special applications. A number of columns have appeared for the separation of sugars, although most are only an amino phase given a special name. There are a series of special columns which are now available for separation of optical isomers. Table 12 provides a listing of many of the packings which fall into the special function category.

Finally, bonded phase ion exchange packings are also available, but have not been very

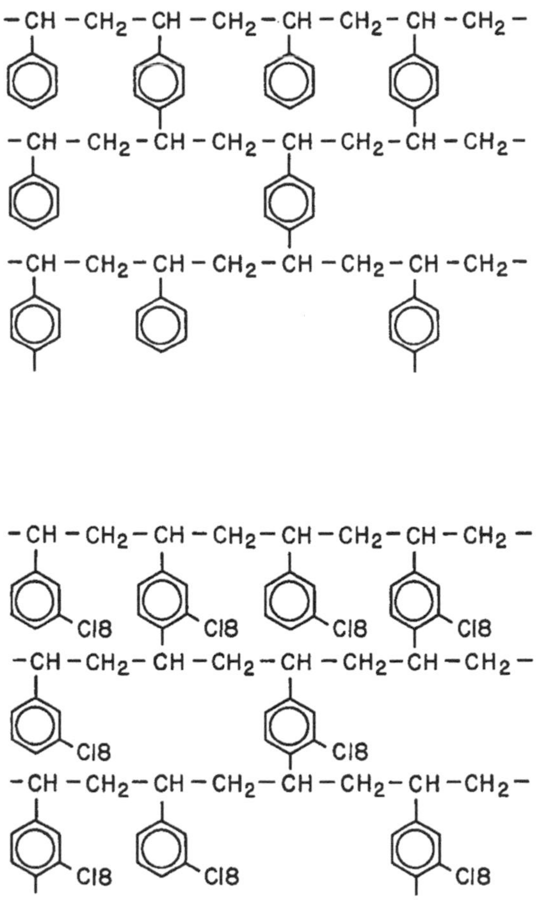

FIGURE 19. Structures of polymeric reverse phase column packing materials.

successful. The pH limitations of these materials severely limit their applications. The capacity of bonded phase ion exchange materials is also relatively low. The majority of compounds amenable to ion exchange chromatography can be handled by other more efficient and less troublesome chromatographic procedures. Table 13 lists a number of bonded ion exchange packing materials.

4. *Stability of Bonded Phase Materials*

Bonded phase packings are stable in most common organic solvents and in aqueous systems in the pH range where silica is stable. Bonded phase materials are generally stable over a pH range of 2.5 to about 7.5. Many methods described in the literature today use pH values above 7.5. In such applications silica precolumns may be used to saturate the mobile phase with silica to minimize degradation of the analytical column.

In addition to hydrolysis of the silica, some bonded phase packings have chemical limitations imposed by the reactivity of the organic group. Normal phase packings generally have very polar functional groups present which may be subject to chemical attack under some conditions. Some of the more troublesome packings in this respect are the bonded amine packing materials, which are quite susceptible to oxidation. An active amine also may form Schiff bases with aldehydes. The types of samples run through a bonded phase column must therefore be selected with some care to avoid materials that might react with the packing material.

Table 10
POLYMER-BASED REVERSED PHASE MATERIALS

Packing type	Structure	Manufacturer[a]	Surface area	Notes
Gamma-C18	Gamma-bonded octadecyl	ES Industries		Gamma-bonded polymer on silica, 20—22% carbon
Gamma-RP-1	Gamma-bonded octadecyl	ES Industries		Gamma-bonded polymer on alumina, 5—8% carbon
PRP-1	Polystyrene gel	Hamilton		
Hitachi Gel 3030	Methacrylate	Hitachi		
Hitachi Gel 3011	Polystyrene divinylbenzene	Hitachi		
ACT-1	C-18 modified polystyrene	Interaction		Branched C18 groups attached to aromatic rings
ACT-2	Amino C-18 modified polystyrene	Interaction		Branched C18 amino alkyl groups attached to aromatic rings
PLRP-S	Polystyrene gel	Polymer Laboratories	100	
PLRP-S 300	Polystyrene gel	Polymer Laboratories	300	
RSPAK DS613	Polystyrene	Shodex		Similar to C18 column
RSPAK DE613	Polymethacrylate	Shodex		Similar to C8 column
RSPAK DM614	Polyester	Shodex		Similar to C3 column
RSPAK DC613	Hydrophillic polystyrene	Shodex		Similar to NH2 packing

[a] Addresses of column manufacturers are given in Table 15.

Table 11
NORMAL PHASE PACKINGS

Support material	Desgination	Functionality	Manufacturer[a]	Shape	Carbon load	Pore size (Å)	Endcapped	Notes
Adsorbosphere	CD	Nitrile	Alltech	S		80		
	NH_2	Amino	Alltech	S		80		
Econosil	CN	Nitrile	Alltech	I		80		
	NH_2	Amino	Alltech	I		80		
Econosphere	CN	Nitrile	Alltech	S		80		
	NH_2	Amino	Alltech	S		80		
RSIL	CN	Nitrile	Alltech	I		80		
	Amino	Amino	Alltech	I		80		
	NO_2	Nitro	Alltech	I	5	80		
Sepralyte	Cyanopropyl	Cyanopropyl	Analytichem					
	Diol	Alcohol	Analytichem					
	Aminopropyl	Amino	Analytichem					
Bakerbond	WP-cyanopropyl	Nitrile	Baker	S		300		
	WP-PEI	Polyethyleneimine	Baker	S		300		
	WP-CBX	Carboxymethyl	Baker	S		300		
	Cyanopropyl	Nitrile	Baker	S	5	170	Yes	
	Diol	Alcohol	Baker	S	3	170	Yes	
	Aminopropyl	Amino	Baker	S		170	Yes	
	Pre Cyanopropyl	Nitrile	Baker	I	5	150	Yes	
	Prep diol	Alcohol	Baker	I	4	150	No	
	Pre aminopropyl	Amino	Baker	I		150	Yes	
Ultrasphere	Cyano	Nitrile	Beckman-Altex	S				
	NH_2	Amino	Beckman-Altex	S				
Ultrasil	NH_2	Amino	Beckman-Altex	I				
Spheri	Cyanopropyl	Nitrile	Brownlee	S				
	Aminopropyl	Amino	Brownlee	S				
	Diol	Alcohol	Brownlee	S				

Table 11 (continued)
NORMAL PHASE PACKINGS

Support material	Desgination	Functionality	Manufacturer[a]	Shape	Carbon load	Pore size (Å)	Endcapped	Notes
Chemcosorb	CN	Nitrile	Chemco-Dychrom	S		100, 300		
	NH_2	Amino	Chemco-Dychrom	S		100		
Zorbax	CN	Nitrile	DuPont	S				
	NH_2	Amino	DuPont	S				
LiChrosorb	Diol	Alcohol	EM Science	I		60		
	CN	Nitrile	EM Science	I		60		
	NH_2	Amino	EM Science	I		60		
Lichrospher II	Diol	Alcohol	EM Science	S		100		
Chromegabond	Amine	Amine	ES Industries	I		60, 100		
	Diamine	Diamine	ES Industries	I		60		
	Triamine	Triamine	ES Industries	I		60		
	Fluoroether	Fluoroether	ES Industries	I		60, 100		
	Fluorodecyl	Fluorodecyl	ES Industries	I		60, 100		
	Fluoropropyl	Fluoropropyl	ES Industries	I		60, 300		
	Diol	Diol	ES Industries	I		60, 100		
	Phenethyl urea	Phenethyl urea	ES Industries	I		60		
	CN	Nitrile	ES Industries	I	7	60, 300		
	DNAP	Dinitroanilino propyl	ES Industries	I		60, 100		Charge transfer type
	TENF	Tetranitroflu-orenimo	ES Industries	I		100, 500		Charge transfer type
Techopak	CN	Nitrile	HPLC Technology					
	CHO		HPLC Technology					
Techsphere	CN	Nitrile	HPLC Technology	S				
	NH_2	Amino	HPLC Technology	S				
Techsil	Amino	Amino	HPLC Technology					
	Nitrile	Nitrile	HPLC Technology					
	NO_2	Nitro	HPLC Technology					
IBM	Cyano	Nitrile	IBM	S	2.2	100	No	IBM same as Apex II series columns

	Amino	Amino	IBM		4	100	
Finepak SIL	NH	Aminopropyl	Jasco	I		95	
Apex I	Amino	Amino	Jones	S		100	
	Cyano	Nitrile	Jones	S		100	
Apex II	Amino	Amino	Jones	S		100	
	Cyano	Nitrile	Jones	S		100	
	Diol	Alcohol	Jones	S		100	
Apex WP	Cyano	Nitrile	Jones	S		300	
	Diol	Alcohol	Jones	S		300	
Separon SI	CN	Cyanoethyl	Laboratory Instrument Works	S		130	Yes
Separon SIL	NH_2	Amino	Laboratory Instrument Works	S		130	
Nucleosil	CN	Nitrile	Macherey-Nagel	S		100	
	NO_2	Nitro	Macherey-Nagel	S		100	
	NH_2	Amino	Macherey-Nagel	S		100	
	Diol	Alcohol	Macherey-Nagel	S		100	
	CN	Nitrile	Macherey-Nagel	S		300	
	$N(CH_3)_2$		Macherey-Nagel	S		100	
	NH_2	Amino	Macherey-Nagel	S		120	
	CN	Nitrile	Macherey-Nagel	S		120	
	Diol	Alcohol	Macherey-Nagel	S		300	
	CN	Nitrile	Macherey-Nagel	S		300	
	Diol	Alcohol	Macherey-Nagel	S		500	
	CN	Nitrile	Macherey-Nagel	S		500	
	Diol	Alcohol	Macherey-Nagel	S		1000	
Polygosil	CN	Nitrile	Macherey-Nagel	I		60	
	NH_2	Amino	Macherey-Nagel	I		60	
	NO_2	Nitro	Macherey-Nagel	I		60	
Microsil	CN	Nitrile	Micromeritics	S		65	
	NH_2	Amino	Micromeritics	S		65	
Nova-Pak	CN	Nitrile	Millipore-Waters	S	3	60	Yes
Resolve	CN	Nitrile	Millipore-Waters	S	4	90	No
μBondapak	CN	Nitrile	Millipore-Waters	I	6	125	No
	NH_2	Amino	Millipore-Waters	I		125	No
Spherisorb	CN	Nitrile	Phase Separations	S	3.5	80	
	NH_2	Amino	Phase Separations	S	2	80	
Bondex	CN	Nitrile	Phenomenex				
	NH_2	Amino	Phenomenex				
	CHO		Phenomenex				

Table 11 (continued)
NORMAL PHASE PACKINGS

Support material	Desgination	Functionality	Manufacturer[a]	Shape	Carbon load	Pore size (Å)	Endcapped	Notes
Maxsil	CN	Nitrile	Phenomenex	I		65		Alternative to Partisil and LiChrosorb
	NH_2	Amino	Phenomenex	I		65		
	NO_2	Nitro	Phenomenex	I		65		
W-Porex	CN	Nitrile	Phenomenex			300		
Ultremex	CN	Nitrile	Phenomenex	S				
	NH_2	Amino	Phenomenex	S				
Spherex	NH_2	Amino	Phenomenex	S		100		
	CN	Nitrile	Phenomenex	S		100		
	Diol	Alcohol	Phenomenex	S		100		
IB-SIL	CN	Nitrile	Phenomenex	S				Similar to IBM packing
	NH_2	Amino	Phenomenex	S				Similar to IBM packing
Hichrom	CN	Nitrile	Regis	S	3.5	80		Spherisorb CN
	NH_2	Amino	Regis	S		80		Spherisorb NH_2
Vydac	501TP Polar	Oxynitrile	Separations Group	S		330		
Hypersil	APS	Amino	Shandon	S	2.2	120	Yes	
	CPS	Nitrile	Shandon	S	2	120	Yes	
Micropak	CN	Nitrile	Varian	I		60		LiChrosorb CN
	NH_2	Amino	Varian	I		60		LiChrosorb NH_2
Partisil	PAC	Amino/cyano	Whatman	I		85		Amino/cyano, 2:1
Partisphere	PAC	Amino/cyano	Whatman	S		120		Amino/cyano, 2:1
YMC-Gel	Amine	Amino	YMC	S		100		
	Cyano	Nitrile	YMC	S		100		
	Diol	Alcohol	YMC	S		100		

[a] Addresses of column manufacturers are given in Table 15.

Table 12
SPECIAL PACKINGS

Support material	Designation	Functionality	Manufacturer[a]	Pore size (Å)	Notes
Alpha Cyclodextrin			Astec		Separation of isomers and enantiomers
Beta Cyclodextrin			Astec		Separation of isomers and enantiomers
Gamma Cyclodextrin				Astec	Separation of isomers and enatiomers
Bakerbond	Chiral Phase DNBPG		Baker		*R-N*-3,5-Dinitrobenzoyl-phenylglycine
	Chiral Phase DNBLEU		Baker		*S-N*-3,5-Dinitrobenzoylleucine
	Chiralpak OT		Baker		Polytriphenylmethyl
	Chiralpak OP		Baker		Methacrylate
	Chiralcil OA		Baker		Cellulose esters
	Chiralcil OB		Baker		Cellulose esters
	Chiralcil OC		Baker		Cellulose carbamate
	Chiralcil OK		Baker		Cellulose ether
	Chiralcil WH		Baker		Ligand exchangers
	Chiralcil WM		Baker		Ligand exchangers
Polypore CA			Brownlee		Oligosaccharides
Polypore PB			Brownlee		Monosaccharides
Polypore H			Brownlee		Organic acids
ORH-801			Interaction		Organic acids
ION-300			Interaction		Organic acids
ARH-601			Interaction		Fast acid
ION-100		Anion	Interaction		
ION-200		Cation	Interaction		
ION-110		Anion	Interaction		Difficult anions
ION-210		Cation	Interaction		Transition metals
Nucleogen	DEAE-60		Macherey-Nagel	60	
	DEAE-500		Macherey-Nagel	500	
	DEAE-4000		Macherey-Nagel	4000	

Table 12 (continued)
SPECIAL PACKINGS

Support material	Designation	Functionality	Manufacturer[a]	Pore size (Å)	Notes
Resovosil	Bovine serum albumin		Macherey-Nagel	300	Separation of optical isomers
μBondapak	Fatty acid		Millipore-Waters		
Vydac	Oligonucleotide	Separations Group			
	Nucleotide		Separations Group		Low capacity ion exchange

[a] Addresses of column manufacturers are given in Table 15.

Table 13
ION EXCHANGE PACKINGS

Support material	Designation	Functionality	Manufacturer[a]	Shape	Pore size (Å)	Endcapped
Adsorbosphere	SAX	Anion	Alltech	S	80	
	SCX	Cation	Alltech	S	80	
RSIL	AN	Quaternary amine	Alltech	I	80	
	CAT	Sulfonic acid	Alltech	I	80	
Sepralyte	Anion	Quaternary amine	Analytichem			
	Weak cation	Carboxylic acid	Analytichem			
	Strong cation	Sulfonic acid	Analytichem			
Bakerbond	Quaternary amine	Anion	Baker	S	170	Yes
	Weak cation	Propyl carboxylic	Baker	S	170	Yes
	Strong cation	Propyl sulfonic	Baker	S	170	Yes
	Prep weak anion	Amino	Baker	I	150	Yes
	Prep strong anion	Quaternary amine	Baker	I	150	Yes
	Weak cation	Carboxylic acid	Baker	I	150	Yes
	Prep strong cation	Propyl sulfonic acid	Baker	I	150	Yes
Ultrasil	SAX	Anion	Beckman	I		
	SCN	Cation	Beckman	I		
Spheri	Anion	Anion	Brownlee	S		
Aquapore	AX-300	Anion	Brownlee	S	300	
	CX-300	Cation	Brownlee	S	300	
Chemcosorb	SCX	Cation	Chemco-Dychrom	S	100, 300	
	SAX	Anion	Chemco-Dychrom	S	100, 300	
Ionospher	Anion	Anion	Chrompack	S		
Zorbax	SAX	Quaternary amine	DuPont	S		
	SCX	Sulfonic acid	DuPont	S		
LiChrosorb	KAT	Cation	EM Science	S		
	AN	Quaternary amine	EM Science	I	100	
Chromegabond	P-SCX	Phenyl sulfonate	ES Industries	I	60, 300	
	A-SCX	Alkyl sulfonate	ES Industries	I	60, 300	
	P-WCX	Phenyl carbonate	ES Industries	I	60	
	A-WXC	Alkyl carbonate	ES Industries	I	60	
	SAX	Strong anion exchange	ES Industries	I	60	

Table 13 (continued)
ION EXCHANGE PACKINGS

Support material	Designation	Functionality	Manufacturer[a]	Shape	Pore size (Å)	Endcapped
	RP-SAX	Hybrid RP/strong anion exchange	ES Industries	I	60	
	RP-SCX	Hybrid RP/strong cation exchange	ES Industries	I	60	
	M-WAX	Mono weak anion exchange	ES Industries	I	60	
	D-WAX	Di weak anion exchange	ES Industries	I	60	
	T-WAX	Tri weak anion exchange	ES Industries	I	60	
Techsil	SAX	Anion	HPLC Technology			
	SCX	Cation	HPLC Technology			
Nucleosil	SB	Quaternary amine	Macherey-Nagel	S	100	
	SA	Sulfonic acid	Macherey-Nagel	S	100	
Polygosil	Weak anion	Dimethylamino	Macherey-Nagel	I	60	
Microsil	SAX	Quaternary amine	Micromeritics	S	65	
	SCX	Sulfonic acid	Micromeritics	S	65	
Spherisorb	SAX	Propyl-trimethyl ammonium	Phase Separations	S	80	
W-Porex	CM	Carboxymethyl	Phenomenex		300	
	DEAE	Diethyl aminoethyl	Phenomenex		300	
	SP	Sulfopropyl	Phenomenex		300	
Spherex	SAX	Anion	Phenomenex	S	100	
	SCX	Cation	Phenomenex	S	100	
Dynamax	150A-AX	Anion	Rainin	I		
	300AX	Anion	Rainin	S		
Vydac	Anion	Quaternary amine	Separations Group	S	300	
	Cation	Sulfonic acid	Separations Group	S	300	
Vydac TP	301 TP-Anion	Anion	Separations Group	S	300	
	401 TP-Cation	Cation	Separations Group	S	300	
Supelcosil	LC-SAX	Quatenarypropyl	Supelco	S		
	LC-SCX	Propylsulfonic	Supelco	S		
	LC-3SAX	Quaternarypropyl	Supelco	S	300	
	LC-3SCX	Propylsulfonic	Supelco	S	300	

Synchropak	CM-300	Cation	Synchrom	S	
	AX-300	Anion	Synchrom	S	
Micropak AX	AX	Difunctional amine	Varian	I	60
Micropak	SAX	Quaternary amine	Varian	I	60
Partisil	SAX	Quaternary amine	Whatman	I	85
	SCX	Phenyl sulfonic acid	Whatman	I	85
Partisphere	SAX	Quaternary amine	Whatman	S	120
	SCX	Phenyl sulfonic acid	Whatman	S	120
	WAX	DEAE	Whatman	S	120
	WCX	Carboxylic acid	Whatman	S	120

[a] Addresses of column manufacturers are given in Table 15.

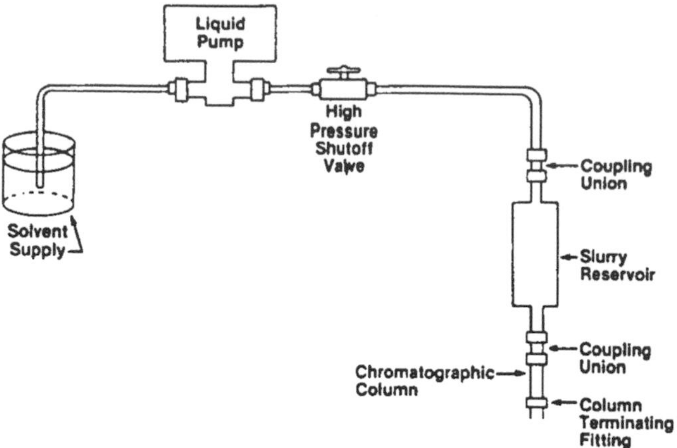

FIGURE 20. Block diagram of apparatus for packing HPLC columns.

B. Packing HPLC Columns

1. Packing Porous Layer Materials

Most porous layer materials are easily packed in a dry state. The tap fill method is the procedure of choice for these materials. Column tubing with a smooth interior wall is cleaned and an appropriate end fitting is attached. The column is then filled by adding small increments of the packing material while rotating and gently tapping on the side of the tube. Between additions of packing material, the column is also tapped in the vertical mode. After the column is filled, an appropriate terminating device with a stainless steel frit is attached. All terminating fittings should be low or zero dead volume fittings.

2. Packing Small Porous Particles

Small particles are especially difficult to pack. They often agglomerate or build up electrostatic charges, which make them very difficult to handle in a dry state. Slurry techniques are usually employed for packing columns with small porous particles.

Many different procedures are described in the literature for the packing of small porous particles into HPLC columns.[28-32] The key to these packing procedures is that of obtaining a tight compact bed with a homogeneous mixture of the packing. All HPLC packings have a range of particle sizes surrounding the mean value. Thus as these types of materials are packed some method for keeping a homogeneous mixture of the entire packing without segregation of the packing by particle size is necessary. Two methods have been used to obtain a homogeneous packing slurry. One method uses a highly viscous solvent which slows the settling of the particles. The second approach uses a balanced density solvent which supports the particles by having the same density as the particle.

To pack good columns, some specialized equipment is required. A typical apparatus for column packing is shown in Figure 20. This type of apparatus requires a high pressure high volume pump. For packing analytical columns of 4 to 4.6 mm I.D., pumping volumes of at least 5 mℓ/min are required with pressure capability of 5000 to 7000 psi. Most people usually prefer air driven pneumatic amplifier pumps, such as those manufactured by Haskel Engineering* for this purpose. The reservoir may be one of those commercially available for this purpose or a fabricated chamber from $^3/_8$- or $^1/_2$-in high pressure tubing with appropriate reducing fittings.

The tubing used to pack such columns should be cleaned and have a smooth or polished interior surface. With the fine particulate packings even very small scratches in the surface

* Haskel Engineering Inc., 100 East Graham Place, Burbank, CA 91502

can have detrimental effects upon the performance of the column. The end fittings for the column should be low or preferably zero dead volume fittings designed for HPLC columns.

Packing HPLC columns properly presents some very severe safety related problems. In packing such columns, high pressures and high flow rates are used. To add to the problem, the pressure changes are very rapid, causing severe shock upon the apparatus, tubing, and fittings. Under such conditions there is always the possibility of a fitting giving way causing rapid depressurization of the system with the possibility of a rapidly accelerating piece of metal breaking loose from the system. When packing HPLC columns, good heavy protective safety shields should be used to protect the operator from any unexpected problems which might develop in the packing of the HPLC column.

Small particulate HPLC packings are especially difficult to pack, and packing problems have become more serious as the nominal size of the particles has decreased. Today, columns of 3-, 5-, and 10-μm particles are common. A key factor in the preparation of stable, efficient columns is the generation of a tightly compacted bed with a uniform distribution of the packing. Several methods have been described for generating such columns. Columns are usually packed with a slurry of the packing material which is mixed and forced into the column at high pressures. Pressures of 5000 to 12,000 psi should be used for column packing. Air-driven pumps are usually used to provide the high pressure and fast flow rates required. Normally the slurry reservoir is pressurized and then the packing is allowed to expand into a column blank. After the column blank has been loaded, the column bed must be "set". Setting is usually done in several steps. First, about 100 mℓ of the slurry solvent is pumped through the column. Next, the slurry solvent is removed by flushing with a solvent such as methanol. Frequently a "slamming" process will be used. In this process, the pressure is raised to maximum value for the pump. Then the pressure is released onto the column, shocking the column while allowing 50 to 100 mℓ of solvent to flow through it. This process is repeated three to four times. Finally, the column is removed from the packing device. A sharp knife is used to remove excess packing from the end of the column, a suitable terminating device with a frit is attached, and the column is ready for testing.

Several types of slurries can be used to pack HPLC columns. Four classes of materials are in use today. Table 14 lists some of the common reagents used with each of the slurry techniques.

Balanced density slurry techniques were one of the first methods used to pack columns with small particulate materials. In this technique, the density of the solvent is adjusted so that it is the same density as the dry packing material. The method is ideal for packings with a wide particle size distribution and for larger size packings (10 μm and larger). This method does, however, use some rather toxic components in the slurry mixtures.[33-38]

The nonbalanced density method of column packing is probably the most widely used technique. Solvents such as methanol and isopropanol can be used to pack silicas. Binary solvents such as a 1:1 mixture of chloroform/methanol or trichloroethylene/ethanol can also used.[33,39-46]

Modified silicas present a different set of packing problems. For alkyl-modified reverse phase packings, a 1:3 mixture of isooctane and chloroform is frequently used. Alkyl silicas often develop electrostatic charges which makes the packing of these substances difficult. To counteract the charge, electrolytes are often added. A common solvent for this purpose is 80 parts methanol and 20 parts 0.1% sodium acetate.[47] Liao and Ponzo have used mixtures of organic acids to interact with the acidic functions on the silica.[48] They have shown that mixtures of heptanoic acid and dichloroacetic acid are an ideal matrix for packing reverse phase materials.

Polar organic-modified silicas are frequently packed as slurries in *n*-propanol, chloroform/pyridine, tetrabromoethane/tetrahydrofuran, or similar solvents.

Two other types of slurries are sometimes employed. A dilute ammonia solution (0.001 *M*) may be used. This tends to neutralize charges and gives a more uniform packing bed.[49]

Table 14
SLURRY TECHNIQUES

Balanced density	Mixtures of Tetrabromoethane Tetrachloroethylene Iodomethane Perchloroethylene
Nonbalanced	Carbon tetrachloride Methanol Acetone Tetrahydrofuran/water Isopropanol Chloroform/methanol Pyridine Organic acids—heptanoic and dichloroacetic Methanol/water/sodium acetate (0.1%)
Ammonia stablilized	0.001 *M* aqueous ammonia
High viscosity	Cyclohexanol Glycerol Ethylene glycol Polyethylene glycol

Finally, high viscosity solvents are sometimes used. Solvents such as cyclohexanol, glycerol, ethylene glycol, and polyethylene glycol may be used. These are most often used where a wide distribution in the size of the particles is expected.[50]

The flowing procedure is one of the common methods used to actually fill the column. Assemble the column and reservoir and fill the column with the solvent used to make the slurry. Fill the reservoir with the slurry and rapidly assemble the system. Pressurize the system to 5000 to 7000 psi and pump about 75 mℓ of solvent through the system. The solvent which is pumped into the system need not be the same solvent that was used to make the slurry. Shut off the pump and allow all flow to stop and the pressure to be dissipated throughout the system. Disconnect the column, scrape the excess packing from the inlet end of the column, and attach an appropriate inlet fitting with a stainless steel frit to the inlet end of the column. Condition the column to the test solvent mixture and evaluate the column for performance criteria before use.

C. Guard Columns

Guard columns are frequently encountered in routine use in HPLC. These columns are inserted between the injection device and the analytical column to protect the expensive analytical column from particulate matter or other substances which remain at the head of the column. Guard columns can be very beneficial to an HPLC system when working with very dirty samples such as crude extracts or biological fluids.

The guard columns are frequently 2 to 3 in. in length and are usually packed with a pellicular material similar to the packing in the analytical column. The guard columns are used for a fixed number of samples and then replaced. The old packing in the guard columns is frequeqntly discarded and the column is repacked for insertion into the HPLC system at some later date.

Another approach to eliminate the need for the guard column has come with the Waters Prep-Pak®. The Prep-Pak® is essentially a minicolumn which can be used to remove particulates and sample materials which have a small retention volume. An appropriate type of solvent is then used to remove the sample from the Prep-Pak®,which is then discarded or reconditioned for the next sample.

D. Column Tubing and Hardware

Several criteria should be considered when choosing the tubing for column bodies and the hardware for column end fittings. It is generally accepted today that smooth wall tubing yields columns with higher efficiencies than can be obtained with rough wall tubing.[51] The actual degree of the polish of the tubing does not appear to be overly important as long as there are no major depressions in the tubing wall. The tubing used for packing most analytical HPLC columns has an internal diameter of 2 to 4.7 mm. Most of the columns in use today are packed in stainless steel tubing with $^1/_4$ in. O.D.

End fittings now in use for HPLC columns usually consist of an inverted nut fitting as shown in Figure 5. A normal compression fitting is used for connection to the column. The inlet to the column uses an inverted compression fitting. The column outlet fitting also contains a fret for retention of the column packing. Unfortunately each manufacturer uses different dimensions for these fittings which results in the problems discussed in Section III.F.

Many of the HPLC applications today use cartridge columns. Cartridge columns are now offered by almost all of the major column manufacturers. These columns usually use a stainless steel sleeve which is packed with the HPLC packing material. Special reusable end fittings are required for terminating the column blanks when attached to the HPLC instrument. This approach can reduce column costs since the terminating fittings do not have to be replaced when the column is replaced. The problem with this approach is that there is no standardization of the hardware, and each manufacturer uses his own design. Thus an entire set of terminating fittings must be purchased from each supplier whose columns are used in the laboratory.

As HPLC column technology has improved, small effects that were insignificant in the early days have become the limiting factors in column performance. One of these problems is mismatch of the metals in an HPLC system. This mismatch often involves the column frits where the metals are not propertly matched with that of the other fittings or column body. This mismatch of metals causes an electrical potential between the different metals which can lead to oxidation of the metal and subsequent movement of metal ions onto the HPLC column. This problem is frequently observed in reverse phase chromatographic columns where the head of the column becomes very black due to the high metal content. In time this causes degradation in column performance.

E. Column Care

Most columns in use today are microparticulate columns. These columns are usually purchased prepacked from a commercial column supplier (Table 15). Since these columns are rather expensive, it is important to take care of the column. Most column manufacturers suggest procedures for maintaining their specific columns. Several general hints for prolonging column life are given below.

To maintain good column efficiency, the column bed should be disturbed as little as possible. Thus, it is important to avoid mechanical shock and vibration of the microparticulate columns. Furthermore, it is generally recommended that flow rates be increased gradually rather than switching a high-pressure stream directly onto the column. Avoiding column shock is more important for smaller particulate packings.

Column life is improved if changes in solvent composition are made gradually. This is especially true in reverse phase chromatography. Thus it is not a good idea to go directly from an organic solvent to all water or vice versa.

Column life can also be affected by the conditions used to store idle columns. It is

Table 15
ADDRESSES OF COLUMN MANUFACTURERS

Ace Glass
1430 Northwest Blvd.
Vineland, NJ 08360

Alltech Associates, Inc.
2051 Waukegan Road
Deerfield, IL 60015

American Burdick & Jackson
American Scientific Products
1430 Waukegan Road
McGraw Park, IL 60085

Amicon
Amicon Corporation
17 Cherry Hill Drive
Danvers, MA 01923

Analytichem
Analytichem International
24210 Frampton Avenue
Harbor City, CA 90710

ASTEC
Advanced Separation Technologies, Inc.
37 Leslie Court
Whippany, NJ 07981

Beckman-Altex
Beckman Instruments Inc.
1716 Fourth Street
Berkeley, CA 94710

Benson Polymeric
P.O. Box 12812
Reno, NV 89510

Bio-Rad Laboratories
2200 Wright Avenue
Richmond, CA 94804

Bioanalytical Systems
2710 Kent Avenue
West Lafayette, IN 47906

Brownlee Labs
2045 Martin Avenue
Santa Clara, CA 95050

Chemco-Dychrom
LC Lab, Inc.
P.O. Box 70116
Sunnyvale, CA 94086

Chrompack
Chrompack, Inc.
P.O. Box 6795
Bridgewater, NJ 08807

Dionex Corporation
1228 Titan Way
Sunnyvale, CA 95086

DuPont
E.I. DuPont de Nemours & Co. Inc.
Clinical and Instrument Systems
Wilmington, DE 19898

EM Science
111 Woodcrest Road
Cherry Hill, NJ 08034

ERBA
ERBA Instruments, Inc.
4 Doulton Place
Peabody, MA 01960

ES Industries
8 South Maple Avenue
Marlton, NJ 08053

Hamilton Company
4970 Energy Way
Reno, NV 89502

Hitachi
Hitachi/NSA
460 E. Middlefield Road
Mountain View, CA 94043

HPLC Technology
P.O. Box 7000-196
Palos Verdes Peninsula, CA 90274

IBM
IBM Instruments
P.O. Box 3020
Wallington, CT 06492

ISCO
4700 Superior Avenue
Lincoln, NE 68504

J.T. Baker Chemical Co.
222 Red School Lane
Phillipsburg, NJ 08865

Jasco Inc.
314 Commerce Drive
Easton, MD 21601

Jones Chromatography
Jones Chromatography USZ, Inc.
P.O. Box 620999
Littleton, CO 80162

Table 15 (continued)
ADDRESSES OF COLUMN MANUFACTURERS

Micromeritics Instrument Corporation
One Micromeritics Drive
Norcross, GA 30093-1877

Millipore Waters Chromatography Division
34 Maple Street
Milford, MA 01757

Perkin-Elmer
Main Avenue
Norwalk, CT 06856

Phase Sep
Phase Separations, Inc.
140 Water Avenue
Norwalk, CT 06850

Phenomenex
6100 Palos Verdes Drive South
Rancho Palos Verdes, CA 90274

Prolabo
12 Rue Pelee, BP 369
Paris Cedex 11 France 75526

Rainin Instrument Co. Inc.
Mack Road
Woburn, MA 01801-4628

Regis Chemical
8110 North Austin
Morton Grove, IL 60053

Shandon
Shandon Southern Instruments
515 Broad Street
Sewickley, PA 15143

Spectra-Physics
Autolab Division
3333 North First Street
San Jose, CA 95134-1995

Supelco, Inc.
Supelco Park
Bellefonte, PA 16823-0048

Synchrom, Inc.
P.O. Box 310
Lafayette, IN 47902-0310

The Anspec Co.
P.O. Box 7730
Ann Arbor, MI 48107

The Separations Group
16695 Spruce Street
Hesperia, CA 92345

Tracor Instruments
6500 Tracor Lane
Austin, TX 78725

Varian Associates
220 Humboldt Court
Sunnyvale, CA 94089

Whatman
9 Bridewell Place
Clifton, NJ 07014

YMC, Incorporated
P.O. Box 492
Mt. Freedom, NJ 07970

recommended that the column beds be kept wet if at all possible. If a column is to be stored, all buffers or salt solutions should be thoroughly washed from the column before storage. Finally, it is generally better to store a column in an organic solvent rather than in water.

One of the biggest sources of column problems is dirty samples. Samples often contain particulate matter or substances which are not eluted from the column under normal operating conditions. Particulate material can plug tubing connections in the sampling valve, connections leading to the column, and the entrance frit of the column. Particulate matter may also damage the sampling valve. Therefore, samples which contain particulate matter should be filtered through a suitable filter before being applied to the chromatographic system. Even with these precautions the entrance frits of the column become plugged with time, causing the pressure drop across the column to increase. When the pressure drop across the column increases by 50% over the original value, it is advisable to replace the frits at the head of the column. This should be done carefully, with the column bed wetted with solvent to avoid disturbing it as much as possible. Some people clean these frits in nitric acid and backflush the frit with solvent. Our experience has indicated that it is not worth the time and effort to clean the frits unless replacement frits are not available.

The problem of components that are not eluted from a column can be difficult to correct. This type of contamination will often be at the head of a column and may cause increased peak tailing as well as peak splitting. Column performance also tends to degrade as the column is used due to changes in the packing material. The degradation process is due in part to hydrolysis of the bonded phase from the silica, erosion of the silica backbone of the packing, and fracture of some of the packing particles in the column. These are nonrecoverable phenomena. There are, however, many instances where contamination of the column is the problem. Loss of performance due to contamination often can be rectified, and the column can be returned to a useful state. However, regeneration of an HPLC column can be very time consuming. Therefore many laboratories find it more economical to discard a column that performs poorly and continue with the work schedule, while other institutions find it advantageous to attempt to recondition the column for further use.

Although many manufacturers have discouraged reversing the direction of flow of the solvent through an HPLC column, this is often a very effective means for restoring the column to a useful condition.[52] The contaminants are generally concentrated on the inlet frit and/or head of the column. By reversing the flow, some of these materials can be washed from the column. Often the contaminated packing has a much smaller influence upon the operation of the column when it is near the outlet than it appears to have when at the inlet of the column. Today the stability of the bed toward flow reversal in the column is quite good. Any state-of-the-art commercial HPLC column should be capable of operation with almost equal efficiency with flow in either direction.

There are many other steps that can be taken to remove contamination from an HPLC column. Rabel[53] has published a set of recommended conditions for reconditioning different types of HPLC columns. Table 16 gives a sequence of solvents which may be used to regenerate a reverse phase HPLC column.

The use of nonpolar solvents to enhance the elution properties of substances from reverse phase columns is an ideal means for removing many substances from a contaminated column. There are, however, other polar impurities which bind to the residual silanols and cannot be removed by this procedure. A wash with a strong acid is used to remove these polar substances from the contaminated column. When reconditioning reverse phase columns, the column should be washed with about 20 to 30 column volumes of each solvent. These solvents can be pumped through the column at flow rates of 2 to 3 mℓ/min. Table 17 provides approximate wash volumes for several popular column sizes.

Hexane or heptane are other solvents that might be added to the group of washing solvents. Additionally, many authors have recommended use of dimethylformamide and dimethylsulfoxide as wash solvents for reverse phase columns.

The most important item in reconditioning reverse phase columns is the removal of all salts from the column. Therefore adequate water washing of reverse phase columns is essential. When changing solvents, one should also assure that the two solvents are miscible.

Table 18 lists a series of washing solvents suggested for normal phase columns. This series of solvents is recommended for silica columns as well as bonded phases like amino, cyano, and alcohol columns. A major problem with normal phase columns is change in the amount of water on the column. For many applications using silica columns, water must be completely eliminated from the solvents or carefully controlled at a well-defined level. Therefore, when regenerating these columns, steps must also be taken to reactivate the column (i.e., reduce the water content of the column.) The series of dry solvents listed in Table 18 are often used to reactivate a silica column. In this step it is critical that the solvents be dry. This is usually accomplished by adding dry molecular sieves to the solvent to capture any water which might be present.

The use of dry solvents is a physical means for drying a column. Bredeweg et al.[54] have described a chemical means for reactivating columns. They use 2,2-dimethoxypropane to

Table 16
SOLVENTS FOR REGENERATING REVERSE PHASE COLUMNS

Water
Methanol
Chloroform
Methanol
Water
0.1 *M* sulfuric acid
Water

Table 17
WASH VOLUMES

Column size (mm)	Wash volume (mℓ)
2.1 × 250	10
4.6 × 150	30
4.6 × 250	50
6.0 × 250	70
8.0 × 250	150

Table 18
WASHING SOLVENTS FOR NORMAL PHASE COLUMNS

(Silica- and Polar-Bonded Phases)

Heptane (hexane)
Chloroform
Ethyl acetate
Acetone
Ethanol
Water
Dry ethanol
Dry acetone
Dry ethyl acetate
Dry chloroform
Dry heptane (hexane)

Table 19
WASH SOLVENTS FOR BONDED ION EXCHANGE COLUMNS

Water
0.1 *M* buffer
Water
0.1 *M* sulfuric acid
Water
Acetone
Water
0.1 *M* disodium EDTA
Water

react with the water to form acetone and methanol. This reagent is used as a 2 to 2.5% solution in chloroform or hexane with a small amount of glacial acetic acid present.

Amino columns may require a special reconditioning step. Since these columns have amino functions present, steps must be taken to assure that the amino groups are intact. If the amino function has been quaternized, a wash with 0.1 *M* ammonium hydroxide is often added to the reconditioning cycle.

Bonded phase ion exchange columns offer a real reconditioning challenge. In these columns one must address the possibility of contamination of the silica gel matrix, contamination of the ionic function, and contamination of the aliphatic or aromatic groups holding the ionic function to the silica gel. Table 19 lists solvents for use in reconditioning bonded ion exchange columns.

Column voiding is a recurring problem in HPLC. This is frequently observed as a sudden loss of efficiency and the appearance of a tailing shoulder on a single component peak. When these symptoms appear, the column will have a significant void area present at its head. Much of the column efficiency can be regained by filling this void with column packing material. If available, this void should be filled with the same material as is present in the column. When carrying out the filling operation, it is advisable to wet both the bulk packing material and the head of the column with methanol to help pack the material into the void. If a match of packing is not available a similar pellicular material may also be used to fill the void. This will certainly be much better than leaving the area void, but is not as good as topping off with the exact packing.

Many reverse phase columns which develop a void will also have a dark area of packing in the top quarter inch of the bed. Much can be gained by removing this quarter inch of packing and topping out the column with new packing material. This procedure will not generate a column with the plate count of the original, but will provide a column suitable for many applications and will greatly extend the usable life of the column. Experience in our laboratories has shown that a column can be topped off three or four times before its performance is degraded too far for further use.

F. Storage of HPLC Columns

The treatment and storage of HPLC columns is a very important problem for many laboratories. The type of operation in a laboratory often dictates the means by which columns are stored and handled. For example, in our laboratory the major cause of column loss is associated with column changes and storage between chromatographic runs. Many laboratories dedicate an HPLC column to a specific assay or solvent system. Other laboratories will use the same column for many different applications. The treatment for these different modes of operation dictate different handling procedures for the columns.

One situation that occurs in many laboratories is that of an HPLC system which is used daily for the same assay. Many people recommend that at the end of each day, the column be washed and stored in a solvent such as methanol. This procedure of course means that the column must be reconditioned for use on the next day of operation. Experience in our laboratory has shown that in most cases such treatment significantly shortens the average column life. The switch of solvents appears to be one of the most detrimental steps for an HPLC column. Our experience has shown that for an assembled system, avoiding solvent changes altogether provides the best column life and overall system performance. The only exceptions are cases where the solvents are very basic (>7.5) or contain high levels of salts (>0.3 to 0.5 *M*). Another exception might be the analysis of biological or other very complex samples where many components are present that are not normally eluted from the column. A regular wash with a strong solvent might be in order in this case. In most cases, exposure of the column to the analytical solvent on a long-term basis has no detrimental effect. Ideally one would also like to maintain a very slow flow (0.1 mℓ/min) through the column during idle periods.

A different mode of operation must be used when an HPLC column is to be removed from the instrument and stored. In this case, it is critical that the column first be washed with an appropriate solvent to remove all salts. The next step depends upon whether the column is dedicated to a single analysis or is used in a variety of procedures. If the column is dedicated to a single assay, it is washed with an organic solvent such as methanol or acetonitrile. The column can be stored in the organic solvent or in water containing enough organic solvent to prevent bacterial growth. Methanol or acetonitrile at a level of 15 to 20% is effective in preventing bacterial growth. The column can now be removed from the system. Columns should be stored wet, and therefore the column should be capped with an appropriate fitting to prevent evaporation of the solvent from the column. Normal phase columns can be treated in a similar manner. These columns are normally stored in hexane or methanol depending upon the nature of the chromatographic system in which they are used.

A more elaborate procedure is required when the HPLC column is used in different assay procedures with different solvent systems. For such cases, it is suggested that the reconditioning steps outlined above be employed before removal of the column for storage. Even with these reconditioning steps, one will often have difficulty regenerating columns that have been used for ion pair chromatography. If at all possible, columns which have been used with an ion pairing reagent should be dedicated to analyses employing that reagent.

The removal and storage of HPLC columns seems to be a very stressful operation for HPLC columns. When a column has been stored, it should be tested to assure that it is functioning properly when it is used again. Evaluation of the column efficiency (plate counts) and the column tailing parameters is a reasonable means for establishing the acceptability of the column for a given assay.

VI. SOLVENT SELECTION IN HPLC

A. Selecting a Solvent System for a Separation

One of the most important steps in the development of an HPLC assay is selection of a suitable solvent system. The selection of stationary phases is rather limited, with only a few basic functionalities being generally available as column packing materials. Therefore, the potential of HPLC rests to a large extent in the selection of a suitable mobile solvent.

Solvents are generally ranked by polarity, but polarity is not a uniquely defined physical property of a substance. Hence the relative polarity of a solvent will be somewhat dependent on the method used to measure it. One of the classical methods of ranking solvents uses adsorption energy as a measure of polarity, with a greater energy corresponding to higher polarity. The classical listing of solvents published by Snyder[24] and reproduced in Table 20 is based upon the energy of adsorption on alumina. Although alumina is not one of the more popular HPLC column materials, the data is useful because the general ordering is similar for other adsorbents. This table is one of the most widely used guides for selection of solvents in HPLC work. The UV cutoff, refractive index, and viscosity for many of the solvents are also included in the table. When using this table for adsorption and normal phase separations, the strength of the solvent increases as one goes down the list, and the energy values become larger. For reverse phase separations, the order is just reversed, with the stronger solvents being near the top of the list.

When selecting a solvent, consideration should be given to the solvent compatibility with the detector of choice for a given problem. The use of the UV detector severely limits the number of solvents which are generally useful for HPLC. The most commonly encountered solvents in HPLC are hexane, isooctane, butyl chloride, methyl t-butyl ether, chloroform, methylene chloride, tetrahydrofuran, acetonitrile, methanol, and water. These solvents generally have acceptable UV cutoffs and have reasonably low viscosities. The low viscosity is important in mass transfer considerations and is a requirement for the development of high column efficiencies.

Solvent polarity is a complex function of many parameters in addition to adsorption energy. A more recent ranking of solvents by Snyder[25] is based on a combination of parameters such as dipole moment, proton acceptor or donor properties, and dispersion forces in the solvent.

The dipole moment is a measure of the permanent charge separation in a molecule and therefore of the electric field surrounding it. A molecule with a large dipole moment interacts more strongly with other charged or dipolar species than does one with a small dipole moment, and therefore acts as a "more polar" substance.

The proton acceptor or donor properties of a molecule are a measure of its ability to hydrogen bond with other molecules. Compounds which display high values for this property will frequently exist in dimeric or higher oligomeric configurations. Water is a good example of such a substance.

The dispersion forces of a molecule are a measure of its ability to polarize another molecule. The parameter is highly dependent on size. Larger molecules are generally affected more by such forces than smaller molecules.

Snyder's polarity index[55] ranks solvents according to a complex theoretical summing of these properties. As a rule, the higher the polarity index, the more polar the solvent. Table 21 lists solvents by increasing polarity index. In general, the order is very similar to the order in Table 20, which is based on adsorption energies. One advantage of the polarity index is that a numerical value is obtained which can be used to estimate the rank of a mixed solvent.

The polarity index offers a convenient guide in the selection of a suitable solvent mixture for a given separation. In quantitative analysis under isocratic conditions, a k' between 2

Table 20
SOLVENT PROPERTIES[a]

Solvent	Adsorption energy[b]	Viscosity	Refractive index	UV cutoff (nm)
Fluoroalkanes	−0.25		1.25	
n-Pentane	0.00	0.23	1.36	210
Isooctane	0.01		1.40	210
n-Decane	0.04	0.92	1.41	210
Cyclohexane	0.04	1.00	1.43	210
Cyclopentane	0.05	0.47	1.41	210
Carbon disulfide	0.15	0.37	1.63	380
Carbon tetrachloride	0.18	0.97	1.47	265
Amyl chloride	0.26	0.43	1.41	225
Xylene	0.26	0.62	1.5	290
Isopropyl ether	0.28	0.37	1.37	220
Isopropyl chloride	0.29	0.33	1.38	225
Toluene	0.29	0.59	1.50	285
n-Propyl chloride	0.30	0.35	1.39	225
Chlorobenzene	0.30	0.80	1.53	
Benzene	0.32	0.65	1.50	280
Ethyl bromide	0.37		1.42	
Ethyl ether	0.38	0.23	1.35	220
Ethyl sulfide	0.38	0.45	1.44	290
Chloroform	0.40	0.57	1.44	245
Methylene chloride	0.42	0.44	1.42	245
Methyl isobutyl ketone	0.43		1.39	330
Tetrahydrofuran	0.45		1.41	220
Ethylene dichloride	0.49	0.79	1.44	230
Methyl ethyl ketone	0.51		1.38	330
1-Nitropropane	0.53		1.40	380
Acetone	0.56	0.32	1.36	330
Dioxane	0.56	1.54	1.42	220
Ethyl acetate	0.58	0.45	1.37	260
Methyl acetate	0.60	0.37	1.36	260
Amyl alcohol	0.61	4.1	1.41	210
Dimethyl sulfoxide	0.62	2.24		
Aniline	0.62	4.4	1.59	
Diethyl amine	0.63	0.38	1.39	275
Nitromethane	0.64	0.67	1.39	380
Acetonitrile	0.65	0.37	1.34	210
Pyridine	0.71	0.94	1.51	305
Isopropanol	0.82	2.3	1.38	210
Ethanol	0.88	1.20	1.36	210
Methanol	0.95	0.60	1.33	210
Acetic acid	Large	1.26	1.37	210
Water	Very large			

[a] Other properties of common solvents are given in Tables 2 and 3 of Volume I, Chapter 2 and Table 11 in Volume 1, Chapter 3.
[b] Energy of adsorption on alumina.

and 5 is usually desired, so a solvent or solvent mixture must be found which gives a k′ value in this range. However, a solvent with the right polarity does not necesssarily offer the selectivity required to separate the compounds of interest. A solution to this problem can frequently be found by trying other solvent mixtures with approximately the same polarity

Although solvent polarity for mixed solvents is not a linear function of composition, the

Table 21
POLARITY INDEX OF SOLVENTS[a]

Solvent	Polarity index	Solvent group	Solvent	Polarity index	Solvent group
n-Decane	−0.3	0	Isopropanol	4.3	2
Isooctane	−0.4	0	Chloroform	4.3	9
n-Hexane	0.0	0	Acetophenone	4.4	6
Cyclohexane	0.0	0	Methyl ethyl ketone	4.5	6
Carbon disulfide	1.0	0	Cyclohexanone	4.5	6
Butyl ether	1.7	1	Nitrobenzene	4.5	7
Carbon tetrachloride	1.7	7	Benzonitrile	4.6	6
Triethylamine	1.8	1	Dioxane	4.8	6
Isopropyl ether	2.2	1	Tetramethyl urea	5.0	3
Toluene	2.3	7	Diethylene glycol	5.0	4
p-Xylene	2.4	7	Ethanol	5.2	2
Chlorobenzene	2.7	8	Pyridine	5.3	3
Phenyl ether	2.8	8	Ethylene glycol	5.4	4
Ethyl ether	2.9	1	Acetone	5.4	6
Ethoxybenzene	2.9	7	Tetramethyl guanidine	5.5	1
Benzene	3.0	7	Methoxyethanol	5.7	4
n-Octanol	3.2	2	Propylene carbonate	6.0	7
Fluorobenzene	3.3	8	Aniline	6.2	6
Benzyl ether	3.3	7	Acetonitrile	6.2	6
Methylene chloride	3.4	5	Methyl formamide	6.2	3
Methoxybenzene	3.5	7	Acetic acid	6.2	4
1-Pentanol	3.6	2	*N,N*-Dimethylacetamide	6.3	3
Ethylene chloride	3.7	5	Dimethyl formamide	6.4	3
Bis(-2-ethoxyethyl)ether	3.9	5	Dimethyl sulfoxide	6.5	6
n-Butanol	3.9	2	Methanol	6.6	2
Isobutanol	3.9	2	Nitromethane	6.8	7
n-Propanol	4.1	2	Formamide	7.3	4
Tetrahydrofuran	4.2	3	Water	9.0	9
Ethyl acetate	4.3	6	Tetrafluoropropanol	9.3	9

[a] Other properties of common solvents are given in Tables 2 and 3 of Volume I, Chapter 2 and Table 11 of Volume I, Chapter 3.

linear approximation is sufficiently accurate to allow its use for estimating the composition of solvent mixtures that will have approximately the same polarity.[56] Thus,

$$\mathrm{PI(SM)} = \%\,\mathrm{SOL(A)} \times \mathrm{PI(A)} + \%\,\mathrm{SOL(B)} \times \mathrm{PI(B)} \qquad (7)$$

where PI(SM) = polarity index of solvent mixture, PI(A) = polarity index of A solvent, and PI(B) = polarity index of B solvent can be used.

Suppose a separation has been developed which offers an appropriate k′ for a particular mixture with a solvent composition of 10% chloroform in hexane. The polarity of this solvent mixture can be calculated by

$$\begin{aligned}\text{polarity of mixture} &= 0.90 \times \mathrm{PI(hexane)} + 0.1 \times \mathrm{PI(chloroform)}\\ &= 0.9 \times (0.0) + 0.1 \times (4.4)\\ &= 0.00 + 0.44\\ &= 0.44\end{aligned}$$

However, the alpha, or separation, between the two components of interest was not satis-

Table 22
SOLVENT GROUPING[a]

Solvent	Solvent group	Polarity index	Solvent	Solvent group	Polarity index
n-Decane	0	−0.3	Formamide	4	7.3
Isooctane	0	−0.4	Methylene chloride	5	3.4
n-Hexane	0	0.0	Ethylene chloride	5	3.7
Cyclohexane	0	0.0	Bis(-2-ethoxyethyl)ether	5	3.9
Carbon disulfide	0	1.0	Ethyl acetate	6	4.3
Butyl ether	1	1.7	Acetophenone	6	4.4
Triethylamine	1	1.8	Methylethylketone	6	4.5
Isopropyl ether	1	2.2	Cyclohexanone	6	4.5
Ethyl ether	1	2.9	Dioxane	6	4.8
Tetramethyl guanidine	1	5.5	Acetone	6	5.4
n-Octanol	2	3.2	Aniline	6	6.2
1-Pentanol	2	3.6	Acetonitrile	6	6.2
n-Butanol	2	3.9	Dimethyl sulfoxide	6	6.5
1-Butanol	2	3.9	Carbon tetrachloride	7	1.7
n-Propanol	2	4.1	Toluene	7	2.3
1-Propanol	2	4.3	*p*-Xylene	7	2.4
Ethanol	2	5.2	Ethoxy benzene	7	2.9
Methanol	2	6.6	Benzene	7	3.0
Tetrahydrofuran	3	4.2	Benzyl ether	7	3.3
Tetramethyl urea	3	5.0	Methoxybenzene	7	3.5
Pyridine	3	5.3	Propylene carbonate	7	6.0
Methyl formamide	3	6.2	Nitromethane	7	6.8
N,N-Dimethylacetamide	3	6.3	Chlorobenzene	8	2.7
Dimethyl formamide	3	6.4	Phenyl ether	8	2.8
Diethylene glycol	4	5.0	Fluorobenzene	8	3.3
Ethylene glycol	4	5.4	Chloroform	9	4.4
Benzyl alcohol	4	5.5	Water	9	9.0
Methoxyethanol	4	5.7	Tetrafluoropropanol	9	9.3
Acetic acid	4	6.2			

[a] Other properties of common solvents are given in Tables 2 and 3 of Volume I, Chapter 2 and Table 11 of Volume I, Chapter 3.

factory. Therefore, another solvent combination with approximately the same polarity is desired. A suitable solvent pair might be hexane and isopropyl ether.

$$0.4 = \% \text{ (hexane)} \times 0.0 + \% \text{ (IPE)} \times 2.2$$

It can be seen that a mixture of about 20% of isopropyl ether in hexane will be required.

In examining the chromatographic properties of many solvents, it is found that certain solvents often offer approximately the same results with respect to selectivity. Thus the chemistry of many solvents are similar, and these solvents can be grouped together. When looking for a solvent, it should then be necessary to try only one from each group, other members of that group being expected to give similar results. Table 22 is a listing of solvents by group and is an important reference for the practicing chromatographer. If a solvent of appropriate polarity (k′) but insufficient selectivity (α) has been found, it is often advantageous to make the second attempt at separation using a solvent mix of similar polarity to the first, but composed of solvents taken from a group far removed in Table 22 from the group containing the first solvent. For example, a solvent composed of 10% chloroform in hexane was used in the example of polarity calculations. Chloroform is a member of group 9, and therefore the next solvent should be in a group far removed from 9. Isopropyl ether would be an appropriate selection since it is a member of group 1. As calculated previously a mixture of 20% isopropyl ether in hexane would have approximately the same polarity as 10% chloroform in hexane and would therefore be the solvent mixture evaluated next. If this change does not offer the appropriate selectivity, solvents from other groups can be tried. Once a solvent from each group has been tried, and still no adequate separation has been obtained, it is usually necessary to use a different column packing to achieve the desired separation.

The particular solvent selected from a group must be compatible with the detector and column being used. A UV detector requires use of a solvent with a UV cutoff below the wavelength at which measurements will be made. The most commonly used solvents in HPLC are hexane, isooctane, butyl chloride, methyl t-butyl ether, chloroform, methylene chloride, tetrahydrofuran, acetonitrile, methanol, and water. These solvents all have relatively low UV cutoffs, cover the full range of polarities, and have reasonably low viscosities. Low viscosity is important in achieving adequate flow without excessive pressure drop across the column and is also a requirement for the development of high column efficiencies.

When a refractive index detector is used, a solvent should be selected whose refractive index differs as much as possible from that of the substance being analyzed. A large difference in the refractive indexes of the sample and solvent give better sensitivity with this type of detector. The solvent selected for use with a fluorescence detector should have a UV cutoff well below the excitation wavelength being used, should not be fluorescent or contain fluorescent impurities, and should not quench the fluorescence of the substance being analyzed. The latter requirement rules out the use of halogenated solvents in many cases. Solvent effects in fluorescence are discussed more completely in Section II.B.2.a.ii of Volume I, Chapter 3. Physical properties of a number of common solvents can be found in Volume I, Chapter 2, Table 4.

B. Solvent Quality

Whenever possible, a grade of solvent should be selected for HPLC which is reasonably constant in physical and chemical characteristics from lot to lot. This is especially important for nonpolar hydrocarbons where a small amount of a polar impurity can greatly affect the behavior of the solvent. In addition, solvents selected for use with a UV detector should not contain impurities that absorb above the cutoff wavelength of the solvent itself, and solvents with low residue levels must be used if constituents of the sample are to be collected from the column eluate for further analytical work.

Today, essentially all solvents which are used in HPLC are available as specialty solvents prepared for HPLC use. These solvents are generally optimized to give good spectral properties as well as having uniform chromatographic properties. These solvents are generally prepared from special distillation procedures. The major problem with such solvents is the stability of certain solvents and thereby the shelf life of the solvent. The solvents causing the most problems are tetrahydrofuran and other ethers as well as the chlorinated solvents.

Preparative liquid chromatography where large volumes of solvents are used is about the only place in liquid chromatography where lower grade solvents are used. In this type of usage, the major concern is one of having a solvent with essentially no nonvolatile residues.

Nonpolar and intermediate polarity solvents frequently have troublesome impurities or preservatives present. Common solvents in this category are (1) the hydrocarbons, such as hexane and isooctane, (2) the chlorinated solvents, such as butyl chloride, methylene chloride, and chloroform, and (3) the ethers, with tetrahydrofuran and methyl t-butyl ether being the most important. Even when buying specialty solvents of these types, care must be taken to insure the absence of troublesome preservatives. This is especially true for tetrahydrofuran, which is very susceptible to peroxide build-up. To avoid this problem, phenolic antioxidants are frequently added. In HPLC, preservative-free tetrahydrofuran is used, and care must be taken with this material because of the peroxide hazard. Such solvents should be stored under nitrogen, and it is not advisable to store such solvents for long periods of time before use. Solvents such as tetrahydrofuran, dioxane, and other ethers should be tested for the presence of peroxides before use. If peroxides are present, steps can be taken to eliminate them before use.

Chlorinated solvents are another group which present impurity problems in HPLC. Chloroform is probably one of the most popular solvents in this group and is also one of the most troublesome. Chloroform usually has ethanol present in significant amounts to serve as a preservative. The presence of ethanol of course means that chloroform will act as a much more polar solvent than expected from its position in the eluotropic solvent series. If chloroform is to be used for HPLC, it should be ordered without preservative. A second problem with chlorinated solvents is the small residual levels of hydrochloric acid that are usually present. The residual hydrochloric acid can lead to problems when using these chlorinated solvents with samples or columns which are acid-sensitive.

Perhaps the most widely used solvent in HPLC is water. This is probably one of the most difficult solvents with which to deal because most water sources are not very pure, and most laboratory distillation and deionization systems produce a quality of water which is not suitable for HPLC use.[57-59] This problem is especially severe in trace work. Perhaps the best solution for this problem is to install a top quality water purification system in the laboratory. The Millipore Milli-Q Apparatus* is currently one of the best systems available for this purpose, and the water is suitable for use in trace level (parts per billion) metal ion analyses as well as trace organic analysis.

A second method for removing impurities from water for trace organic analysis is to insert a reverse phase trapping column in the water line between the pump and the mixing chamber. The reverse phase column, which is usually packed with a rather coarse, totally porous reverse phase packing such as Waters C18® on Porasil B®, removes or reduces most trace organic components from the water. One problem with such a system is that the trap column must be reconditioned frequently by pumping methanol or acetonitrile through it to remove the trapped organic substances. The trap column must then be flushed of the organic solvent with water before returning it to use.

C. Water-Saturated Solvents

Many applications of adsorption chromatography employ an organic solvent partially saturated with water. A typical specification might be a 50% water-saturated solvent. These solvents are prepared by adding excess water to the organic solvent and stirring overnight. The organic layer is then separated from the water layer, and this serves as the water-saturated solvent.

The anhydrous organic solvent is usually prepared by passing the solvent over an open column bed of activated alumina. The alumina column takes the water out of the solvent, giving a very dry solvent. A second alternative for drying the solvent is to treat it with a molecular sieve which will remove water. The dry solvent is then mixed with the water-saturated solvent in the appropriate ratio called for in the procedure. A 1:1 ratio of the water-saturated solvent to the dry solvent would give the 50% water-saturated-solvent mentioned above. Such solvents are used in adsorption chromatography to control the activity of the silica column.

D. Buffer Systems

Buffering systems are very important in several aspects of HPLC. They are essential in ion exchange chromatography and are also used when ion suppression and ion pairing techniques are employed.

The use of silica-based ion exchange packings is limited to the pH range from about 2.5

* Millipore Corporation, Beford, MA 01730.

Table 23
SUBSTANCES USED IN COMMON HPLC BUFFERS[a]

Acetate
Formate
Phosphate
Borate
Carbonate
Triethylamine
Diethylamine
Perchlorate
Nitrate

[a] Specific compositions of several buffers are given in Table 3 and Figures 6 and 7 of Volume I, Chapter 2.

to 7.5 or 8.0. A wider range can be used with polymeric resin ion exchangers, but the physical strength of these materials limits their use to relatively low column head pressures. The buffers most frequently used in the 2.5 to 8.0 pH range are listed in Table 23. The concentration of the buffers usually range from 0.005 *M* to 0.5 *M*, with the more dilute ones being encountered most frequently. Compositions of a number of common buffers are given in Table 3 and Figures 6 and 7 in Volume I, Chapter 2.

In ion pairing and ion suppression techniques, buffers which have reasonable solubility in the organic phase are frequently required. Dilute solutions of ammonium carbonate or phosphate, or buffers made from ammonium acetate or ammonium formate can be used in these cases. The acetate and formate buffers have good solubility in the organic solvents commonly encountered in these applications.

Care must be exercised when using mixed aqueous-organic solvent systems containing buffers or other salts since the miscibility of the solvents may be altered by the salt. Neither separation of the phases nor precipitation of the salts must occur in the solvent composition range to be used in a chromatographic separation.

The nitrate and perchlorate ions were included in the list of buffers in Table 23. These ions are used in some buffer systems, but are more commonly added to a solvent system to reduce peak tailing. Nitrate and perchlorate, like many other inorganic ions, absorb UV light and must not be used in solvent systems when detection is to be made at short UV wavelengths.

Pyridine and triethyl amine formate and acetate have also become popular HPLC buffers. These materials are sufficiently volatile to allow lyophilization from a sample. This property is extremely valuable in HPLC applications.

When using buffered solvents, it is necessary to select a very good grade of salt having no impurities which will interfere with the UV detectors. Probably the most commonly used buffers in HPLC are the acetate and phosphate buffer systems. The ammonium salts of these materials are probably most desirable because they have better solubility in mixed organic/aqueous solvents than other salts. Unfortunately, most reagent grades of these salts are not sufficiently uniform in their UV properties to assure a consistent supply of these materials of acceptable quality. Some specialty HPLC grades of the common buffering salts are now becoming available. At present, the best way of preparing these buffers is to start with the appropriate acid and base, make an initial dilution of one material, titrate the solution to the appropriate pH with the other substance, and, finally, dilute the solution to the appropriate volume. This procedure generally provides solutions with much lower levels of particulate matter and generally good spectral characteristics.

E. Final Solvent Workup

Before a solvent is used with a chromatographic instrument, some precautions should be exercised. Some manufacturers recommend filtering all solvents though a 0.5 μm (approximate) filter before use. This is especially advantageous for aqueous solvents. Solvent degassing is also desirable when working with polar solvents and especially with aqueous solvents. Degassing can be accomplished by oiling or pulling a vacuum on the solvent.

Care must be used when filtering flammable solvents which have very low electrical conductivities. Static charges can build up on the filtering apparatus causing a discharge. The problem arises when the apparatus is touched or additional solvent is added. The discharge can be hot enough to ignite solvent vapors. When filtering such solvents, it is advisable to limit the amount of solvent at any one time and to use stainless steel solvent reservoirs, funnels, and receivers which are connected to one another with a grounding strap.

VII. DATA HANDLING IN HPLC

A. Types of Data Available

Most applications of HPLC in pharmaceutical analysis are carried out to either identify or quantitate a compound. A measurement of time and/or peak area is usually required to accomplish this.

Such measurements usually require a time measurement and/or a peak integration procedure. These measurements can be carried out using a ruler, a planimeter, and electronic integrator, or a multi-instrument computer data system. The most highly automated system available in a laboratory will normally be used for most applications. It should, however, be pointed out that one type of system may be best suited for one problem while another type may be better for another problem. For instance, the digital integrator may be very good for the integration of the main peak in a determination of raw material purity, while a simple peak height measurement might be superior for a trace component assay where the baseline is very noisy.

The time of occurrence of the peak maximum is often used for identification of a compound or determination of its presence or absence in a sample. For major component identification the technique is useful within the usual limitations of any chromatographic identification procedure. The major limitation is, of course, the possibility that a given peak could be caused by the elution of some other compound with similar retention characteristics. Therefore, when chromatographic identification techniques are employed, care should be taken to determine that no other likely impurity, degradation product, or intermediate would have the same retention characteristics as the component in question. Matching retention volumes in a second, very different chromatographic system increases the likelihood that the identification is correct, but other physical data such as NMR or mass spectra are essential whenever a firm identification is required.

Quantitative analysis of a sample may be done using either peak height or peak area measurements. Peak areas are proportional to the amount of material eluting from the column as long as the solvent flows at a constant rate. Peak heights are proportional to the amount of material only when peak widths are constant and are strongly affected by the sample injection techniques, but are not as sensitive to fluctuations in flow rate as peak areas. Peak areas were almost always used for quantitation in early applications of HPLC. However, the use of automated injection techniques (especially in conjunction with reverse phase columns) has greatly reduced the problem of peak width variations, so that peak height measurements are now preferred over area measurement in many applications of HPLC.

Several studies have examined the merits of peak area and peak height measurements for quantitative HPLC. These studies showed that peak height measurements offer better precision and accuracy when one is dealing with peaks that are not fully resolved. Poor resolution may be caused by incomplete separation of major components or the presence of small peaks eluting on the tail of a major peak. For well-resolved HPLC peaks, peak area determinations almost always give better precision.[60-62]

B. Peak Area Normalization

The peak area normalization technique is often used for complex mixtures where many similar components are present. It is also frequently used in attempts to evaluate absolute purity of a sample. The procedure is to sum the areas under all peaks in the chromatogram and then calculate the percentage of the total area that is contributed by the compound of interest. The principal problem with this approach in HPLC is that common HPLC detectors are not mass detectors. Thus, the difference in response of each component must be considered if the analysis is to be accurate. However, the response factors are not known in most applications of this technique. The technique further assumes that all of the sample is eluted from the column. This can be a dangerous assumption if the sample is likely to contain substances with vastly different chemical and physical properties, as is often the case in reaction mixtures where the product is considerably different from the reactants.

C. Internal Standard Method

In the internal standard method, a known amount of a substance with a k′ different from any component present in the sample is added to each sample and standard. Following this initial addition, which must be precise, all further work is concerned with the ratio of the response of the sample to the response of the internal standard peak.

$$\text{Area Ratio} = \text{Area Sample/Area Internal Standard} \tag{8}$$

The concentration of a sample solution is calculated from the area ratios using the following equation:

$$\text{Sample Concentration} = \frac{\text{Area Ratio of Sample}}{\text{Area Ratio of Standard}} \times \text{Conc of Standard} \tag{9}$$

Changes in the volume of the sample solution or small differences in injection volume are of no real concern, as they would affect sample and internal standard proportionately.

There are several problems with the use of the internal standard method in HPLC. One of the major problems is selection of a suitable internal standard. Several guidelines should be observed in this selection. To be useful as an internal standard, a compound must have a k′ different from all other possible sample constituents and yet elute from the column in a reasonable time. Elution of the internal standard should not extend the time required for the analysis by more than a factor of 2 to 3. The search for a suitable compound can be very time-consuming, and success is not guaranteed. To serve as an internal standard, a compound should also show changes in its chromatographic behavior of the same type and magnitude as occur in the main constituent when small variations in mobile phase composition or column aging occur. A substance which moves from a tailing peak to a leading peak with such parameter variations is not a viable internal standard compound. The easiest way to meet these criteria is often to select a related compound, such as a homologue of the major sample component, as the internal standard.

The need for internal standards has been greatly reduced by the use of improved HPLC injectors with loop valves. Loop injectors allow injection of precisely controlled volumes of sample onto the column. The precision of the injections, coupled with the high performance of modern HPLC instruments, has led to a situation where the largest source of variation in an HPLC determination is often the integration of the peaks. The precision and accuracy of peak integrations is often limited by the effectiveness of the algorithm used to select peak start and finish times. When the variance associated with peak integration dominates the method variance, a more precise analytical method may be achieved using external rather than internal standardization since external standardization requires only one integration per sample rather than two.

D. External Standard Method

The external standard method employs separate injections of a fixed volume of the sample and standard solutions. The peaks are integrated and the concentration of the sample solution calculated by direct comparison of the sample and standard peaks using the following equation:

$$\text{Sample Concentration} = \frac{\text{Peak Area in Sample}}{\text{Peak Area in Standard}} \times \text{Conc of Standard} \qquad (10)$$

In applications where the samples are expected to have a relatively broad range of concentrations, a series of standards covering the anticipated range of sample values are analyzed. Values for the samples are then calculated from the regression line through the standard values. Good precision using the external standard method requires good reproducibility in the volume of sample injected.

With the use of precise sampling valves, the external standard approach has become the principal means for quantitation. The smaller loop valves which deliver a fixed volume of sample offer the best results for this type of work. The precision of injection and peak integration using the external standard method in a well behaved HPLC system utilizing automated sampling valves is routinely in the range of 0.5 to 1% or better. One of the biggest sources of variation in these systems is variation of the sample volume with small changes in temperature while the samples are waiting for injection. Therefore, any automated sampling device used for quantitative HPLC with external standardization should be positioned to minimize temperature fluctuations in the sample during operation.

VIII. DERIVATIZATION IN HPLC

Even though the use of chemical derivatization was slow to gain a foothold in HPLC, it has now become a significant factor in the practice of HPLC. In HPLC, chemical derivatization can occur prior to the separation process or follow the separation when a postcolumn reactor is used.

Precolumn chemical derivatization is primarily used to increase detector response for the compound of interest. Although of lesser importance, chemical derivatization can be used to alter the extraction efficiency of a sample from a difficult sample matrix. It can also be used to alter or enhance the HPLC separation. In this respect, an amine or other very polar group might be derivatized to decrease column interaction and tailing. Finally, optical isomers might be derivatized to form diastereomers which can be separated on conventional (as opposed to chiral) column packings.

Precolumn derivatization is preferred over postcolumn derivatization because it imposes fewer restrictions on the sample solvent and reagents since the reaction is not carried out in the presence of the chromatography solvent. With precolumn derivatization, extractions or other forms of sample cleanup can be used prior to injection to avoid potential chromatographic interferences. Precolumn reaction presupposes the stability of the derivative under the conditions used in the subsequent separation and analysis. Precolumn techniques present less problems in instrument design, for they do not require large amounts of extra-column volume in which the chemical reaction is allowed to occur. Thus the separation efficiency is maintained with the precolumn derivatization technique.

Postcolumn derivatization is used almost exclusively to enhance the sensitivity of the detector to the compound of interest. When dealing with fast reactions, a mixing tee can be used to add the reagents to the flowing stream followed by a short section of tubing between the tee and detector to facilitate mixing and completion of the chemical reaction. For slower reactions, the mixture of reagents, sample, and mobile phase may be passed through a packed bed. This facilitates mixing and reaction of the chemicals without causing a large degradation of the separation because of band spreading. A third type of postcolumn reactor uses an air segmented stream. The segmented stream decreases the potential of band spreading and is used when long reaction or mixing times are required.

Chemical derivatizations to produce fluorescent analogs are by far the most important reactions in HPLC. Of these, the *o*-phthalaldehyde reaction for amines and amino acids has become the most important. Of lesser importance are the reactions to produce ultraviolet chromophors for detection. The advent of optical isomer separations has produced yet another class of reactions to yield diastereomers. Each of these groups of reactions will be discussed in the following sections.

Reverse phase HPLC is generally the mechanism of choice for separating the materials formed in precolumn derivatizations. It is also the usual choice when postcolumn derivatization procedures are used. Although reverse phase HPLC is the separation mechanism of choice, both normal phase and adsorption chromatographic procedures appear in the literature for separation of many of the derivatives discussed in the following sections. The choice of an HPLC procedure for a derivatized sample ultimately depends upon the chemical stability and chromatographic behavior of the derivative of the target compound and any potential interferences that might be present in the sample.

A. Derivatives for Fluorometric Detection

The formation of fluorescent derivatives to enhance detection sensitivity is by far the largest application of chemical derivatization in HPLC. Lingeman et al.[63] have provided a very good review of the use of fluorescence in HPLC. This includes naturally fluorescent compounds as well as descriptions of many procedures that can be used to form fluorescent products. Table 24 is a listing of fluorescent labels commonly used in HPLC and the functionalities with which they react.

1. o-Phthalaldehyde (OPA)

The reaction of OPA with primary amines is probably the most widely used reaction today. This reagent is convenient for use with postcolumn reactors. One advantage is that the reagents have no fluorescence in their unreacted forms. OPA is reacted with a primary amino function along with a mercaptan such as 2-mercaptoethanol, 3-mercapto-1-propanol, or ethanethiol in a basic media (pH 10) to form an isoindole which is fluorescent.[154,155]

$$C_6H_4(CHO)_2 + RNH_2 + HSCH_2CH_2OH \xrightarrow{[OH^-]} \text{isoindole}(SCH_2CH_2OH)\text{N-R} + 2\,H_2O \qquad (11)$$

The generalized procedure involves dissolving about 400 mg of OPA in 400 ml of 0.4 *M* boric acid buffer adjusted to pH 10.4 with potassium hydroxide. Then 2 ml of 2-mercaptoethanol is added. The reactions are fast (1 to 2 min) at room temperature.[156] The reagent is quite stable, but the stability of the derivative is somewhat dependent upon the thiol used. The use of 2-ethanethiol and 3-mercapto-1-propanol gives more stable products.[157] Detection limits for these products are in the 1- to 10-ng range.

Table 24
FLUORESCENCE LABELS USED IN HPLC

Reagent	Abbreviation	Functionality	Ref.
N-(9-Acridinyl) maleimide	NAM	Sulfhydryl	64, 56
(D)-(L)-1-Aminoethyl-4-dimethylaminonaphthalene	DANE	Carboxyl	66, 67
4-Amino-7-nitrobenzo-2-oxa-1,3-diazole	NBD-amine	Hydrolytic enzymes	68
9-Aminophenanthrene		Carboxyl	69
2-Aminopropionitrile-fumarate-borate	AFB	Carbohydrate	70
1-Anilinonaphthylmaleimide	ANM	Sulfhydryl	71
1- and 9-anthroylnitrile		Hydroxyl	72
p-(9-Anthroyloxy)phenacyl bromide		Carboxyl, imide, phenol, sulfhydryl	73
9-Anthryldiazomethane	ADAM	Carboxyl, imide, phenol, sulfhydryl	74, 75
p-(2-Benzimidazoyl) phenylmaleimide	BIPM	Sulfhydryl	76
Benzoin		Guanidine	77, 78
p-(2-Benzoxazolyl)phenylmaleimide	BOPM	Sulfhydryl	79
Boc-aminomethyl-/Boc-aminophenylisothiocyanate		Amine	80
2-Bromoacetonaphthon (naphthacyl bromide)		Carboxyl, imide, phenol, sulfhydryl	81
1-Bromoacetylpyrene		Carboxyl, imide, phenol, sulfhydryl	82
4-Bromomethyl-7-acetoxycoumarin	Br-Mac	Carboxyl, imide, phenol, sulfhydryl	83
4-Bromomethyl-6,7-dimethoxycoumarin	Br-Mdmc	Carboxyl, imide, phenol, sulfhydryl	84
4-Bromomethyl-7-methoxycoumarin	Br-Mmc	Carboxyl, imide, phenol, sulfhydryl	85, 86
N-Chlorodansylamide	NCDA	Sulfhydryl	87, 88
9-(Chloromethyl)anthracene	9-CIMA	Carboxyl, imide, phenol, sulfhydryl	89
4-Chloro-7-nitrobenzo-2-oxa-1,3-diazole	NBD-Cl	Amine (primary, secondary), phenol	90, 91
2-Cyanoacetamide		Reducing hydroxyl	92, 93
2-Dansylaminoethanol	DAE	Carboxyl	94
2-Dansylethylchloraformate	Dns-ECF	Amine	95
9,10-Diaminophenanthrene	DAP	Carboxyl	96, 97
4-Diazomethyl-7-methoxycoumarin	D-Mmc	Carboxyl, imide, phenol, sulfhydryl	98
5-Di-*n*-Butylaminonaphthalene-1-sulfonyl chloride	Bns-Cl	Amine (primary, secondary), hydroxyl, phenol, sulfhydryl	99
N,N'-Dicyclohexyl-/*N,N'*-diisopropyl-*O*-(7-methoxycoumarin) methyl-isourea	DCCI/DIDI	Carboxyl	100, 101
4,5-Dimethoxy-1,2-diaminobenzene	DBB	Aldehyde	102, 103
N-(7-Dimethylamino-4-methyl-3-coumarinyl)-maleimide	DACM	Sulfhydryl	104
5-Dimethylaminonaphthalene-1-sulfonyl-aziridine	Dns-A	Sulfhydryl	105, 106
5-Dimethylaminonaphthalene-1-sulfonyl-cadaverine	Dns-C	Carboxyl	107
5-Dimethylaminonaphthalene-1-sulfonyl chloride	Dns-Cl	Amine (primary, secondary), hydroxyl, phenol, sulfhydryl	108, 109

Table 24 (continued)
FLUORESCENCE LABELS USED IN HPLC

Reagent	Abbreviation	Functionality	Ref.
5-Dimethylaminonaphthalene-1-sulfonyl-hydrazine	Dns-H	Carbonyl	110, 111
4-Dimethylamino-1-naphthoylnitrile	dMA-NN	Hydroxyl (primary, secondary)	112
4-Dimethylamino-1-naphthylisothiocyanate		Amine	113
2-Diphenylacetyl-1,3-indandione-1-hydrazone	DIH	Aldehyde	114
1,2-Diphenyl-ethylenediamine	DPE	Reducing hydroxyl	115
9-Fluorenylmethylchloroformate	FMOCCI	Amine	116
Fluoresceinisothiocyanate		Amine	117
4-Fluorobenzo-2-oxa-1,3-diazole-7-sulfonate	SBD-F	Sulfhydryl	118, 119
4-Fluro-7-nitrobenzo-2-oxa-1,3-diazole	NBD-F	Amine (primary, secondary), phenol	120, 121
4-Fluro-7-sulfamoylbenzo-2-oxa-1,3-diazole	NH2-SBD-F	Sulfhydryl	122
Glycinamide		Hydroxyl	123
4-Hydrazino-7-nitrobenzo-2-oxa-1,3-diazole	NBD-H	Carbonyl	124
4′-Hydrazino-2-stilbazole	4H2S	Carbonyl	125, 126
9-(Hydroxymethyl)anthracene	HMA	Carbonyl	127
4-Hydroxymethyl-7-methoxycoumarin	Hy-Mmc	Carboxyl	128, 129
9-Isothiocyanatoacridine		Amine	130
7-Methoxycoumarin-3/4-carbonyl azide	3-MCCA/4-MCC	Hydroxyl	131
2-Methoxy-2,4-diphenyl-3(2H)-furanone	MDPF	Amine (primary, secondary)	132, 133
4-(6-Methylbenzothiazol-2-yl)phenylisocyanate	Mbp	Amine	134, 135
(+)/(−)-2-Methyl-1,1′-binaphthalene-2′-carbonylnitrile		Hydroxyl	136
5-Methylphenylaminonaphthalene-1-sulfonyl chloride	Mns-Cl	Amine (primary, secondary) hydroxyl, phenol, sulfhydryl	137, 138
Monobromo-trimethyl-ammoniobimane		Sulfhydryl	139, 140
1,2-Naphthoylenebenzimidazole-6-sulfonyl chloride	NBI-SO2Cl	Amine (primary, secondary) hydroxyl, phenol, sulfhydryl	141, 142
1-Naphthylamine		Carboxyl	143
2-Naphthylchloroformate	NCF	Amine	144
Naphthylisocyanate	NIC	Amine	145
o-Phenylenediamine		α-Ketocarboxyl	146, 147
Phenylisothiocyanate		Amine	148
4-Phenylspiro[furan-2(3H)-1′-phthalan]-3,3′-dione	Flur	Amine (primary, secondary)	138, 149
o-Phthalaldehyde	OPA	Amine (primary, secondary)	150, 151
N-(1-Pyrene)maleimide	PM	Sulfhydryl	152
N-Succinimidyl-2-naphthoxyacetate	SNA	Amine	153

2. Fluorescamine

Another popular fluorescent reagent is fluorescamine (4-phenylspiro[furan-2(3H)-1′-phthalan]-3,3′dione). This compound reacts with nucleophilic groups such as alcohols, primary and secondary amines, and water. However, fluorescamine only forms fluorescent products in its reactions with primary amines.[158,159]

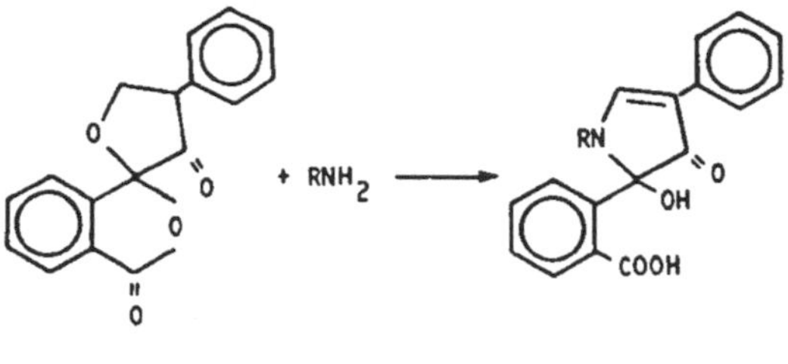

(12)

This derivatization is run in aqueous borate or phosphate buffer in the pH range of 7.5 to 9. Fluorescamine dissolved in ethanol, acetonitrile, or acetone is then added. This reagent is generally used in postcolumn reactors. When used in precolumn derivatization of compounds such as amino acids that have a free carboxylic acid group present, a lactone formed by reaction of the free carboxylic acid of the product with the proximal hydroxyl group[160-162] may exist in equilibrium with the initial product.

A closely related compound, 2-methoxy-2,4-diphenyl-3(2H)-furanone (MDPF), reacts with both primary and secondary amines. Primary amines form pyrrolinones. Secondary amines form a nonfluorescent aminodienone which reacts further with ethanolamine to yield a fluorescent product.[132,133]

3. Dansyl Reactions

Another popular family of compounds which form fluorescent derivatives belong to the dansyl (i.e., 5-dimethylaminonaphthalene-1-sulfonyl) group.

a. Dansyl Chloride (Dns-Cl)

The most important compound of this family is Dns-Cl. Dns-Cl reacts with primary and secondary amines as well as phenols, alcohols, and thiols.

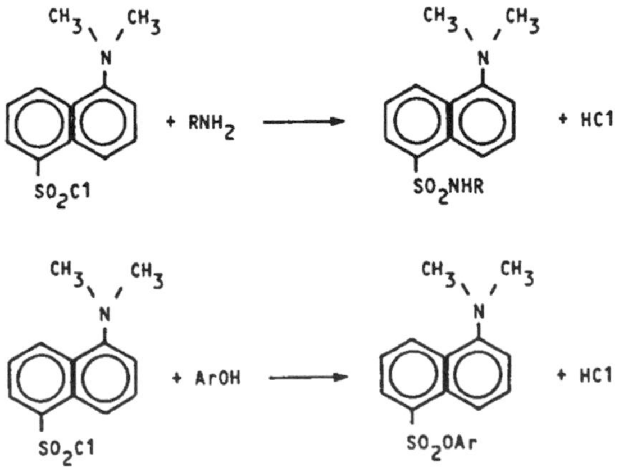

(13)

These reactions are performed in a slightly alkaline medium (pH 9). Sodium carbonate or triethylamine is usually added to the sample along with a solution of Dns-Cl. The sample may be heated slightly. The sensitivity is typically in the subnanogram or picogram range.[163-166]

b. Bansyl Chloride (Bns-Cl)

A series of modified dansyl reagents have also evolved to fulfill very specific needs. Bns-Cl has a 5-dibutylamino group substituted for the dimethylamino group. This compound is more lipophilic and offers an advantage if the derivatives are to be extracted from the aqueous reaction mixture. The detection limit is again in the low nanogram range.[99]

c. Mansyl Chloride (Mns-Cl)

Where extreme detection sensitivity is desired, Mns-Cl is often used. This reagent has a 5-methyl-phenylamino group present. This structure increases the fluorescence efficiency of the derivative.[138]

d. Dansylaziridine (Dns-A)

The aziridine group is yet another modification that is made to the reagent. This group is substituted for the chlorine on the sulfonic acid. This modification produces a reagent which is selective for sulfhydryl groups.

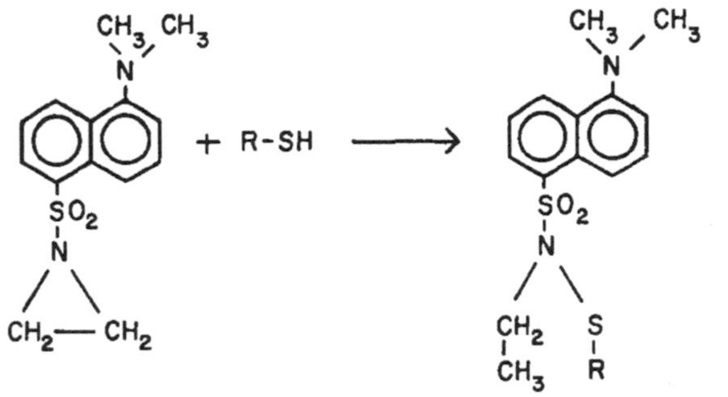

CH₃ CH₃ N — naphthalene — SO_2–N(CH₂—CH₂) + R-SH ⟶ CH₃ CH₃ N — naphthalene — SO_2–N(CH₂CH₃)(S–R) (14)

This reaction is run in phosphate buffers at a pH of 8.2. The mixture is heated at 60°C for 1 h.[105,106,156]

e. Dansylhydrazine (Dns-H)

Finally, a hydrazine moiety may be substituted to give Dns-H. This reagent is used with compounds containing a reactive carbonyl group to form fluorescent hydrazones. It is useful for ketosteroids[110,166-172] and sugars.[111,173,174] The reaction is usually run in methanol with a two-fold excess of Dns-H. The reaction is heated to about 70°C for 15 min.[156]

4. 4-Bromomethyl-7-Methoxycoumarin (BrMmc)

Methoxycoumarins have been used for preparation of fluorescent compounds with acidic function. The general reagent is BrMmc. The reaction is

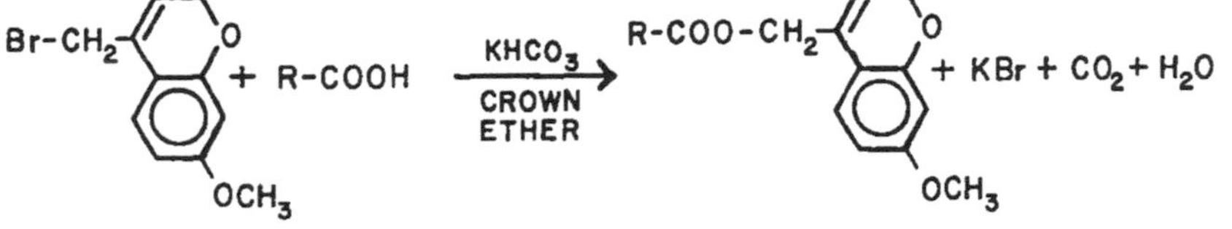

$$\text{Br-CH}_2\text{-(7-OCH}_3\text{-coumarin)} + \text{R-COOH} \xrightarrow[\text{CROWN ETHER}]{\text{KHCO}_3} \text{R-COO-CH}_2\text{-(7-OCH}_3\text{-coumarin)} + \text{KBr} + CO_2 + H_2O \quad (15)$$

This reaction is run by dissolving about 3 μmol of acid in acetone. Next, 1 mg of crown ether is added along with 9 μmol of BrMmc and 25 mg of solid potassium carbonate. The solution is refluxed for about 1 h.[85,86,128,156,175-177] The detection limit is about 7 pmol.[178]

5. 7-Chloro-4-Nitrobenzo-2-Oxa-1,3-Diazole (NBD Chloride)

NBD chloride is a reagent which reacts with thiols and amines.[138,179,180]

$$\text{NBD-Cl} + RSH \longrightarrow \text{NBD-SR} + HCl \tag{16}$$

The reaction takes place in a neutral buffer. About 100 μmol of sample at a pH of 7 is treated with 50 m*M* sodium citrate and 1 m*M* EDTA. NBD chloride is added and the reaction is complete in about 2 h at room temperature.

B. Derivatives for Ultraviolet/Visible Detection

The favored detector in HPLC is still the ultraviolet detector. Therefore it is often desirable to have compounds which can be monitored by these devices. The following series of reagents have been developed and are commercially available for use in HPLC.

Development of visible chromophores also finds some use. The most widely used application of this general technique is the postcolumn reaction of amino acids to enable their quantitation at low concentrations.

1. N-Succinimidyl-p-Nitrophenylacetate (SNPA)

SNPA reacts with primary and secondary amines and amino acids to form the corresponding *p*-nitrophenylacetamides.

$$O_2N\text{-}C_6H_4\text{-}CH_2-\overset{O}{\overset{\|}{C}}-O-N(\text{succinimide}) + HN{<}^{R}_{R'} \longrightarrow O_2N\text{-}C_6H_4\text{-}CH_2-\overset{O}{\overset{\|}{C}}-N{<}^{R}_{R'} + HON(\text{succinimide}) \tag{17}$$

The reaction is run by treating 1 to 5 mg of sample with 4 ml of tetrahydrofuran and 50 mg of SNPA. The mixture is heated at 60°C for about 1 h.[156]

2. 3,5-Dinitrobenzoyl Chloride (DNBC)

DNBC reacts with alcohols and phenols to form the 3,5-dinitrobenzoate esters. DNBC also reacts with primary and secondary amines to form the 3,5-dinitrobenzamides.

$$(O_2N)_2C_6H_3-\overset{O}{\overset{\|}{C}}-Cl + HOR \longrightarrow (O_2N)_2C_6H_3-\overset{O}{\overset{\|}{C}}OR + HCl \tag{18}$$

$$(O_2N)_2C_6H_3-\overset{O}{\overset{\|}{C}}-Cl + HN{<}^{R}_{R'} \longrightarrow (O_2N)_2C_6H_3-\overset{O}{\overset{\|}{C}}-N{<}^{R}_{R'} + HCl$$

This reaction is run by treating 1 to 5 mg of sample with 4 mℓ of tetrahydrofuran, 40 mg DNBC, and 3 drops of pyridine. The reaction mixture is heated at 60°C for about 1 h.[156,181]

3. p-Nitrobenzyloxyamine Hydrochloride (PNBA)

PNBA reacts with aldehydes and ketones to form the *p*-nitrobenzyloximes.

$$O_2N-C_6H_4-CH_2-ONH_2 \cdot HCl + R-\overset{O}{\overset{\|}{C}}-R' \xrightarrow{Et_3N} O_2N-C_6H_4-CH_2ON=C\begin{smallmatrix}R\\R'\end{smallmatrix} + HCl \quad (19)$$

About 0.5 mg of sample and 10 mg PNBA are dissolved in 5 $\mu\ell$ of pyridine and heated at 60°C for 30 min.[156]

4. O-p-Nitrobenzyl-N,N′-Diisopropylisourea (PNBDI)

PNBDI reacts with carboxylic acids to form the *p*-nitrobenzyl esters. The reaction is usually run in methylene chloride with mild heat.

$$O_2N-C_6H_4-CH_2OC\begin{smallmatrix}=NCH(CH_3)_2\\NCH(CH_3)_2\end{smallmatrix} + RCOOH \longrightarrow O_2N-C_6H_4-CH_2O\overset{O}{\overset{\|}{C}}-R + O=C\begin{smallmatrix}NHCH(CH_3)_2\\NHCH(CH_3)_2\end{smallmatrix} \quad (20)$$

A sample of about 3 μmol is treated with 125 $\mu\ell$ of methylene chloride and 9 μmol PNBDI in 125 $\mu\ell$ of methylene chloride. The vial is sealed and heated for 2 h at 80°C.[156]

5. p-Bromophenacyl Bromide (PBPB)

PBPB is used to treat carboxylic acids.

$$Br-C_6H_4-\overset{O}{\overset{\|}{C}}-CH_2-Br + R-COOH \longrightarrow Br-C_6H_4-\overset{O}{\overset{\|}{C}}-CH_2-O\overset{O}{\overset{\|}{C}}-R + HBr \quad (21)$$

The reaction is performed by titrating 3 to 4 μmol of acid in methanol with methanolic KOH to the phenolphthalein end point. The methanol is stripped and 5 μmol of PBPB and 0.25 μmol of 18-crown-6-(1,4,7,10,13,16-hexaoxacyclooctadecane) in 5 ml acetonitrile are added. The solution is heated at 80°C for 15 min.[156,182]

C. Derivatives for Separation of Optical Isomers

The separation of optical isomers has gained widespread attention in pharmaceutical analysis. The activity of a drug as well as its absorption, metabolism, and elimination may be dictated by the configuration of the isomer used. Thus it is often desirable to resolve the optical isomers of a drug when it is analyzed. One of the preferred methods for the analysis of optical isomers is to react the substance with an appropriate chiral reagent to form diastereomers which can then be resolved on conventional HPLC columns. Souter[183] has published a book that describes many procedures for the separation of stereoisomers by both HPLC and GC.

1. 1-Fluoro-2,4-Dinitrophenyl-5-L-Alanine Amide (FDAA or Marfey's Reagent)

Marfey's reagent is being used for the separation of amino acids.[184] When this substance is reacted with amino acids, the D-amino acids have strong intramolecular bonding which allows them to be selectively retained with respect to the L-derivatives on reverse phase columns. About 5 μmol of sample is dissolved in 100 $\mu\ell$ of solvent and 200 $\mu\ell$ of a 1% solution of FDAA in acetone is added. Next, 40 $\mu\ell$ of 1.0 *M* sodium bicarbonate is added and the tube is sealed. It is then heated for 1 h at 40°C. The solution is acidified and examined by reverse phase HPLC.[185]

2. 2,3,4,6-Tetra-O-Acetyl-β-D-Glucopyranosyl Isothiocyanate (TAGIT)

TAGIT is a reagent which has been used to react with primary and secondary amino groups to give thioureas.[186]

3. O,O-Dibenzoyl Tartaric Acid Anhydride

O,O-Dibenzoyl tartaric acid anhydride is an interesting reagent for the resolution of alkanolamines by formation of a monoalkanolamine monoester.[187]

Up to several milligrams of an alkanolamine is dissolved in a dry aprotic solvent such as tetrahydrofuran. Excess trichloroacetic acid and the chiral reagent are added. The tube is sealed and heated at 50°C for several hours.

IX. MICRO-HPLC TECHNIQUES

Micro-HPLC techniques, which have been variously called microbore and microcolumn techniques by different authors, offer an exciting expansion of HPLC technology. The use of small diameter columns allows very high efficiency separations, analysis of small quantities of material, low volume flow rates, and easy interface with a variety of detectors. A range of types and sizes of columns fall into the micro-HPLC category, and with the many advantages of small diameter columns, microtechniques are rapidly making inroads into the practice of liquid chromatography. Micro-HPLC has already matured to the point that several books devoted to the subject are now available.[188-190]

Microcolumn HPLC has many potential advantages over conventional HPLC.[191,192] There has been a significant driving force to explore these techniques. Micro-HPLC techniques have the potential for producing high column efficiencies which make it ideal for separation of complex mixtures. Microcolumn techniques considerably reduce the dilution effects common in liquid chromatography. Therefore the mass sensitivities are greatly increased over those typical for normal HPLC. This offers considerable advantages for dealing with small or very limited samples. The reduction in the dilution in micro-HPLC also makes it an ideal selection for trace analysis problems.[193]

The low volume flow rates used in micro-HPLC offer several advantages. This characteristic greatly reduces solvent consumption and the associated problems of solvent disposal. The low flows also allow use of more exotic solvents and reagents to facilitate the separation, since the total use is a relatively small volume. Finally, low flow rates allow the interface of a wide variety of detectors with HPLC. Microsystems can be easily interfaced to mass spectrometers and laser-based instruments for measurement of fluorescence, light scattering, and optical rotatory dispersion.

Micro-HPLC techniques fall into several categories depending upon the column types employed. Early micro-HPLC techniques used packed columns with diameters up to about 2 mm. Table 25 gives some of the characteristics of micro-HPLC column types in use today.

The size and type of column also dictate the flow rates and amount of sample which can be injected onto such a column.[194] The optimal sample mass for samples used with conventional HPLC columns is in the range of 10 to 100 μg of sample. Small bore packed columns can tolerate 1 to 10 μg of material. Finally, the open capillaries will only tolerate about 100 ng of sample.

The flow rates through the various types of columns also vary.[195] Table 26 shows some typical flow rates.

The small columns and low flow rates put great demands upon the design of micro-HPLC systems. The system should have very good flow control at the very low flow rates required

Table 25
TYPES OF COLUMNS FOR MICRO-HPLC

Type	Size	Packing	Column construction
Small bore packed column	0.2—1 mm I.D.,	Small particulate packings, usually 3- to 5-μm packings	Stainless steel Glass lined Stainless steel Fused silica
Packed capillary	40—80 μm I.D.	Small embedded particulate packings 1—30 μm	Fused silica Glass
Open tubular capillary	15—50 μm I.D.	A bonded phase on the wall	Fused silica Glass

Table 26
TYPICAL FLOW RATES IN HPLC COLUMNS

Column types	Flow rates
Standard packed	1 mℓ/min
Small bore packed	1—20 μℓ/min
Packed capillary	0.5—2 μℓ/min
Open tubular capillary	<1 μℓ/min

of these systems. The low flow rates and overall small system volume make it imperative that the entire chromatographic system have extremely small dead volume.[196] Special valves with small internal loops are usually employed. The columns used in these systems are usually inserted directly into the valve in a position almost adjacent to the valve rotor. This is done to minimize the dead volume associated with the sampling valve.[197]

The detectors used with micro-HPLC techniques must also be optimized.[198,199] The ultraviolet detector is still the major detector used with micro-HPLC. With the use of these detectors, the column must be brought almost adjacent to the detector cell. The cell volume must also be reduced considerably from conventional HPLC detectors. For many applications where fused silica columns are used, the detector cell is included as a small section of the tubing at the end of the column. The small volumes generally allow for high concentrations of components in a very narrow band. Thus the sensitivities may be better with microsystems because of the reduced volume in which the sample component resides: mass sensitivities for micro-HPLC systems are typically much greater than obtainable with conventional HPLC.

Micro-HPLC techniques are an ideal approach for use with mass spectrometers. The low flow rates make transport detectors very useful devices for these systems. Depending upon the solvents and the nature of the mass spectrometers, the column eluent may also be introduced directly into the mass spectrometer.[200-203] The high mass-to-volume ratio of the eluent peak gives good sensitivity with the mass spectrometers.

Other specialized detectors have also been described for use with micro-HPLC systems. Various laser-based detectors for fluorescence[204,206] and light scattering techniques[207,208] have been described. Laser-based systems have also been described for sensitive detection based upon optical rotation.[209,211] These are extremely interesting devices, for micro-HPLC systems are finding many applications in dealing with biological samples.

The micro-HPLC systems also make the moving wire flame ionization detectors more practical.[212] With the low flow rates, the entire column eluent can be used with this type of detector.

Micro-HPLC is a specialty branch of HPLC that is just beginning to be widely accepted. Unfortunately, there is little instrumentation currently available to support the technique.

Usually suitable components need to be procured from multiple sources to construct the instrumentation. Almost none of the very specialized detectors described is commercially available. Therefore these devices must be fabricated when one has a special application that demands it use. The high efficiencies, high detector sensitivities, and advantages especially in interfacing HPLC to mass spectrometers still make this a very attractive adjunct of HPLC.

X. PREPARATIVE HPLC

Preparative HPLC uses HPLC techniques for the isolation and purification of useful quantities of a chemical substance. The differences between preparative and analytical HPLC arise from the need to separate and collect larger quantities of material in preparative work than can be accommodated using normal analytical procedures. Since open column chromatography was practiced for preparative purposes long before the advent of HPLC, it was only natural to expect HPLC techniques to be quickly implemented in preparative applications. Today, both adsorption and reverse phase HPLC are widely used in the preparative mode. Reverse phase preparative HPLC techniques are being coupled with biotechnology to produce very pure biological products for pharmaceutical and other biomedical uses.

Preparative HPLC can be used to isolate very small amounts of material (milligrams or less), or it can be used for plant-scale operations (kilograms to tons). The equipment used for these extremes may be very different.

Verzele and Geeraert[213] divide preparative HPLC into five major types. Type 1 by their definition utilizes analytical columns and instrumentation for isolation of a small amount of a pure component. This is often used where a small amount of material is needed for identification purposes. Type 2 is an extension of type 1 techniques using slightly larger columns. Columns up to about 1 cm in diameter may be used in type 2 separations. Quantities of pure material up to several hundred milligrams can be isolated using this technique. Many HPLC column manufacturers offer columns approaching 1 cm in diameter packed with the same materials as used in their analytical columns. These larger columns are sold as preparative or semipreparative columns. Whatman and Waters offer special columns for this type of application. Whatman offers the Magnum 9 columns for this purpose. Waters offers a special series of radial compression columns which fit into this category. Both type 1 and type 2 separations can be performed in almost any laboratory. Most analytical HPLC equipment fulfills the requirements for conducting preparative HPLC on this scale.

Verzele's type 3 form of HPLC makes use of long columns. In this mode of operation, gram amounts of material may be easily isolated. A disadvantage of type 3 operation is that the pressure across the column limits the flow rates that are achievable and therefore limits the utility of these columns.

Type 4 begins to require elaborate special equipment for its practice. This mode is ideal for producing gram to kilogram quantities of material. Separation of such large quantities requires the use of larger diameter columns than type 1 to 3 separations. Columns of 5 to 10 cm in diameter are common. The columns may be packed with the small 5- to 10-μm particulate packings, or the packings may approach 30 to 50 μm in diameter. These columns require very special hardware to distribute the sample and flow uniformly over the column bed. These types of columns also require large flows of solvent. Flow rates of 500 mℓ/min to 1 ℓ/min are common in this type of column. There are commercial pieces of equipment designed specifically for this mode of HPLC. The Water Prep 500 series instruments, Jobin-Yvon Industries Chromatospac systems, and YMC preparative instruments all fit into this category. Preparative HPLC on this scale also fits into pilot plant operations, and many specialized instruments have been developed to fulfill the need for preparative HPLC in pilot operations.

Type 5 preparative HPLC by Verzele's classification corresponds to full-scale production plant operation. There are many examples of the commercial application of preparative HPLC. It is a very critical technique in the production of many of the products of biotechnology research. Most, if not all, of the recombinant DNA products on the market today use preparative-scale HPLC techniques in their production.

Many questions are still being explored regarding preparative HPLC techniques. The question of the selection of the best k′ values for the components being separated has not been fully resolved.[214] Furthermore, questions abound as to the appropriate load of sample to place on a column.[33,214,215] Some systems are operated in a mass overload mode,[216] and other systems are operated in a volume overload mode. The definition of a column overload requires that the k′ associated with elution of the peak be shifted by more than 10%.

People often argue that samples in preparative HPLC should be kept concentrated. Counter to this argument is the solubility of the sample. It is very important to keep the sample in solution in the chromatographic process. The precipitation of purified, concentrated materials in column frits, detectors, and chromatographic plumbing is a common problem in the practice of preparative HPLC.

Careful consideration must be given to the selection of systems and approaches for preparative separations. In preparative HPLC, resolution is usually sacrificed for throughput. There are, however, beneficial effects that help to offset the losses in performance that would be expected from heavy column loading. Impurities eluting before the major peak tend to be pushed further out in front of that peak. Thus, by judiciously cutting the front of the peak, one can often eliminate most of the early eluting components. Likewise, later eluting components will frequently be retarded even further by interaction with the high concentration of the major component eluting ahead of it. Thus, if one uses only the center portion of a major peak, one may obtain a reasonably pure sample in only one pass through the chromatographic column.

The separations problem should be thoroughly evaluated before deciding upon a tentative preparative procedure. If the major component of a sample is to be purified, the center of the main peak may be collected as it should be the purest portion of the peak. The front and tail portions of the peak can then either be discarded or collected and reprocessed in a similar manner. If a minor component is to be isolated, multistep approach might be more effective. In this case, the sample is run through a chromatographic system with poor resolution to collect only the general area in which the component elutes. This process will greatly concentrate the desired component. The concentrated sample can then be fractionated using a chromatographic system with much higher resolution to yield a purified sample.

Preparative HPLC makes use of many of the same types of packing materials used in regular HPLC. Reverse phase packings and silicas are the most widely used materials in preparative applications. Ion exchange and exclusion chromatography are also used in certain cases in both semiquantitative and plant-scale separations. The selection of the separation mode depends upon the nature of the sample. Both analytical and preparative HPLC often use the same chromatographic systems.

Solvents are a very real problem in preparative HPLC. Often cheaper solvents are employed in preparative HPLC than would normally be used in analytical HPLC. The contamination of solvents by UV absorbing species is not often a problem in preparative HPLC. The major problem with the lower quality solvents used in preparative HPLC is nonvolatile residues. Since large volumes of solvent are often evaporated from the sample following the chromatographic separation, the solvents must be essentially free of all nonvolatile residues if high product purity is to be achieved. Selection of buffers and chromatographic reagents (e.g., ion pair reagents) must also be done with care. Ideally, reagents are used that can be removed using an evaporative technique. Regular evaporative removal or lyophilization may be used to remove the solvent. Nonvolatile salts and reagents can present a very difficult problem in the final isolation of the separated compound.

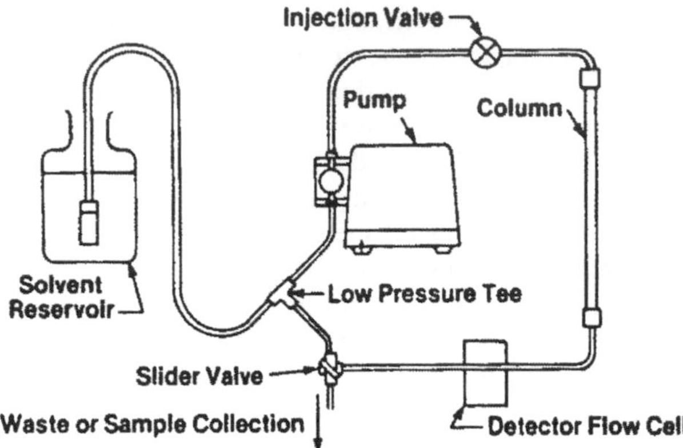

FIGURE 21. Instrument configuration for recycle with single head pump.

Another chromatographic step is often used to remove chromatographic reagents from the sample. This procedure is also used to concentrate the sample. Reverse phase chromatography is often very effective for this purpose. In this approach, the sample in the original chromatographic solvent is diluted with a weaker solvent such as water. The diluted sample is now pumped onto the column followed by a water rinse. The polarity of the solvents is such that the sample of interest has little or no mobility on the column. All of the salts and reagents can then be removed from the sample by flushing them through the column with the water rinse. Following the water rinse, a strong solvent such as methanol is pumped through the column. The methanol will elute the sample component of interest in a sharp, concentrated band. The sample is now concentrated in a volatile solvent which can be easily handled by conventional means.

Preparative HPLC has also made use of some specialized multicolumn pass techniques. These specialized techniques are discussed in the following sections.

A. Recycle HPLC

Recycle HPLC is used when sufficient separation of two components cannot be achieved under normal HPLC conditions. In recycle HPLC, the column eluate is directed to waste or some other appropriate collection vessel until the components of interest begin to elute. At this point the column eluate is rerouted so that it returns to the pump and is pumped back onto the column. The eluate is pumped back onto the column until the whole band has been returned to the head of the column. This process is repeated as often as necessary to achieve the desired separation. The maximum number of times the material can be recycled through the column is limited by the increase in peak width which occurs with each cycle and eventually results in the band being spread through the whole column. The number of cycles used is generally far less than the maximum number possible.

Recycle HPLC places some very stringent requirements on the HPLC system. Since the column eluate is run through the column a second time, the external volume of the HPLC system must be reduced to a minimum to avoid large peak spreading. This implies that the optimum system for recycle would be a multihead reciprocating pump of low displacement volume. In such a system, the column eluate can serve as the solvent supply for the pump, which then drives the eluting sample back onto the head of the column for the next cycle.

In applications where a single head reciprocating pump is used, the recycle apparatus must be designed as shown in Figure 21. Here the column eluate which is to be recycled is actually pumped into the solvent supply line for the pump. The column eluate displaces some of the solvent which is adjacent to the pump head. In such a system all of the tubing and fittings should be of a small volume type to avoid excessive peak spreading.

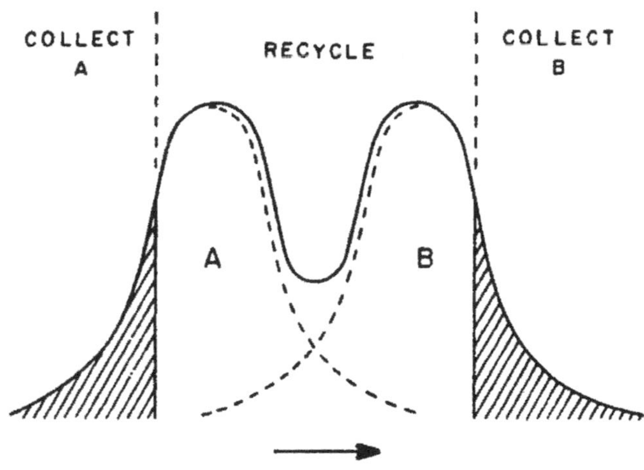

FIGURE 22. Shave-recycle chromatogram.

B. Shave-Recycle HPLC

The shave-recycle method is often used in the collection of samples in preparative HPLC. Figure 22 shows chromatograms obtained using shave-recycle HPLC. The leading edge of the first peak will usually be fairly pure first component. Therefore, this material is collected until a point is reached in the chromatogram where the column eluate starts to contain the second constituent. At this point, the column eluate is recycled. The eluate is recycled until a point is reached on the tailing edge of the second constituent. The tailing edge of this peak should be reasonably pure in the second material. This material is now collected as it elutes from the column. This process can be continued until all of the material has been collected or until the amount still remaining on the column becomes so small so as to make it impractical to continue the process. Using this technique, one gains the advantage of decreasing the column loading and therefore improving the resolution with each pass through the column.

C. Peak Detection in Preparative HPLC

The large samples used in preparative HPLC will saturate the UV detector if it is set to monitor at the absorption band maximum. If a UV detector is used, it is often set at a wavelength other than the absorption maximum in order to keep the absorption of the eluting peaks within the detector range. The refractive index detector, which has only marginal sensitivity for most analytical work, can also be used to good advantage here. The refractive index detector is also sensitive to a wider variety of compounds than a UV detector, which can be especially important in detecting substances coeluting with the compound of interest in preparative work.

XI. SUPERCRITICAL FLUID CHROMATOGRAPHY (SFC)

SFC is a form of HPLC in which the mobile phase is a fluid above its critical temperature and pressure. The critical fluid above its critical temperature and pressure. The critical constants for several solvents commonly used in SFC are listed in Table 27. SFC fits between GC and liquid chromatography, and can be used to achieve separations that are difficult or unsatisfactory with either GC or liquid chromatography. Although the potential benefits of using supercritical fluid mobile phases in chromatography have been known for many years,[217-222] practical development of the technique did not begin until the early 1980s. The application of SFC has grown rapidly since that time due to improved understanding of the physical

Table 27
PHYSICAL PROPERTIES OF SOLVENTS FOR SUPERCRITICAL CHROMATOGRAPHY

Solvent	Boiling point (°C)	Critical temperature (°C)	Critical pressure (atm)	Critical density (g/cm³)
Ethane	−88.6	32.3	48.1	0.203
Propane	−42.1	96.7	41.9	0.217
n-Butane	0.5	152.0	37.5	0.23
n-Pentane	36.1	196.6	33.3	0.232
Cyclohexane	80.7	280.0	40.2	0.273
Ethylene	−103.7	9.2	49.7	0.218
Benzene	80.1	288.9	48.3	0.302
Toluene	110.6	320.0	40.6	0.292
o-Xylene	144.4	357.0	35.0	0.284
Methanol	64.7	240.5	78.9	0.272
Ethanol	78.5	243.0	63.0	0.276
2-Propanol	82.5	235.3	47.0	0.273
Fluoromethane	−78.4	44.6	59.9	0.3
Trifluoromethane	−82.2	25.9	46.9	0.52
Chlorotrifluoromethane	31.2	28.0	38.7	0.579
Dichlorodifluoro-methane	−29.8	111.8	40.7	0.56
Carbon dioxide	−78.5	31.3	72.9	0.448
Nitrous oxide	−88.6	36.5	71.7	0.45
Ammonia	−33.4	132.4	112.5	0.235
Water	100.0	374.2	218.3	0.315
Sulfur hexafluoride	−63.8 (subl)	45.5	37.1	0.74
Xenon	−107.1	16.6	58.4	1.10

process involved in practical SFC separations coupled with advances in column and detector technology. The advantages of SFC are a direct result of the unusual combination of physical properties characteristic of supercritical fluids.

Viscosities, densities, and diffusion coefficients in supercritical fluids lie between those of liquid and vapor phases of the same composition. Substances in the supercritical state have relatively large compressibilities, and therefore their physical properties depend strongly upon the pressure. At the pressures typically used in SFC, diffusion coefficients are 10 to 100 times greater than in the corresponding liquid, while viscosities are much lower than in the liquid. Low mobile phase viscosities minimize the pressure drop across the column and allow higher linear velocities to be achieved. Large diffusion coefficients allow rapid equilibration of the analyte between the stationary and mobile phases, resulting in high efficiencies, narrow peaks, and fast optimum linear velocities through the column.

The densities of supercritical fluids under the conditions used in SFC typically range from 0.1 to 0.8 of their liquid densities. Densities of several fluids at the critical point are included in Table 27. The solvent strength of a supercritical fluid is roughly proportional to density[222] and approaches that of the liquid at higher pressures. The increase in solvent strength with pressure is greatest at the fluid's critical point.[223] Good solvent properties make it possible to use SFC to separate high molecular weight and polar compounds that cannot be separated by GC because of their low vapor pressure, and to achieve the separation with a speed and efficiency greater than that achievable with liquid chromatography. Since solvent strength depends upon pressure, it is also possible to use pressure changes to assist in effecting a separation.[224-231]

A typical chromatograph for supercritical fluid work consists of a high-pressure pump with a pressure control system, column oven with precision temperature control, sample inlet system, column, and detector. A block diagram of a typical supercritical fluid chromatograph is shown in Figure 23. Most of the early work with SFC was done using homemade

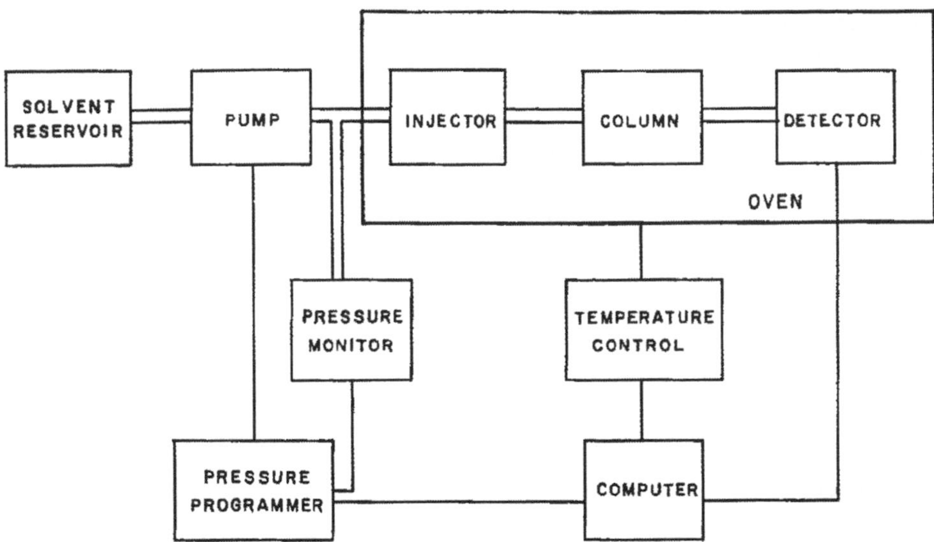

FIGURE 23. Block diagram of a typical supercritical fluid chromatograph. The fluid is pressurized in the pump at subcritical temperatures, then heated above its critical temperature in the oven. The injector is usually in the oven so that the sample is injected into the supercritical phase.

instruments (e.g., see References 228 and 232 to 237), but a number of companies now sell instruments for this purpose including Lee Scientific, Suprex, Computer Chemical Systems, Brownlee Labs, ISCO, and JASCO.[238] An SFC system for chemical process monitoring has also been introduced by Combustion Engineering.[239]

The mobile phase in SFC is usually pumped as a liquid. Although it is possible to pump a supercritical fluid, the high compressibility of the fluid makes it difficult to maintain good control of the flow rate and hold pressure pulsations to a tolerable level. Pumping a compressible fluid also generates heat that needs to be removed if the state of the fluid is to be well controlled. Syringe pumps are preferred for SFC because they can produce high pressures without pulsation and are very easily controlled. However, other types of pumps have been used on occasion.[240-242] When the mobile phase is liquid at room temperature and atmospheric pressure, the pump is connected to a solvent reservoir in the same manner as for conventional HPLC. For solvents like carbon dioxide, the solvent is stored in a pressure vessel in the liquid state and the pump is connected through an eductor tube placed near the bottom of the pressure vessel. If the critical temperature of the liquid is very low, it may be necessary to cool both the solvent reservoir and the pump chamber. A pressure sensor is placed at the pump output and used in a feedback loop to control the pump. Inclusion of a microprocessor in the feedback loop can be used to operate in constant pressure, constant density, programmed pressure, or programmed density mode.[238]

After leaving the pump, the liquid is heated above its critical temperature, usually by passing it through a heat exchanger in the column oven. Separation is often done at reduced temperatures (i.e., fluid temperature/critical temperature) in the range of 1.02 to 1.40[243] The temperature and pressure are maintained somewhat above their critical values because control is very difficult to maintain near the critical point. The temperature control in the oven should be at least as good as that used for gas chromatography.

The sample injector is usually a standard HPLC injection valve. Sample loop volumes of 5 to 30 $\mu\ell$ are commonly used with packed columns, while volumes less than 1 $\mu\ell$ are typical in capillary column SFC. The injector may be located between the pump and the oven[237,244,245] so that the sample is injected into the liquid stream, but injection in the supercritical phase is more common.[236,238,240,242,246] Injection into the supercritical phase is preferred because the potential for plugging the injector is reduced by the high solubility of most sample components in the fluid phase relative to the liquid phase.

The pioneering work on SFC was done with packed column[217,227] and some packed column work is still done.[240,242] A serious limitation on packed columns in SFC arises from the compressibility of the mobile phase and the large pressure gradient existing across the column. In a conventional HPLC separation the mobile phase is essentially incompressible and column efficiency increases with decreasing particle size of the packing material. In SFC, the opposite effect is observed.[227,228] A decrease in particle size increases the pressure drop across the column and, because of the large compressibility factor, produces a decrease in column efficiency. Gouw and Jentoft[228] compared this effect of an increased pressure gradient in SFC to the effect of a decreasing temperature program in GC.

The pressure drop across an open tubular capillary column is much less than across a packed column, so the effect of mobile phase compressibility is much less. As a result, most commercial instruments for SFC are designed for capillary work. Typical columns are 50 to 100 μm in diameter and contain a chemically bonded stationary phase with nominal phase thicknesses ranging from 0.25 to 1.0 μm. The stationary phases must be bonded and cross-linked to prevent dissolution in the mobile phase, since the common mobile phases are good solvents for the more common stationary phases. The stationary phases used in SFC are the same types of polymers used in capillary column GC. Capillary columns intended for SFC and coated with methyl-, methylphenyl-, *n*-octyl-, biphenyl-, and cyanopropyl-polysiloxanes are all available commercially. Other stationary phases can be used, including certain liquid crystal phases for separations of isomers.[248]

One of the greatest advantages of SFC over both GC and HPLC is the ease of interfacing the column with a variety of detectors. Most detectors that can be used in either GC or HPLC can also be used in SFC. This versatility is due to the fact that the effluent can be kept pressurized and in the fluid state in a detector designed for liquid phase work, or allowed to expand into a vapor through a restrictor at the end of the column for measurement with a gas phase detector. Liquid chromatographic detectors that have been used for SFC include ultraviolet photometers,[224,249-251] fluorometers,[236] refractive index detectors,[252] and scintillation detectors.[251] GC detectors have included the flame ionization detector,[218,224,238,253-257] flame photometric detector,[258] mass spectrometer,[237,243,259-263] and Fourier transform infrared spectrometer.[264-266] The flame ionization detector is the most widely used of these detectors. The selection of a detector should be based upon the sensitivity and selectivity required for the analysis, and on the fluid used for the mobile phase.

The flame ionization detector and ultraviolet photometers both offer reasonable sensitivity, but relatively poor selectivity. Of course, the flame detector could not be used if the mobile phase were pentane or another fluid that caused a detector response. The greatest sensitivity is achievable with the fluorometric, flame photometric, and mass spectrometric detectors. The fluorometric and mass spectrometric detectors have the additional advantage of being useful with almost any of the common mobile phases. Mass spectrometric and Fourier transform infrared detection offer the greatest selectivity, although the sensitivity of the infrared method is not adequate for many analytical problems.

The most commonly used mobile phase in SFC is carbon dioxide. Carbon dioxide is ideal because it has a low critical temperature and pressure, is not flammable or toxic, is readily available and inexpensive, and is relatively inert and compatible with most detectors. The low critical constants are important in separations of labile materials because they allow the separation to be carried out under very mild conditions. Carbon dioxide has very low polarity in the supercritical state, although small quantities of other solvents can be added to modify its behavior.[243,267-269] The behavior of mixed solvents in SFC is still a subject of active research. Nitrous oxide, pentane, and other hydrocarbons are also nonpolar. Most polar solvents have critical temperatures that are too high to be useful for organic analysis.

Ammonia is an exception, but is very difficult to handle experimentally.

Only a few applications of SFC to problems in pharmaceutical analysis have been reported. Richter[257] analyzed erythromycin A using supercritical carbon dioxide on an SE-54 (5% phenyl polymethylphenylsiloxane) stationary phase in an 80-μm open tubular capillary column. The analysis was performed at 40°C with the density of the mobile phase programmed from 0.225 to 0.83 g/mℓ. A flame ionization detector was used. The analysis of polyethylene glycol was also reported in this paper.

White and Houck[270] used supercritical carbon dioxide to analyze a series of mono-, di-, and triglycerides on both DB-5 (95% dimethyl-5% diphenyldisoloxane) and DB-225 (50% cyano-propyl-methyl 50%-methylphenylpolysiloxane) 100 μm capillary columns with a flame ionization detector. Pressure programming from approximately 150 to 300 atm was employed. Column temperatures ranged from 90°C for the lower molecular weight glycerides to 150°C for the higher molecular weight triglycerides. SFC analysis of triglycerides has also been reported by Chester[255] and by Rawdon and Norris.[271] Rawdon and Norris used a packed column with a mixture of carbon dioxide/methanol, 99:1, as mobile phase and an ultraviolet detector. One advantage of SFC for analysis of this family of compounds is that the separation is achieved at low-enough temperatures that degradation does not occur on column. Pyrolysis of the analyte is a serious problem when attempting to analyze the higher molecular weight glycerides by GC.

Smith et al. employed SFC with mass spectrometric and flame ionization detection to analyze a series of trichothecene mycotoxins.[272] SE-54 (5% phenyl polymethylphenylsiloxane) capillary columns were used with carbon dioxide as the mobile phase, temperatures up to 100°C, and pressures ranging from 75 to 350 atm. Electron impact ionization and chemical ionization using methane, isobutane, and ammonia were all used to obtain mass spectra.

Markides et al. reported the determination of several pesticides including parathion, chloropyrifos, and larvin by capillary SFC.[258] The separation was accomplished on a poly(methyl-*n*-octylsiloxane) column using carbon dioxide and nitrous oxide as mobile phases. A flame photometric detector was used to achieve very good sensitivity in these measurements (0.25 ng in sulfur mode and 0.5 ng in phosphorus mode when using carbon dioxide as the mobile phase).

XII. DEVELOPING AN HPLC PROCEDURE

The various components of a liquid chromatograph have been discussed in the previous sections. The analytical potential of HPLC, however, lies in the chemist's ability to combine these parts into a procedure to answer a particular question or solve a given problem.

Consideration must be given to the nature of the problem in the development of any HPLC procedure. Elaborate and sophisticated procedures may be quite acceptable in a methods development laboratory or other laboratories where experienced chromatographers will be performing the assay. However, such procedures may be unacceptable if the procedure will normally be run in a quality control environment where personnel not fully understanding HPLC techniques may be running the assay and where rapid turnaround is important. These situations require simple, straightforward procedures to assure the routine acquisition of good and reliable HPLC data.

Four fundamental questions that should be asked each time work is started on a new problem are given in Table 28.

The very first question that should be asked is whether or not HPLC is the best approach to the problem. The answer will depend upon the complexity of the separation problem, solubility and volatility of the compound, precision and accuracy requirements, etc. If HPLC is the method of choice, several additional questions must be answered to determine the particular HPLC technique that will be used. A problem which requires an answer based upon a very limited number of samples might make use of long and involved assays. One

Table 28
BASIC SEPARATION QUESTIONS

Is HPLC the logical choice of method, or is there a simpler or more direct method that can be used?
Is this a single short-term problem, or will the method be adapted as a routine control procedure?
Will the method be transferred to other corporate and regulatory laboratories?
Is the method convenient for the laboratory which will be the recipient of the samples?

potential solution is to use a solvent gradient to elute the sample components from the column. Such a procedure might be very adequate and less time-consuming than trying to optimize conditions for an isocratic assay. However, if a large number of samples must be run in a routine manner, optimization of the method to reduce both the instrument time and the chemist's time might be in order.

The second question is related to the first and centers on the final disposition of the method. If the method is to be transferred to other corporate or regulatory laboratories, it is certainly advisable to keep it as simple as possible. Standard, commercially available instruments and supplies should be used if at all possible.

If the HPLC solution is to result in a method which will be run by a production or control laboratory, it is advisable to adapt procedures which will be convenient for such a laboratory. If several possible solutions to the problem are available using different column and solvent combinations, it is advisable to select combinations which are used or similar to those already used in the receiving laboratory. Such as choice can drastically reduce the time required to perform the analyses in the laboratory. This is especially beneficial if the laboratory has many different types of samples coming in for assay on a limited number of instruments.

After the basic philosophy for solution of the problem has been adopted, the remaining questions are concerned with the nature of the sample. Table 29 lists several characteristics of the sample that should be considered before any work proceeds. If the sample is a complex mixture of components, a very efficient column system will be required, and consideration should be given to the use of a gradient technique. However, if only one or two components of the mixture are of interest, a simple procedure resolving these compounds may be satisfactory even though other components are not resolved. This situation often arises with complex formulations and with chemical reaction mixtures where only the quantities of the principal product and residual starting materials are required.

Raw materials frequently have very tight purity specifications; therefore, an HPLC procedure is required which separates all likely intermediates, impurities, and degradation products from the principal component. The narrow acceptance limits on purity dictate that the precision for the method be very good. If the method relative standard deviation becomes much larger than 1.5 or 2%, the number of replicate assays required to generate an answer at an acceptable confidence level becomes so great that the method becomes impractical for routine use.

Less demanding procedures are often acceptable if the problem involves the determination of an active component in a formulated product. In this instance, a separation which discriminates between the active component, excipients present in the formulation, and likely degradation products is sufficient. The control limits for formulations are frequently broader than for raw materials, and less precise assays may be satisfactory for this application. However, when estimating the assay precision required in a particular situation, it must be kept in mind that it is the sum of the variances arising from manufacturing and analysis that must be compared to the control range rather than the assay variance alone.

The last major type of sample commonly analyzed by HPLC are those requiring assay of one or more trace contaminants in a product. Procedures which offer good efficiency are required for this type of sample in order to differentiate small amounts of material from baseline noise. Column and solvent conditions must be chosen to allow introduction of large samples onto the chromatographic column.

Table 29
SAMPLE CHARACTERISTICS TO CONSIDER IN HPLC METHOD DEVELOPMENT

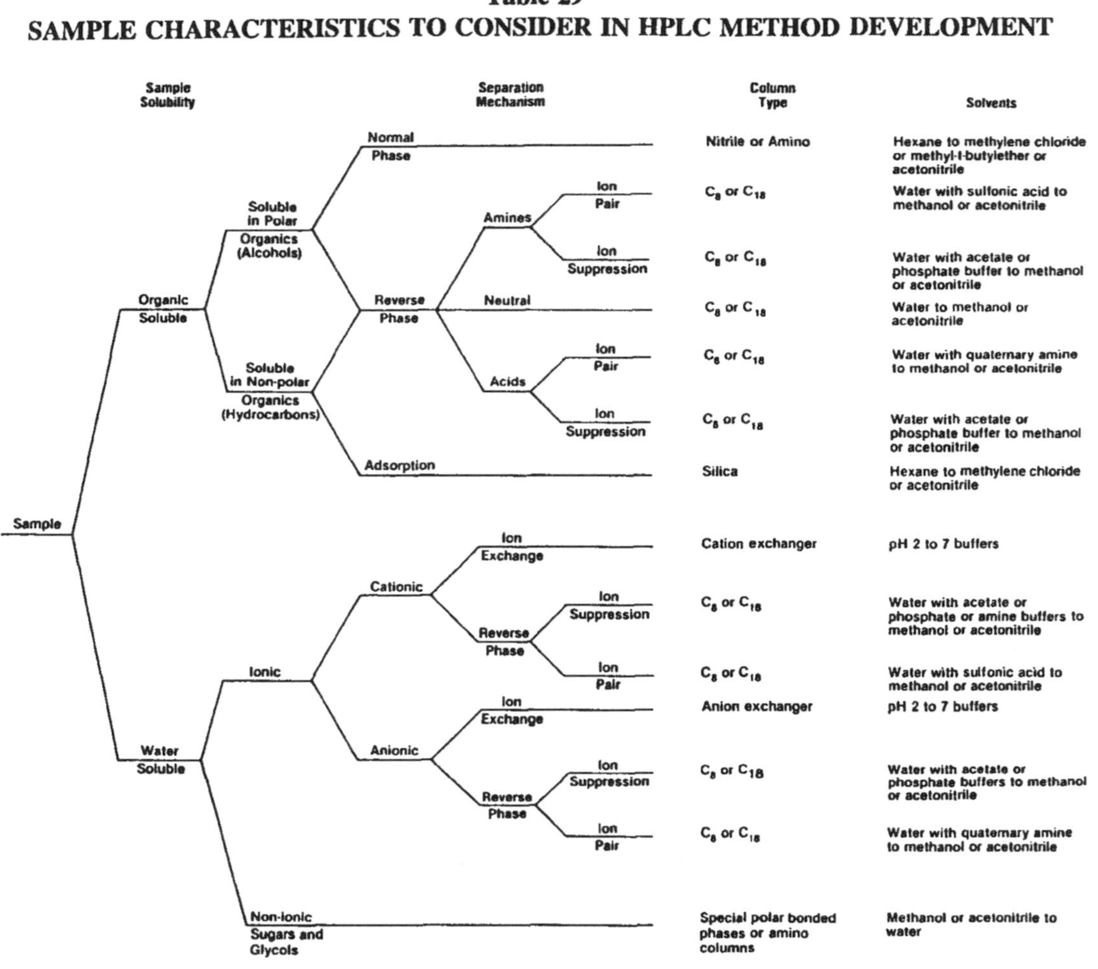

Sample	Sample Solubility		Separation Mechanism			Column Type	Solvents
Sample	Organic Soluble	Soluble in Polar Organics (Alcohols)	Normal Phase			Nitrile or Amino	Hexane to methylene chloride or methyl-t-butylether or acetonitrile
		Soluble in Polar Organics (Alcohols) / Soluble in Non-polar Organics (Hydrocarbons)	Reverse Phase	Amines	Ion Pair	C_8 or C_{18}	Water with sulfonic acid to methanol or acetonitrile
				Amines	Ion Suppression	C_8 or C_{18}	Water with acetate or phosphate buffer to methanol or acetonitrile
				Neutral		C_8 or C_{18}	Water to methanol or acetonitrile
				Acids	Ion Pair	C_8 or C_{18}	Water with quaternary amine to methanol or acetonitrile
				Acids	Ion Suppression	C_8 or C_{18}	Water with acetate or phosphate buffer to methanol or acetonitrile
		Soluble in Non-polar Organics (Hydrocarbons)	Adsorption			Silica	Hexane to methylene chloride or acetonitrile
	Water Soluble	Ionic	Cationic	Ion Exchange		Cation exchanger	pH 2 to 7 buffers
			Cationic	Reverse Phase	Ion Suppression	C_8 or C_{18}	Water with acetate or phosphate or amine buffers to methanol or acetonitrile
			Cationic	Reverse Phase	Ion Pair	C_8 or C_{18}	Water with sulfonic acid to methanol or acetonitrile
			Anionic	Ion Exchange		Anion exchanger	pH 2 to 7 buffers
			Anionic	Reverse Phase	Ion Suppression	C_8 or C_{18}	Water with acetate or phosphate buffers to methanol or acetonitrile
			Anionic	Reverse Phase	Ion Pair	C_8 or C_{18}	Water with quaternary amine to methanol or acetonitrile
		Non-Ionic Sugars and Glycols				Special polar bonded phases or amino columns	Methanol or acetonitrile to water

Information on the chemical nature of the sample is very valuable in the initial selection of HPLC conditions. Information on the chemical structures of the sample components, and especially on the functional groups present, is very helpful in selecting the proper column for a separation. Solubility data is also helpful in selecting the mode of HPLC to use and, in addition, give insight into likely choices for the eluting solvent. Some information on the physical properties of the sample is also required since HPLC detection generally involves measurement of some physical property of the compound as it elutes from the column. Since UV detection is often the method of choice, knowledge of the UV and visible spectrum of the compound are very helpful in selecting the best wavelength. If the UV absorption does not offer acceptable sensitivity, other properties of the compound may be useful. Finally, knowledge of the chemical functionality and reactivity of the molecule can give valuable insights as to the nature of potential derivatives which might be formed to make detection easier.

Knowledge of the stability of the compound is also important in developing procedures for handling it. If certain conditions are likely to cause decomposition of a compound, it is certainly advisable to avoid those conditions in the chromatographic examination of the compound.

Column selection is the first step in development of an HPLC procedure. As mentioned earlier, the nature of the sample will dictate the column to be used.

In pharmaceutical analysis, one seldom deals with high molecular weight compounds; therefore, methods other than exclusion chromatography are generally used. The first question in column selection is the solubility of the sample. If the sample is soluble in very polar organic solvent such as alcohols, both normal phase and reverse phase columns utilizing bonded phase packing may be used. The type of column adopted for an assay will ultimately be the one which gives the best resolution of the principal components of the sample. However, a reverse phase column would be a good starting point in method development if the sample matrix contained highly polar materials such as sugars and starches, since these materials would not be retained by the column, thus assuring better column stability and easing the requirements for pre-column sample preparation. For similar reasons, a normal phase column might be the best starting point if the sample matrix is composed largely of materials less polar than the compounds of interest.

If the compounds are nonpolar and soluble in hydrocarbon solvents, either a reverse phase or an adsorption column can be used. Some adsorption columns are probably the most efficient columns available, but the constant problems of maintaining proper adjustment of the column activity limits their use in routine analyses to cases where their high efficiency is essential. For example, the separation of a group of closely related steroids often requires the efficiency of an adsorption column. Column activity is usually controlled by addition of methanol containing a given amount of water to the eluant. An equilibrium is achieved between the water and methanol in the solvent and that adsorbed on the column packing. The difficulties in using these columns arise because very small changes in the quantity of water and methanol adsorbed on the column can cause large changes in the k' values for the components of the sample. Very close control of the composition of the solvent and drying of the samples are required to maintain stable column characteristics. For these reasons, a reverse phase column will be the best choice for separation of relatively nonpolar compounds provided adequate separation can be achieved. Probably better than three fourths of all pharmaceutical applications of HPLC utilize the reverse phase mode of separation.

Another class of compounds are those which are water soluble. This class is subdivided into ionic and nonionic compounds or electrolytes and nonelectrolytes. This distinction separates the sugars and polyols from the ionic substances. The natural choice for separation of ionic substances is ion exchange chromatography. The typical ion exchange packing for HPLC is a surface modified silica and is therefore limited to operation between pH 2.5 and

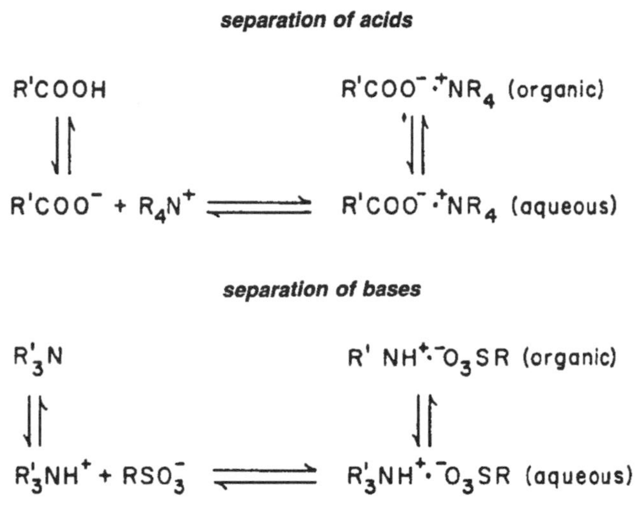

separation of acids

$$\begin{array}{ccc} R'COOH & & R'COO^- \cdot {}^+NR_4 \text{ (organic)} \\ \downarrow\uparrow & & \downarrow\uparrow \\ R'COO^- + R_4N^+ & \rightleftharpoons & R'COO^- \cdot {}^+NR_4 \text{ (aqueous)} \end{array}$$

separation of bases

$$\begin{array}{ccc} R'_3N & & R'\,NH^+ \cdot {}^-O_3SR \text{ (organic)} \\ \downarrow\uparrow & & \downarrow\uparrow \\ R'_3NH^+ + RSO_3^- & \rightleftharpoons & R'_3NH^+ \cdot {}^-O_3SR \text{ (aqueous)} \end{array}$$

FIGURE 24. Equilibria involved in ion pair separations.

7.5. This range drastically limits the utility of the column for separating weak acids by anion exchange. Ion exchange columns have not offered outstanding performance for other reasons as well. The efficiencies of such columns are usually not very good, and their ion exchange capacities are quite low. If high column efficiencies are not required, the possibility of using a polymeric ion exchange resin in the 200 to 400 mesh size range with a low head pressure (0 to 500 psi) should be considered for the separation. Such resins can be used over a wide pH range and have a relatively high capacity. When a suitable ion exchange system cannot be worked out, several alternative approaches are available for use with ionic substances.

One very useful technique for weakly ionic compounds is ion suppression. This technique makes use of the following equilibrium to suppress dissociation so that the compound can be chromatographed as a neutral substance.

$$AB \rightleftharpoons A^+ + B^- \tag{22}$$

When dealing with a weak organic acid, the addition of an acid to the eluant will force much of the weak acid into its nonionic form,

$$RCOO^- + H^+ \rightleftharpoons RCOOH \tag{23}$$

The nonionic form is of course soluble in organic solvents, and a column for the separation can now be selected from those suitable for this class of compounds. When using the ion suppression technique, the reverse phase option is again generally the mode of choice. To control retention in this system, buffers are added to the mobile phase. Typical buffers are listed in Table 23. The pH range is again limited by the silica column and extends from 2.5 to 7.5. This technique offers a viable approach to separation of acids with pK_a greater than 2 and for bases with pK_a of less than 8. Stronger acids and bases remain ionic in the pH range of 2 to 8 and will either not be retained or will exist in a mixed ionic-nonionic form which leads to badly skewed peaks. Other techniques must therefore be utilized for strong acids and bases.

Another means of handling ionic materials is by ion-pair chromatography. This technique is also applicable to the stronger acids and bases. Although the actual phenomena occurring on the column in this technique are not well understood, the equilibriums in Figure 24 represents some of the more important processes underlying this mode of separation. The figure shows acids being separated using a quaternary amine, and bases using a sulfonic acid because these are common reagent choices. Other reagent choices and a more detailed discussion of the principles of ion pair formation and extraction are given in Volume I, Chapter 2, Section IV. The essential feature is that the ion pair favors the organic phase and therefore can be chromatographed on a reverse phase column.

For basic samples, alkyl sulfonates are used with pentane through octane, sulfonic acids being the most common reagents. The pH of the solvent system must be such that both the analyte and the reagent are in their ionic forms. Thus, for basic compounds, a pH in the range of 7 to 8 is usually required. Phosphate buffers are usually used in this pH range.

When dealing with acidic samples in ion-pair chromatography, quaternary amines are generally utilized as the pairing entity. The common pairing reagent is the tetrabutylammonium ion. For these systems, the pH of the mobile phase is usually low and in the range of 3 to 4. Again, phosphate buffers are commonly used.

Another class of compounds is the water soluble nonelectrolytes. This class of compounds includes primarily the sugars and polyols. This class of substances presents very difficult problems for the HPLC chromatographer. Some success has recently been achieved with this class of compounds through use of rather polar column packings such as the bonded phase amine columns, which are often used for normal phase HPLC. Several specialty columns are also offered by column manufacturers for use with this class of compounds.

Once a potential mode of separation has been chosen based upon the solubility and functional groups of the compound, a specific column must be selected. As discussed in Section V, there are many sources for the different types of column packings used today, but for routine analysis, it is probably advisable to purchase a prepacked column, especially if the methodology is eventually going to be transferred to other laboratories.

However, packing materials with similar functionality are frequently very different when obtained from different manufacturers. Therefore, in HPLC analysis, it is very important that the column used be specified both by type and manufacturer.

Several different size columns are also available. The typical analytical HPLC column appears to be standardized on $^1/_4$ in. columns with internal diameters of 4 to 4.7 mm. The length of such columns is usually 12 to 30 cm.

Once a column has been selected, the next task is to find a suitable solvent. The choice of a solvent is probably one of the most critical decisions that must be made in the development of an HPLC procedure. The common types of solvents for each of the different sample classifications are also included in Table 29. Also, the previous section on solvents in HPLC gives insights into different methods for selection of solvents.

Various solvent optimization procedures are now being followed in the normal development of HPLC methods.[273-285] Some of these procedures attempt to develop an idealized separation where all key sample components are separated in a minimum amount of time. Other procedures concentrate more on developing a separation for all of components without regard for time. At least three solvents are used in any of these optimization procedures. Each of these solvents should be from different solvent groups discussed in Section VI.A. One of the major limitations of these procedures is that they require the use of authentic samples of all of the components to be separated. Unfortunately, in pharmaceutical development efforts with new compounds, degradation products and other impurities are usually not available when the initial method development for a compound occurs. The technique is still valuable, for the resulting separation may reveal the presence of yet unrecognized components in the sample.

For development of a reverse phase separation, several reverse phase solvents with varied properties are first selected to use in the optimization procedure. A solvent triangle is often part of the optimization strategy. A typical solvent triangle for reverse phase separations is shown in Figure 25. In reverse phase systems, methanol is often used as a proton acceptor,

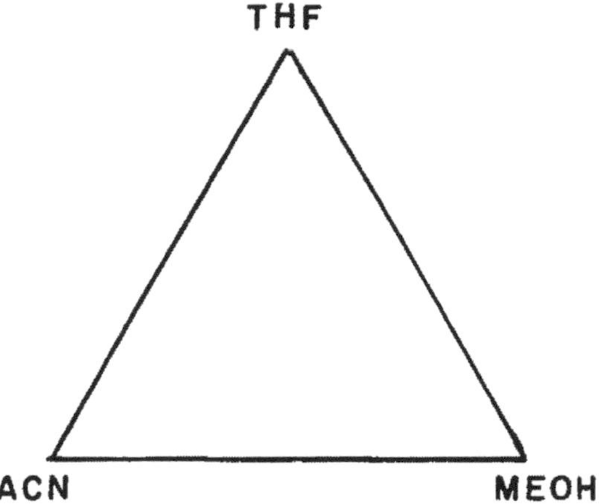

FIGURE 25. Solvent triangle for reverse phase separations.

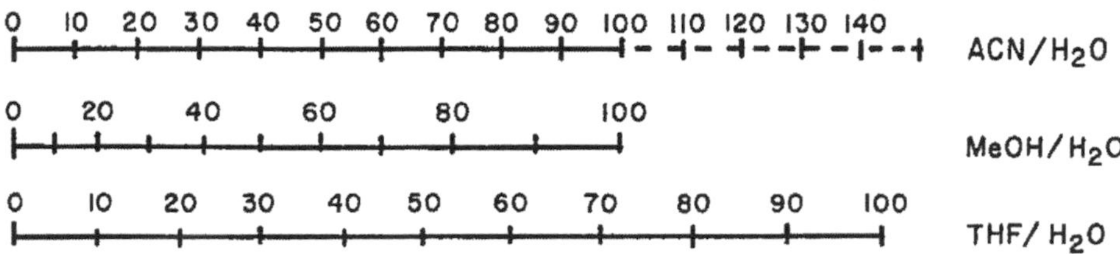

FIGURE 26. Iso-eluotropic series for reverse phase separations.

acetonitrile as a proton donor, and tetrahydrofuran to interact with dipoles. For reverse phase separations, these three solvents are mixed with water which serves as the moderator to adjust the solvent strength.

After an appropriate reverse column has been selected, one of the solvents is chosen to determine the approximate solvent strength that will be required for the separation. Methanol is frequently the solvent that is chosen for this purpose. The approximate methanol/water ratio required to elute the test compound can be found by conducting a gradient elution from 100% water to 100% methanol. Alternatively, if one prefers to use isocratic chromatography to determine the approximate solvent mixture, one would start with methanol and gradually add water until the sample component moves from the solvent front to a position corresponding to a k′ value between 3 and 10.

After the solvent strength for methanol has been determined, one can then use an iso-eluotropic guide for the selection of the solvent compositions at the other apexes of the triangle. Figure 26 has been prepared by Schoenmakers et al.[286] for the selection of mixtures of the other two solvents that will have the same solvent strength.

If one had a chromatographic run in which the methanol concentration required to give the desired elution of the sample was 50% methanol and 50% water, one could then use the guide to establish the composition of the solvents at the other two apexes of the triangle. To use the chart in Figure 26 to find the other two compositions, one would first find the position along the methanol/water line which corresponds to the 50% point. A perpendicular line is then drawn through this point so that the new line intersects with the other two lines in the figure. The points at which the lines cross provide estimates of the solvent composition of acetonitrile and tetrahydrofuran that will be required. For the example, one would estimate that the acetonitrile system should consist of 40% acetonitrile and 60% water. For the

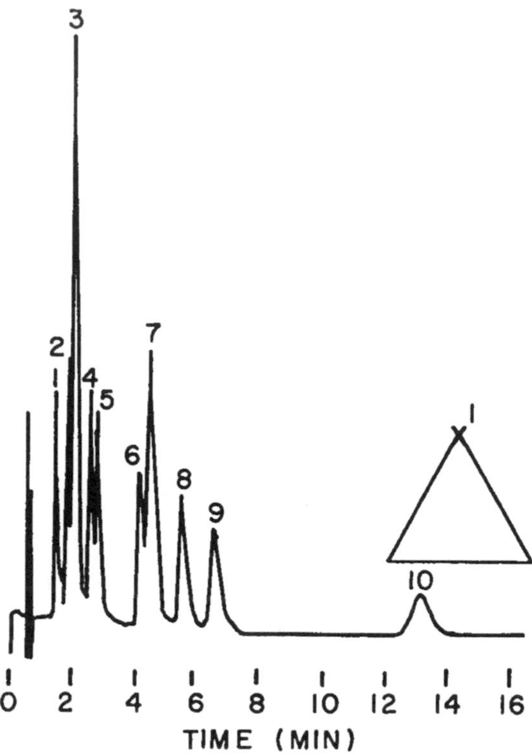

FIGURE 27. Separation of ten phenols using a two-component mobile phase as one step in the systematic development of an optimized chromatographic procedure.
Mobile phase: 1% aqueous acetic acid/methanol, 60:40
Column: Zorbax C8, 4.6 mm I.D. × 15 cm
Flow rate: 3.0 cm^3/min
Temperature: 50°C
Detector: ultraviolet @ 254 nm

tetrahydrofuran system, one would estimate the solvent composition to be 30% tetrahydrofuran and 70% water. These are then the solvent compositions that are required at the apex of the solvent triangle.

To optimize the solvent composition, a series of seven experiments would then be run as shown in the solvent triangle in Figure 25. In the first step of the optimization procedure, each of the solvents at the apexes of the triangle is run. Next, binary mixtures of two of the solvents are run. The composition of these are determined by taking half of the organic portion of the solvent at each apex and adding water to provide 100%. In the case of the earlier example, solvents of 25% methanol, 20% acetonitrile, and 55% water would be used for point 4. The compositions at points 5 and 6 are established in a like manner. Finally, point 7 uses a mixture of all three of the solvents. In our example, it would be 16.6% methanol, 13.3% acetonitrile, 10% tetrahydrofuran, and 60.1% water. Each of the sample components would then be run in each of the seven solvent systems.

An example of the application of this procedure is the development of a system for separation of a series of phenols. The first experiment showed that a solvent system of 40% methanol and 60% water was needed. Each of the components as run and identified in the chromatogram (see Figure 27)

The iso-eluotropic series was then used to select the solvent for the second system which was 28% acetonitrile and 72% water. Each of the components was run in this system. The

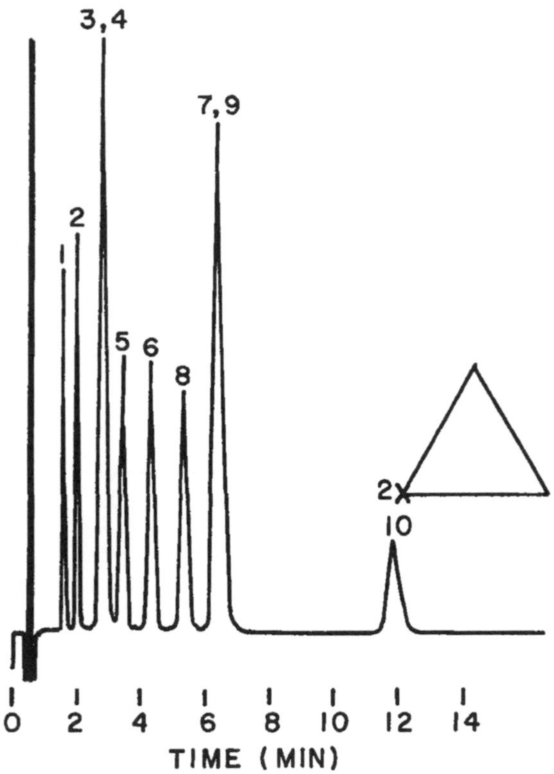

FIGURE 28. Separation of ten phenols using a two-component mobile phase as one step in the systematic development of an optimized chromatographic procedure.
Mobile phase: 1% aqueous acetic acid/acetonitrile, 72:28
Column: Zorbax C8, 4.6 mm I.D. × 15 cm
Flow rate: 3.0 cm^3/min
Temperature: 50°C
Detector: ultraviolet @ 254 nm

resulting separation of the mixture is shown in Figure 28. Note that in this separation the positions of peaks 7 and 8 have changed relative to their position in the water/methanol system. Figure 29 shows the result for the third set where a solvent composition of 30% tetrahydrofuran and 70% water was used. The next set of experiments used mixtures of two organic modifiers. Figure 30 shows the result for the system where 20% methanol, 14% acetonitrile, and 66% water were used. Figures 31 and 32 show the composition and results for the other two binary solvent systems. Finally, Figure 33 gives the results for the system including all three solvents that were used. The solvent composition for this experiment was 13% methanol, 9% acetonitrile, 10% tetrahydrofuran, and 68% water.

The chromatograms of the mixtures for all seven systems are now examined. The best separations appear to be in the methanol and acetonitrile systems. Components 2 and 3, 4 and 5, and 6 and 7 are not fully resolved in the methanol system. The acetonitrile system separates the pairs that were not adequately resolved in the methanol system, but tends to merge peaks 3 and 4, as well as peaks 7 and 9. Therefore, it would be desirable to improve the resolution between 2 and 3, 4 and 5, and 6 and 7 while not merging peaks 3 and 4 and 7 and 9. To accomplish his goal, one might try a system with the addition of 5% acetonitrile. Experiments with several compositions around this area finally led to the system shown in Figure 34 where 34% methanol, 3% acetonitrile, and 63% water were used.

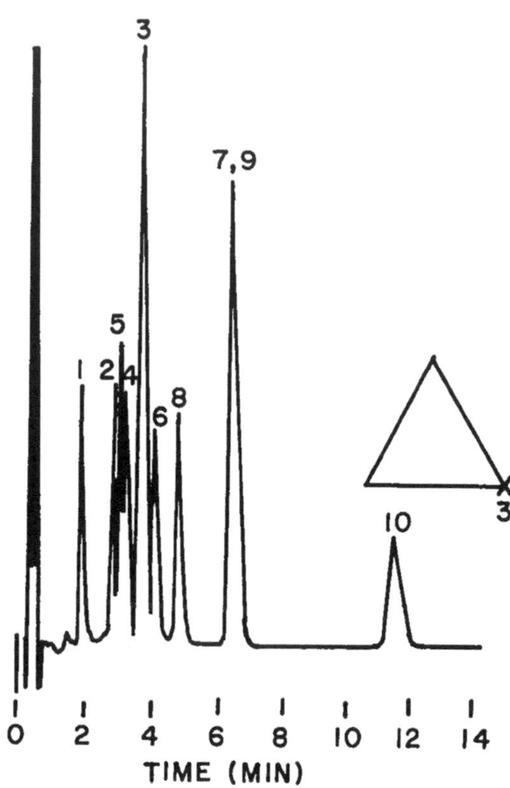

FIGURE 29. Separation of ten phenols using a two-component mobile phase as one step in the systematic development of an optimized chromatographic procedure.
Mobile phase: 1% acetic acid/tetrahydrofuran, 70:30
Column: Zorbax C8, 4.6 mm I.D. × 15 cm
Flow rate: 3.0 cm³/min
Temperature: 50°C
Detector: ultraviolet @ 254 nm

The approach just described is a case where visual optimization was used. The same information could have been obtained through use of overlapping resolution mapping.[273] This is a mathematical approach which can be run on a computer to make the same decisions purely on mathematical grounds. Both approaches usually give results which are very similar.

If one had a broad range of polarities in the sample matrix, it might be useful to modify this approach and use gradients for the screening runs with each of the seven solvent mixtures. In any case, it is necessary to establish the elution position of each of the sample components. This can be achieved by individual runs with each component. An alternative is to use different concentrations of each of the components in a sample mixture. This would allow the identities of the peaks to be established by measuring the areas of the peaks in each chromatogram. This approach can significantly reduce the number of chromatograms required for method development.

The same approach can be used to develop ion pairing systems. In ion pairing systems, method development may include variations in pH and ion pairing reagent as well as the solvent components.

Normal phase HPLC separations can also be developed using a similar solvent optimization scheme. In normal phase systems the common solvents are methylene chloride, methyl t-butyl ether, and acetonitrile. The moderator solvent is usually hexane. The procedure is the

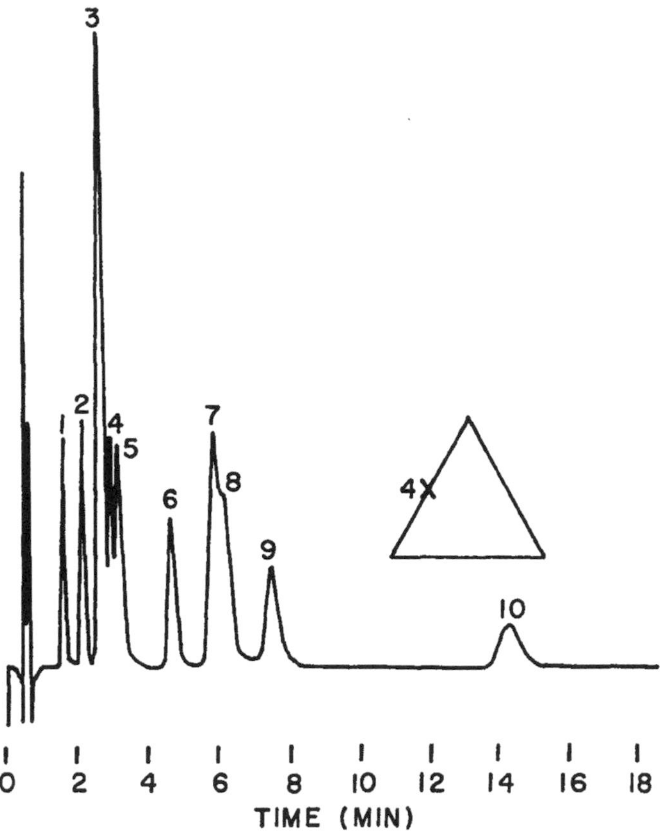

FIGURE 30. Separation of ten phenols using a three-component mobile phase as one step in the systematic development of optimized chromatographic procedure.
Mobile phase: 1% aqueous acetic acid/methanol/acetonitrile, 66:20:14
Column: Zorbax C8, 4.6 mm I.D. × 15 cm
Flow rate: 3.0 cm^3/min
Temperature: 50°C
Detector: ultraviolet @ 254 nm

same as described above except that a different iso-eluotropic series must be used. Table 30 shows one approximation for normal phase solvents on a cyano column.[287] This information can then be used to estimate the strength of the solvents needed for the normal phase experiments.

When a suitable isocratic solvent system has been established, the method should be thoroughly evaluated to determine if it offers accurate and precise answers for the components in question. The utility for such methods lies primarily in raw material assays or principal component assays for formulated products where only a limited number of components must be separated. The separation of complex mixtures will usually require greater resolution than can be achieved by isocratic procedures.

Once a separation procedure has been established, it is wise to develop a system suitability test to determine if a particular column will resolve all of the critical components, or if a newer, more efficient column is required. The test should be based upon the most difficult separation that is required for the analysis. An adequate value for the resolution of these two components should then be selected and specified in the HPLC procedure. This estab-

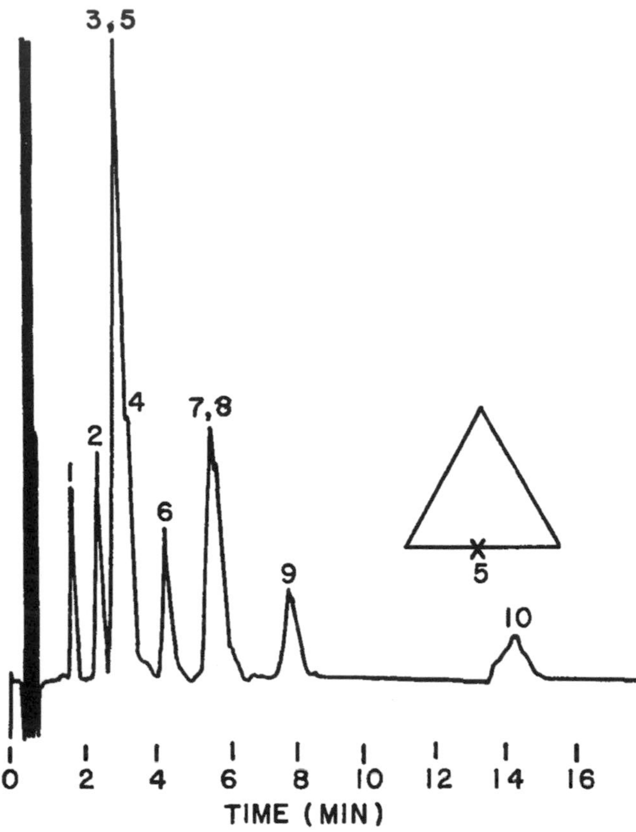

FIGURE 31. Separation of ten phenols using a three-component mobile phase as one step in the systematic development of an optimized chromatographic procedure.
Mobile phase: 1% aqueous acid/acetonitrile/tetrahydrofuran, 71:14:15
Column: Zorbax C8, 4.6 mm I.D. × 15 cm
Flow rate: 3.0 cm^3/min
Temperature: 50°C
Detector: ultraviolet @ 254 nm

lishes a convenient test to perform on the column before an assay is run to determine whether or not column performance is adequate.

Previous sections discussed the different techniques for introducing the sample onto the HPLC column. The questions remain of which solvent to use to dissolve the sample and what volume of the solution to apply to the column. In early HPLC work it was often assumed that the sample volume should be small. Later work has shown that this premise is not altogether true, and that the maximum acceptable volume depends upon the choice of sample solvent.

Good solubility of the sample is the primary criterion in selection of a sample solvent, but the stability of the sample in the proposed solvent and compatibility of the solvent with the detector are also very important. When using automated HPLC analysis, it is often desirable to store sample solutions for 16 to 24 h on the autoinjector. The long time period between sample dissolution and actual chromatographic analysis requires that the solvent be chosen so that the sample shows no significant changes over these time periods.

The most desirable situation is obtained when the sample solvent and the column solvent are identical. Under these conditions, the column is not disturbed by the addition of the

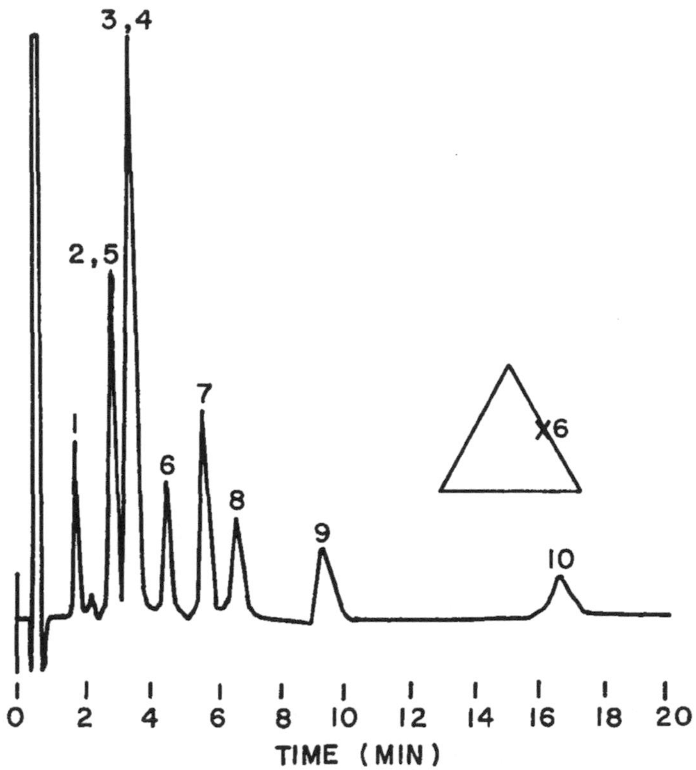

FIGURE 32. Separation of ten phenols using a three-component mobile phase as one step in the systematic development of an optimized chromatographic procedure.
Mobile phase: 1% aqueous acetic acid/methanol/tetrahydrofuran, 65:20:15
Column: Zorbax C8, 4.6 mm I.D. × 15 cm
Flow rate: 3.0 cm^3/min
Temperature: 50°C
Detector: ultraviolet @ 254 nm

sample solvent, and relatively large sample volumes can be used. The size sample which can be injected is the volume which will not significantly affect the elution efficiency of the peak of interest. As a rule of thumb, the acceptable sample volume can be up to one half of the peak volume when the eluant is used as the sample solvent. The peak volume is the volume of solvent which elutes from the column during the time that the sample peak is eluting.

Another common situation is the one in which the sample solvent is considerably different from the column solvent. Ideally, the sample solvent is a weaker solvent than the eluant. This situation allows application of reasonably large sample volumes often on the order of 50 to 100 $\mu\ell$ for typical analytical columns. The case in which the sample is in a stronger solvent than the column eluant can cause variations in the retention volumes of the compound as the sample volume changes. In addition, a portion of the sample may come through the column with the sample solvent band. To minimize these effects, small sample volumes should be used. The exact volume of course depends on the individual chromatographic system; however, when dealing with this situation, common maximum sample volumes are probably 10 to 50 $\mu\ell$ when working on analytical microparticulate columns. The most commonly used sample volumes are in the range of 10 to 20 $\mu\ell$, which would be an acceptable volume for any sample solvent.

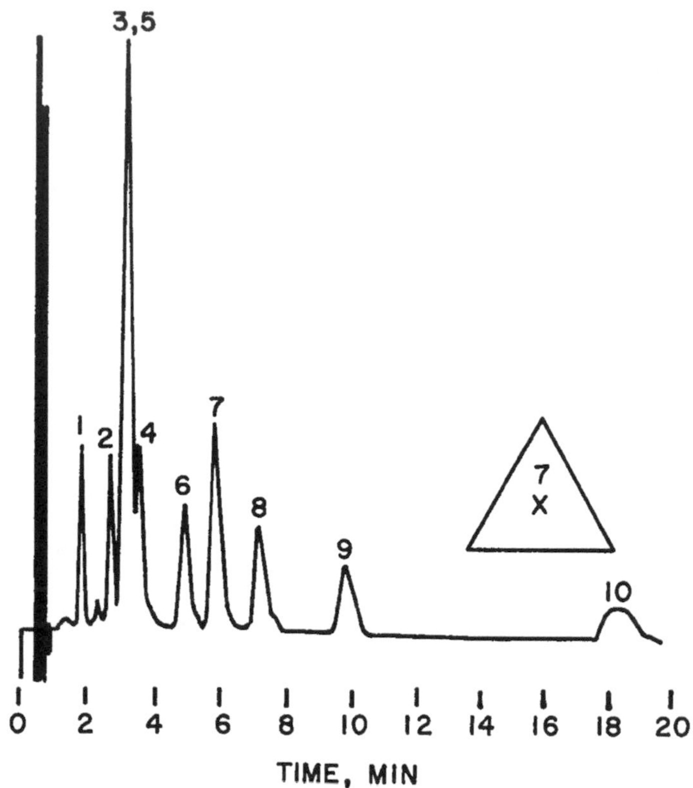

FIGURE 33. Separation of ten phenols using a four-component mobile phase as one step in the systematic development of an optimized chromatographic procedure.
Mobile phase: 1% aqueous acetic acid/methanol/acetonitrile/tetrahydrofuran, 65:13:9:10
Column: Zorbax C8, 4.6 mm I.D. × 15 cm
Flow rate: 3.0 cm^3/min
Temperature: 50°C
Detector: ultraviolet @ 254 nm

HPLC is a rather forgiving technique. If sensitivity is a problem, compromises of sample volume with its associated effects can be made and very usable methods still obtained. Changes in peak retention volume with sample size should remain constant when sample size is fixed, and usable data can be generated with only a little extra care. Such compromises are commonly made in trace analysis work to obtain adequate detection levels for the compound of interest.

Another form of sample introduction of special utility in trace work may use the column as a trap for the sample. If the sample is in a weak solvent, large volumes of sample solution may be run through the column, and the sample components will be held at the head of the column. Finally, a gradient is employed to elute the sample from the column. This is a common procedure when working with very dilute solutions.

Another critical selection in the development of an HPLC method is the detector. The choice of a detector rests upon a knowledge of the physical properties of the compound. The UV detector is usually the detector of choice when the sample components contain a UV chromophore, and is probably used in 95% of the applications of HPLC in pharmaceutical analysis. A major problem with this detector is that it limits the solvents that can be used to those with a UV cutoff below the wavelength at which the column is to be monitored.

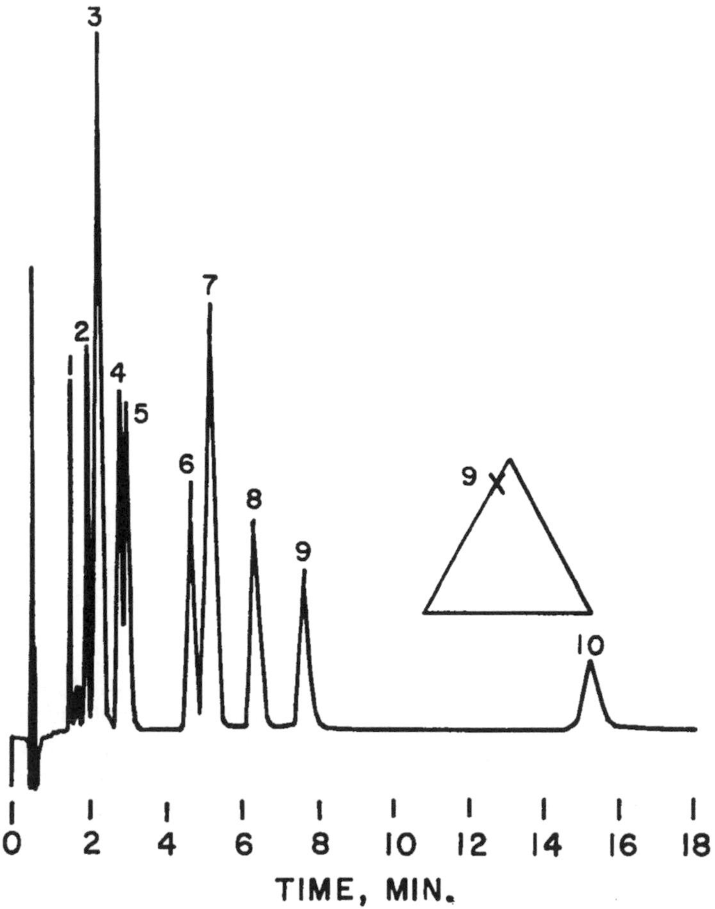

FIGURE 34. Separation of the ten phenols in the optimized, three-component mobile phase developed using the systematic optimization scheme.
Mobile phase: 1% aqueous acetic acid/methanol/acetonitrile, 63:34:3
Column: Zorbax C8, 4.6 mm I.D. × 15 cm
Flow rate: 3.0 cm^3/min
Temperature: 50°C
Detector: ultraviolet @ 254 nm

This problem can sometimes be overcome by monitoring on the long wavelength tail of the absorption band, but oftentimes compromises must be made in solvent selection in order to achieve the desired sensitivity.

As discussed in Section III.E, there is a host of other detectors available. Many of these devices can offer a good solution to a specific problem. If one is working with a separation problem where UV detection is presenting problems, other physical properties of the compound should be examined to determine if any other detection device may offer a better approach to the detection problem.

After all of the chromatography decisions have been made and a sample is run through the chromatographic system, the one remaining problem is how best to handle the resulting data.

In most applications of HPLC, a quantitative estimation of a component is required from the experiment, and some form of peak measurement is therefore required. Either peak height or peak area measurements may be used. For most applications, either of these

Table 30
SOLVENT STRENGTH MONOGRAPH FOR POLAR-BONDED PHASE SEPARATIONS

$\alpha\epsilon^\circ$	%V B (cyano-silica)				%V B (amino-silica)		
B	MC	ACN	MTBE	MeOH	MC	ACN	MTBE
A	hex	hex	hex	MC	hex	hex	hex
0.00	0	0	0		0	0	0
0.02	17	6	12		5	2	13
0.04	42	15	29		11	4	28
0.06	100	27	48	0	19	6	45
0.08		48	73	12	30	9	59
0.10		84	100	31	47	13	100
0.12				57	75	18	
0.14				100	100	25	
0.16						34	
0.18						45	
0.20						65	
0.22						100	

Note: MC = methylene chloride, ACN = acetonitrile, MTBE = methyl t-butyl ether, MeOH = methanol, hex = hexane or FC113.

techniques will give quite adequate results. Peak height data is very much affected by injection technique and therefore is most applicable where completely automated sample injection is used. Nonetheless, peak height measurements may be preferred when dealing with a badly tailing peak using electronic integration devices, since a badly skewed peak will invariably cause problems in the determination of the peak finish. The integration procedure in which the peak finish is instrumentally established can cause large variations in the included peak area. When such a situation exists, one finds that the variation in peak heights is much less than the variation in peak areas.

The second situation in which peak height data may be better than area data is where partially unresolved peaks occur on the leading or tailing edge of the peak of interest. Small peaks in these locations will generally have less effect on peak heights than on peak areas.

For most HPLC applications, the most sophisticated data system available in the laboratory will be used. There are, however, cases where such devices can be very troublesome unless a totally interactive mode of data reduction is available. One such case is in trace analysis work where sensitivity is a problem and very noisy detector signals must be processed, or where one must deal with partially unresolved components. Under these circumstances it may be advantageous to forget the sophisticated data processors and resort to a simple peak height measurement by manual methods.

XIII. SYSTEM SUITABILITY

As HPLC became popular for quantitative analysis of chemicals and pharmaceuticals, the need for reproducibility between operators and laboratories rose. Reproducibility has been difficult to achieve in HPLC because there are many packings which have similar descriptions, but behave considerably differently depending upon the manufacturer. Furthermore, most HPLC packings and/or columns deteriorate with use. The deterioration may be caused by hydrolysis of the bonded phase from the silica backbone of the packing, fracture of the silica particles, or decay of the bed with subsequent formation of a column void. The user is always faced with the problem of knowing if the column is behaving in a manner similar to the one used for method development, and if it is capable of providing acceptable results.

When an HPLC procedure is used, the entire HPLC system must be performing adequately. The performance of the system is usually influenced mostly by the column and the solvent used in the separation. Ideally one would like to develop simple criteria to assure that the system is performing adequately. Oftentimes one or more of the fundamental chromatographic parameters are used to measure the acceptability of the system performance. Separation factor, resolution, number of theoretical plates, and peak tailing are parameters which are used for this purpose.

Separation factor (α) and resolution (R) can be used for system evaluation provided there are at least two components in the sample. These quantities are defined in Equations 3 and 6, respectively. The pair of components which is most difficult to separate should be used for column evaluation when more than two components are present in a sample. One of the components would normally be the major compound of interest. Having selected an appropriate pair, a minimum value for α or R can be specified. Having specified a suitable value, one can verify that the criterion is met before a set of samples is run. Likewise, the data can be rejected if the sample set was run but the suitability parameter was not satisfactory.

When dealing with pharmaceutical analysis, the separation efficiency (α or R) may not be an easy test to run. The nature of the test requires the presence of a second component. Most raw materials or pharmaceutical preparations do not have a major second component (impurity or degradation product) present. One method of overcoming the absence of the second component is to use a reference sample which has been degraded or altered in such a way so as to produce at least two major peaks. From a chromatographic viewpoint, this is a very desirable means for addressing the problem. However, from a practical standpoint, maintaining a reference sample for such use may not be very desirable. The maintenance of a reference sample becomes very difficult when analyses are being run at different locations or when there are multiple sources for the drug material.

A second solution lies in the introduction of a second commercially available and structurally related component. This may be an acceptable solution when one is dealing with a large family of commercially available compounds. In this case, the second compound can be spiked into the sample to provide the appropriate separation data. In some instances internal standards may be used in the assay. The internal standard may, in fact, serve as the second component of the sample, although the internal standard often is not structurally similar to the compound of interest.

Peak shape is yet another parameter of chromatographic interest which can be used as a control for system suitability. One of the major methods for monitoring peak shape utilizes the concept of number of theoretical plates, which is a normalized measure of peak width. Unfortunately, the methods used for determination of the number of theoretical plates in a column are far from standardized in HPLC. The models used for the determination of theoretical plate assume a Gaussian peak shape. The general relationship which has evolved for the determination of the number of theoretical plates N was given in Equation 4. The factor of 16 in Equation 4 was the result of measuring peak width by the distance between the points where tangents to the peak sides cross the baseline. Equation 4 can be generalized for use with other measures of peak width:

$$N = a\frac{T_R^2}{W^2} \tag{24}$$

values of the constant a for use with other measures of peak width are summarized in Table 31.

The use of N for column evaluation is very popular, although there are fundamental difficulties in its use. The expressions for determining N are not strictly applicable to asymmetrical peaks. The most popular relationships for estimating N involve the measure-

Table 31
PARAMETER VALUES FOR ESTIMATING THE NUMBER OF THEORETICAL PLATES[a]

w	a	Description of method
w_1	4	Width measured at inflection points
$w_{1/2}$	5.51	Width measured at half the peak height
w_{tan}	16	Width measured at the base between tangents to the points of inflection
w_{α}	9	Width at 3ϵ[b]
	16	Width at 4ϵ[b]
	25	Width at 5ϵ[b]

[a] See Equation 24.
[b] ϵ = standard deviation of the peak.

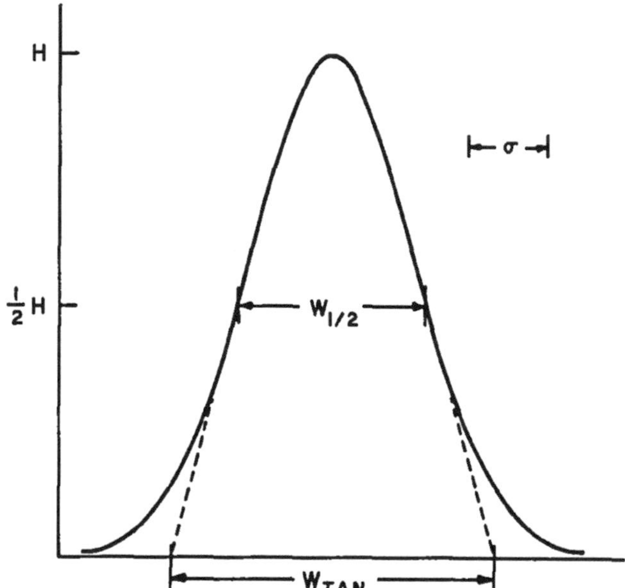

ment of peak width at half height or the tangent width at the baseline. These are the two easiest forms to use when manual methods are used for the calculation. The half-height method, however, has poor sensitivity toward peak tailing, for the width measurement is made at a position on the peak which is not greatly influenced by tailing. The tangent method is also strongly biased by the upper parts of the peak which greatly influence the construction of the tangent lines. The other methods and positions for measurement of peak width are more appropriate when a computer system is available for processing the data.

When large data systems are available for processing the peak data, other means might be appropriate for the determination of N. The analysis of moments is often used for the determination of N. A generalized expression for the number of theoretical plates is

$$N = \frac{T_R^2}{\delta_s^{\,2}} \tag{25}$$

where T_R is the retention time of the peak in seconds and δ_s is the variance of the peak in seconds. With moment analysis, the zeroth moment is the area of the peak. The first moment is the position of the peak mean, and the second moment is the peak variance. Thus by dividing the first moment by the second moment and squaring, an estimate of N can be obtained.

More recently, many people have touted the use of asymmetry-based methods for the determination of system plate numbers. Foley and Drosey[288] have developed an expression which can be used for determination of N.

$$N = \frac{41.7(T_R/w_{0.1})}{(A/B + 1.25)} \tag{26}$$

In this equation, T_R is the retention time at the peak maxima. The value $w_{0.1}$ is obtained by measuring the peak width at 10% of the peak height. A is the distance from the peak maximum to the position on the peak envelope measured at 10% of the peak height. B is the distance from the front of the peak envelope to the peak maximum position.

The use of peak moments and calculations that include peak asymmetry appears to give the best estimate of actual column behavior. The *United States Pharmacopeia XXI* utilizes a tangent approach for determination of column plate numbers. HPLC manufacturers use many different means for determination of N for their columns.

If the plate count for an HPLC column is used as a system suitability test, one can use any of the techniques for determination of N as long as there is always consistency in how the calculation is done. It does, however, require that realistic limits be set for the various tests used in the method.

Asymmetry-based methods for determination of theoretical plates include a parameter to account for peak tailing. Often the tailing characteristics of a peak are in their own right a good handle for control of the overall system. A tailing factor can be defined as

$$T = w/2f \tag{27}$$

The terms are shown in Figure 35. The value f is defined as the distance from the front of the peak envelope to the point of the peak maximum, and w is the distance across the entire peak. The major difficulty with this approach is in deciding at what percentage of the total peak height the measurement will be made. The USP currently specifies that the measurement be made at 5% of the peak height. At the 5% level, noise and other factors affecting determination of the baseline influence the number to a great degree. Thus it may be difficult to obtain a good number for the tailing factor. More recently Foley and Drosey[288] have recommended that the measurements be made at 10% of the peak height. In practice this gives a more useful and reproducible number.

The use of peak tailing as a system suitability test requires that efforts be made during the development stages of a method to establish how much tailing occurs when a good column is used. Peak tailing can be sensitive to column voiding and degradation of the column packing. Thus, peak tailing can be a very good tool in determining the adequacy of a column setup or in establishing when a column has deteriorated to the point that it should not be used.

Many people use statistical parameters based upon all of their data or a part of their data in each run to determine the acceptability of the data set. The slope, intercept, and scatter of data points on a multipoint calibration curve can give good insight into how the run performs, especially when compared to historical values for these parameters. If the current values deviate from average historical values for a parameter, one should explore why the variation occurred and determine if the data set should be accepted or rejected. Many people use the RSD of a set of data points to determine the acceptability of a set results. Often the

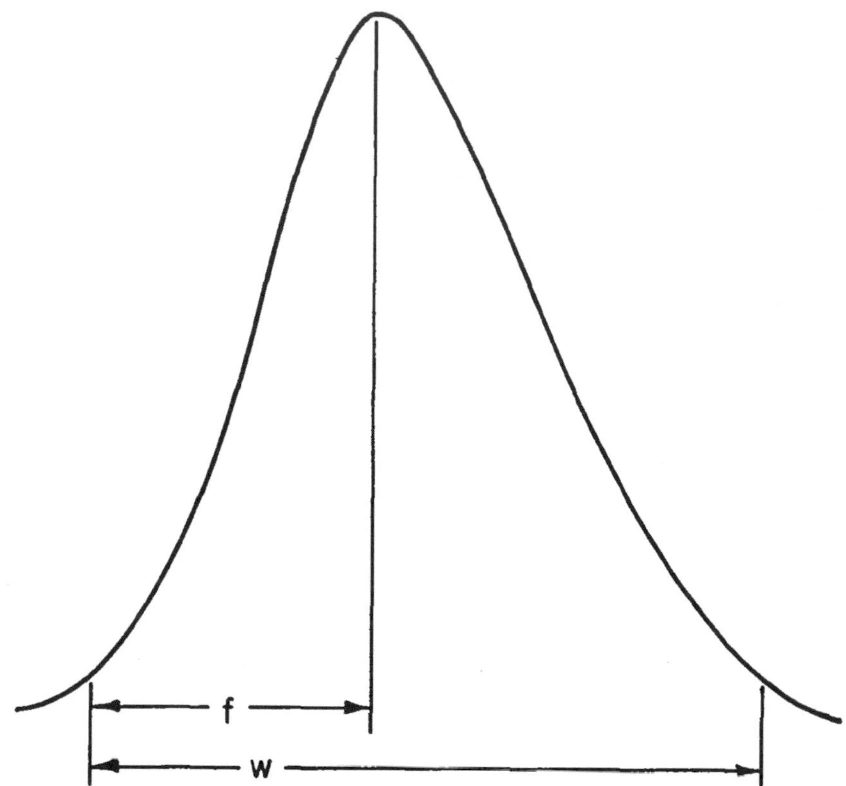

FIGURE 35. Definition of the parameters used to define peak tailing (see Equation 27).

RSD is determined from replicate measurements on a single sample or standard. The problem with this approach is that only a very small portion of the run may truly be monitored. It may also require the preparation of several assay samples which have no use other than for determining the variance of the system at this particular time. Perhaps a better approach is to evaluate the RSD for the results on each sample. This can be accomplished easily when the analysis protocol requires several replicate measurements to establish the assay value for the sample. The advantage of monitoring each data set is that it gives a picture of the behavior of the system throughout the sample run.

Maintaining the historical performance data for an assay can be very valuable. The knowledge of performance values for key parameters such as slope of calibration line, number of theoretical plates, and peak tailing, along with the acceptable operating ranges for these values, allows one to use these as acceptance criteria for each run. When a key parameter falls outside the acceptable range (or approaches its limiting value), corrective action can be taken before time is lost because of system failure. Thus, this type of information gives one the opportunity to control the immediate system as well as to forecast the approach of problems.

XIV. TRACE ANALYSIS BY HPLC

HPLC can be used to determine trace level components in a sample. Applications include analysis of trace components in natural product extracts, residual intermediates or synthetic by-products in a drug substance, and determination of drug levels in the body. HPLC procedures have been successfully used for all types of trace analysis, but sacrifices in what would normally be called good HPLC technique must often be made to find a realistic solution for such problems.

For trace analysis work, detector systems are required which provide the ultimate in detection sensitivity for a specific compound. Sometimes, the material of interest may be fluorescent or possess chemical functionalities which can be easily derivatized to produce a fluorescent product. These cases are the easiest to handle because fluorescence provides a convenient and sensitive means for detecting a compound. A good fluorometric detector is required to effectively take advantage of the fluorescent character of the analyte. A fluorometer is required which permits selection of the optimum excitation and emission wavelengths in order to maximize the signal-to-noise ratio for the analysis.

The more typical situation is one in which the compound has a UV absorption band which can be used for detection purposes. If the absorption maximum for the trace level compound is near 254 nm, the best detector is probably a good, single-wavelength 254 nm device. As the absorption maximum moves away from 254 nm, the variable wavelength detectors become the detectors of choice. If the absorption maximum of the compound misses 254 nm by more than about 10 nm, the variable detector should be used.

Use of a variable wavelength detector for trace analysis below 220 nm requires a properly maintained instrument. Deuterium lamps are not inherently long lived, and they accumulate operating hours rapidly when used in HPLC detectors. The lamps generally become noisy at the low wavelengths with age. Thus, frequent source replacement may become necessary in trace analysis applications. Most photometric detectors have one or more mirrors in the light path. The mirrors may become dirty or etched, decreasing the amount of light passed through the system. As the light energy decreases, the noise level in the detector increases. Therefore, any detector which is going to be used for trace level analysis work must be cleaned, tested, and have faulty components replaced on a regular basis. It is also important to use very high quality reagents and solvents to reduce the background absorbance of the solvent as much as possible, and to avoid extraneous peaks or retention volume shifts due to trace impurities in the reagents. The noise level can also be reduced in some cases by thermostating both the columns and detectors.

After the spectral properties of the trace level component have been established and a suitable detection device has been selected, a separation technique must be developed. Even though the trace level component may be separated from other sample components when run at normal concentrations, the separation may be inadeqate for trace level monitoring. A very good separation is required for monitoring extremely low levels of a substance in a sample. For trace level analysis, the analyte should ideally elute early in the chromatogram and far ahead of major sample components. An early eluting component will generally provide a sharper peak, and the peak can be measured against a lower background signal if major sample components elute much later.

Trace components which elute after the major components can be very difficult to detect and quantitate. It is common to overload the column greatly with the major sample component when analyzing for a trace component. This will often cause the peak from the major component to tail badly. The trace level material must therefore be well separated from the major component. This presents a problem in isocratic systems because as the retention volume increases, the peak becomes much broader. As the peak becomes broader, the sensitivity of detection for that component decreases.

Isocratic trace level analysis procedures are of course desirable, but, in reality, they are often impractical. The solubility of the sample is often low enough that large volumes must be injected to obtain the required sensitivity for the trace level component. It is often necessary to use a strong column solvent in order to dissolve enough sample to allow analysis. The use of large volumes and strong solvents complicate the situations for isocratic analysis. The large volumes and strong solvents tend to push both the trace level components and the major sample component through the column much quicker than would be expected under normal operating conditions. Under such conditions, the trace level constituent will often not be adequately resolved from the major sample components, although the preliminary work with smaller sample volumes indicated that the system should work.

To overcome these problems, gradient techniques are frequently employed in trace level analysis. When using gradient techniques, the column is first eluted with a very weak solvent. In fact, the initial solvent should not cause any significant movement of the trace level component on the column. If the sample solvent is sufficiently weak, several milliliters of the sample solution can be pumped through the system and the sample concentrated at the head of the column. A solvent gradient is then run to elute the materials from the column. By using a gradient, the different materials can be resolved better than under isocratic conditions. Most compounds will also elute in sharper peaks when gradient elution is employed, thus enhancing the sensitivity of the method.

Column selection for trace component analysis imposes requirements that are not present in normal HPLC. The choice of column size can be important in trace analysis. Standard analytical columns are frequently overloaded by the large samples required for trace analysis. Fortunately, standard columns may still work quite well, even with overloading, for the assay of the minor components in the range of 500 to 1000 ppm. When sample size and volume compromise the performance of an analytical column to the extent that it does not provide adequate separation, semipreparative columns prepared with the same microparticulate packings may offer a viable solution to the problem. The larger semipreparative columns will generally handle the larger samples required for trace analysis and frequently provide better resolution than standard analytical columns. A disadvantage of larger diameter columns is that much faster flow rates must be used, and therefore much larger volumes of solvent must be used. They also compromise the detector limit, since the material of interest is now diluted into a larger solvent volume.

After developing a suitable separation procedure, the analyst is still faced with the problem of quantitation. Most trace level analyses utilize the lower end of the detector sensitivity, and the signals are therefore quite noisy. Processing such signals through digital integrators and various computer data systems which use fixed algorithms for peak selection and integration will not provide adequate quantitation. The chromatographer generally has two alternatives for processing such data. The first and most desirable method utilizes an interactive data system which allows the chromatographer to make the critical decisions on peak start and finish and location of the baseline. The data system can then use this information to calculate the peak area or peak height required for quantitation. If such a data system is not available, the next best method is to use a ruler and manually draw appropriate baselines on the chromatogram and manually measure the peak height for the peak of interest.

Standardization of the analysis is usually carried out by the external standard procedure employing samples of the trace level components at known concentrations. When working with very low levels of the trace constituents, the sample matrix may affect the quantitative results. Therefore, the trace impurity should be added to an authentic sample of the major substance if at all possible.

Procedures which concentrate the sample at the head of the column may not achieve complete elution of the material as the analysis proceeds. Thus it is not unusual to see a peak elute at the position of the major sample component on subsequent blank runs. The ghost peaks may be observed for several runs after a column has been used in this fashion. When injecting large sample volumes, it is also not unusual to see shoulders appearing on the leading or tailing edge of the peak. This is not necessarily a sign of a defective column, but is often a function of the sample volume and injection.

XV. PURITY EVALUATION WITHOUT REFERENCE STANDARDS

HPLC has become one of the primary tools used in characterizing purity of reference standards and other samples for which there are no suitable reference materials available. In this application HPLC is used to detect and estimate the level of impurities in the sample rather than for quantitating the main component directly. Estimating impurity levels requires a powerful chromatographic system which separates many potential contaminants from the sample. Ideally, several systems with different selectivities and different column/sample interactions would be used. It would also be advantageous to monitor the eluate at several different wavelengths. This would provide the best set of circumstances for observing as many of the foreign materials in the sample as possible. Unfortunately, there exist an almost infinite set of conditions which could be run in such circumstances. A compromise procedure is generally used in which one or two systems are used that generally work well for the particular type of compounds being analyzed. The sample is frequently monitored at only one or two optimum wavelengths when UV detectors are used, although this may change since multiwavelength UV detectors are now commercially available. The main peak from the sample can be very far off scale since the real interest lies in the foreign materials present in the sample. In such procedures, sample concentrations are generally increased by a factor of at least 50 to 100 times the concentration used in the routine analysis of the major sample component. These sample concentrations may have to be moderated if sample solubility becomes a problem, or if the sample contains a high level of some impurity. A more realistic picture of the sample may be obtained by using a concentration which will just keep the major impurity on scale.

In this type of work, maximum sensitivity and maximum resolution are required. Thus small volumes of a concentrated sample are preferred in this type of work. The same criteria that were used to select appropriate sample volumes for normal HPLC procedures are also applicable here.

Selection of the proper chromatographic system is extremely important in this type of work. Systems developed for routine analysis are usually designed to separate known impurities and degradation products from the main compound. In routine methods, however, many of the impurities and foreign sample components may elute in the solvent front or so far out in the chromatogram that they cannot be detected due to excessive band broadening. These problems can be overcome by the use of gradient elution. The gradient will allow the resolution of many sample components and provide suitably sharp peaks for all components which elute under the conditions reached in the gradient.

Quantitation of the impurity bands is a major problem in purity evaluation. Identities of some of the impurities may be known allowing them to be evaluated by direct comparison of the peak area with that of an authentic sample of known concentration chromatographed under the same conditions. Unfortunately, there will generally be a whole series of trace level components present whose identities are not known. A number of assumptions must be made in order to estimate the concentration of the unidentified imipurities in the sample.

Since the unidentified trace materials often have chromatographic characteristics similar to those of the main component, and since they usually arise from the same manufacturing process, the assumption is often made that the trace components will have spectral properties similar to those of the main component. Quantitation of the trace components can then be accomplished by using dilute solutions of the major component to obtain a response cure for the impurities. The response of the diluted samples is then compared to each of the trace level peaks to ascertain relative quantities of each material present.

It is important to include a blank run with an injection of the sample solvent in the data set, when using the procedure just outlined. Often there will be peaks appearing from the sample or column solvents. In trace analysis, these background peaks may be of a similar magnitude to those peaks caused by substances actually present in the sample. The blank run determines which peaks are from the sample and which peaks are from other sources and should be ignored. When more sophisticated data systems are available, the background chromatogram can be subtracted directly from the sample chromatogram. This will produce a residual chromatogram which is free of anomalies generated by the chromatographic system.

Very useful information relevant to the evaluation of reference materials and other special samples can be obtained by using these HPLC techniques. The HPLC procedures can provide information that is almost impossible to obtain in any other fashion. However, it should be remembered that HPLC may not provide a totally accurate picture of the sample. There may be many components present in a sample which are not observed by the HPLC procedure. For example, residual solvents and inorganic materials frequently are missed completely. There is always the possibility that a major impurity is not separated from the major sample component. Also, the simplifying assumptions made to permit quantitation of the impurities may produce very erroneous results if the spectral properties of the impurity are significantly different from the spectral properties of the major sample component. HPLC data should therefore be coupled with other suitable techniques to arrive at the most realistic purity value for the sample.

XVI. TROUBLESHOOTING AND MAINTENANCE

With the advent of Good Laboratory Procedures (GLP) it has become imperative that good instrument maintenance records be kept. A good running log of instrument and a record of all problems and repairs can be a very useful tool in the proper use and maintenance of a HPLC instrument.

Most HPLC instrumentation consists of a complex array of electronic and mechanical parts. In routine operation, the majority of problems with HPLC instrumentation are related to the mechanical operation of the device. These failures frequently appear as unusual recorder outputs from the instrument. Most of these problems can be solved by a person having a modest ability in disassembling and reassembling mechanical devices. The ability to recognize problems and to correct them as they arise in the laboratory generally allows more reliable results to be generated with fewer major problems, and therefore with more efficient use of both instruments and operators.

There are many different types of HPLC instrumentation available today. Many typical problems give similar characteristics regardless of the instrument used. Other problems may be peculiar to one instrument or to one type of instrument.

Typical problems seen in day-to-day operation of HPLC instrumentation

Symptom	Possible cause	Possible remedy
No column flow or no column pressure	No solvent in column supply	Resupply solvent reservoir
	A plugged filter in the solvent supply reservoir or pump	Replace in-line filters before pump
	An air bubble lodged in a reciprocating pump chamber or check valve	Purge pump with solvent
	Leak in sample valve prior to column	Tighten valve by placing more tension on seal or replace valve seal
	Leak in plumbing prior to column	Locate and repair leak
Column pressure and/or flow erratic	Leak around pump seal	Replace pump seal
	Leak in sampling valve	Increase tension on seal of valve by tightening valve or replace seal

	An air bubble lodged in a reciprocating pump chamber or check valve	Purge pump with solvent
	Foreign material in pump check valve	Wash pump head with water or acid as recommended in pump manual, or replace check valve
High column head pressures	Inlet frit on column plugged	Backwash frit or replace
	Foreign particle restricting flow through tubing in system	Loosen fittings at various points in system until problem is isolated; clean particle from system
	Plugged lines at the detector, especially in systems containing salts	Disassemble and repair detector cell
		Some aqueous systems support biological growth; if present, discard solvent; clean HPLC system and prepare fresh solvent
	Collapse of column bed	Replace column
Peak splitting, usually on tailing edge, or peak doublet	Void area in column	Fill column with packing material or with a pellicular material, or replace column
	Large void area in HPLC column	Reduce void areas in HPLC system
	Carry-over of sample in switching of HPLC sampling valve	Repair valve
	Mismatch of column solvent and sample solvent	Change sample solvent to make it more similar to the column solvent
Loss of resolution	Column overloaded	Reduce amount of sample
	Sample volume too large	Inject smaller volume of more concentrated sample solution
	Large void area in HPLC column	Fill column with packing material Replace column
	Bad column	Replace column
	Strongly retained material remaining on column	Regenerate column
	Peak Shape and Retention Volumn Problems	
Flat tops	Column overload	Reduce amount of sample
	Detector overload	Monitor at less efficient wavelength or reduce amount of sample
Badly skewed peaks	Poor geometry of tubing, such as going from large to small bore connections	Replace all lines on fittings with material of uniform inner diameter
	Large void area in HPLC column	Fill column with packing material or replace column
	Bad column	Replace column
Increased retention time	Leak in system prior to column in sample valve or connecting fitting	Isolate position of leak and repair
	Low flow rate	Check output of pump and correct if improper flow rate is being delivered On multihead pumps, one pump chamber may not be delivering solvent; purge bubble from chamber A check valve may be leaking; clean or replace leaking check valve May also be an electronic problem in flow monitor or pump control circuits
Decreased retention time	Loss of stationary phase from column	Replace column
	Column not fully equilibrated to new solvent after solvent change	Equilibrate column

	Detector Problems	
Spikes on detector signal	Air bubbles passing through cell	Degas solvent prior to use Apply a small back pressure to detector cell; usually a needle valve is used and adjusted until problem is alleviated
	Bad source lamp (lamp flickers)	Replace source lamp
	Detector will not zero	
High background	Dirt or foreign material in sample cell	Disassemble and repair detector cell
	Past cutoff of solvent	Select higher wavelength before solvent cutoff or select different solvent
	Solvent contains UV absorbing impurities	Use different lot or brand of solvent which is free of such impurities
	Faulty lamp	Replace lamp
	Detector cells not balanced	Disassemble, clean, and rebuild detector cell
	If photoresistor detector arrangement, photoresistors are out of balance	Replace photoresistor assembly
Low background	Dirt or foreign material present in reference cell	Disassemble, clean, and rebuild cell
	Leak into reference cell or moisture present in reference cell	Dry cell and, if necessary, disassemble, replace, gaskets, and rebuild cell
	Faulty lamp	Replace lamp
	If photoresistor detector arrangement, photoresistors are out of balance	Replace photoresistor assembly
Noisy baseline	Contamination or foreign material present in sample or reference cell	Wash cells with a series of miscible solvents Clean or replace detector cells
	Defective or aging UV lamp	Replace UV lamp
	Small bubbles present in cell	Increase back preessure on cell until bubble is removed
	Recorder or instrument grounding problem	Check recorder and instrument ground; eliminate ground loops
	Dirty solvents being used	Use a different lot, brand or better grade of solvent
Baseline drift	Contaminated or bleeding column	Regenerate or replace column
	Leak between sample and reference cell	Rebuild detector cell assembly
	Gradual change in column temperature	Use a temperature controlling device
	Gradual change in composition of column solvent	Take preecautions to avoid evaporation or other itiems causing potential changes in the column solvent
	Recorder Problems	
Steps present on chromatogram or baseline shifts	Recorder gain and/or damping control improperly adjusted	Adjust recorder gain and damping-refer to recorder instrument manual
	Large dead band in recorder	Have recorder repaired
	Instrument and/or recorder improperly grounded	Eliminate ground loops and have everything attached to a good earth ground
	Low level AC signal being fed to recorder	Locate source of AC signal and eliminate

XVII. APPLICATIONS

HPLC is beginning to make inroads into the officially accepted procedures for many drugs. For example, a method for folic acid is undergoing tests for inclusion in the *U.S. Pharmacopeia* as an official test for this substance. HPLC is being used in the laboratories of many pharmaceutical companies for routine assay of new drugs as well as for substitutes for older, more troublesome assays for marketed drugs. Many of these applications are just beginning to find their way into the scientific literature. HPLC is also very widely used in studies of drug levels in biological fluids.

There are many possible solutions to most HPLC separation problems. The methods used by one group to solve their assay problems often gives valuable insights into potential methods for another group's problems. For this reason, a listing of some recent publications on HPLC analysis of drugs is included here. The reader is referred to the original literature for a more detailed discussion of each application.

Table 32
PHARMACEUTICAL APPLICATIONS OF HPLC

Compound	Solvent system	Column	Detector	Ref.
Acetaminophen	225 mg tetramethylammonium hydroxide 750 mℓ water 125 mℓ methanol 125 mℓ acetonitrile 1 mℓ acetic acid	L1	UV/280	289, p. 14
Acetylcholine chloride	1.03 g sodium heptanesulfonate 700 mℓ water 10 mℓ methanol Adjust pH to 4.0 with acetic acid or ammonium hydroxide 50 mℓ acetonitrile Dilute to 1 ℓ with water	L1	RI	289, p. 18
Acetylcysteine	6.8 g potassium dihydrogen phosphate Dilute to 1 ℓ with water	L1	UV/214	289, p. 19
Acyclovir	0.25 mℓ acetonitrile 0.9 g heptanesulfonic acid 999.75 mℓ 0.01 *M* potassium dihydrogen phosphate (pH 2.3)	μBondapak C18 (L1)	UV/254	290
Alclometasone	2 methanol 1 0.05 *M* potassium dihydrogen phosphate	L1	UV/254	291, p. 1804
Allopurinol	0.05 *M* ammonium dihydrogen phosphate	L1	UV/254	291, p. 1806
Alprazolam	850 acetonitrile 80 chloroform 50 ℓ butanol 20 water 0.5 acetic acid	L3	UV/254	291, p. 1807
Alprostadil	1 mℓ water	L3	UV/254	292, p. 1697

	7.5 mℓ t-amyl alcohol 1000 methylene chloride			
Amcinonide	80 methanol 20 water	L1	UV/254	289, p. 34
Amiloride	71 water 25 methanol 4 pH 3 buffer (136 g potassium dihydrogen phosphate/ℓ to pH 3 with phosphoric acid)		UV/286	289, p. 36
Aminobenzoic acid	300 mℓ methanol 10 mℓ acetic acid 690 mℓ water	L1	UV/254	289, p. 39
Aminosalicylic acid	17.04 g disodium hydrogen phosphate 16.56 g sodium dihydrogen phosphate monohydrate 9.2 g tetrabutyl ammonium hydroxide 1700 mℓ water 300 mℓ methanol	L1	UV/254	289, p. 47
Ampicillin	200 mℓ methanol 250 mℓ 0.02 *M* phosphate buffer pH 6.00 0.25 mℓ alkyldimethylamine (ADMA C10)	μBondapak C18 (L1)	UV/220	293
Anthralin	395 mℓ water 550 mg sodium heptane sulfonate 5 mℓ acetic acid 600 mℓ acetronitrile	L1	UV/365	289, p. 68
Apomorphine	34 methanol 66 buffer (0.02 *M* disodium hydrogen phosphate and 0.03 *M* citric acid, pH 3.2)	μBondapak C18 (L1)	Electrochemical	294
Aspartame	25% methanol 75% aequous 1% triethylammonium acetate, pH 4.5	Cyclobond I	UV/214	295
Aspirin	2 g sodium heptanesulfonate 850 mℓ water 150 mℓ acetonitrile pH to 3.4 with acetic acid	L1	UV/280	289, p. 78

Table 32 (continued)
PHARMACEUTICAL APPLICATIONS OF HPLC

Compound	Solvent system	Column	Detector	Ref.
	225 mg tetramethylammonium hydroxide 750 mℓ water 125 mℓ methanol 125 mℓ acetonitrile 1 mℓ acetic acid	L1	UV/280	289, p. 14
Azlocillin	40 methanol 60 0.01 *M* acetate, pH 4.8	μBondapak C18 (L1)	UV/220	296
Bamipine	20 methanol with 0.02% ammonium hydroxide 80 dichloromethane	μPorasil (L3)	UV/258	297
Beclomethasone dipropionate	3 acetonitrile 2 water	L1	UV/254	289, p. 91
Benzodiazepines	550 methanol 250 water 200 0.1 *M* pH 7.2 phosphate buffer	Hypersil ODS (L1)	UV/240	298
Benzoyl peroxide	5 acetonitrile 10 water	L1	UV/254	289, p. 100
Benztropine mesylate	350 mℓ 0.005 *M* pH 3.0 octylamine phosphate buffer 650 acetronitrile	L7	UV/259	289, p. 103
Betamethasone	2 water 1 acetronitrile	L1	UV/254	299, p. 1975
Bleomycin	960 mg sodium pentane sulfonate 1000 mℓ 0.08 *N* acetic acid	L1	UV/254	289, p. 126
Bupivacaine	65 acetonitrile	L1	UV/263	289, p. 137

	35 pH 6.8 phosphate buffer (1.94 g potassium dihydrogen phosphate, 2.48 g dipotassium hydrogen phosphate in 1 ℓ water)			
Butalbital	3100 water 725 acetonitrile 4 phosphoric acid	L1	UV/214	291, p. 1820
Butaperazine	100 chloroform 3.5 methanol	μBondapak amine	UV/280	300
Butorphanol tartrate	3 0.05 *M* ammonium acetate in water 1 acetonitrile pH to 4.1 with acetic acid	L11	UV/254	289, p. 142
Calcifediol	6 heptane 6 water-saturated heptane 3 methylene chloride 5 ethyl acetate	L3	UV/254	289, p. 144
Carbamazepine	55 water 40 methanol	Resolve C-18	UV/254 5 acetonitrile	301
Carbidopa	5 methanol 95 0.05 *M* sodium dihydrogen phosphate, pH 2.7	L1	UV/280	289, p. 160
Carboprost tromethamine	992 mℓ methylene chloride 0.5 mℓ water 7 mℓ 1,3-butanediol	L3	UV/254	289, p.162
Cefazolin	1 acetonitrile 9 pH 3.6 buffer (0.9 g disodium hydrogen phosphate and 1.298 g citric acid monohydrate in 1 ℓ water)	L1	UV/254	289, p. 174
Cefonicid	1 0.2 *M* ammonium phosphate 2.5 methanol 16.5 water	L1	UV/254	299

Table 32 (continued)
PHARMACEUTICAL APPLICATIONS OF HPLC

Compound	Solvent system	Column	Detector	Ref.
Cefoperazone	1.2 triethylamine 2.8 1 *N* acetic acid 120 acetonitrile 876 water	L1	UV/254	289, p. 175
	1 Methanol 1 water	μBondapak C18 (L1)	UV/228	302
Cerforanide	18 mℓ 10% aqueous tetrabutylammonium hydroxide 8.6 mℓ 11 *N* potassium hydroxide 700 water 200 methanol Titrate to pH7 and dilute to 1 ℓ with water	L1	UV/254	299
Cefoxitin	800 water 10 acetic acid 190 acetonitrile	L1	UV/254	299
Ceftazidime	20 mℓ acetic acid 200 water 120 acetonitrile Dilute to 2 ℓ and titrate to pH 4.0	μBondapak C18 (L1)	UV/254	303
Ceftizoxime	1 acetonitrile 9 pH 3.6 buffer (1.42 g citric acid monohydrate and 1.73 g disodium hydrogen phosphate in 1 ℓ water)	L1	UV/254	299, p. 1980
	13 acetonitrile 87 0.02 *M* phosphate (pH 2.6)	μBondapak Phenyl (L11)	UV/254	304
Ceftriaxone	4 g tetrabutylammonium bromide 500 mℓ acetonitrile 440 water	L1	UV/270	299

	55 pH 7 phosphate buffer (13.6 g dipotassium hydrogen phosphate and 4 g dipotassium hydrogen phosphate diluted to 1 ℓ) 5 citrate buffer (25.8 g sodium citrate/1) 250 water			
Cefuroxime	10 0.1 *M* pH 3.4 acetate buffer (50 mℓ 0.1 *M* sodium acetate diluted to 1 ℓ with 0.1 *M* acetic acid) 1 acetonitrile	L1 (hexyl)	UV/254	289
Cephalexin in cephradine	48.6 g anhydrous sodium sulfate 1900 mℓ water 20 mℓ 1 *N* acetic acid, pH to 4.7, and diulte to 2 ℓ	L6	UV/254	289, p. 182
Cephradine	3 acetic acid (4%) 15 sodium acetate (3.86%) 200 methanol 782 water	L1	UV/254	289
Chlorambucil	500 methanol 1 acetic acid Dilute to 1 ℓ with water	L1	UV/254	289, p. 187
Chloramphenicol	500 mℓ methanol 500 buffer (0.68 g sodium acetate trihydrate to pH 4 in 500 mℓ water)	L1	UV/254	289, p. 191
	172 methanol 27 water 1 acetic acid	L1	UV/280	289, p. 191
Chlordiazepoxide	5 water 4 tetrahydrofuran 1 methanol	L1	UV/254	289, p. 193
Chlorothiazide	9 0.1 *M* disodium hydrogen phosphate 1 acetonitrile to pH3	L1	UV/254	289, p. 199

Table 32 (continued)
PHARMACEUTICAL APPLICATIONS OF HPLC

Compound	Solvent system	Column	Detector	Ref.
Chlorthalidone	3 0.01 *M* diammonium hydrogen phosphate 2 methanol	L7	UV/254	289, p. 210
Chlorzoxazone	Water methanol gradient	L1	UV/280	291, p. 1827
Cholecalciferol	3 *N* amyl alcohol 997 hexane	L3	UV/254	289, p. 213
Cimetidine	75 acetonitrile 15 methanol 15 buffer (0.01 *M* pH 3.5 phosphate)	L18	UV/229	299, p. 1981
	20 acetonitrile 80 water with 0.005 *M* octanesulfonic acid	μBondapak C18 (L1)	UV/229	305
Cinnarizine	9 methanol 1 pH 7 phosphate buffer	Ultrasphere RP-18 (L1)	UV/254	306
Cinoxacin	100 mℓ sodium borate (38.1 g sodium borate, add water to 1 ℓ) 0.426 g sodium sulfate Dilute to 1 ℓ with water	L12	UV/254	289, p. 219
Ciprofloxacin	5 acetonitrile 95 0.025 *M* phosphoric acid with tetrabutylammonium hydroxide	Spherisorb ODS II (L1)	Fluorometric	307
Citric Acid	0.2% sulfuric acid	L17	RI	292, p. 1712
Clemastine fumarate	83 methanol 17 pH 7 buffer (1.44 g disodium hydrogren phosphate and 0.62 g potassium dihydrogen phospate in 1 ℓ water)	L7	UV/220	289, p. 221

Clomiphene citrate	80 hexane 20 alcohol-free chloroform 0.1 triethylamine	L3	UV/302	289, p. 231
Clonidine	4 water 3 methonol 3 acetonitrile	L1	UV/254	289, p. 233
Clotrimazole	3 methanol 1 dipotassium hydrogen phosphate (4.35 g/ℓ)	L1	UV/254	291, p. 1829
Copovithane	30 acetonitrile 70 water	μBondapak C18 (L1)	UV/340	308
Cortisone Acetate	95 butyl chloride 95 butyl chloride water saturated 14.7 tetrahydrofuran 7 methanol 6 acetic acid	L3	UV/254	289, p. 251
Crotamiton	3 acetronitrile 2 water	L1	UV/254	289, p. 254
Cyclophosphamide	70 water 30 acetonitrile	L1	UV/195	289, p. 263
Cyclosporine A	45 acetonitrile 55 water	Zorbax CN (L10)	UV/210	309
Cyclothiazide	54 water 27 acetonitrile 27 methanol 0.2 acetic acid	L1	UV/280	289, p. 266
Dactinomycin	46 acetonitrile 25 0.04 *M* sodium acetate 25 0.07 *M* acetic acid	L1	UV/254	289, p. 271

Table 32 (continued)
PHARMACEUTICAL APPLICATIONS OF HPLC

Compound	Solvent system	Column	Detector	Ref.
Danazol	6 acetonitrile 4 water	L1	UV/254	289, p. 272
Dapsone	100 mℓ isopropyl alcohol 100 mℓ acetonitrile 100 mℓ ethyl acetate to 1 ℓ with pentane	L3	UV/254	289, p. 273
Daunorubicin	62 water 38 acetonitrile to a pH 2.2 with phosphoric acid	L1	UV/254	289, p. 275
Desoximetasone	65 methanol 35 water 1 acetic acid	L7	UV/254	289, p. 284
Dexamethasone	7 water 3 acetonitrile	L7	UV/254	289, p. 288
	52 methanol 45 0.05 *M* phosphate buffer (6.9 g sodium dihydrogen phosphate to 1 ℓ with water)	L1	UV/254	289, p. 292
Dextroamphetamine Sulfate	1.1 g sodium heptanesulfonate in 525 mℓ water 25 mℓ acetic acid (14 mℓ glacial acetic acid to 100 mℓ with water) 450 mℓ methanol Adjust to pH 3.3 with acetic acid	L1	UV/254	299, p. 1832
Dextromethorphan	70 acetonitrile 30 buffer (0.007 *M* docusate sodium 0.007 *M* ammonium nitrate)	L1	UV/280	291, p. 1832

Dextrose	800 acetonitrile 200 0.0067 *M* sodium dihydrogen phosphate to pH 6.5	L8	RI	289, p. 354
Diazepam	70 acetonitrile 30 water	Zorbax ODS (L1)	UV/242	289, p. 307; 310
Diazoxide	80 0.01 *M* sodium pentanesulfonate 20 methanol 1 acetic acid	L11	UV/254	289, p. 310
Dichlorphenamide	1 acetonitrile 1 buffer (1 part of 0.02 *M* sodium dihydrogen phosphate and 1 part disodium hydrogen phosphate)	L1	UV/280	289, p. 312
Didemnin B	57 acetonitrile 43 50 m*M* potassium dihydrogen phosphate 0.01% triethylamine	RSil C3 (L1)	UV/220	311
Diflorasone diacetate	350 butyl chloride water saturated 125 methylene chloride water saturated 15 acetic acid 10 tetrahydrofuran	L3	UV/254	299, p. 1986
Diflunisal	55 water 23 methanol 10 acetonitrile 2 acetic acid	L1	UV/254	299, p. 1987
Digitoxin	450 acetonitrile 550 water	L1	UV/281	289, p. 326
Dihydrotachysterol	100 isooctane 1 isopropyl alcohol	L3	UV/254	299, p. 1988
Dimenhydrinate	0.83 g diammonium hydrogen phosphate in 165 mℓ water 935 mℓ methanol	L1	UV/254	289, p. 335

Table 32 (continued)
PHARMACEUTICAL APPLICATIONS OF HPLC

Compound	Solvent system	Column	Detector	Ref.
Dinoprost	496 methylene chloride 3.5 1,3-butanediol 0.25 water	L3	UV/254	291, p. 1835
Diphenhydramine hydrochloride	60 acetonitrile 40 water 0.5 triethylamine titrate to pH 6.5	L10	UV/254	291, p. 1836
Dipipanone	70 acetonitrile 30 1% ammonium acetate 0.05 *M* diethylamine	Spherisorb ODS (L1)	UV/230	312
Dipryidamole	750 mℓ methanol 250 mℓ (250 mg disodium hydrogen phosphate in 250 mℓ water to pH 4.6 with phosphoric acid)	L1	UV/288	289, p. 346
Docusate Potassium	180 mℓ water 6 mℓ acetic acid 8 mℓ 25% tetrabutyl ammonium hydroxide in methanol QS to 1 ℓ with methanol	L1	RI	289, p. 351
Docusate	65 acetonitrile 35 phosphoric acid (1 in 2500)	L14	UV/214	289, p. 352
Dopamine Hydrochloride	260 mℓ acetonitrile 1740 mℓ (0.1% octanesulfonate and 1% acetic acid in water)	L1	UV/280	289, p. 354
Doxorubicin	69 water 31 acetonitrile to pH 2 with phosphoric acid	L1	UV/254	289, p. 358

Dyclonine	600 acetonitrile 400 (potassium dihydrogen phosphate and 0.45 mℓ heptylamine in 350 mℓ water, pH to 3 and dilute to 400 mℓ)	L13	UV/254	289, p. 364
Dyphylline	1.4 g potassium dihydrogen phosphate in 1350 mℓ water QS to 2 ℓ with methanol	L1	UV/254	289, p. 366
Emepronium bromide		LiChrosorb RP-8 (L1)	UV/258	313
Epinephrine	980 mℓ 0.12 *M* sodium dihydrogen phosphate 20 mℓ methanol 1 mℓ 0.1 *M* disodiumedetate 1.2 mℓ 0.5 *M* sodium heptanesulfonate pH to 3.4	L1	Electrochemical	289, p. 376
Equilin	35 acetonitrile 65 water	L1	UV/280	289, p. 380
Ergocalciferol	3 *N* amyl alcohol 997 hexane	L3	UV/254	289, p. 381
Ergoloid mesylate	80 water 20 acetonitrile 2.5 triethylamine	L1	UV/280	289, p. 383
Ergonovine maleate	20 acetonitrile 80 0.5 *M* phosphate buffer (6.8 g potassium dihydrogen phosphate, dissolve in water and titrate to 2.1; dilute to 1 ℓ)	L1	UV/312	289, p. 385
Erythromycin	50 acetonitrile 32 20 m*M* sodium perchlorate 8 20 m*M* ammonium acetate 10 methanol	Sepralyte diphenyl (L)	Electrochemical	314
Estradiol	1 acetonitrile 1 water	L1	UV/280	299, p. 1991

Table 32 (continued)
PHARMACEUTICAL APPLICATIONS OF HPLC

Compound	Solvent system	Column	Detector	Ref.
Estradiol cypionate	0.8 g ammonium nitrate in 300 mℓ water	L1	UV/280	289, p. 399
Estrone	700 mℓ acetonitrile 980 methylene chloride 20 methanol 1.5 water	L3	UV/280	289, p. 403
Estropipate	200 acetonitrile 200 methanol 600 buffer (13.6 g potassium dihydrogen phosphate/ℓ water)	L1	UV/213	291, p. 1839
Ethinyl Estradiol	1 water 1 acetonitrile	L1	UV/280	291, p. 1840
Ethynodiol Diacetate	95 cyclohexane 5 ethylacetate	L3	RI	289, p. 416
Flunisolide	3 water 2 acetonitrile	L7	UV/254	289, p. 433
Fluocinolone	77 water 13 acetonitrile 10 tetrahydrofuran	L1	UV/254	291, p. 1841
Fluocinolone Acetonide	60 methanol 40 water	L1	UV/254	289, p. 434
Fluorometholone	60 methanol 40 water	L1	UV/254	289, p. 439
Fluoxymesterone	475 butyl chloride 475 butyl chloride water saturated 40 tetrahydrofuran	L3	UV/254	289, p. 443

	35 methanol 30 acetic acid			
	58 water 42 acetonitrile	L1	UV/254	291, p. 1842
Flurandrenolide	60 methanol 40 water	L1	UV/240	289, p. 445
Fluspirilene	50 chloroform 50 methanol	μPorasil (L3)	UV/254	315
Folic acid	35.1 g sodium perchlorate 1.4 g potassium dihydrogen phosphate 7 mℓ 1 *M* potassium hydroxide 40 mℓ methanol Dilute to 1 ℓ with water and titrate to pH 7.2	L1	UV/254	289, p. 449
Furosemide	35 methanol 65 buffer (0.01 *M* sodium dihydrogen phosphate, pH 3.5)	Zorbax ODS (L1)	UV/235	316
Gemfibrozil	10 mℓ acetic acid 750 mℓ methanol QS to 1 ℓ with water	L1	UV/276	289, p. 457
Glutethimide	2 acetonitrile 3 pH 4 acetate buffer (0.82 g sodium acetate in 1 ℓ water)	L1	UV/254	299, p. 1994
Guaifenesin	60 water 40 methanol 1.5 acetic acid	L1	UV/276	299, p. 1994
Guanabenz	57 water 43 methanol 0.3 phosphoric acid	L1	UV/245	292, p. 1726

Table 32 (continued)
PHARMACEUTICAL APPLICATIONS OF HPLC

Compound	Solvent system	Column	Detector	Ref.
Halcinonide	1 acetonitrile 1 water	L1	UV/254	292, p. 1727
Hexylcaine hydrochloride	65 water 20 0.006 *M* octylammonium phosphate 15 acetonitrile	L10	UV/254	289, p. 490
Hydralazine hydrochloride	95 tetramethylammonium nitrate (1.36 g/950 mℓ water plus 2 mℓ acetic acid) 5 methanol	L1	UV/254	291, p. 1843
	0.015 *M* potassium dihydrogen phosphate 0.1% acetic acid 2% methanol 97.9% water	μBondapak C18 (L1)	UV/256	317
Hydrochlorothiazide	9 0.1 *M* sodium dihydrogen phosphate (pH 3)	L1	UV/254	299, p. 1998
Hydrocortisone	95 butyl chloride 95 butyl chloride water saturated 14 tetrahydrofuran 7 methanol 6 acetic acid	L3	UV/254	289, p. 499
	75 water 25 acetonitrile	L1	UV/254	289, p. 499
Hydroquinone	55 methanol 45 water	L1	UV/280	289, p. 512
Hydroxyzine hydrochloride	45 0.01 *M* sodium phosphate 55 acetonitrile	L1	UV/230	299, p. 2000

Hydroxyzine pamoate	940 acetonitrile 25 water 25 acetic acid 25 trolamine	L3	UV/254	289, p. 521
Indenolol	2 methanol 1 acetate buffer pH 7	μBondapak CN (L10)	UV/250	318
Indomethacin	1 0.01 *M* sodium dihydrogen phosphate 1 0.01 *M* disodium hydrogen phosphate 1 water 1 acetonitrile	L1	UV/254	289, p. 532
Insulin	74 0.1 *M* sodium dihydrogen phosphate 26 acetonitrile	L13	UV/214	299, p. 2004
Isoetharine mesylate	8 mℓ acetic acid 100 mℓ sodium sulfate QS to 1 ℓ with water	L9	UV/254	289, p. 562
Isoniazid	4.4 g docusate sodium in 600 mℓ methanol 400 water Titrate to pH2	L1	UV/254	291, p. 1852
Isoprenaline (isoproterenol)	5 methanol 95 0.01 *M* perchloric acid	Ultrasphere ODS (L1)	UV/280	319
Isoproterenol hydrochloride	1.76 g sodium heptanesulfonate in 800 mℓ water 200 mℓ methanol Titrate to pH 3.0 with phosphoric acid	L1	UV/280	299, p. 2009
Ketoconazole	7 (1 in 500 diisopropylamine in methanol) 3 ammonium acetate (1 g/200 mℓ)	L1	UV/225	292, p. 1737
Leucovorin calcium	15 tetrabutylammonium hydroxide (25% solution) 835 water 125 acetonitrile Titrate to pH 7.5	L1	UV/254	292, p. 1738

Table 32 (continued)
PHARMACEUTICAL APPLICATIONS OF HPLC

Compound	Solvent system	Column	Detector	Ref.
Leukotrienes	67 methanol 33 water 0.02 acetic acid	μBondapak C18 (L1)	UV/280	320
Levothyroxine sodium	65 water 35 acetonitrile 1 phosphoric acid	L10	UV/225	291, p. 1856
Levonorgestrel	350 acetonitrile 150 methanol 450 water	L7	UV/215	292, p. 1739
Lidocaine	50 mℓ pH 7 phosphate buffer 400 mℓ water 550 acetonitrile	L1	UV/261	289, p. 595
Liothyronine sodium	700 water 300 acetonitrile 5 phosphoric acid	L10	UV/225	289, p. 599
Meclizine hydrochloride	45 methanol 55 buffer (0.69 g sodium dihydrogen phosphate in 100 mℓ water, titrated to a pH of 4.0)	L9	UV/230	289, p. 625
Meclocycline	85 0.001 *M* ammonium edetate 15 tetrahydrofuran	L1	UV/340	289, p. 626
Medroxyprogesterone acetate	700 butyl chloride 300 hexane 80 acetonitrile	L3	UV/254	289, p. 627

Mefloquine	67 50 m*M* sodium sulfate titrated to pH 2.84 33 acetonitrile	Nucleosil C18 (L1)	UV/220	321
Megestrol acetate	55 acetonitrile 45 water	L1	UV/280	289, p. 629
Melphalan	50 methanol 49 water 1 acetic acid	L1	UV/254	289, p. 631
Metaraminol	7 methanol 3 0.0032 *M* hexanesulfonate, pH3	L7	UV/264	299, p. 2012
Metaproterenol	10 formic acid 990 water	L1	UV/278	291, p. 1862
Methadone	Water methanol gradient	L1	UV/280	289, p. 649
	0.1% sodium octanesulfonate 60 acetonitrile 40 potassium dihydrogen phosphate titrate to pH 3.5	Spherisorb CN (L10)	UV/259	322
Methocarbamol	75 pH 4.3 buffer 25 methanol	L1	UV/274	289, p. 662
Methoxsalen	35 acetonitrile 100 water	L1	UV/254	299, p. 2013
Methyldopa	6.8 g potassium dihydrogen phosphate in 750 mℓ water, pH to 3.5 QS to 1 ℓ	L1	UV/280	289, p. 673
Methylprednisolone	475 butyl chloride 475 butyl chloride water saturated 70 tetrahydrofuran 35 methanol 30 acetic acid	L3	UV/254	289, p. 681

Table 32 (continued)
PHARMACEUTICAL APPLICATIONS OF HPLC

Compound	Solvent system	Column	Detector	Ref.
Methyprylon	60 methanol 40 water	L1	UV/280	289, p. 687
Metoclopramide	65 acetonitrile 35 buffer (0.03 *M* sodium acetate, pH 7.4)	Techsphere CN (L10)	UV/280	323
Metoprolol	961 mg pentanesulfonic acid 82 mg anhydrous sodium acetate 550 mℓ methanol 470 mℓ water 0.57 mℓ acetic acid	L1	UV/254	289, p. 690
Metronidazole	930 pH 4 phosphate buffer 70 methanol	L1	UV/320	289, p. 691
Minoxidil	700 methanol 300 water 10 acetic acid 3 g/1 docusate sodium	L1	UV/254	291, p. 1867
Moxalactam disodium	19 0.10 *M* ammonium acetate 1 methanol	L1	UV/254	289, p. 702
Nadolol	700 methanol 1300 water 5.84 g sodium chloride 1 0.1 *M* hydrochloric acid	L16	UV/220	291, p. 1869
Nafimidone	2 m*M* tetrabutylammonium bromide 25 acetonitrile 75 0.04 *M*pH 7 phosphate buffer	Hypersil ODS (L1)	UV/254	324

Nalorphine	800 potassium dihydrogen phosphate (1 g in 500 mℓ water) 200 acetonitrile	L7	Fluorescence	299, p. 2014
Naloxone hydrochloride	70 acetonitrile 30 0.01 *M* potassium dihydrogen phosphate	Micropak MCH (C18) (L1)	UV/254	325
Naproxen	50 methanol 50 pH 6.5 citrate buffer	Lichrosorb RP8 (L1)	UV/235	326
Nifedipine	50 water 25 acetonitrile 25 methanol	L1	UV/235	299, p. 2018
Nitroglycerin	1 water 1 methanol	L1	UV/220	299, p. 2019
Minocycline	550 0.2 *M* ammonium oxalate 250 dimethylforamide 200 0.1 *M* disodium edetate Titrate to pH 6.2 with tetrabutylammonium hydroxide	L7	UV/280	291, p. 1865
Norepinephrine bitartrate	1.1 g sodium heptanesulfonate in 800 mℓ water 200 methanol Titrate to pH of 3	L1	UV/280	299, p. 2020
Nylidrin hydrochloride	1 0.1 *M* diammonium hydrogen phosphate Titrate to pH of 7 4 methanol	L1	UV/276	289, p. 751
Pentamidine	1800 mℓ acetonitrile 2200 mℓ water 8 mℓ 10% tetramethylammonium chloride 4 mℓ phosphoric acid	Ultrasphere octyl (L1)	Fluorescene	328
Phenacemide	50 acetonitrile 50 0.05 *M* acetate buffer (pH 4.2)	Ultrasphere octyl (L1)	UV/254	329

Table 32 (continued)
PHARMACEUTICAL APPLICATIONS OF HPLC

Compound	Solvent system	Column	Detector	Ref.
Phendimetrazine tartrate	65 acetonitrile 35 pH 7.5 phosphate buffer	L15	UV/210	289, p. 814
Phenobarbital	2 methanol 3 pH 4.5 buffer	L1	UV/254	289, p. 818
Phenolphthalein	50 water 50 methanol 1 acetic acid	L1	UV/280	299, p. 2025
Phenylephrine hydrochloride	4 methanol 1 water	L1	UV/280	289, p. 828
Phenytoin	55 methanol 45 water	L1	UV/254	292, p. 1755
Physostigmine	1 acetonitrile 1 0.05 *M* ammonium acetate	L1	UV/254	291, p. 1878
Phytonadione	2000 hexane 1.5 amyl alcohol	L3	UV/254	291, p. 1879
	95 dehydrated alcohol 5 water	L1	UV/254	289, p. 836
Piperacillin	450 methanol 100 0.2 *M* sodium dihydrogen phosphate 3 tetrabutyl ammonium hydroxide (1 in 10) QS to 1 ℓ with water and titrate to pH 5.5	L1	UV/254	289, p. 841

	400 mℓ methanol 500 mℓ water 94 mℓ disodium hydrogen phosphate Titrate to pH 6.5	μBondapak C18 (L1)	UV/254	330
Pirenzepine	70 acetonitrile 40 methanol 15 5% acetic acid	μBondapak C18 (L1)	UV/285	331
Piroxicam	45 methanol 55 buffer (7.72 g citric acid, 5.35 g sodium dihydrogen phosphate in 1 ℓ)	L1	UV/254	291, p. 187
Plicamycin	650 0.01 *M* phosphoric acid 350 acetonitrile	L1	UV/278	299, p. 2026
Pramoxine hydrochloride	27 acetonitrile 17 water 1 pH 7.5 buffer (3.5 g dipotassium hydrogen phosphate/100 mℓ water)	L1	UV/224	292, p. 1756
Prazepam	85 methanol 15 water 0.5 acetic acid	L1	UV/254	289, p. 867
Prazosin	700 methanol 300 water 10 acetic acid	L3	UV/254	289, p. 868
Prednisolone	95 butyl chloride 95 butyl chloride water saturated 14 tetrahydrofuran 7 methanol 6 acetic acid	L3	UV/254	289, p. 870
Prednisone	1.36 g potassium dihydrogen phosphate in 600 mℓ water 400 mℓ methanol	L1	UV/254	299, p. 2028

Table 32 (continued)
PHARMACEUTICAL APPLICATIONS OF HPLC

Compound	Solvent system	Column	Detector	Ref.
Prilocaine	2.16 sodium octanyl sulfate 37 mg disodium edetate 950 water 20 acetic acid 20 acetonitrile 10 methanol	L10	UV/254	289, p. 876
Probenecid	50 1% acetic acid with 0.05 *M* sodium dihydrogen phosphate (pH 3) 50 1% acetic acid in acetonitrile	L11	UV/254	299, p. 2029
Prochlorperazine	2.07 g sodium dihydrogen phosphate in 500 mℓ water 5 1 *M* tetrabutylammonium hydroxide 1500 methanol Titrate to pH 4.8	L1	UV/254	292, p. 1757
	45 buffer (4.3 g sodium octane sulfate, 4 mℓ acetic acid in 1 ℓ water) 40 acetonitrile 15 methanol	L1	UV/254	292, p. 1758
Progesterone	28 isopropyl alcohol 100 water	L2	UV/254	289, p. 889
Promazine	25 5% aqueous ammonium acetate 75 methanol	Spherisorb ODS (L1)	UV/240	332
Prostaglandin E2	70 cloroform 30 Hexane	Ultrasphere CN (L10)	UV/254	333
Pseudoephedrine	17 alcohol 3 ammonium acetate (1 g in 250 mℓ water)	L3	UV/254	289, p. 914

Pyridoxine	1.2 g sodium heptanesulfonate 470 mℓ methanol 20 mℓ acetic acid 2 ℓ water Titrate to pH of 3	L1	UV/280	299, p. 2030
Quinestrol	4 acetonitrile 1 water	L1	UV/281	289, p. 925
Quinidine gluconate	860 water 100 acetonitrile 20 10% diethylamine solution 20 methanesulfonic acid solution (35 mℓ methanesulfonic acid, 20 mℓ acetic acid diluted to 500 mℓ	L1	UV/235	289, p. 927
Quinine	860 water 100 acetonitrile 20 10% diethylamine solution 20 methanesulfonic acid solution (35 mℓ methanesulfonic acid, 20 mℓ acetic acid diluted to 500 mℓ with water)	L1	UV/235	289, p. 929
Resorcinol	55 water 7 acetonitrile 6 methanol	L1	UV/280	289, p. 941
Ritodrine hydrochloride	63 water 25 methanol 10 ammonium acetate (1.8 g ammonium acetate, 120 acetic acid to 1 ℓ with water) 2 0.25 *M* sodium heptanesulfonate	L7	UV/254	289, p. 947
Salicylic acid	225 mg tetramethylammonium hydroxide pentahydrate 770 mℓ water 150 methanol 150 acetonitrile 1 acetic acid	L1	UV/280	289, p. 952

Table 32 (continued)
PHARMACEUTICAL APPLICATIONS OF HPLC

Compound	Solvent system	Column	Detector	Ref.
Sodium nitroprusside	30 acetonitrile 70 buffer (1.36 g potassium dihydrogen phosphate, 5.2 mℓ 25% tetrabutylammonium hydroxide in methanol, dilute to 1 ℓ and titrate to pH 7.1)	L11	UV/210	291, p. 1888
Spironolactone	55 acetonitrile 45 0.02 *M* diammonium hydrogen phosphate	L1	UV/254	289, p. 981
Succinylcholine	1 1*N* tetrabutylammonium chloride in methanol 9 water	L3	UV/214	289, p. 985
Sulfadiazine	87 water 12 acetonitrile 1 acetic acid	L1	UV/254	289, p. 989
Sulfadoxine	4 1% aqueous acetic acid 1 acetonitrile	L1	UV/254	289, p. 991
Sulfathiazole	78 water 22 acetonitrile 10 1 *M* tetrabutylammonium hydroxide (pH 7.7)	L1	UV/280	289, p. 998
Sulfinpyrazone	65 acetonitrile 35 tetrahydrofuran	L1	UV/235	292, p. 1762
Tamoxifen citrate	320 mℓ water 2 mℓ acetic acid 1.08 g sodium octanesulfonate 678 mℓ methanol	L11	UV/254	289, p. 1011
Tazadolene Succinate	4 mℓ triethylamine 700 mℓ water	Zorbax C8 (L1)	UV/254	334

	Adjust to pH 2.5 250 mℓ acetonitrile 50 mℓ tetrahydrofuran			
Terbutaline Sulfate	8 methanol 92 0.02 *M* potassium dihydrogen phosphate (pH 3.6)	μBondapak Phenyl	UV/278	335
Terfenadine	1 acetonitrile 1 0.05 *M* sodium acetate (pH 5.0)	μBondapak C18 (L1)	UV/225	336
Tetracycline	5.2 g diammonium hydrogen phosphate 760 mℓ water 5 mℓ ethanolamine 240 mℓ acetonitrile 60 mℓ dimethylformamide	μBondapak C18 (L1)	UV/254	337
Theophylline	64 water 35 methanol 1 acetic acid	L1	UV/254	289, p. 1042
Thioridazine	850 acetonitrile 150 water 1 triethylamine	L1	UV/265	299, p. 2036
Thiothixene	1400 ethanolamine solution (0.5 mℓ ethanolamine in 3780 mℓ methanol) 200 water	L4	UV/254	289, p. 1057
Ticarcillin	9 acetonitrile 91 0.2 *M* sodium acetate (pH 2.7)	HC ODS (L1)	UV/220	338
Timolol Maleate	2 methanol 3 pH 2.8 phosphate buffer (22.08 g sodium dihydrogen phosphate in 2 ℓ water titrated to pH 2.8)	L1	UV/295	289, p. 1064
Tioconazole	440 acetonitrile 400 methanol 280 water 2 ammonium hydroxide	L1	UV/219	219, p. 1895

Table 32 (continued)
PHARMACEUTICAL APPLICATIONS OF HPLC

Compound	Solvent system	Column	Detector	Ref.
Tolazamide	475 hexane 475 hexane water saturated 20 tetrahydrofuran 15 alcohol acetic acid	L3	UV/254	289, p. 1067
Tolnaftate	1 water 3 acetonitrile	L1	UV/254	289, p. 1072
Triamcinolone	60 methanol 40 water	L1	UV/254	289, p. 1076
Triamcinolone Acetonide	70 water 20 acetonitrile	L1	UV/254	289, p. 1077
Triamcinolone hexacetonide	5 isopropanol 95 methylene chloride	L4	UV/254	289, p. 1081
Triazolam	850 acetonitrile 80 chloroform 50 butyl alcohol 20 water 0.5 acetic acid	L3	UV/254	299, p. 2037
Trimethoprim	84 1% aqueous acetic acid 16 acetonitrile	L1	UV/254	289, p. 1096
Trimetrexate	60 water with 0.02% phosphoric acid and 0.08% triethylamine 40 acetonitrile	μBondapak C18 (L1)	UV/241	339
Triprolidine hydrochloride	17 alcohol 3 ammonium acetate solution (1 g in 250 mℓ water)	L3	UV/254	289, p. 1098

Verapamil	55 methanol 45 0.1 *M* ammonium acetate	MOS Hypersil (L1)	Fluorescence	340
Vidarabine	2.2 g docusate sodium in 10 mℓ acetic acid and 500 mℓ methanol diluted to 1 ℓ with water	L1	UV/254	289, p. 1115
Vincristine sulfate	5 mℓ diethylamine 295 water Titrate to pH 7.5 Dilute to 1 ℓ with methanol	L7	UV/297	289, p. 1116
Vitamin A	1 methyl t-butyl ether 99 hexane	LiChrosorb SI-60 (L3)	UV/313	341
Warfarin	64 methanol 36 water 1 acetic acid	L1	UV/280	291, p. 1899

REFERENCES

1. **Johnson, E. L. and Stevenson, R.,** *Basic Liquid Chromatography,* 2nd ed., Varian Aerograpah, Walnut Creek, CA, 1978.
2. **Brown, P. R.,** *High Pressure Liquid Chromatography: Biochemical and Biomedical Applications,* Academic Press, New York, 1973.
3. **Snyder, L. R. and Kirkland, J. J.,** *Introduction to Modern Liquid Chromatography,* 2nd ed., Wiley-Interscience, New York, 1979.
4. **Deyl, Z., Macek, K., and Janak, J.,** *Liquid Column Chromatography,* Elsevier, New York, 1975.
5. **Parris, N. A.,** *Instrumental Liquid Chromatography,* Elsevier, New York, 1976.
6. **Heftmann, E.,** *Chromatography: A Laboratory Handbook of Chromatographic and Electrophoretic Methods,* Van Nostrand Reinhold, New York, 1975.
7. **Krstulovic, A. M. and Brown, P. R.,** *Reversed-Phase High-Performance Liquid Chromatography,* Wiley-Interscience, New York, 1982.
8. **Hearn, M. T. W., Ed.,** *Ion-Pair Chromatography,* Marcel Dekker, New York, 1985.
9. **Perry, S. G., Amos, R., and Brewer, P. I.,** *Practical Liquid Chromatography,* Plenum Press, New York, 1972.
10. **Little, J.,** ASTM Committee E-19, 25th Annu. Conf. on the Practice of Chromatography, Chicago, October 1986.
11. **Munk, M. N.,** Refractive index detectors, in *Liquid Chromatographic Detection,* Vickrey, T. M., Ed., Marcel Dekker, New York, 1983.
12. **Reeve, D. R. and Crozier, A.,** *J. Chromatogr.,* 137, 271, 1977.
13. **Arpino, P. J.,** Mass spectrometric detectors, in *Liquid Chromatography Detectors,* Vickrey, T. M., Ed., Marcel Dekker, New York, 1983, 243.
14. **Niessen, W. M. A.,** *Chromatographia,* 21, 342, 1986.
15. **Pretorius, V. and Van Rensburg, J. F. J.,** *J. Chromatogr. Sci.,* 11, 355, 1973.
16. **Slais, K. and Krejci, M.,** *J. Chromatogr.,* 91, 181, 1974.
17. **Armstrong, D. W.,** *Sep. Purif. Methods,* 14, 213, 1985.
18. **Armstrong, D. W. and Henry, S. J.,** *J. Liq. Chromatogr.,* 3, 657, 1980.
19. **Arunyanart, M. and Cline Love, L. J.,** *Anal. Chem.,* 56, 1557, 1984.
20. **Cline Love, L. J., Habarta, J. G., and Dorsey, J. G.,** *Anal. Chem.,* 56, 1133A, 1984.
21. **Arunyanart, M. and Cline Love, L. J.,** *Anal. Chem.,* 57, 2837, 1985.
22. **Dorsey, J. G., Khaledi, M. G., Landy, J. S., and Lin, J. L.,** *J. Chromatogr.,* 316, 183, 1984.
23. **Landy, J. S. and Dorsey, J. G.,** *J. Chromatogr. Sci.,* 22, 68, 1984.
24. **Barford, R. A. and Sliwinski, B. J.,** *Anal. Chem.,* 56, 1554, 1984.
25. **Berthod, A., Girard, I., and Gonnet, C.,** *Anal. Chem.,* 58, 1362, 1986.
26. **Berthod, A., Girard, I., and Gonnet, C.,** *Anal. Chem.,* 58, 1359, 1986.
27. **Goldberg, A.,** *Comparison of Reversed Phase Packings, Liquid Chromatography Technical Report,* DuPont Instruments, Wilmington, DE, 1980.
28. **Majors, R. E.,** *Anal. Chem.,* 44, 1722, 1972.
29. **Majors, R. E.,** *Am. Lab.,* 7, 13, 1975.
30. Slurry Packing Vydac TP Separations Materials, The Separations Group, Hesperia, CA, 1979.
31. **Little, C. J., Dale, A. D., Ord, D. A., and Marten, T. R.,** *Anal. Chem.,* 49, 1311, 1977.
32. **Bahalyar, S., Yuen, J., and Henry, D.,** *How to Pack Liquid Chromatography Columns,* Spectra Physics Technical Bulletin 114-76, San Jose, CA.
33. **Snyder, L. R. and Kirkland, J. J.,** *Introduction to Modern Liquid Chromatography,* 2nd ed., Wiley-Interscience, New York, 1979.
34. **Kirkland, J. J.,** *J. Chromatogr. Sci.,* 9, 206, 1971.
35. **Majors, R. E.,** *Anal. Chem.,* 44, 1722, 1972.
36. **Kirkland, J. J.,** *Chromatographia,* 8, 661, 1975.
37. **Strubert, W.,** *Chromatographia,* 6, 50, 1974.
38. **Cassidy, R. M., LeGay, D. S., and Frei, R. W.,** *Anal. Chem.,* 46, 340, 1974.
39. **Webber, T. J. N. and McKerrell, E. H.,** *J. Chromatogr.,* 122, 243, 1976.
40. **Coq, B., Gonnett, C., and Rocca, J. L.,** *J. Chromatogr.,* 108, 249, 1975.
41. **Kirkland, J. J. and Antle, P. E.,** *J. Chromatogr. Sci.,* 15, 137, 1977.
42. **Bristow, P. A., Brittain, P. N., Riley, C. M., and Williamson, B. E.,** *J. Chromatogr.,* 131, 57, 1977.
43. **Cox, G. B., Loscombe, C. R., Slucutt, M. J., Sugden, K., and Upfield, J. A.,** *J. Chromatogr.,* 117, 269, 1976.
44. **Linder, H. R., Keller, H. P., and Frei, R. W.,** *J. Chromatogr. Sci.,* 14, 234, 1976.
45. **Little, C. J., Dale, A. D., Ord, D. A., and Marten, T. R.,** *Anal. Chem.,* 49, 1311, 1977.
46. **Chang, S. H., Gooding, K. M., and Regnier, F. E.,** *J. Chromatogr.,* 125, 103, 1976.

47. **Knox, J. H.,** *Intensive Course on High-Performance Liquid Chromatography,* Wolfson Liquid Chromatography Unit, Department of Chemistry, University of Edinburgh, Edinburgh, 1977, 163.
48. **Liao, J. C. and Ponzo, J. L.,** *J. Chromatogr. Sci.,* 20, 14, 1982.
49. **Kirkland, J. J.,** *J. Chromatogr. Sci.,* 10, 593, 1972.
50. **Endele, R., Halasz, I., and Unger, K.,** *J. Chromatogr.,* 99, 377, 1974.
51. **Anderson, J. M.,** *J. Chromatogr. Sci.,* 22, 332, 1984.
52. **Vendrell, J. and Aviles, F. X.,** *J. Chromatogr.,* 356, 420, 1986.
53. **Rabel, F. M.,** *J. Chromatogr. Sci.,* 18, 394, 1980.
54. **Bredeweg, R. A., Rothman, L. D., and Pfeiffer, C. D.,** *Anal. Chem.,* 51, 2061, 1979.
55. **Snyder, L.,** *J. Chromatogr.,* 92, 223, 1974.
56. Waters Associates Liquid Chromatography School Notes, Waters Associates Inc., Milford, MA.
57. **Gabler, R., Hegde, R., and Hughes, D.,** *J. Liq. Chromatogr.,* 6, 2565, 1983.
58. **Reust, J. B. and Meyer, V. R.,** *Analyst,* 107, 673, 1982.
59. **Bristol, D. W.,** *J. Chromatogr.,* 188, 193, 1980.
60. **Aiken, R. L., Fritz, G. T., Marmion, R. M., Michel, K. H., and Wolf, T.,** *J. Chromatogr. Sci.,* 119, 338, 1981.
61. **McCoy, R. W., Aiken, R. L., Pauls, R. E., Ziegel, E. R., Wolf, T., Fritz, G. T., and Marmion, D. M.,** *J. Chromatogr. Sci.,* 22, 425, 1984.
62. **McCoy, R. W., Aiken, R. L., Pauls, R. E., Ziegel, E. R., Wolf, T., Fritz, G. T., and Marmion, D. M.,** *J. Chromatogr. Sci.,* 24, 273, 1986.
63. **Lingeman, H., Underberg, W. J. M., Takadate, A., and Hulshoff, A.,** *J. Liq. Chromatogr.,* 8, 789, 1985.
64. **Takahashi, H., Yoshida, T., and Meguro, H.,** *Bunseki Kagaku,* 30, 341, 1980.
65. **Anzai, N., Kimura, T., Chida, S., Tanaka, T., Takahashi, H., and Meguro, H.,** *Yakugaku Zasshi,* 101, 1002, 1981.
66. **Goto, J., Goto, N., Hikichi, A., Nishimaki, T., and Nambara, T.,** *Anal. Chim. Acta,* 120, 187, 1980.
67. **Goto, J., Goto, N., and Nambara, T.,** *J. Chromatogr.,* 239, 559, 1982.
68. **Sato, E., Miyakawa, M., and Kanaoka, Y.,** *Chem. Pharm. Bull. Jpn.,* 32, 336, 1984.
69. **Ikeda, M., Shimada, K., and Sakaguchi, T.,** *J. Chromatogr.,* 305, 261, 1984.
70. **Kato, T. and Kinoshita, T.,** *Bunseki Kagaku,* 31, 615, 1984.
71. **Kanoaka, Y., Machida, M., Machida, M. I., and Sekine, T.,** *Biochim. Biophys. Acta,* 317, 562, 1975.
72. **Goto, J., Goto, N., Shamsa, F., Saito, M., Komatsu, S., Suzaki, K., and Nambara, T.,** *Anal. Chim. Acta,* 147, 397, 1983.
73. **Watkins, W. D. and Peterson, M. B.,** *Anal. Biochem.,* 125, 30, 1982.
74. **Barker, S. A., Monti, J. A., Christian, S. T., Benington, F., and Morin, R. D.,** *Anal. Biochem.,* 107, 116, 1980.
75. **Nimura, N. and Kinoshita, T.,** *Anal. Lett.,* 13, 191, 1980.
76. **Kanoaka, Y., Machida, M., Ando, K., and Sekine, T.,** *Biochim. Biophys. Acta,* 207, 269, 1970.
77. **Ohkura, Y. and Kai, M.,** *Anal. Chim. Acta,* 106, 89, 1979.
78. **Goto, J., Saito, M., Chikai, T., Goto, N., and Nambara, T.,** *J. Chromatogr.,* 276, 289, 1983.
79. **Miners, J. O., Fearnly, I., Smith, K. J., Birkett, D. J., Brooks, P. M., and Whithouse, M. W.,** *J. Chromatogr.,* 275, 89, 1983.
80. **L'Italien, J. J. and Kent, S. B. H.,** *J. Chromatogr.,* 283, 149, 1984.
81. **Distler, W.,** *J. Chromatogr.,* 192, 240, 1980.
82. **Kamada, S., Maeda, M., and Tsuji, A.,** *J. Chromatogr.,* 272, 29, 1983.
83. **Tsuchiya, H., Hayashi, T., Naruse, H., and Takagi, N.,** *J. Chromatogr.,* 234, 247, 1982.
84. **Farinotti, R., Siard, Ph., Bourson, J., Kirkiacharian, S., Valeur, B., and Mahuzier, G.,** *J. Chromatogr.,* 269, 81, 1983.
85. **Crozier, A., Zaerr, J. B., and Morris, R. O.,** *J. Chromatogr.,* 238, 157, 1982.
86. **Iwamoto, M., Yoshida, S., Chow, T., and Hirose, S.,** *Yakugaku Zasshi,* 103, 967, 1983.
87. **Murayama, K. and Kinoshita, T.,** *Anal. Lett.,* 14, 1221, 1981.
88. **Murayama, K. and Kinoshita, T.,** *Anal. Lett.,* 15, 123, 1982.
89. **Korte, W. D.,** *J. Chromatogr.,* 243, 153, 1982.
90. **Ahnoff, M., Grundevik, I., Arfwidsson, A., Fonselius, J., and Persson, B. A.,** *Anal. Chem.,* 53, 485, 1981.
91. **Johnson, L., Lagerkvist, S., Lindroth, P., Ahnoff, M., and Martinsson, K.,** *Anal. Chem.,* 54, 939, 1982.
92. **Honda, S., Suzuki, S., Takahashi, M., Kakehi, K., and Ganno, S.,** *Anal. Biochem.,* 134, 34, 1983.
93. **Honda, S., Matsuda, Y., Takahashi, M., Kakehi, K., and Ganno, S.,** *Anal. Chem.,* 52, 1079, 1980.
94. **Goya, S., Takadate, A., and Iwai, M.,** *Yakugaku Zasshi,* 101, 1164, 1981.
95. **Takadate, A., Iwai, M., Fujino, H., Tahara, K., and Goya, S.,** *Yakugaku Zasshi,* 103, 962, 1983.
96. **Lloyd, J. B. F.,** *J. Chromatogr.,* 189, 359, 1980.

97. **Kanaoka, Y., Yonemitsu, O., Tanizawa, K., and Ban, Y.,** *Chem. Pharm. Bull. Jpn.*, 12, 799, 1964.
98. **Takadate, A., Tahara, T., Fujino, H., and Goya, S.,** *Chem. Pharm. Bull. Jpn.*, 30, 4120, 1982.
99. **Kamimura, H., Sasaki, H., and Kawamura, S.,** *J. Chromatogr.*, 225, 115, 1981.
100. **Goya, S., Takadate, A., and Fujino, H.,** *Yakugaku Zasshi*, 102, 63, 1982.
101. **Goya, S., Takadate, A., Fujino, H., and Tanaka, T.,** *Yakugaku Zasshi*, 100, 744, 1980.
102. **Ishida, J., Yamaguchi, M., Kai, M., Ohkura, Y., and Nakamura, M.,** *J. Chromatogr.*, 305, 381, 1984.
103. **Ishida, J., Kai, M., and Ohkura, Y.,** *Proc. 104th Annu. Meet. of Pharm. Soc. Jpn.*, Sendai, March 1984, 564.
104. **Machida, M., Ushijima, N., Machida, M. I., and Kanoaka, Y.,** *Chem. Pharm. Bull. Jpn.*, 23, 1385, 1975.
105. **Lankmayr, E. P., Budna, K. W., Mueller, K., and Nachtmann, F.,** *Fresenius Z. Anal. Chem.*, 295, 371, 1979.
106. **Lankmayr, E. P., Budna, K. W., Mueller, K., Nachtmann, F., and Rainer, F.,** *J. Chromatogr.*, 222, 249, 1981.
107. **Lingeman, H., Underberg, W. J. M., and Hulshoff, A.,** submitted for publication.
108. **Werkhoven-Goewie, C. E., Brinkman, U. A.Th., and Frei, R. W.,** *Anal. Chim. Acta*, 114, 147, 1980.
109. **Sommadossi, J. P., Lemar, M., Necciari, J., Sumirtapura, Y., Cano, J. P., and Gaillot, J.,** *J. Chromatogr.*, 228, 205, 1982.
110. **Chayen, R., Dvir, R., Gould, S., and Harell, A.,** *Anal. Biochem.*, 42, 283, 1971.
111. **Alpenfels, W. F., Mathews, R. A., Madden, D. E., and Newson, A. E.,** *J. Liq. Chromatogr.*, 5, 1711, 1982.
112. **Goto, J., Komatsu, S., Goto, N., and Nambara, T.,** *Chem. Pharm. Bull. Jpn.*, 29, 899, 1981.
113. **Ichikawa, H., Tanimura, T., Nakajima, T., and Tamura, Z.,** *Chem. Pharm. Bull. Jpn.*, 18, 1493, 1970.
114. **Swarin, S. J. and Lipari, F.,** *J. Liq. Chromatogr.*, 6, 425, 1983.
115. **Noda, H., Mitsui, A., and Ohkura, Y.,** *Proc. of 104th Annu. Meet. of Pharm. Soc. Jpn.*, Sendai, March 1984, 558.
116. **Anson, M. H. and Boning, A. J., Jr.,** *Anal. Lett.*, 12, 25, 1979.
117. **Maeda, H., Ishida, N., Kawauchi, H., and Tuzimura, K.,** *J. Biochem.*, 65, 777, 1969.
118. **Imai, K., Toyo'oka, T., and Watanabe, Y.,** *Anal. Biochem.*, 128, 471, 1983.
119. **Toyo'oka, T. and Imai, K.,** *J. Chromatogr.*, 282, 495, 1983.
120. **Imai, K. and Watanabe, Y.,** *Anal. Chim. Acta*, 130, 377, 1981.
121. **Watanabe, Y. and Imai, K.,** *Anal. Chem.*, 55, 1786, 1983.
122. **Toyo'oka, T., Miyano, H., and Imai, K.,** *Proc. of 104th Annu. Meet. of Pharm. Soc. Jpn.*, Sendai, March 1984, 558.
123. **Seki, T. and Yamaguchi, Y.,** *Anal. Lett.*, 15, 1111, 1982.
124. **Guebitz, G., Wintersteiger, R., and Frei, R. W.,** *J. Liq. Chromatogr.*, 7, 839, 1984.
125. **Mitzutani, S., Wakuri, Y., Yoshida, N., Nakajima, T., and Tamura, Z.,** *Chem. Pharm. Bull. Jpn.*, 17, 2340, 1969.
126. **Hirata, T., Kai, M., Kohashi, K., and Ohkura, Y.,** *J. Chromatogr.*, 226, 25, 1981.
127. **Lingeman, H., Hulshoff, A., Underberg, W. J. M., and Offerman, F. B. J. M.,** *J. Chromatogr.*, 290, 215, 1984.
128. **Duenges, W. and Seiler, N.,** *J. Chromatogr.*, 145, 483, 1978.
129. **Goya, S., Takadate, A., Fujino, H., and Irikura, M.,** *Yakugaku Zasshi*, 101, 1064, 1981.
130. **Sinsheimer, J. E., Hong, D. D., Stewart, J. T., Fink, M. L., and Burckhalter, J. H.,** *J. Pharm. Sci.*, 60, 141, 1971.
131. **Goya, S., Takadate, A., Irikura, M., Suehiro, T., and Fujino, H.,** Proc. 104th Annu. Meet. Pharm. Soc. Jpn., Sendai, March 1984, 582.
132. **Nakamura, H., Takagi, K., Tamura, Z., Yoda, R., and Yamamoto, Y.,** *Anal. Chem.*, 56, 919, 1984.
133. **Nakamura, H. and Tamura, Z.,** *Anal. Chem.*, 52, 2087, 1980.
134. **Tocksteinova, D., Churacek, J., Slosar, J., and Skalik, L.,** *Mikrochim. Acta*, 1, 507, 1978.
135. **Wintersteiger, R., Gamse, G., and Packa, W.,** *Fresenius Z. Anal. Chem.*, 312, 455, 1982.
136. **Goto, J., Goto, N., and Nambara, T.,** *Chem. Pharm. Bull. Jpn.*, 30, 4597, 1982.
137. **Cory, R. P., Becker, R. R., Rosenebluth, R., and Isenberg, T.,** *J. Am. Chem. Soc.*, 90, 1643, 1968.
138. **Seeiler, N. and Demisch, L.,** Fluorescent derivatives, in *Handbook of Derivatives for Chromatography*, Blau, K. and King, G. S., Eds., Heyden, London, 1977.
139. **Fahey, R. C., Newton, G. L., Dorian, R., and Kosower, E. M.,** *Anal. Biochem.*, 121, 357, 1981.
140. **Newton, G. L., Randel, D., Dorian, R., and Fahey, R. C.,** *Anal. Biochem.*, 114, 383, 1981.
141. **Tochsteinova, D., Slosar, J., Urbanek, J., and Churacek, J.,** *Mikrochim. Acta*, 11, 193, 1979.
142. **Jandera, P., Pechova, H., Tocksteinova, D., Churacek, J., and Kralovsky, J.,** *Chromatographia*, 16, 275, 1982.

143. Ikeda, M., Shimada, K., and Sakaguchi, T., *J. Chromatogr.*, 272, 251, 1983.
144. Guebitz, G., Wintersteiger, R., and Hartinger, A., *J. Chromatogr.*, 218, 51, 1981.
145. Wintersteiger, R., Wenninger-Weinzierl, G., and Pacha, W., *J. Chromatogr.*, 237, 399, 1982.
146. Steinberg, S. M. and Bada, J. L., *Mar. Chem.*, 11, 299, 1982.
147. Liao, J. C., Hoffman, N. E., Barboriak, J. J., and Roth, D. A., *Clin. Chem.*, 23, 802, 1977.
148. Edman, P., *Acta Chem. Scand.*, 4, 283, 1950.
149. Samejima, K., *J. Chromatogr.*, 96, 250, 1974.
150. Stobaugh, J. F., Repta, A. J., Sternson, L. A., and Garren, K. W., *Anal. Biochem.*, 135, 495, 1983.
151. Felix, A. M. and Terkelson, G., *Arch. Biochem. Biophys.*, 157, 177, 1973.
152. Wu, C. W., Yarbrough, L. R., and Wu, F. Y. H., *Biochemistry*, 115, 2863, 1976.
153. Falter, H., Jayasimhula, K., and Day, R. A., *Anal. Biochem.*, 67, 359, 1975.
154. Hodgin, J. C., *J. Liq. Chromatogr.*, 2, 1047, 1979.
155. Gehrke, C. W., *J. Chromatogr.*, 162, 293, 1979.
156. Regis Product Literature, Regis Chemical Company, Morton Grove, IL.
157. Simons, S. S., Jr. and Johnson, D. F., *Anal. Biochem.*, 82, 250, 1977.
158. Bernardo, S. De., Weigele, M., Toome, V., Manhart, K., Leimgruber, W., Boehlen, P., Stein, S., and Udenfriend, S., *Arch. Biochem. Biophys.*, 162, 390, 1974.
159. Stein, S., Boehlen, P., Imai, K., Stone, J., and Udenfriend, S., *Fluoresc. News*, 7, 9, 1973.
160. McHugh, W., Sandman, R. A., Haney, W. G., Sood, S. P., and Wittmer, D. P., *J. Chromatogr.*, 124, 376, 1976.
161. Szokan, G., *J. Liq. Chromatogr.*, 5, 1493, 1982.
162. Furukawa, H., Sakaikiba, E., Kamel, A., and Ito, K., *Chem. Pharm. Bull. Jpn.*, 23, 1625, 1975.
163. Frei, R. W., Santi, W., and Thomas, M., *J. Chromatogr.*, 116, 365, 1976.
164. Schwedt, G. and Bussemas, H. H., *Fresenius Z. Anal. Chem.*, 283, 23, 1977.
165. Saeki, Y., Uehara, N., and Shirakawa, S., *J. Chromatogr.*, 145, 221, 1978.
166. Brown, N. D., Sweett, R. B., Kintzios, J. A., Cox, H. D., and Doctor, B. P., *J. Chromatogr.*, 164, 35, 1979.
167. Kawasaki, T., Maeda, M., and Tsuji, A., *J. Chromatogr.*, 163, 143, 1979.
168. Kawasaki, T., Maeda, M., and Tsuji, A., *Yakugaku Zasshi*, 100, 925, 1980.
169. Kawasaki, T., Maeda, M., and Tsuji, A., *J. Chromatogr.*, 226, 1, 1981.
170. Kawasaki, T., Maeda, M., and Tsuji, A., *J. Chromatogr.*, 232, 1, 1982.
171. Kawasaki, T., Maeda, M., and Tsuji, A., *J. Chromatogr.*, 233, 61, 1982.
172. Kawasaki, T., Maeda, M., and Tsuji, A., *J. Chromatogr.*, 272, 261, 1983.
173. Mopper, K. and Johnson, L., *J. Chromatogr.*, 256, 27, 1983.
174. Takeda, M., Maeda, M., and Tsuji, A., *J. Chromatogr.*, 244, 347, 1982.
175. Duenges, W., *Prae-Chromatographische Mikromethoden*, Dr. Alfred Huethig Verlag, New York, 1979, 106.
176. Grushka, E., Lam, S., and Chassin, J., *Anal. Chem.*, 50, 1398, 1978.
177. Lam, S. and Grushka, E., *J. Chromatogr.*, 158, 207, 1978.
178. Hayashi, K., Kawase, J., Yoshimura, K., Ara, K., and Tsuji, K., *Anal. Biochem.*, 136, 314, 1984.
179. Kissinger, P. T., Felix, L. J., Miner, D. J., Preddy, C. R., and Shoup, R. E., Detectors for trace organic analysis by liquid chromatography: principles and applications, in *Contemporary Topics in Analytical and Clinical Chemistry*, Vol. 2, Hercules, D. M., Hieftje, G. M., Snyder, L. R., and Evenson, M. A., Eds., Plenum Press, New York, 1978.
180. Seilere, N ., *J. Chromatogr.*, 143, 221, 1977.
181. Carey, M. A. and Persinger, H. E., *J. Chromatogr. Sci.*, 10, 537, 1972.
182. Taylor, J. T. and Freeman, S., *Chromatogr. Newsl.* 9, 1, 1981.
183. Souter, R. W., *Chromatographic Separations of Stereoisomers*, CRC Press, Boca Raton, FL, 1985.
184. Marfey, P., *Carlsberg Res. Commun.*, 49, 591, 1984.
185. Pierce Product Literature, Pierce Chemical Company, Rockford, IL.
186. Sedman, A. J. and Gal, J., *J. Chromatogr.*, 278, 199, 1983.
187. Lindner, W., Leitner, C., and Uray, G., *J. Chromatogr.*, 316, 605, 1984.
188. Kucera, P., Ed., *Microcolumn High-performance Liquid Chromatography*, Elsevier, Amsterdam, 1984.
189. Scott, R. P. W., Ed., *Small Bore Liquid Chromatography Columns*, Wiley-Interscience, New York, 1984.
190. Novotny, M. and Ishii, D., Eds., *Microcolumn Separation Methods*, Elsevier, Amsterdam, 1985.
191. Scott, R. P. W., *J. Chromatogr. Sci.*, 23, 233, 1985.
192. Eckers, C., Cuddy, K. K., and Henion, J. D., *J. Liq. Chromatogr.*, 6, 2383, 1983.
193. van der Wal, Sj., *J. Liq. Chromatogr.*, 9, 1815, 1986.
194. Wainer, I. W., *Liq. Chromatogr.*, 3, 876, 1985.
195. Reese, C. E. and Scott, R. P. W., *J. Chromatogr. Sci.*, 18, 479, 1980.
196. Rabel, F. M., *J. Chromatogr. Sci.*, 23, 247, 1985.
197. Martin, M., Eon, C., and Guiochon, G., *J. Chromatogr.*, 108, 229, 1975.

198. **Hirata, Y. and Novotny, M.,** *J. Chromatogr.*, 186, 521, 1979.
199. **Hirata, Y., Lin, P. T., Novotny, M., and Wightman, R. M.,** *J. Chromatogr.*, 181, 787, 1980.
200. **Lee, E. D. and Henion, J. D.,** *J. Chromatogr. Sci.*, 23, 253, 1985.
201. **Scott, R. P. W.,** *Trace Organic Analysis: A New Frontier in Analytical Chemistry*, National Bureau Standards Special Publication 519, U.S. Government Printing Office, Washington, D.C., 1979, 637.
202. **Henion, J.,** *Microcolumn Separation Methods*, Novotny, M. and Ishii, D., Eds., Elsevier, Amsterdam, 1985, 243.
203. **Henion, J.,** *Microcolumn High-Performance Liquid Chromatography*, Kucera, P., Ed., Elsevier, Amsterdam, 1984, 260.
204. **Folestad, S., Galle, B., and Josefsson, B.,** *J. Chromatogr. Sci.*, 23, 273, 1985.
205. **Folestad, S., Johnson, L., Josefsson, B., and Galle, B.,** *Anal. Chem.*, 54, 925, 1982.
206. **Hershberger, L. W., Callis, J. B., and Christian, G. D.,** *Anal. Chem.*, 51, 1444, 1979.
207. **Jorgenson, J. W., Smith, S. L., and Novotny, M.,** *J. Chromatogr.*, 142, 233, 1977.
208. **Stolyhwo, A., Colin, H., Martin, M., and Guiochon, G.,** *J. Chromatogr.*, 288, 253, 1984.
209. **Yeung, E. S., Steenhoek, L. E., Woodruff, W. D., and Kuo, J. C.,** *Anal. Chem.*, 52, 1399, 1980.
210. **Bobbitt, D. R. and Yeung, E. S.,** *Anal. Chem.*, 56, 1577, 1984.
211. **Yeung, E. S.,** *J. Pharm. Biomed. Anal.*, 2, 255, 1984.
212. **Veening, H., Tock, P. P. H., Kraak, J. C., and Poppe, H.,** *J. Chromatogr.*, 352, 345, 1986.
213. **Verzele, M. and Geeraert, E.,** *J. Chromatogr. Sci.*, 18, 559, 1980.
214. **Bombaugh, K. J. and Almquist, P. W.,** *Chromatographia*, 8, 109, 1975.
215. **Scott, R. P. W. and Kucera, P.,** *J. Chromatogr.*, 119, 467, 1976.
216. **Fallick, G. J.,** *Am. Lab.*, 5, 19, 1973.
217. **Klesper, E., Corwin, A. H., and Turner, D. A.,** *J. Org. Chem.*, 27, 700, 1962.
218. **Myers, M. N. and Giddings, J. C.,** *Sep. Sci.*, 1, 761, 1966.
219. **Sie, S. T. and Rijnders, G. W. A.,** *Sep. Sci.*, 2, 699, 1967.
220. **Sie, S. T. and Rijnders, G. W. A.,** *Sep. Sci.*, 2, 729, 1967.
221. **McLaren, L., Myers, M. N., and Giddings, J. C.,** *Science (Washington, D. C.)*, 159, 197, 1968.
222. **Giddings, J. C., Myers, M. N., McLaren, L., and Keller, R. A.,** *Science (Washington, D. C.)*, 162, 67, 1968.
223. **Smith, R. D. and Udseth, H. R.,** *Anal. Chem.*, 55, 2266, 1983.
224. **Van Wasen, U., Swaid, I., and Schneider, G. M.,** *Angew. Chem. Int. Ed. Engl.*, 19, 575, 1980.
225. **Sie, S. T. and Rijnders, G. W. A.,** *Anal. Chim. Acta*, 38, 31, 1967.
226. **Jentoft, R. E. and Gouw, T. H.,** *J. Chromatogr. Sci.*, 8, 138, 1970.
227. **Novotny, M., Bertsch, W., and Zlatkis, A.,** *J. Chromatogr.*, 61, 17, 1971.
228. **Gouw, T. H. and Jentoft, R. E.,** *J. Chromatogr.*, 68, 303, 1972.
229. **Conaway, J. E., Graham, J. A., and Rogers, L. B.,** *J. Chromatogr. Sci.*, 16, 102, 1978.
230. **Fjeldsted, J. C., Jackson, W. P., Peaden, P. A., and Lee, M. L.,** *J. Chromatogr. Sci.*, 21, 222, 1983.
231. **Smith, R. D., Chapman, E. G., and Wright, B. W.,** *Anal. Chem.*, 57, 2829, 1985.
232. **Jentoft, R. E. and Gouw, T. H.,** *Anal. Chem.*, 44, 681, 1972.
233. **Sie, S. T. and Rijnders, G. W. A.,** *Sep. Sci.*, 1, 459, 1966.
234. **Nieman, J. A. and Rogers, L. B.,** *Sep. Sci.*, 10, 517, 1975.
235. **Van Lenten, F. J. and Rothman, L. D.,** *Anal. Chem.*, 48, 1430, 1976.
236. **Peaden, P. A., Fjeldsted, J. C., Lee, M. L., Springston, S. R., and Novotny, M.,** *Anal. Chem.*, 54, 1090, 1982.
237. **Smith, R. D., Fjeldsted, J. C., and Lee, M. L.,** *J. Chromatogr.*, 247, 231, 1982.
238. **Later, D. W., Richter, B. E., Felix, W. D., Anderson, M. R., and Knowles, D. E.,** *Am. Lab.*, 18(8), 108, 1986.
239. **Levy, G. B.,** *Am. Lab.*, 18(8), 62, 1986.
240. **Gere, D. R., Board, R., and McManigill, D.,** *Anal. Chem.*, 54, 736, 1982.
241. **Schmitz, F. P., Hilgers, H., and Klesper, Ee.,** *J. Chromatogr.*, 267, 267, 1983.
242. **Klesper, E. and Leyendecker, D.,** *Int. Lab.*, 16(9), 18, 1986.
243. **Smith, R. D., Wright, B. W., and Yonker, C. R.,** Characterization of supercritical fluid solutions using supercritical fluid chromatography and mass spectrometry, in *Theories and Applications of Supercritical Gas Extraction*, NTS, Inc., 1987.
244. **Wright, B. W., Udseth, H. R., Smith, R. D., and Hazlett, R. N.,** *J. Chromatogr.*, 314, 253, 1984.
245. **Smith, R. D., Udseth, H. R., and Hazlett, R. N.,** *Fuel*, 64, 810, 1985.
246. **Novotny, M., Springston, S. R., Peaden, P. A., Fjeldsted, J. C., and Lee, M. L.,** *Anal. Chem.*, 53(3), 407A, 1981.
247. **Klesper, E.,** *Angew. Chem. Int. Ed. Engl.*, 17, 738, 1978.
248. **Rokushika, S., Naikwadi, K. P., Jadhav. A. L., and Hatano, H.,** *J. High Resolut. Chromatogr. Chromatogr. Commun.*, 8, 480, 1985.

249. **Gouw, T. H. and Jentoft, R. E.,** *Adv. Chromatogr.*, 13, 1, 1975.
250. **Klesper, E.,** *Angew. Chem. Int. Ed. Engl.*, 17, 738, 1978.
251. **Jentoft, R. E. and Gouw, T. H.,** *Anal. Chem.*, 48, 2195, 1976.
252. **Asche, W.,** *Chromatographia*, 11, 411, 1978.
253. **Sie, S. T., Van Beersum, W., and Rijnders, G. W. A.,** *Sep. Sci.*, 1, 459, 1966.
254. **Bartmann, D.,** *Ber. Bunsenges. Phys. Chem.*, 76, 336, 1972.
254a. **Fjeldsted, J. C., Kong, R. C., and Lee, M. L.,** *J. Chromatogr.*, 279, 449, 1983.
255. **Chester, T. L.,** *J. Chromatogr.*, 299, 424, 1984.
256. **Chester, T. L., Innis, D. P., and Owens, G. D.,** *Anal. Chem.*, 57, 2243, 1985.
257. **Richter, B. E.,** *J. High Resolut. Chromatogr. Chromatogr. Commun.*, 8, 297, 1985.
258. **Markides, K. E., Lee, E. D., Bolick, R., and Lee, M. L.,** *Anal. Chem.*, 58, 740, 1986.
259. **Randall, L. G. and Wahrhaftig, A. L.,** *Anal. Chem.*, 50, 1705, 1978.
260. **Randall, L. G. and Wahrhaftig, A. L.,** *Rev. Sci. Instrum.*, 52, 1283, 1981.
261. **Smith, R. D., Felix, W. D., Fjeldsted, J. C., and Lee, M. L.,** *Anal. Chem.*, 54, 1883, 1982.
262. **Smith, R. D., Kalinoski, H. T., Udseth, H. R., and Wright, B. W.,** *Anal. Chem.*, 56, 2476, 1984.
263. **Nishioka, M., Whiting, D. G., Campbell, R. M., and Lee, M. L.,** *Anal. Chem.*, 58, 2251, 1986.
264. **Pentoney, S. L., Shafer, K. H., and Griffiths, P. R.,** *J. Chromatogr. Sci.*, 24, 230, 1986.
265. **Pentoney, S. L., Shafer, K. H., Griffiths, P. R., and Fuoco, R.,** *J. High Resolut. Chromatogr. Chromatogr. Commun.*, 9, 168, 1986.
266. **Olesik, S. V., French, S. B., and Novotny, M.,** *Anal. Chem.*, 58, 2256, 1986.
267. **Blilie, A. L. and Greibrokk, T.,** *Anal. Chem.*, 57, 2239, 1985.
268. **Hirata, Y. and Nakata, F.,** *J. Chromatogr.*, 295, 315, 1984.
269. **Yonker, C. R. and Smith, R. D.,** *J Chromatogr.*, 361, 25, 1986.
270. **White, C. M. and Houck, R. K.,** *J. High Resolut. Chromatogr. Commun.*, 8, 293, 1985.
271. **Rawdon, M. G. and Norris, T. A.,** *Am. Lab.*, 16, 17, 1984.
272. **Smith, R. D., Udseth, H. R., and Wright, B. W.,** *J. Chromatogr. Sci.*, 23, 192, 1985.
273. **Glajch, J. L., Kirkland, J. J., Squire, K. M., and Minor, J. M.,** *J. Chromatogr.*, 199, 57, 1980.
274. **Drouen, A. C. J. H., Billiet, H. A. H., and de Galan, L.,** *J. Chromatogr.*, 352, 127, 1986.
275. **Engelhardt, H. and Elgass, H.,** *Chromatographia*, 22, 31, 1986.
276. **Dezaro, D. A., Dvoin, D., Horn, C., and Hartwick, R. A.,** *Chromatographia*, 20, 87, 1985.
277. **Goewie, C. E.,** *J. Liq. Chromatogr.*, 9, 1431, 1986.
278. **Issaq, J. J., Muschik, G. M., and Janini, G. M.,** *J. Liq. Chromatogr.*, 6, 259 1983.
279. **Issaq, H. J., McNitt, K. L., and Goldgaber, N.,** *J. Liq. Chromatogr.*, 7, 2535, 1984.
280. **Haky, J. E., Young, A. M., Domonkos, E. A., and Leeds, R. L.,** *J. Liq. Chromatogr.*, 7, 2127, 1984.
281. **Costanzo, S. J.,** *J. Chromatogr. Sci.*, 24, 89, 1986.
282. **Smith, H. K., Switzer, W. L., Martin, G W., Benezra, G. W., Wilson, W. P., and Dixon, D. W.,** *J. Chromatogr. Sci.*, 24, 70, 1986.
283. **Sabati, L. G., Diaz, A. M., Tomas, Z. M., and Gassrot, M. M.,** *J. Chromatogr. Sci.*, 21, 439, 1983.
284. **Schoenmakers, P. J., Drouen, A. C. J. H., Billiet, H. A. H., and de Galan, L.,** *Chromatographia*, 15, 688, 1982.
285. **Snyder, L. R. and Dolan, J. W.,** *Am. Lab.*, 18(8), 37, 1986.
286. **Schoenmakers, P. J., Billiet, H. A. H., and de Galan, L.,** *J. Chromatogr.*, 218, 261, 1981.
287. **Massart,** *Anal. Chem.*, 56, 2662, 1984.
288. **Foley, J. D. and Drosey, J. G.,** *Anal. Chem.*, 55, 730, 1983.
289. *U. S. Pharmacopeia XXI*, Rockville, MD, 1985.
290. **Bouquet, S., Regnier, B., Quehen, S., Brisson, A. M., Courtois, P., and Fourtillan, J. B,** *J. Liq. Chromatogr.*, 8, 1663, 1985.
291. *U. S. Pharmacopeia XXI, Suppl. 2*, Rockville, MD, 1985.
292. *U. S. Pharmacopeia XXI, Suppl. 1*, Rockville, MD, 1985.
293. **Hikal, A. H. and Jones, A. B.,** *J. Liq. Chromatogr.*, 8, 1455, 1985.
294. **Yang, R. K., Hsieh, J. Y.-K., Kendler, K. S., and Davis, K. L.,** *J. Liq. Chromatogr.*, 7, 191, 1984.
295. **Issaq, H. J., Weiss, D., Ridlon, C., Fox, S. D., and Muschik, G. M.,** *J. Liq. Chromatogr.*, 9, 1791, 1986.
296. **Fasching, C. E. and Peterson, L. R.,** *J. Liq. Chromatogr.*, 6, 2513, 1983.
297. **Kountourellis, J. E., Raptouli, A., and Georgakopoulos, P. P.,** *J. Chromatogr.*, 362, 439, 1986.
298. **Gill, R., Law, B., and Gibbs, J. P.,** *J. Chromatogr.*, 356, 37, 1986.
299. *U. S. Pharmacopeia XXI, Suppl. 3*, Rockville, MD, 1986.
300. **Molokhia, A. M., El-Hoofy, S., and Al-Rahman, S.,** *J. Liq. Chromatogr.*, 7, 1643, 1984.
301. **Kumps, A.,** *J. Liq. Chromatogr.*, 7, 1235, 1984.
302. **Dupont, D. G. and DeJager, R. L.,** *J. Liq. Chromatogr.*, 4, 123, 1981.
303. **Hwang, P. T. R., Drexler, G., and Meyer, M. C.,** *J. Liq. Chromatogr.*, 7, 979, 1984.
304. **LeBel, M., Ericson, J. F., and Pitkin, D. H.,** *J. Liq. Chromatogr.*, 7, 961, 1984.

305. **Boutagy, J., More, D. G., Munro, I. A., and Shenfield, G. M.,** *J. Liq. Chromatogr.*, 7, 1651, 1984.
306. **Putteemans, M., Bogaert, M., Hoogewijs, G., Dryon, L., Massart, D. L., and Vanhaelst, L.,** *J. Liq. Chromatogr.*, 7, 2237, 1984.
307. **Gau, W., Ploschke, H. J., Schmidt, K., and Weber, B.,** *J. Liq. Chromatogr.*, 8, 485, 1985.
308. **Rosenblum, M. G., Hortobagy, G. N., Wingender, W., and Hersh, E. M.,** *J. Liq. Chromatogr.*, 7, 159, 1984.
309. **Shihabi, A. K., Scaro, J., and David, R. M.,** *J. Liq. Chromatogr.*, 8, 2641, 1985.
310. **Pakuts, A. P., Matula, D., and Matula, T. I.,** *J. Liq. Chromatogr.*, 6, 2557, 1983.
311. **Hartshorn, J. N., Tong, W. P., Stewart, J. A., and McCormack, J. J.,** *J. Liq. Chromatogr.*, 9, 1489, 1986.
312. **Cathapermal, S. and Caddy, B.,** *J. Chromatogr.*, 351, 249, 1986.
313. **Gordon, S. M., Freeston, L. K., Collins, A. J., and Bain, R.,** *J. Chromatogr.*, 322, 246, 1985.
314. **Duthu, G. S.,** *J. Liq. Chromatogr.*, 7, 1023, 1984.
315. **Hikal, A. H. and Al-Shoura, H. A.,** *J. Liq. Chromatogr.*, 5, 2205, 1982.
316. **Lovett, L. J., Nygard, G., Dura, P., and Khalil, S. K. W.,** *J. Liq. Chromatogr.*, 8, 1611, 1985.
317. **Das Gupta, V.,** *J. Liq. Chromatogr.*, 8, 2497, 1985.
318. **Mohamed, M. E., Aboul-Enein, H. Y., Babhair, S. S., and El-Hofi, S.,** *J. Liq. Chromatogr.*, 6, 715, 1983.
319. **Dolezalova, M.,** *J. Chromatogr.*, 361, 421, 1986.
320. **Jubiz, W., Nolan, G., and Kaltenborn, K. C.,** *J. Liq. Chromatogr.*, 8, 1519, 1985.
321. **Arnold, P. J. and Stetten, O. V.,** *J. Chromatogr.*, 353, 193, 1986.
322. **Adams, P. S. and Haines-Nutt, R. F.,** *J. Chromatogr.*, 329, 438, 1985.
323. **Nygard, G., Lovett, L. J., and Khalil, S. K. W.,** *J. Liq. Chromatogr.*, 9, 157, 1986.
324. **Taylor, R. B., Durham, D. G., Shiviji, A. S. H., and Reid, R.,** *J. Chromatogr.*, 353, 51, 1986.
325. **Tawakkol, M. S., Mohamed, M. E., and Hassan, M. M. A.,** *J. Liq. Chromatogr.*, 6, 1491, 1983.
326. **Loenhout, J. W. A., van Ginneken, C. A. M., Ketelaars, H. C. J., Kimenai, P. M., Tan, Y., and Gribnau, F. W. J.,** *J. Liq. Chromatogr.*, 5, 549, 1982.
327. **Lebelle, M. J., Lauriault, G., and Wilson, W. L.,** *J. Liq. Chromatogr.*, 3, 1573, 1980.
328. **Lin, J. M.-H., Shi, R. J., and Lin, E. T.,** *J. Liq. Chromatogr.*, 9, 2035, 1986.
329. **Foda, N. H. and Jun, H. W.,** *J. Liq. Chromatogr.*, 9, 803, 1986.
330. **Meulemans, A., Mohler, J., Decazes, J. M., Dousset, I., and Modai, A.,** *J. Liq. Chromatogr.*, 6, 575, 1983.
331. **Babhair, S. A.,** *J. Liq. Chromatogr.*, 7, 2401, 1984.
332. **Tebbett, I. R.,** *J. Chromatogr.*, 356, 227, 1986.
333. **Carignan, G. and Lodge, B. A.,** *J. Liq. Chromatogr.*, 8, 1431, 1985.
334. **Franks, T. J., Stodola, J. D., and Walker, J. S.,** *J. Chromatogr.*, 353, 379, 1986.
335. **Das Gupta, V.,** *J. Liq. Chromatogr.*, 9, 1065, 1986.
336. **Gupta, S. K., Gwilt, P. R., Lim, J. K., and Waters, D. H.,** *J. Chromatogr.*, 361, 403, 1986.
337. **Hon, J. Y. C. and Murray, L. R.,** *J. Liq. Chromatogr.*, 5, 1973, 1982.
338. **Aravind, M. K., Miceli, J. N., and Kauffman, R. E.,** *J. Liq. Chromatogr.*, 7, 2887, 1984.
339. **Ackerly, C. C., Hartshorn, J., Tong, W. P., and McCormack, J. J.,** *J. Liq. Chromatogr.*, 8, 125, 1985.
340. **Lim, C. K., Rideout, J. M., and Sheldon, J. W. S.,** *J. Liq. Chromatogr.*, 6, 887, 1983.
341. **Van Antwerp, J. and Lepore, J.,** *J. Liq. Chromatogr.*, 5, 571, 1982.

Chapter 5

THE DETERMINATION OF ISOMERIC PURITY

R. W. Souter

TABLE OF CONTENTS

I. INTRODUCTION

Isomeric compounds are those which have the same empirical formula, but different structures. They can be divided into two general classes: positional isomers and stereoisomers. The former, also called constitutional isomers, differ in the sequence of atoms and bonds, while the latter have identical sequences and differ only in their spatial arrangement. Stereoisomers include optical isomers (enantiomers and diastereomers) as well as geometric isomers and conformational isomers.

Positional isomers are relatively simple, and typical examples include the hexanes (5-isomers) and ethylphenols (*ortho*-, *meta*-, and *para*-). Positional isomers typically differ from one another in both chemical and physical properties, although the differences are sometimes quite small.

Geometric isomers occur as a result of restricted rotation about a bond. These are the familiar *cis*- and *trans*- (or E and Z) isomers. Geometric isomers have significantly different chemical and physical properties. Conformational isomers exist as different arrangements in space made possible by rotation about one or more single bonds, and interconversion of these isomers occurs without breaking bond.[1] An example of conformational isomers would be the *syn-, and anti-*, and *gauche* forms of 1,2-dichloroethane.

Optical isomers are compounds which meet one of the following criteria:

1. The molecule is not superimposable on its mirror image.
2. At least one part of the molecule is not superimposable on its mirror image.

Molecules and parts of molecules that are not superimposable on their mirror images are called "chiral molecules" and "chiral centers", respectively. The most common chiral center encountered in organic chemistry is a carbon atom bearing four chemically distinct substituent groups.

Optical isomers that are mirror images of one another are called enantiomers. Enantiomers have identical physical and chemical properties with two exceptions:

1. Enantiomers rotate the plane of polarization of light by equal but opposite amounts.
2. The reactions of enantiomers with other chiral molecules may occur with different rate and equilibrium constants.

Optical isomers that are not mirror images of one another are called "diastereomers." Diastereomers differ significantly in chemical and physical properties. Diastereomers may also rotate the plane of polarized light, but *meso*-diastereomers possess no optical rotation. Mislow[2] has given a very good discussion of the types, chemistry, and nomenclature of stereoisomers.

It is very important that one be able to accurately assess the isomeric purity of substances, since isomeric impurities may have undesirable pharmacological, toxicological, or other effects. An isomeric impurity may be carried through a synthesis and react preferentially at one step to yield an undesirable level of another impurity. In many cases in the pharmaceutical field, only one isomer of a series produces the desired effect, while the others may have unwanted side effects. For example, α-d-propoxyphene hydrochloride is a widely prescribed analgesic,[3-5] the α-1 form has clinically useful antitussive properties,[6] and the β-d and β-1 isomers are substantially inactive.[7] Simple enantiomers also provide some very interesting examples of differences in interaction with the organism: (−) epinephrine is over ten times as active a vasoconstrictor as (+) epinephrine, and (−) amino acids are either tasteless or bitter, while (+)

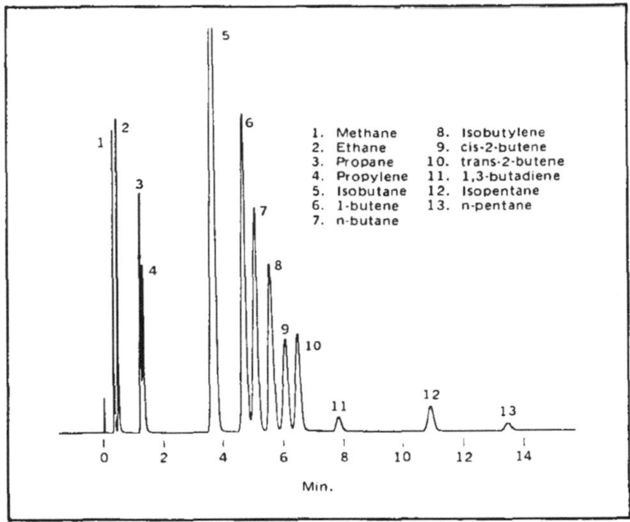

FIGURE 1. Separation of some hydrocarbon positional isomers and related hydrocarbons. 80/100 mesh Carbopack® C/0.19% picric acid, 2 ft. × 1/8 in. O.D. s.s.; column temperature: 1 min hold 30°C to 100°C @ 4°C/min; flow rate: 50 mℓ/min nitrogen @ 53 psi; detector: FID. (From *Chromatography/Lipids*, Vol. 9, No. 3, Supelco, Supelco Park, Bellefonte, Pa. July 1975. With permission.)

amino acids are sweet.[8] These vast differences in effects of various stereoisomers emphasize the need for accurate assessment of the isomeric purity of agricultural, pharmaceutical, and other biochemicals.

Because of their different physical and chemical properties, positional isomers and some stereoisomers yield relatively easily to isomeric purity analysis. Optical stereoisomers are inherently more difficult to study, but such work is critically necessary.

II. POSITIONAL ISOMERS

A. Chromatography

1. Gas Chromatography (GC)

The simplicity of operation, sensitivity, and especially the selectivity offered by gas chromatography (GC) have made it a widely used technique for qualitative and quantitative analyses. Different physical and chemical properties of positional isomers often make them amenable to direct separation by this technique.

Homologous series of isomers of various chemical classes may be separated easily on Carbopack® C packing materials by proper choice of operating conditions. Carbopack® C is a graphitized carbon[9] and, as shown in Figures 1 to 4, various surface modifications of the Carbopack® C under proper operating conditions elute hydrocarbon, alcohol, cresol, and butylbenzene isomers with varying degrees of resolution. Separations of phenols and alcohols on these packing materials should be attempted only on glass columns to avoid tailing of the peaks. The separation of the butylbenzene isomers shown in Figure 4 indicates that the Carbopack® C/0.1% SP-1000 has excellent separating abilities since this difficult separation usually requires a capillary column.[10]

Carbopack® B/3% Carbowax® 20M/0.5% phosphoric acid was developed for separation of volatile carboxylic acids at trace levels. Symmetrical peaks are obtained

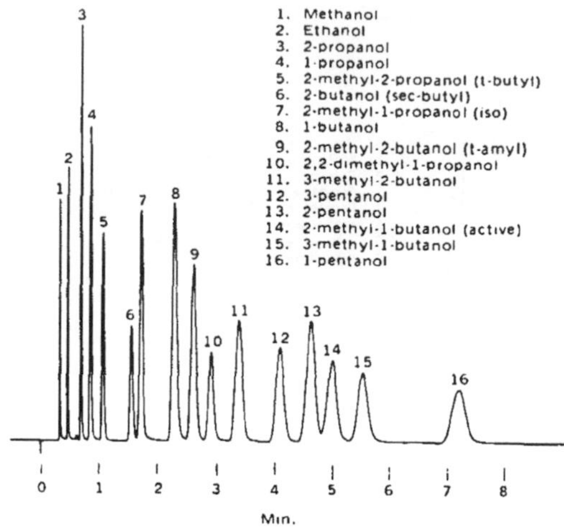

FIGURE 2. Separation of some alcohol positional isomers and related alcohols. 80/100 mesh Carbopack® C/0.2% Carbowax 1500®, 6 ft × 2 mm I.D. glass; column temperature: 125°C; flow rate: 20 mℓ/min nitrogen @ 31 psi; sample size: 0.02 μℓ; detector: FID. (From *Chromatography/Lipids*, Vol. 9, No. 3, Supelco, Supelco Park, Bellefonte, Pa., July 1975. With permission.)

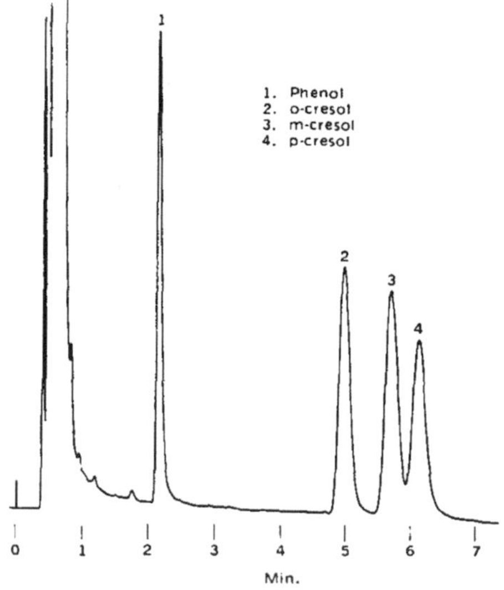

FIGURE 3. Separation of *o*-, *m*- and *p*-cresols and phenol. 80/100 Carbopack® C/0.1% SP-1000, 6 ft × 2 mm I.D. glass; column temperature: 225°C; flow rate: 20 mℓ/min nitrogen; sample size: 0.5 μℓ approximately 0.1% each in methyl ethyl ketone solution; detector: FID. (From *Chromatography/Lipids*, Vol. 9, No. 3, Supelco, Supelco Park, Bellefonte, Pa., July 1975. With permission.)

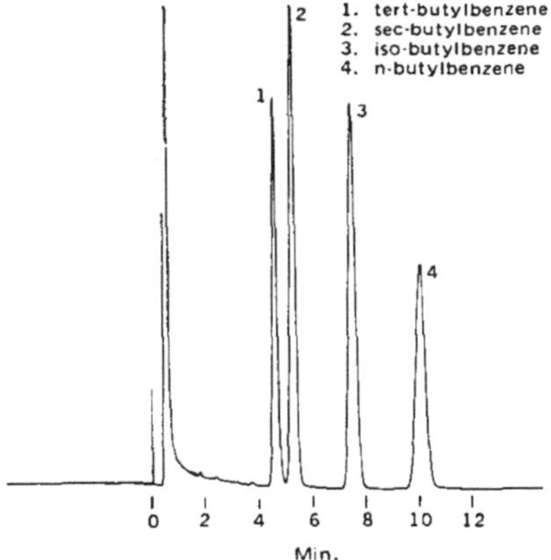

FIGURE 4. Separation of butylbenzene positional isomers. 80/100 mesh Carbopack® C/0.1% SP-1000, 6 ft × 1/8 in. O.D. s.s.; column temperature: 225°C; flow rate: 20 mℓ/min nitrogen @ 42 psi; sample size: 1.0 μℓ; detector: FID. (From *Chromatography/Lipids*, Vol. 9, No. 3, Supelco, Supelco Park, Bellefonte, Pa., July 1975. With permission.)

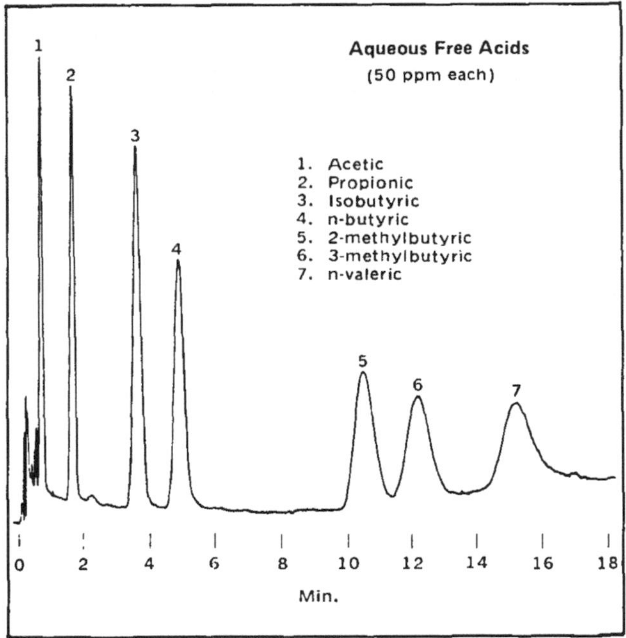

FIGURE 5. Separation of some carboxylic acid positional isomers and related compounds at the ppm level. 60/80 mesh Carbopack® B/3% Carbowax® 20 M/0.5% phosphoric acid: 30 in. × ¼ in. O.D. × 4 mm I.D. glass; column temperature: 160°C — inlet and detector: 200°C; flow rate, 60 mℓ/min nitrogen; sample size: 1 μℓ; detector: FID; 16×10^{-12} AFS. (From Catalog No. 11, Supelco, Supelco Park, Bellefonte, Pa., 1977, 4. With permission.)

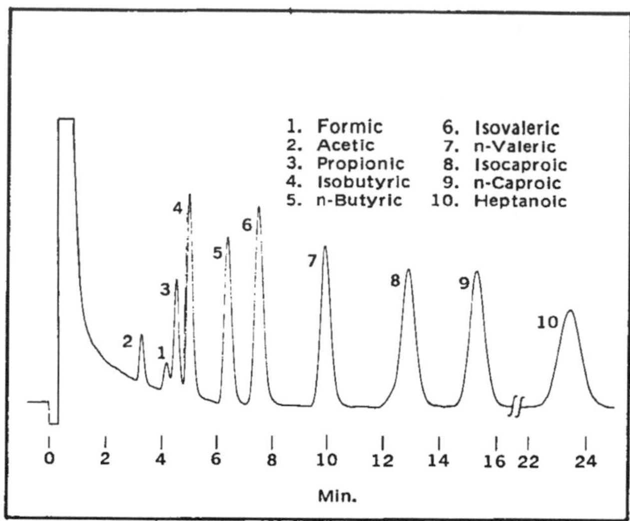

6A

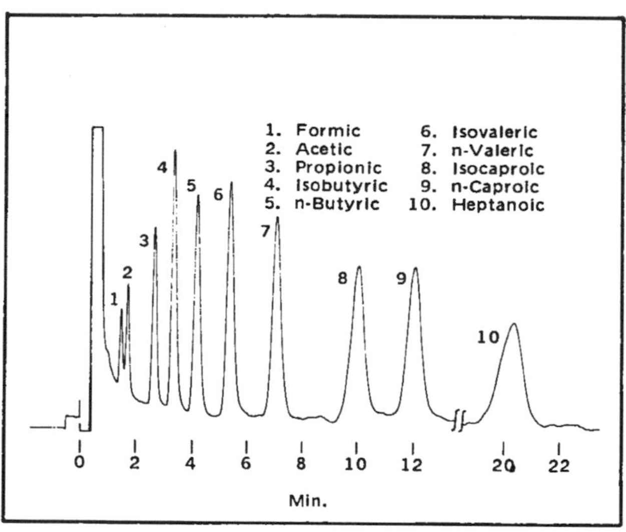

6B

FIGURE 6. Separation of some volatile carboxylic acid positional isomers and related acids. (A) 10% SP-1000/1% H_3PO_4 on 100/120 Chromosorb W AW,® 6 ft × 4 mm I.D. glass; column temperature: 147°C; flow rate, 86 mℓ/min, helium; sample size: 14 μℓ; instrument: Fisher® Model 2400, TC detector. (B) 15% SP-1220/1% H_3PO_4 on 100/120 Chromosorb W AW®, 6 ft × 4 mm I.D. glass; column temperature: 145°C; flow rate: 70 mℓ/min; sample size: 14 μℓ; instrument: Fisher® Model 2400, TC detector. (From Catalog No. 11, Supelco, Supelco Park, Bellefonte, Pa., 1977, 9. With permission.)

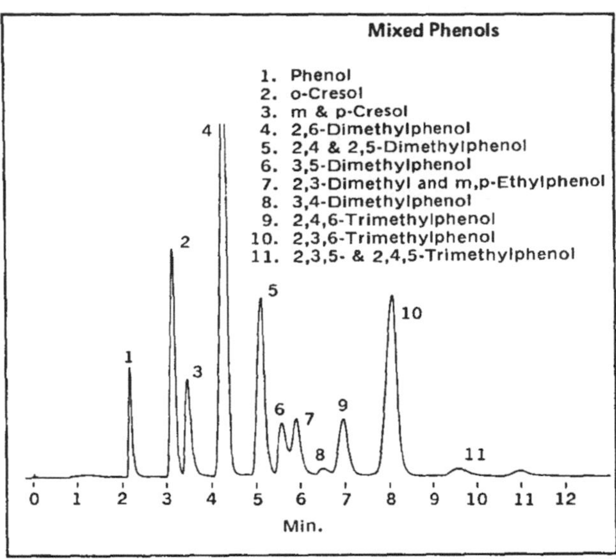

FIGURE 7. Separation of some mixed phenol positional isomers. 10% SP-2100 on 100/120 Supelcoport®, 6 ft × 2 mm ID glass; temperature: 130°C — inlet and detector: 175°C; flow rate: 20 mℓ/min nitrogen; sample size: 0.05 μℓ; detector: FID. (From Catalog 11, Supelco, Supelco Park, Bellefonte, Pa., 1977, 20. With permission.)

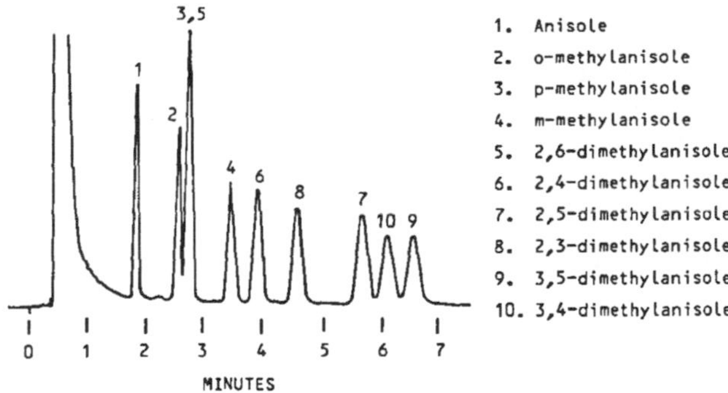

FIGURE 8. Separation of some anisole positional isomers. 5% SP-1200/1.75% Bentone 34® on 100/120 Supelcoport®, 6 ft × 1/8 in SS; column temperature: 150°C; flow rate: 20 mℓ/min nitrogen. (From Catalog 11, Supelco, Supelco Park, Bellefonte, Pa., 1977, 20. With permission.)

at the part-per-million level for such isomers as isobutyric and *n*-butyric or 2- and 3-methylbutyric acids when glass columns are used, as shown in Figure 5. More details are available from work by DiCorcia and Samperi.[11] Isomeric carboxylic acids used to identify anaerobic fermentation cultures may be separated without tailing directly as the acids or as methyl esters on the columns shown in Figure 6. Of these packing materials available from Supelco,[12] the SP-1220 allows shorter analysis times, but both give good peaks without tailing. Using other readily available column packings,[12] mixtures of phenol positional isomers and of methyl- and dimethylanisoles may be separated as indicated in Figures 7 and 8, respectively.

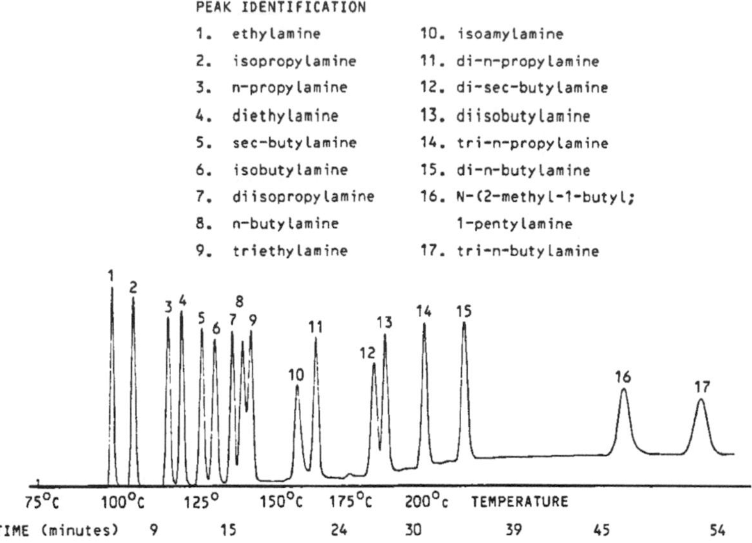

FIGURE 9. Separation of amines on a stock pretested packing. 28% Pennwalt 223® + 4% KOH on 80/100 Gas-Chrom R®; column: 10 ft × ¼ in. O.D. copper; column temperature: progressing from 75 to 200°C at 4°/min; detector: TC at 150 mA; helium carrier: 45 mℓ/min; sensitivity: 1×10^{-7} AFS. (From Catalog No. 20, Applied Science Laboratories, State College, Pa., 1977, 24. With permission.)

A very popular column packing used for separation and analysis of amine isomers is Pennwalt 223® with potassium hydroxide on Gas-Chrom R®. It is available as either 10 or 28% Pennwalt 223® with 4% KOH on 80/100 mesh Gas Chrom R®.[13] It was designed specifically for the analysis of amines and yields symmetrical peaks, as shown in Figure 9, where several sets of isomers are resolved. Since the maximum operating temperature with this phase is only 225°C, it may not be useful for direct separation of some higher-boiling amine isomers.

The proper choice of stationary phase to eliminate peak tailing and to yield symmetrical peaks for the compound class under consideration is very important, and the reader is referred to the section on column packings in the gas chromatography chapter of this text. Further examples of separations of isomeric pyridines, esters, aldehydes, ketones, phenols, alkyl halides, thiols, and sulfides have been collected by Littlewood.[14]

Bentone 34® is a montmortillonite clay in which the naturally occurring inorganic ions have been replaced by dimethyldioctadecyl ammonium ions. Its surface is nearly neutral, but polar compounds will be adsorbed to some degree, causing peak tailing. It may be used successfully for aromatic positional isomer separations when coated on a conventional support, most usually combined with a stationary phase like the low polarity ester SP 1200.[12] Varying the amount of Bentone 34® in the packing material will shift the peak positions, and this may be used to enhance some separations. The Bentone® is thermally stable to 180°C, but should not be used with strong acids or bases as these will destroy its structure. Figure 8 shows one separation effected with the Bentone 34®, and a large number of other examples, including xylenes and dichlorobenzenes, are available.[15]

In addition to standard gas chromatographic phases, special liquid phases may be employed for specific separation purposes. Liquid crystals may be used in the separation of positional isomers. Some types of compounds, including many with rodlike shapes and polar linkages at the ends of the molecule, form ordered liquid mesophases

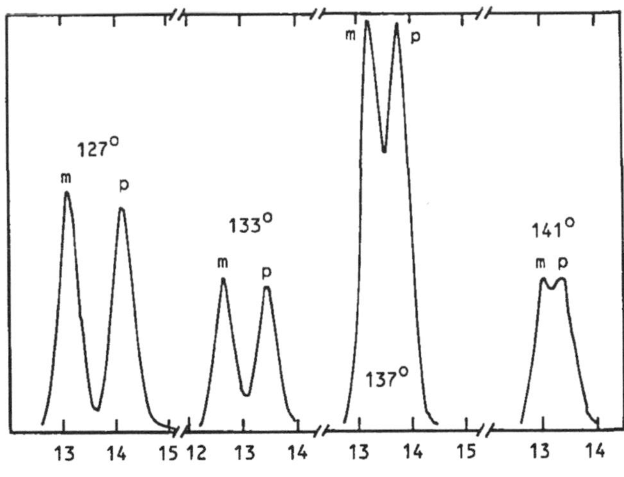

FIGURE 10. Separation of *m*- and *p*-chlorotoluene on 1.5 m 15% *p*-azoxyanisole on 60/80 mesh Chromosorb W® with nitrogen carrier gas flow of 30 to 40 m*l*/min. (From Dewar, M. and Schroeder, J., *J. Am. Chem. Soc.*, 86, 5237, 1964. With permission.)

when melted. There are three types of mesophases: smectic, nematic, and cholesteric.[16] The smectic phase, for example, consists of a two-dimensional array in which the rod-shaped molecules are arranged in parallel in layers. The separation mechanism for positional isomers can be understood from the separation of *p*- and *m*-dichlorobenzene. Since the *para*- isomer is more elongated, it can align itself in the layers more readily than the *meta*- isomer and thus be longer retained. Thus, when liquid crystals are used as stationary liquid phases, the solute retention mechanism involves, in part, a fitting of the molecules in the lattice structure. The ability to fit depends on molecular shape. Figure 10 shows the separation of *m*- and *p*-chlorotoluene on *p*-azoxyanisole, which has a 120 to 135°C nematic range. The two isomers are nearly baseline resolved at 127°C, but are almost unseparated at 141°C. In passing from the nematic to the isotropic (disordered normal liquid) state the separation decreases because an isotropic liquid has no ability to discriminate by shape. Note also that the *para*- isomer elutes after the *meta*-.

Kelker and von Schivizhoffen have reviewed the use of liquid crystals as stationary phases in gas chromatography and have described the fundamental desirable properties of such phases.[17] Using liquid crystals as stationary phases, Dewar and Schroeder have separated various disubstituted benzenes,[18,19] while Dewar and co-workers showed that a nematic liquid crystalline mixture could be used to change the relative retention times and increase the useful gas chromatography temperature range.[20] Andrews et al.[21] reported the use of 4,4′-di-*n*-alkoxyazoxybenzene homologues and mixtures as gas chromatographic liquid phases for shape separations. They concluded that in choosing the components for such "superselective" mixtures: (1) at least one component, A, should have a high nematic → isotropic transition point, (2) the other components should be compatible with A such that the mixture has a stable, well-ordered nematic mesophase, and (3) the mixture should ideally have a high A concentration and a low freezing point so that both molecular ordering and low temperature aid the separation.[21]

Richmond[22] extended the GC studies of liquid crystal phases with benzene positional isomers to include *ortho*- as well as *meta*- and *para*- substituted compounds. Although the average HETP on the tested columns was rather large for conventional GC columns, the α values calculated for the *meta-/para-* pairs were large (they averaged nearly 1.2), and the separations were baseline or nearly so in most cases. One of the stationary phases studied, 4,4′-biphenylene [bis*para*-(heptyloxy) benzoate], had a very wide smectic (150 to 211°C) as well as nematic (211 to 316°C) operating temperature range. Recently Cook and Spangelo[23] separated isomeric mono-substituted phenols by GC on liquid crystal phases and discussed the separations in light of molecular shapes and interactions. For further references involving uses of liquid crystalline phases for positional isomer separations by GC, the recent work by Grushka and Solsky[24] should be consulted.

For laboratories which are so equipped, capillary columns may offer facile isomer separations in cases where previously a specialized liquid phase might have been necessary. Both wall-coated open tubular (WCOT) and support-coated open tubular (SCOT) columns and appropriate GC instrumentation are now readily available. Many laboratories coat their own glass capillary columns. Capillary columns offer a separating power much better than the usual packed columns while permitting the analysis to be done in a shorter time. SCOT columns have the advantages over the WCOT columns that their sample capacities are higher and that they have less stringent requirements concerning instrumentation.

Even though high-efficiency capillary column separations were achieved more than a decade ago, the versatility and application for quantitative analysis of such columns is subject to argument. Because of insufficient accuracy and precision often related to unreliable sampling methods, and due to instrument-installation problems, relatively few quantitative analyses have been reported using capillary GC techniques. Mattsson and Nygren[25] have reported on the determination of a variety of polychlorinated biphenyls and chlorinated pesticides in sewage sludge by use of a capillary column, but results were reported with a relatively wide variation. Only very recently was the quantitative determination of isomeric chlorophenols reported.[26] Capillary columns can be a powerful tool for quantitative analysis of substances (especially isomers) not satisfactorily resolved on packed analytical columns. The ethylphenols are widely used as starting materials and intermediates in organic syntheses, but suffer from incomplete resolution of the *meta-* and *para-* isomers on typical packed columns.[27] Souter and Bishara[28] recently reported the quantitative analyses of isomeric ethylphenols using a SCOT column. Figure 11 demonstrates the difference in resolving power of a typical packed column vs. the SCOT column. Both accuracy and precision for assays of a low level of the *para-* isomer in the presence of the *meta-* depend immensely on the resolution between them.

Mostecky et al.[29] have used capillary gas chromatography to examine isomer content of various petrochemical fractions. Isomeric methyl- and ethyl- as well as dimethylnaphthalenes were resolved and 2-, 3-, and 4-methylbiphenyls were also separated using three different capillary columns. Glass capillary columns were recently used by Fell and Lee[30] in their determination of urinary monohydric and dihydric phenols. *o*-, *m*-, and *p*- Cresols were resolved as well as catechol and resorcinol and 3- and 4-methylcatechol when the compounds were chromatographed as the trimethylsilyl derivatives. Apon and Nicolaides[31] used capillary gas chromatography/mass spectrometry to determine the positions of methyl branches in fatty acid methyl esters.

Highly toxic polychlorinated dibenzo-*p*-dioxin (PCDD) isomers are produced from chlorinated phenols during the manufacturing processes of a series of agricultural and industrial chemicals.[32] A total of 75 different PCDDs exist, and different isomers have significantly different toxicological properties.[33] Buser[34] recently used a glass, OV-61-coated capillary column for the separation and gas chromatography/mass spectrometry identification of a number of the PCDD isomers, including the ten hexachlorodi-

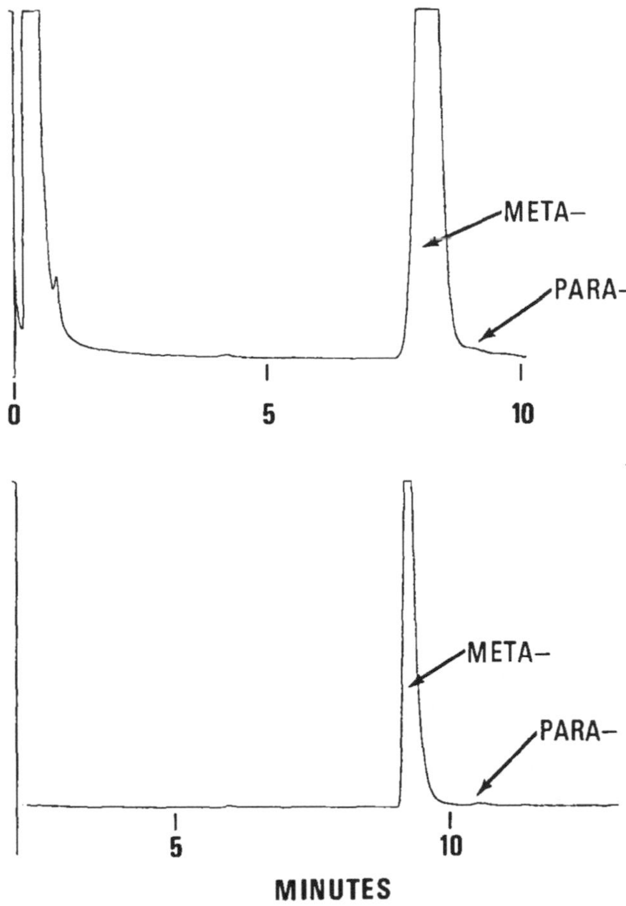

FIGURE 11. Gas chromatograms for approximately 0.5% *p*-ethylphenol in *m*-ethylphenol. Upper: 4 ft × 1/8 in. I.D. packed with 3.8% UC-W-98 on 80/100 mesh Gas Chrom Z® operated at 70°C (with helium flow 60 mℓ/min. Lower: 50 ft × 0.02 in. I.D. SCOT coated with OV-17 operated at 115°C with helium about 5 mℓ/min. (From Souter, R. W. and Bishara, R. H., *J. Chromatogr.*, 140, 245, 1977. With permission.)

benzo-*p*-dioxins shown in Figure 12. The separating abilities of capillary columns is well demonstrated by this example for positional isomers.

Hoshika and Takata[34] separated a variety of carbonyl compounds, including positional isomeric aldehydes and ketones as the 2,4-dinitrophenylhydrazones on glass capillaries. Isomeric butyl methyl ketones as well as butyr- and valeraldehydes were well resolved.

2. Thin-Layer Chromatography (TLC)

Although the applications of TLC have been, for the most part, qualitative, the technique serves well for separations of the components of mixtures. Its popularity may be attributed primarily to its marked simplicity in equipment and technique. Because of the variation in physical and chemical properties, most positional isomers may be relatively easily separated by TLC if the proper combination of adsorbent and solvent system is found. The chromatographic (TLC) behavior of isomers based on their chemical constitution has been discussed by Petrowitz.[35]

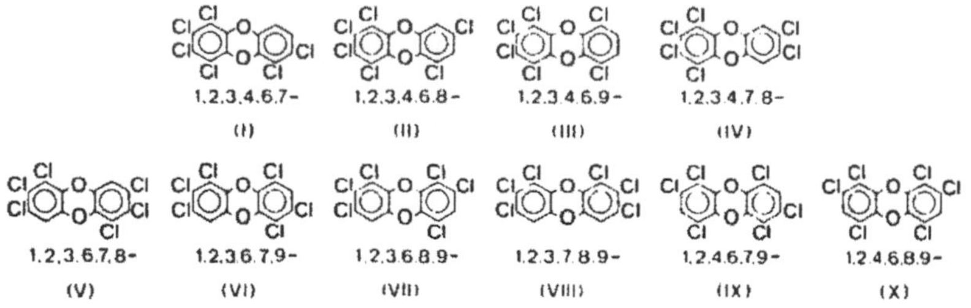

1,2,3,4,6,7- (I) 1,2,3,4,6,8- (II) 1,2,3,4,6,9- (III) 1,2,3,4,7,8- (IV)

1,2,3,6,7,8- (V) 1,2,3,6,7,9- (VI) 1,2,3,6,8,9- (VII) 1,2,3,7,8,9- (VIII) 1,2,4,6,7,9- (IX) 1,2,4,6,8,9- (X)

FIGURE 12. Hexachlorodibenzo-*p*-dioxins separated by capillary column GC on OV-61. (From Buser, H., *J. Chromatogr.*, 114, 95, 1975. With permission.)

Jakovljevic et al.[36] recently described the TLC separation of some chloro- and methyl- substituted 2-aminopyridines. Chloroaminopyridines are useful as fungicides and as intermediates in the preparation of heterocycles with herbicidal properties.[37-39] The chromatogram is shown in Figure 13.

Bishara et al.[40] reported thin layer separations of the pesticide DDT from isomers and related compounds using Aluminum Oxide E plates and 33 different solvent systems. Figure 14 shows some of the isomers which could be resolved in various of the solvent systems.

One of the 14 carcinogens for which OSHA has recently established exposure limits is 4-aminobiphenyl, which has been shown to cause bladder cancer in man.[41] Conversely, the 2-aminobiphenyl isomer exhibits no carcinogenic activity, since it is metabolized to harmless compounds.[42] Jakovljevic et al. reported the TLC separation of these isomers in benzene on silica gel,[43] and R_f values were 0.42 and 0.28 for 2-aminobiphenyl and 4-aminobiphenyl, respectively.

LeRosen and co-workers[44] studied steric factors in the TLC separations of disubstituted benzenes. A systematic study of *ortho*- and *para*- isomers indicated that adsorption of the *ortho*- isomer was dependent on internal hydrogen bond formation, steric hindrance, inductive effects, solvent effects, and the spatial arrangement of the substituent groups around the benzene ring. The same factors, with the exception of internal hydrogen bonding, were found to influence the behavior of *para*- isomers. TLC separations of isomeric halo- and nitro- derivatives of aniline were studied by Fishbein[45] who discussed the order of R_f values for mono- , di- , and tri- substituted isomers on silica gel.

A key work in the TLC separation of primary aromatic amine positional isomers is that of Gillio-Tos et al.;[46] some hR_f values ($hR_f = 100 \times R_f$) are presented in Table 1. Petrowitz has extensively studied adsorption affinity of heterocyclic compounds; he established that the differences in R_f values between α- and β- substituted derivatives are usually small in comparison with the differences between α- or β- and γ-substituted derivatives.[47] The hR_f values of the individual pyridine compounds are quoted in Table 2 in order of decreasing adsorption.[48] The TLC tables compiled by Zweig and Sherma[49] contain examples of positional isomer separations for compounds of various chemical classes.

Just as capillary columns can increase resolution over packed columns in gas chromatography, so multiple developments of a plate can increase resolution in TLC. This may sometimes be useful in instances where a selective solvent system cannot be found. Table 3 shows how relative resolution is related to the number of plate developments. An apparatus is available[50] for automatic redevelopment of thin layer plates. Programmed multiple development (PMD) retains the capabilities of TLC while providing excellent resolution and sensitivity as a result of uniformly narrow spots over the entire

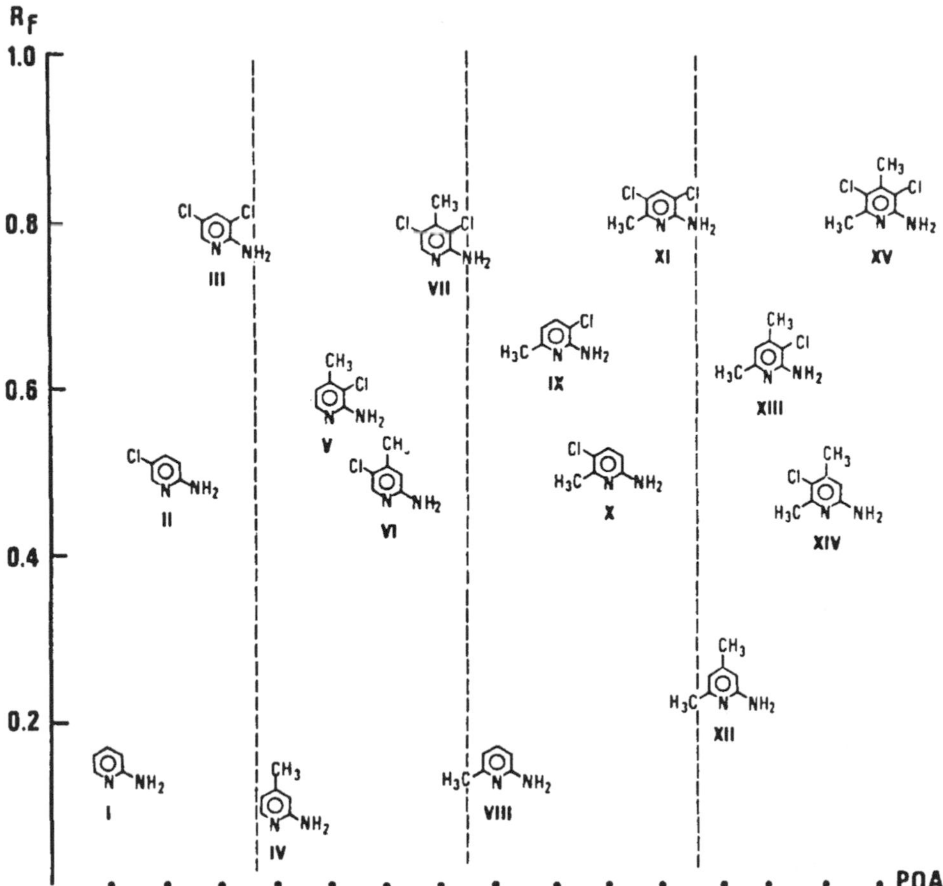

FIGURE 13. Thin-layer chromatogram of aminopyridines. Silica Gel 60 F_{254} developed with chloroform-dioxane (60:40), 10 μg spot. The circle in the pyridine represents both the unsaturation of the ring and the actual position of the developed spot (POA, point of application). The following 15 compounds were studied: 2-aminopyridine (I); 2-amino-5-chloropyridine (II); 2-amino-3,5-dichloropyridine (III); 2-amino-4-methylpyridine (IV); 2-amino-3-chloro-4-methylpyridine (V); 2-amino-4-methyl-5-chloropyridine (VI); 2-amino-3,5-dichloro-4-methylpyridine (VII); 2-amino-6-methylpyridine (VIII); 2-amino-3-chloro-6-methylpyridine (IX); 2-amino-5-chloro-6-methylpyridine (X); 2-amino-3,5-dichloro-6-methylpyridine (XI); 2-amino-4,6-dimethylpyridine (XII); 2-amino-3-chloro-4,6-dimethylpyridine (XIII); 2-amino-5-chloro-4,6-dimethylpyridine (XIV); 2-amino-3,5-dichloro-4,6-dimethylpyridine (XV). (From Jakovljevic, I. M., Bishara, R. H., and Kress, T. J., *J. Chromatogr.*, 134, 238, 1977. With permission.)

length of the chromatogram.[51] PMD has found a number of useful applications where improved resolution was needed.[52]

3. High-Pressure Liquid Chromatography (HPLC)

HPLC has been applied in a number of instances for the separation of positional isomers. As an analysis technique, HPLC may offer advantages over GC in some cases with respect to sensitivity, interference, and ability to handle fragile or nonvolatile substances.

Figure 15 illustrates separations of alkyl and aromatic nitro compounds on Varian Micropak® liquid chromatography columns. The nitropropane isomers were chromatographed on 10-μm silica; the more sterically hindered 2-nitropropane eluted before the straight-chain isomer, since it was less able to interact with the silica surface.

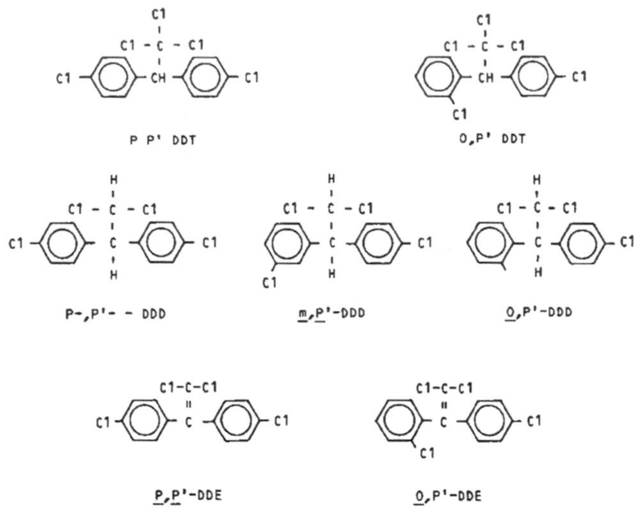

P P' DDT

O,P' DDT

P-,P'- - DDD

m,P'-DDD

O,P'-DDD

P,P'-DDE

O,P'-DDE

FIGURE 14. Structure of DDT and related compounds separated by TLC on Aluminum Oxide (Type E, F_{254}) Plates. (From Bishara, R. H., et al., *J. Chromatogr.*, 64, 135, 1972. With permission.

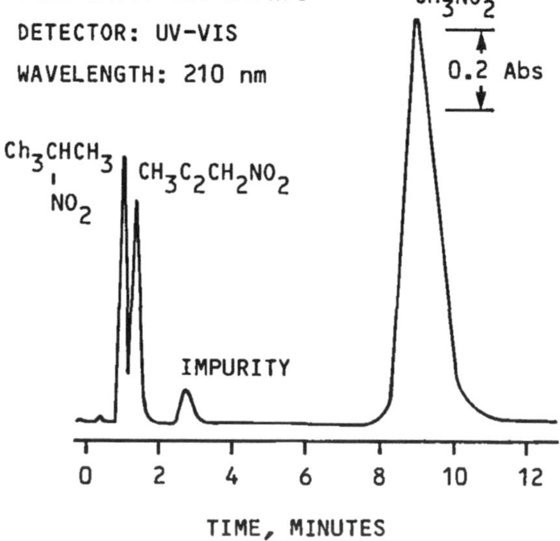

FIGURE 15. Separation of alkyl nitro- and of nitroaniline isomers on HPLC columns. (From Majors, R. E., *Liquid Chromatography at Work -18 and -19*, Varian Instr. Div., Walnut Creek, Calif., 1976. With permission.)

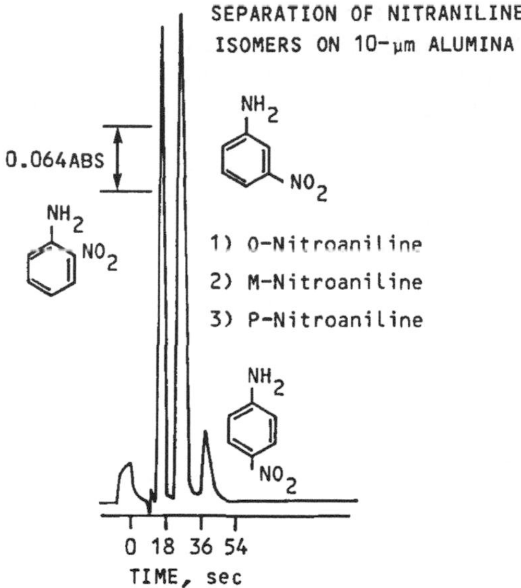

COLUMN: MICROPAK WITH 10 μM LICHROSORB ALOX T
DIMENSION: 15 cm x 2 mm
MOBILE PHASE: 40% CH_2Cl_2 IN HEXANE
FLOW RATE: 100ml/hr, PRESSURE: 325 psi
SAMPLE SIZE: 1 μl
SAMPLE CONCENTRATION: 1 mg/ml IN METHYLENE CHLORIDE
DETECTOR: UV

FIGURE 15. (Continued)

For the nitroaniline isomers, a Micropak® alumina column eluted the *ortho-* isomer first since intramolecular electronic attractions are strongest, and it is absorbed least on the alumina. In the *para-* isomer, both functional groups are free to interact, so it elutes last. Micropak® alumina and silica columns possess unique and different selectivities for different classes of compounds. Alumina is considered more selective for acidic compounds, sulfides, unsaturated molecules, and halogenated compounds.[53]

Nitroaniline positional isomers may also be completely resolved on a μ Bondapak®-NH_2 column, as shown in Figure 16. This column may circumvent problems sometimes encountered with silica gel columns due to changing water content.

In the selection of a packing to separate positional isomers, the nature of the group that is positioned differently in the different isomers is of utmost importance. If the difference is in the position of a functional group, silica or alumina would be first choices, or if the difference is in the position of a nonpolar group (like methyl), a C_{18} packing would probably be first choice. Silica is a good general-purpose adsorbent available in a wide variety of forms. It has high linear capacity and high efficiency. Columns packed with alumina normally provide slightly lower efficiencies than corresponding silica columns. Water content of a polar adsorbent must be held constant for repeatable separations. The separation of chloroaniline isomers on μ Porasil® (silica) is shown in Figure 17.

The difference in selectivities of column packing materials for positional isomers is demonstrated by Figure 18. A comparison of peaks 2 and 3 going from the amine

Table 1
HR$_f$[a] VALUES OF PRIMARY AROMATIC AMINES ON SILICA GEL G LAYERS

Amine	Solvent[b] A	B	C
o-Toluidine	62	17	84
m-Toluidine	54	10	83
p-Toluidine	40	5	80
o-Aminophenol	34	—	80
m-Aminophenol	29	—	75
p-Aminophenol	6	—	62
o-Aminobenzoic acid	62	44	98
m-Aminobenzoic acid	50	12	95
p-Aminobenzoic acid	59	29	97
o-Anisidine	60	15	81
m-Anisidine	51	9	80
p-Anisidine	11	2	58
o-Nitroaniline	69	52	93
m-Nitroaniline	64	36	92
p-Nitroaniline	58	29	91
o-Phenylenediamine	—	—	63
m-Phenylenediamine	—	—	53
p-Phenylenediamine	—	—	40
o-Bromoaniline	81	69	95
m-Bromoaniline	70	44	93
p-Bromoaniline	61	27	89
o-Chloroaniline	78	66	96
m-Chloroaniline	68	40	94
p-Chloroaniline	60	22	89

[a] The hR$_f$ values must be regarded only as guide values.

[b] Solvents: A: Dibutyl ether-ethyl acetate-acetic acid (50 + 50 + 5); B: dibutyl ether-*n*-hexane-acetic acid (80 + 16 + 4); C: *n*-butanol-acetic acid-water (40 + 10 + 50) (upper phase).

From Stahl, E., *Thin-Layer Chromatography*, 2nd ed., Springer-Verlag, New York, 1969, 501. With permission.

column to the nitrile column shows that elution order is reversed, and the separation is not very good on the nitrile column.

Paired-ion chromatography (PIC) may also be successfully applied for HPLC separations of isomers. PIC allows compounds which are strongly ionic to be separated by reverse phase chromatography. A large organic counter-ion is added to the mobile phase to form a reversible ion-pair complex with the ionized sample. The complex acts as an electrically neutral, nonpolar compound. Details of the technique are available.[54] In Figure 19, a PIC separation of phthalic acid isomers is accomplished by use of a tetrabutylammonium phosphate reagent to pair with acids.

Recently, HPLC was used for a sensitive, specific assay of sulfonylureas in human plasma.[55] Different retention times were reported for the 2-hydroxy- and the 3-hydroxychlorpropamides, metabolites of chlorpropamide.

Table 2
STRUCTURES AND hR_f VALUES OF VARIOUS PYRIDINE DERIVATIVES

Solvent[a]	α-substituted pyridine	hR_f	β-substituted pyridine	hR_f	γ-substituted pyridine	hR_f
E	—COOH	2	—COOH	6	—OH	0
A	—COOH	4	—COOH	6	—OH	2
E	—OH	6	$—CH_2OH$	13	—COOH	5
A	—OH	20	$—CH_2OH$	39	—COOH	5
E	$—CH_2OH$	18	$—NH_2$	18	$—NH_2$	5
A	$—CH_2OH$	45	$—NH_2$	45	$—NH_2$	14
E	$—NH_2$	27	—OH	23	$—CH_2OH$	4
A	$—NH_2$	50	—OH	53	$—CH_2OH$	39
E	$—CH_3$	30	—CHO	33	$—CH_3$	27
A	$—CH_3$	54	$—CH_3$	55	$—CH_3$	48
E	—CHO	51	$—CH_3$	35	—CHO	36
A	—CHO	67	—CHO	58	—CHO	56
E	—Hal	61	—Hal	57		
A	—Hal	70	—Hal	70		

Note: Layer: silica gel G.

[a] Solvents: E, ethyl acetate; A, acetone.

From Stahl, E., Ed., *Thin-Layer Chromatography,* 2nd ed., Springer-Verlag, New York, 1969, 504. With permission.

Table 3
RESOLUTION IN MULTIPLE-DEVELOPMENT TLC

	Optimum R_f		
Number of developments	Single development[a]	Total[b]	Relative resolution[c]
1	0.33	0.33	1.00
2	0.29	0.50	1.31
3	0.26	0.61	1.52
5	0.22	0.71	1.73
10	0.16	0.82	2.00

[a] Position of band after first development.
[b] Position of band at completion of multiple development.
[c] R_s divided by R_s for single development (R_f, 0.33).

From Karger, B. V., Snyder, L. R., and Horvath, C., *An Introduction to Separation Science,* John Wiley & Sons, New York, 1973, 294. With permission.

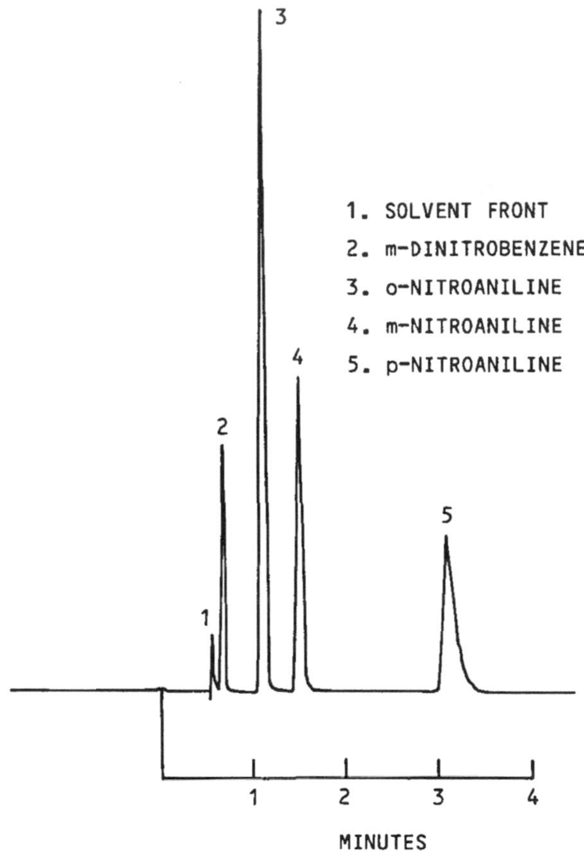

FIGURE 16. Separation of nitroanilines by HPLC. Column: μ Bondapak NH_2® (Waters) 300 mm × 4 mm I.D., mobile phase: hexane-methylene chloride (65:35); flow rate: 6 mℓ/min; ambient temperature. (From Vivilecchia, R. V., Cotter, R. L., Limpert, R. J., Thimot, N. Z., and Little, J. N., *J. Chromatogr.*, 99, 407, 1974. With permission.)

B. Spectroscopy

1. Nuclear Magnetic Resonance (NMR)

NMR is widely applicable to determination of isomeric impurities in substances. Reasonably good quantitation is possible by use of an internal standard added to the tube containing the sample solution and by careful integration. The internal standard should ideally absorb as a singlet only, should be pure material, and the absorption should be positioned in the spectrum such that it does not interfere with any sample peaks, ^{13}C satellites, or spinning side bands.

In many cases, different positional isomers will have separate distinct absorptions which can be used for quantitation. The absorption-per-proton is calculated for the internal standard (IS) and for each isomer, and using the latter, the molecular weights, and the sample weights, the weight percent of each isomer may be found as follows:

$$100 \times \frac{\text{wt IS, mg}}{\text{wt SA, mg}} \times \frac{\text{APP, SA}}{\text{APP, IS}} \times \frac{\text{MW SA}}{\text{MW IS}}$$

where APP = absorption per proton or integration count per proton and MW = molecular weight.

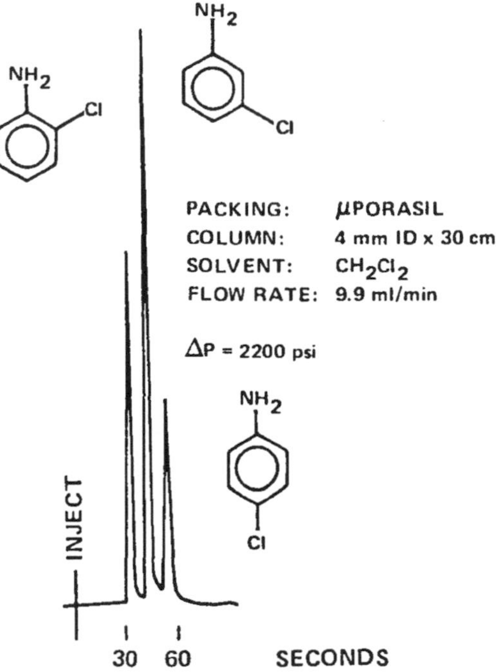

FIGURE 17. Separation of chloroaniline isomers on μPorasil® (Waters). (From LC School reference material, Waters Assoc., Milford, Mass., 1977, GT-3. With permission.)

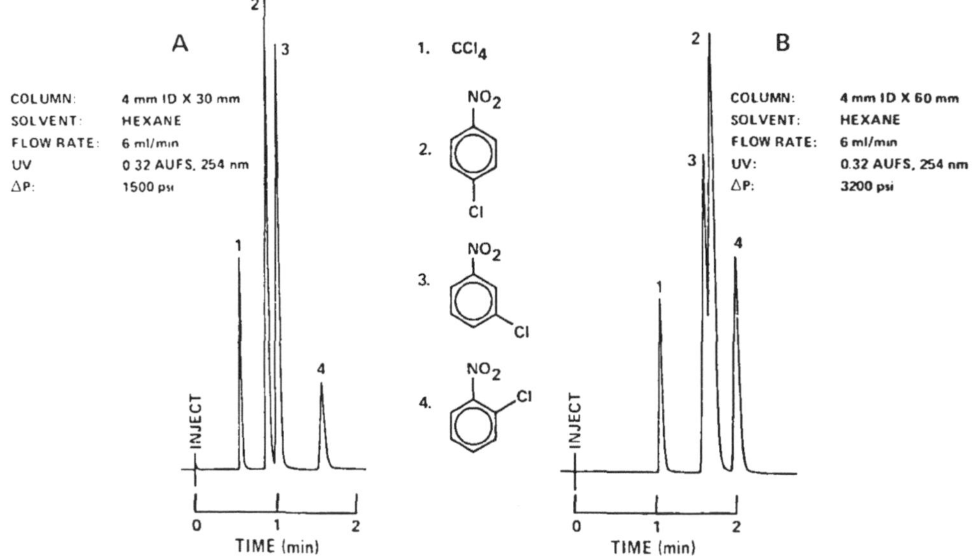

FIGURE 18. Separation of nitrochlorobenzenes: comparison on μBondapak NH_2® (A) and on μBondapak CN® (B). (From LC School reference material, Waters Assoc., Milford, Mass., 1977, PGS-11. With permission.)

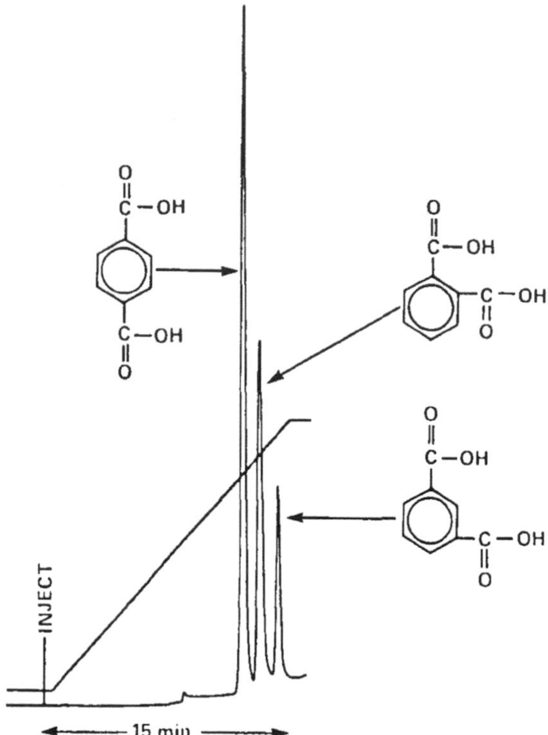

FIGURE 19. PIC separation of phthalic acid isomers. Column: μBondapak C_{18}®, 4 mm I.D. × 30 cm; solvent: (A) water (PIC Reagent A), (B) methanol (PIC reagent A) 5% B → 40% B using gradient program No. 6 with Model 660 programmer 15 min; flow rate: 2.0 mℓ/min; detection by μV @ 254 nm. (From LC School reference material, Waters Assoc., Milford, Mass. 1977, LS-17. With permission.)

Often two or more sample absorptions may be used for comparative calculations. In quantitation by NMR, accuracy of area measurements used in the calculations may be maximized by using (1) the best signal-to-noise ratio attainable to ensure negligible background noise contribution to the integration, (2) care to avoid saturation, (3) several rapid integration sweeps to calculate an average ratio, and (4) proper phase adjustment for easy-to-measure integrals.

Mixtures of alkene structural isomers have been assayed by NMR.[56,57] The extent of isomerization of pure 5-methyl-tetrahydropyridine to the 3-methyl isomer after treatment with a hot acetic acid-hydrochloric acid mixture was readily followed by observing the 3-methyl resonance signal in the total mixture. This type of NMR method is simple and one knows exactly what one is measuring.

Cerfontain et al.[59] recently described the quantitative NMR determination of isomeric arenesulfonic acids using a set of linear equations obtained from areas of the various absorptions of unknown mixtures. Toluene -2,5-, -3,4-, and -3,5-disulfonic acids as well as 3-bromo- or 3-fluorobenzenesulfonic acids in mixtures could be determined. The analysis could in most cases be performed without requiring reference standards of the sulfonic acids.

In spite of practical problems involved with its setup and use, carbon-13 (^{13}C) NMR has greater potential than ^{1}H NMR for the study of organic systems. Carbon resonances of organic compounds cover a chemical shift range of 600 ppm, and the molecular backbones of the molecules may be directly studied. It is well known that some carbon atoms of isomeric organic molecules may give widely different ^{13}C chemical shifts.[60] For instance, *ortho-* , *meta-* , and *para-* carbon atoms of aromatic systems show substantially different chemical shifts for many substituent groups.[61] Smith studied the ^{13}C NMR spectrum of a mixture of the three isomers of dianisylmethane, but quantitation was not discussed.[62] Retcofsky and Friedel have presented extensive data for 2- , 3- , and 4- substituted pyridines.[63,64]

Quantitation by ^{13}C NMR has been slow in developing because significant differences in relaxation times of ^{13}C nuclei in different environments often result in signal areas which are not proportional to the number of nuclei present.

2. IR and UV Spectroscopy and Mass Spectrometry

Just as isomers may be distinguished and quantitated by NMR, so may other spectroscopic techniques be applied. The appropriate sources should be consulted, however, for an explanation of the particular problems involved in quantitation.

Streitweiser and Fahey[65] studied the nitration of Fluoranthene

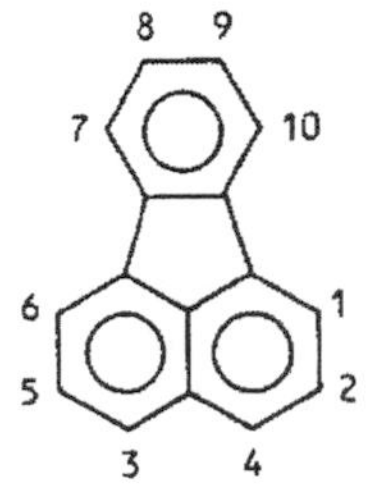

8 9 7 10 6 1 5 2 3 4

which can produce five possible mononitro-isomers (1,2,3,7,8). IR spectra of chloroform solutions of nitration mixtures were taken and compared with similar spectra of the pure isomers. The multicomponent isomer system was quantitated by least squares using a computer program which was given IR intensities at seven selected wavelengths. Multicomponent analysis in absorption spectroscopy is discussed in detail by Bauman.[66]

Mixtures of three isomeric toluenesulfonic acids (*o-* , *m-* , and *p-*) in concentrated sulfuric acid have been quantitatively determined by multicomponent spectrophotometric analysis.[67] The determination was performed by subjecting the absorbances of the unknown mixture and of its constituents (gathered at several wavelengths) to a least squares computer treatment. The *para-* isomer content was determined with excellent accuracy, but the accuracies of determination of the *meta-* and *ortho-* isomers were less satisfactory due to spectral similarities. A similar treatment was used for quantitation of arylsulfonic acids by UV spectrophotometry.[68]

Mantel and Stiller[69] recently reported the simultaneous determination of benzoic acid and its monohydroxy- isomers by measuring absorbances of their mixtures at several wavelengths and that of the Fe(III) complex of the *o-*hydroxy-benzoic acid at another wavelength. Solution of a simple set of two equations yielded individual acid concentrations with an error of ±5 to 10%.

Quantitative analysis by mass spectrometry is handled quite similarly to that by IR or UV absorption spectrometry. Pure samples of each compound must be available in order to obtain mass spectra of each component. From an inspection of the individual mass spectra of compounds known or suspected to be present in the mixture, peaks are chosen for the analysis on the basis of intensity and freedom from interference. If at all possible, single-component peaks are selected. Height of the single-component peak is used for the quantitation, or, if the sample has no single-component peaks, simultaneous linear equations may be used. Gifford et al.[70] analyzed a mixture of isomeric butanols. A large number of compounds may be analyzed by mass spectrometry without extensive sample pretreatment, and very small samples may be used.

Isomeric xylenes yield virtually identical mass spectra since they all form the same methyltropylium ion.[71] Similarly, the mass spectra of methylethylbenzenes shown in Figure 20 fail to exhibit appreciable differences. It is easier to distinguish between similar isomers in heterocyclic compounds, as shown by the mass spectra of 3-ethyl-4-methylpyridine and 3-ethyl-5-methylpyridine, which differ in the M/(M − 1) ratio (see Figure 20). The difference is even more pronounced if the ethyl group is attached to the 2- or 4- position as in 2-ethyl-6-methylpyridine.[72] Examples of complex molecules which are positional isomers are available.[73]

Single-ion monitoring[74,75] may also be used for the quantitation of isomers by mass spectrometry if the isomers produce distinct ions and if pure samples are available. This is very useful in cases of bad or complete overlap of gas chromatographic peaks in gas chromatography/mass spectrometry experiments.

III. STEREOISOMERS

A. Optical Rotation

Enantiomers have optical rotations that are equal in magnitude, but opposite in direction. As optical rotations are easily measured, it was natural that optical rotation was one of the earliest techniques used to determine the purity of mixtures of enantiomers. The optical rotation of a solid or liquid depends on the substance, the layer thickness, the temperature, and the wavelength. For a solution of an optically active substance, the rotation also depends on the concentration of the optically active solute, on the nature of the solvent, and possibly on the nature and concentration of optically inactive co-solutes. Experimental conditions should be fully defined, and results can be reported as specific rotation, the number of degrees of rotation observed if a 1-dm tube is used, and the compound being examined is present to the extent of 1 g/cc. To calculate specific rotation from observations with other conditions,

$$[\alpha] = \text{specific rotation} = \frac{\text{observed rotation}}{\text{length (dm)} \times \text{g/cc}}$$

Specific rotation is not useful as a first-time, absolute assessment of the optical purity of a compound, since one may have isolated a partially resolved enantiomer (which even after repeated crystallization does not change its rotation). A conclusion that a compound is optically pure may be strengthened if both of its enantiomers can be prepared and their rotations are equal and opposite.

If no interaction exists, specific rotations $[\alpha]_1$ and $[\alpha]_2$ of two optically active components in a mixture are additive:[76]

$$[\alpha] = C_1[\alpha]_1 + (1 - C_1)[\alpha]_2$$

If mixtures of optically active species of interest are prepared and concentration vs. specific rotation is plotted, this calibration curve may be used to analyze unknown mixtures of the same isomers. In at least one case, however, specific rotation is not a linear function of enantiomer concentration.[77] The applications of optical rotation are discussed more fully in Volume I, Chapter 6.

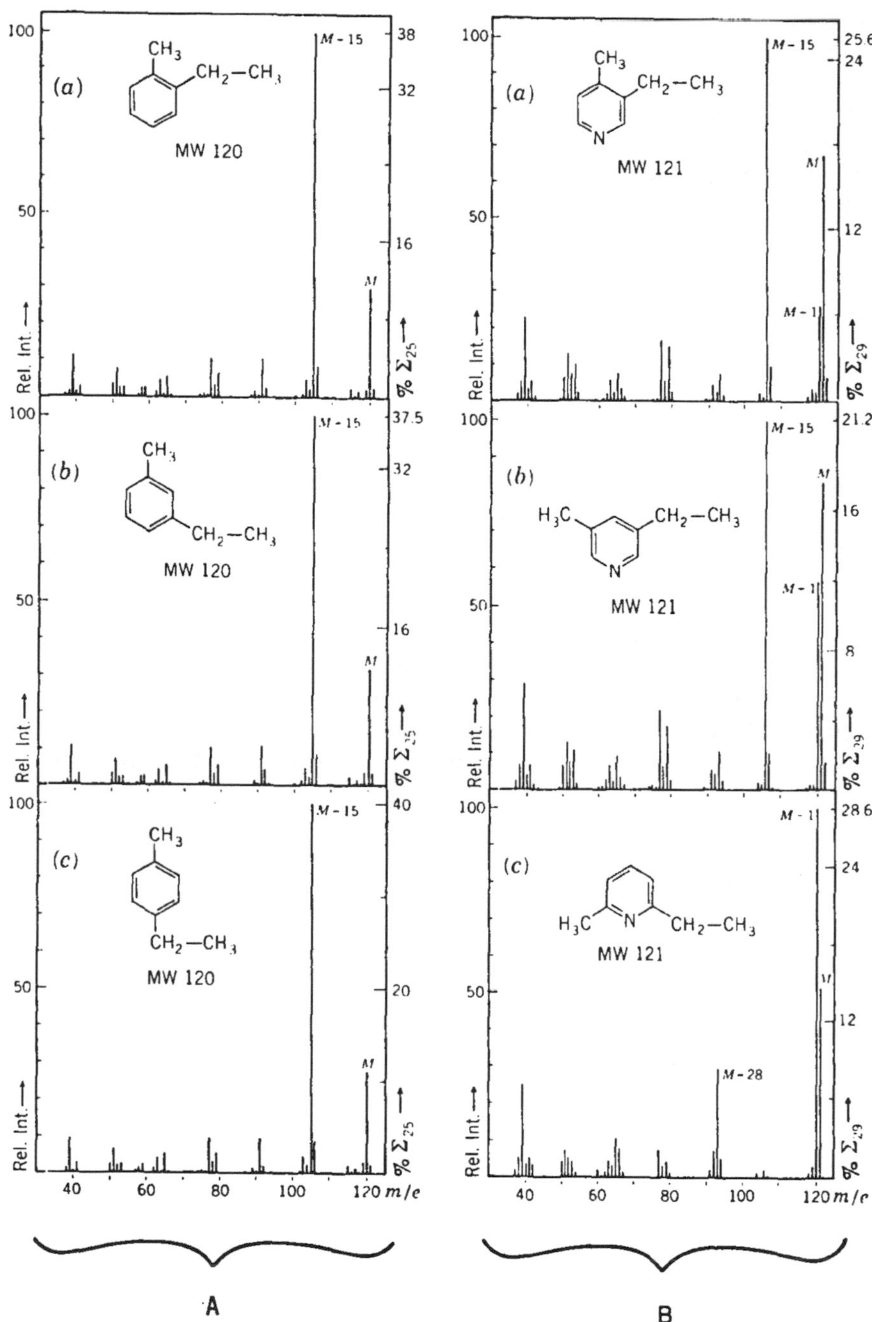

FIGURE 20. (A) Mass spectra of isomeric methylethylbenzenes. (a) 1,2-isomer, (b) 1,3-isomer, and (c) 1,4-isomer. (B) Mass spectra of isomeric pyridines. (a) 3-ethyl-4-methylpyridine, (b) 3-ethyl-5-methylpyridine, and (c) 2-ethyl-6-methylpyridine. (From Biemann, K., *Mass Spectrometry*, McGraw-Hill, New York, 1962, 152. With permission.)

B. Chromatography

Chromatographic techniques have, by far, been the most useful and most widely applied for analysis of stereoisomeric purity. Almost from the first, it was recognized that chromatographic methods offered distinct advantages, including small sample size, independence from the magnitude of specific rotation, and independence from other optically active species initially present. Research of recent years has produced a number of significant advances in chromatographic separation techniques for resolution of optical isomers. Generally, the resolution of enantiomers by chromatographic means has been achieved by either conversion of the racemate to a mixture of diastereomers by a suitable chemical reaction with a chiral reagent or through the use of a chiral sorbent. Diastereomers may generally be separated on the basis of their different physical properties on an achiral sorbent. Geometric isomers also separate relatively easily.

Separation mechanisms and correlations between structure and selectivity for stereoisomer separations have been discussed for gas chromatography[78,79] and liquid column chromatography,[80] as well as for thin-layer chromatography.[81,82]

Conversion of enantiomeric mixtures into diastereomers for their chromatographic separation is still more practical than direct separation on chiral adsorbents or liquid phases. Formation of diastereomers is discussed in detail by Wilen,[83] and resolving agents are readily available in good optical and chemical purity.

Regardless of whether it is done directly or indirectly, the chromatographic separation of stereoisomers provides a basis for development of a quantitative analysis of the isomeric purity. Such factors as completeness of reaction with a derivatizing agent, precision of determination, and accuracy as established by comparison with the results of independent methods, should all be considered in designing a method for assay of stereoisomeric purity.

1. Gas Chromatography (GC)

Examples of separations of geometrical isomers by GC are plentiful in the literature and will not be discussed here extensively. Product literature from Applied Science Laboratories,[84,85] and Supelco[86] discuss separations of *cis-trans* isomers. Recently a bound-layer cation exchanger, the sulfobenzyl derivative of Porasil C®, was synthesized and used as a gas chromatographic stationary phase.[87] It showed marked selectivity for alkenes relative to alkanes when converted to the Ag^+ form, and was useful for the analytical separation of several *cis-trans* isomeric alkenes.

Observations dealing with gas chromatographic resolution of diastereomers began to surface in the literature about 1960. Gil-Av and Nurok completely resolved a series of racemic secondary alcohols as lactic acid derivatives on a 150-ft capillary column coated with polypropylene glycol.[88] Later Gil-Av used a similar approach to resolve racemic amino acids[89] which were chromatographed as the *N*-trifluoroacetyl esters of 2-*n*-alkanols. The latter technique was applied to determination of the configuration of the amino acids in two antibiotics of the Vernamycin B group.[90]

Much of the work on diastereomer GC separations has derived from attempts to separate amino acid enantiomers. Weygand et al.[91] showed that capillary columns could be used to resolve some methyl esters of *N*-trifluoroacetyldipeptides. For the first time, it was possible to accurately quantitate the racemization which occurred during peptide synthesis. A wide variety of derivatizing agents and operating conditions for determination of enantiomeric amino acids have been described and are summarized in Table 4.[92-107]

Another class of compounds extensively subjected to study of enantiomer separations is amines. *N*-Trifluoroacetyl-L-prolyl chloride was used by Gordis[108] to separate the enantiomers of amphetamine on a packed column. Gordis also correlated the res-

Table 4
DETERMINATION OF AMINO ACID ENANTIOMERS BY GAS CHROMATOGRAPHY OF DIASTEREOMERIC DERIVATIVES

Application	Ref.
Alanine, valine, leucine, proline, methionine and phenylalanine as *N*-trifluoroacetyl-L-prolyl methyl esters on a packed column	92
Polyfunctional D,L amino acids separated as the *N*-trifluoroacetyl-L-prolyl methyl esters on a packed column — in some instances, silyl amino acid esters were coupled with the resolving agent	93
Separation of enantiomers of threonine and of allothreonine using *N*-trifluoroacetyl-L-prolyl chloride of improved optical purity	94
Assay for free D-amino acids in the house fly by packed column GC of *N*-trifluoroacetyl-L-alanyl esters	95
S-prolyl dipeptide derivatives for quantitative estimation of R- and *S*-leucine enantiomers	96
15 amino acids on 22 stationary phases; *N*-trifluoroacetyl-L-prolyl methyl esters; reaction conditions for quantitative derivatization; relation between structure and separation	97
Alanine, valine, leucine, proline enantiomer separations; influence of ester and of *N*-perfluoroacyl group on separations	98
Resolution of racemic amino acids by conversion to *n*-butyl esters; use of *N*-trifluoroacetyl-L-prolyl, *N*-trifluoroacetyl-L-hydroxyprolyl, *N*-trifluoroacetyl-L-4-thiazolidine carbonyl and *N*-trifluoroacetyl-L-pyroglutamyl resolving agents	99
Resolution of 21 racemic amino acids as the *N*-trifluoroacetyl-L-prolyl *n*-butyl esters on a polar and a nonpolar stationary phase; relation between structures and separation factors	100
New chiral reagents for separation of amino acid enantiomers; camphor-related resolving agents with rigid, skeletal structures and high volatilities	101
Comparison of three techniques for separation of enantiomers of leucine; contains a number of excellent application references for amino acid enantiomer determinations	102
Assignment of amino acid peaks in chromatogram from marine sediment; *N*-trifluoroacetyl 2-butyl esters	103
Resolution of a variety of amino acids as *N*-perfluoroacyl-L-menthyl esters on seven stationary phases	104
Resolution of DL-isovaline as *N*-trifluoroacetyl-DL-isovalyl-L-leucine isopropyl ester	105
Separation of amino acids including histidine, arginine, and tryptophan on capillary columns as pentafluoropropionyl-(+)-3-methyl-2-butyl esters	106
Separation of amino acid diastereomers on an optically active GC stationary phase	107

olution of amphetamine enantiomers with measurements of optical rotation. Gunne[109] used GC of the same derivative for determination of D- and L-amphetamine in human urine. Excretion of amphetamine enantiomers was followed for up to 60 hr after the administration of a known dose. Wells studied the application of the same technique to determination of amphetamine salts and commercial tablet dosage forms.[110] Recently Souter[111,112] studied the GC resolution of a variety of structurally related enantiomeric primary amines, including several amphetamines. New amino acid resolving agents were applied[111] for the separation of the amine enantiomers and, as shown in Figure 21, 1-methyl-3-phenylpropylamine was better resolved with *N*-trifluoroacetyl-L-leucyl chloride than with the commercially available *N*-trifluoroacetyl-L-prolyl chloride. The use of L-azetidine and L-thioproline ring systems for resolving agents analogous to *N*-trifluoroacetyl-L-prolyl chloride was also studied,[112] and better separations were also observed for some amines. Very recently, Gal[113] used (−)-α-methoxy-α-(trifluoromethyl)phenylacetyl chloride as a resolving agent for amphetamine and eight related amines. This chiral acylating reagent is suitable for the determination of the optical composition of compounds extracted from biological fluids. It was used in the determination of enantiomeric composition of 2,5-dimethyoxy-4-methylamphetamine excreted in rat urine after drug administration.

Table 5 describes some other applications of resolutions of enantiomeric amines as diastereomeric derivatives.[114-118]

Although their importance is every bit as great as amines, the separation of alcohol

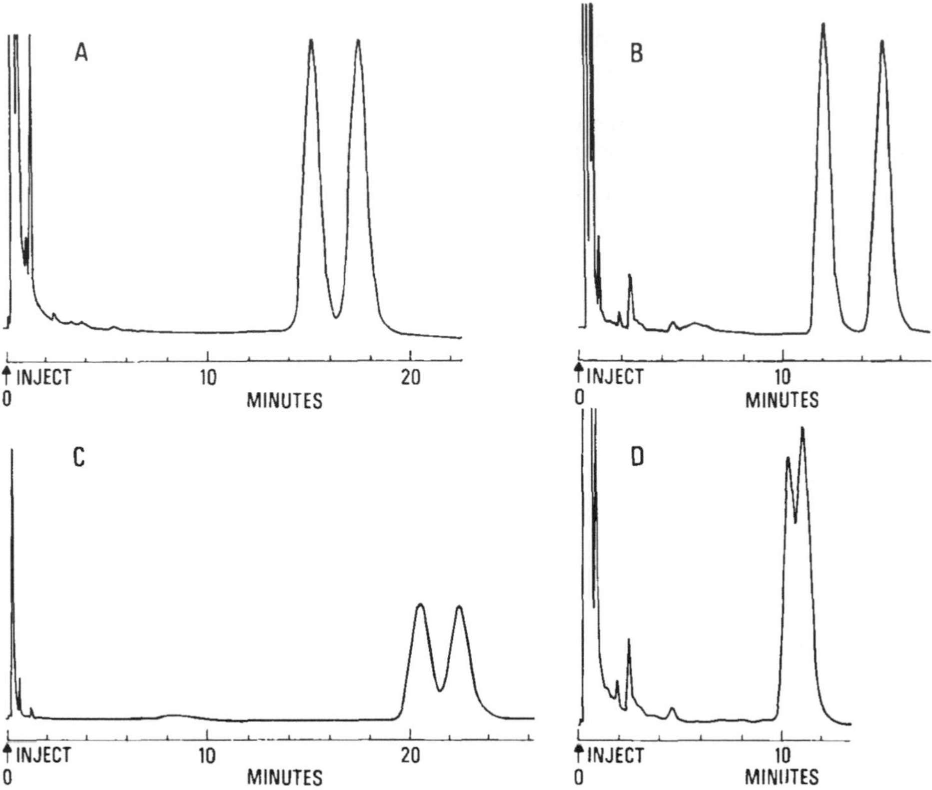

FIGURE 21. Comparison of separations of *N*-trifluoroacetyl-L-prolyl- and *N*-trifluoroacetyl-L-leucylamine diastereomers at 200°C on a 6 ft glass column packed with 5% DEGS on 70/80 mesh Anakrom AB® with a flow rate of 60 m*l*/min of helium. (A) *N*-trifluoroacetyl-L-leucyl-D,L-1-methyl-3-phenylpropylamine, α = 1.153; (B) *N*-trifluoroacetyl-L-prolyl-D,L-α-ethylphenethylamine, α = 1.243; (C) *N*-trifluoroacetyl-L-prolyl-D,L-1-methyl-3-phenylpropylamine, α = 1.096; (D) *N*-trifluoroacetyl-L-leucyl-D,L-α-ethylphenethylamine, α = 1.078. (From Souter, R. W., *J. Chromatogr.*, 108, 265, 1975. With permission.)

Table 5
RESOLUTION OF AMINE ENANTIOMERS AS DIASTEREOMERIC DERIVATIVES BY GC

Application	Ref.
N-trifluoroacetyl-L-prolyl derivatives of racemic amines and *N*-acyl-1-phenyl-2-aminopropane diastereomers on packed columns	114
Resolution of racemic cyclic amines as *N*-trifluoroacetyl-L-prolyl derivatives; structural correlations to separations	115
Analysis of steric purity of aliphatic amines containing two asymmetric centers — resolved as *N*-trifluoroacetyl-(*S*)-prolylamides	116
Separation of amines as chrysanthemoylamide derivatives	117
Use of *N*-pentafluorobenzoyl-*S*-(−)-prolyl-1-imidazolide as an electron capture-sensitive reagent for the determination of enantiomeric composition	118

enantiomers has been less well studied. Simple 2-alkanols were resolved as (+)mandelic acid esters by Cross et al.,[118] and the separations were discussed in light of the compound structures. Other aliphatic alcohols have been resolved with R-(+)-1-phenylethylisocyanate, which was synthesized and then used to prepare diastereomeric carbamates.[119] Relative retention times were used to assign the absolute configuration of a homologous series of secondary alcohols. Anders and Cooper resolved several phenylalkylcarbinols as 3 β-acetoxy-Δ^5-etienates and as menthyl carbonates.[120] (−)-Menthyl chloroformate was used to separate both amino and hydroxyl compounds: the phenylalkylcarbinols were again included.[121] Hamberg[122] used (R)-1-phenylethylisocyanate to form *N*-(1-phenylethyl)urethanes from hydroxyl compounds with the $-CH(OH)CH_3$ group, and these were separated by gas chromatography. Both ω 2-hydroxyacids and 2-alkanols from tissues, etc. were resolved using very small samples. The influence of the structure of the resolving agent on the separations of racemic alcohols has been described.[123,124] Racemates may also be resolved by coinjection of the racemate and an appropriate, volatile resolving agent onto a nonoptically active column;[125] bicyclic ketones and alcohols were studied with a variety of resolving agents and polar GC columns.

Optical isomers of insecticidal pyrethroids have been determined by GC. The pyrethroids have two geometric and two optical isomers arising from the chrysanthemoyl structure (D-*trans*- , L-*trans*- , D-*cis*, and L-*cis*). Isomers of these pyrethroidal esters have very different toxicities to insects. The chrysanthemic acid obtained during hydrolysis of the corresponding pyrethroids was esterified with D- or L-2-octanol and the diastereomeric ester derivatives were resolved on a packed GC column.[126] (+)-α-Methylbenzylamine has also been used to react with chrysanthemoyl chloride (produced from the chrysanthemic acid from the pyrethroid's hydrolyses) to form diastereomeric amides which were separated on a capillary column.[127] Recently these techniques were applied to determination of geometric and optical isomers of a new pyrethroid: DL-*cis*- and DL-*trans*- isomers were separated and quantitated as one step, and the optical isomers were hydrolyzed to the corresponding acids, which were derivatized to diastereomeric esters with D- or 1-2-octanol, and determined by GC.[128]

A representative sample of other determinations of enantiomers as diastereomers by gas chromatography is presented in Table 6.[129-137]

At the Sixth International Symposium on gas chromatography in Rome, Gil-Av et al.[138] reported some preliminary results on optical isomer separations on chiral gas chromatographic stationary phases. Eighteen pairs of *N*-trifluoroacetyl-α-amino acid esters were resolved on glass capillary columns coated with *N*-trifluoroacetyl D- or L-isoleucine lauryl ester and with *N*-trifluoroacetyl L-phenylalanine cyclohexyl ester. The separation of antipodes was theorized as involving readily reversible association between the enantiomers and the asymmetric solvent molecules.

The direct resolution of enantiomers on optically active stationary phases is advantageous in many cases since (1) the chiral stationary phase may be reused whereas formation of diastereomers consumes a supply of resolving agent, and (2) derivatization of enantiomers with achiral reagents is less involved than diastereomer formation in many cases. Although much of the research work with chiral stationary phases has been aimed at mechanistic and structural studies, a variety of types of compounds have been separated, and these will be discussed.

By far most work has been done with amino acid separations. Recently several stationary phases have been developed with higher operating maximum temperatures than were previously available. Much of the work with amino acid separations on chiral GC stationary phases is summarized in Table 7.[139-161] Besides the peptide stationary phases studied so extensively for amino acid separations, the other major class of optically active stationary phases useful for direct gas chromatographic separation of

Table 6
MISCELLANEOUS DETERMINATIONS OF ENANTIOMERIC PURITY BY GC OF DIASTEREOMERIC DERIVATIVES

Application	Ref.
Study of mechanism of separation of diastereomeric esters on a packed column	129
Study of solute structure on separation of diastereomeric esters and amides on a packed column	130
Resolution of racemic carbohydrate diastereomers — determination of configuration on a capillary column	131
Resolution of racemic aldehydes and ketones via diastereomeric acetals using 2,3-butanediol on a capillary column	132
Separation of diastereomeric amides and esters on a capillary column	133
Resolution of diastereomeric ketones on glass capillary columns	134
Separation of vicinal diastereomers (bromomethoxy- and dibromo alkanes)	135
Separation of (−) menthyl and methyl esters of diastereomeric isoprenoid acids of geological interest — pristanic and phytanic acids	136
Chromatographic separability of some diastereomeric carbamates; GC and NMR correlations with separability and stereochemistry	137

enantiomers is the carbonyl-bis-(amino acid esters) the first example of which was reported by Feibush and Gil-Av.[162] This phase was demonstrated to have a significant resolving power for chiral primary amines as *N*-trifluoroacetyl derivatives. The nature of this phase was studied in detail by Corbin and Rogers,[163] Feibush et al.,[164] Rubenstein et al.,[165] and by Lochmüller and Souter.[167-171] The effect of substituent changes on the stationary phase was studied in detail by Lochmüller and Souter,[170] who reported results of separations of amine enantiomers on five different chiral carbonyl-bis-(amino acid ester) phases.

The discovery that enantiomers could be resolved with very large α values on a smectic liquid crystalline phase of carbonyl-bis-(D-leucine isopropyl ester) was recently reported by Lochmüller and Souter.[171] These authors showed that several carbonyl-bis-(L-valine esters) displayed liquid crystalline behavior and were useful as highly selective stationary phases for GC resolution of enantiomeric amines.[170] Differences observed for the number of theoretical plates (N) for each peak in a racemate separation on smectic chiral GC phases may be related to differences in diffusion coefficients of the chiral solutes in the chiral medium.[170] A chromatogram demonstrating this anomalous peak shape is shown in Figure 22.

Recently the status of enantiomeric analysis by GC was summarized by Gil-Av,[172] who commented on some relatively recent physicochemical methods for optical purity determination. Very recently *N*-acyl derivatives of chiral amines were shown to be applicable for separation of amines (aromatic, aliphatic, cyclic), amino acids, and acids.[173]

2. High-Pressure Liquid Chromatography (HPLC)

Most separations accomplished so far for enantiomers by HPLC have been by conversion of the enantiomers to diastereomers, with appropriate resolving agents. The use of HPLC for such work has been expanding steadily because of its high sensitivity and versatility and because the low volatility of many samples precludes use of GC.

Helmchen and Strubert[174] showed that diastereomeric amides formed from reactions of racemic amines with optically pure *O*-methylmandelyl chloride were generally separable by HPLC, and the method described appeared particularly well suited for detection of trace amounts of optical impurities.

Furukawa et al.[175] used HLPC to approach the problem of resolution of amino acid enantiomers. Several amino acids were chromatographed as *N*-D-10-camphorsulfonyl

Table 7
DETERMINATION OF AMINO ACID ENANTIOMERS BY DIRECT RESOLUTION ON CHIRAL GC STATIONARY PHASES

Application	Ref.
Separation of *N*-trifluoroacetyl-(±)-alanine t-butyl ester on packed column coated with *N*-trifluoroacetyl-L-valyl-L-valine cyclohexyl ester	139
Separations of *N*-trifluoroacetyl-α-amino acid esters on di- and tripeptide phases of L-valine (capillary columns)	140
Perfluoroacyl esters of α-amino acid esters separated on *N*-trifluoroacetyl-L-phenylalanyl-L-leucine cyclohexyl ester capillary column	141
N-Trifluoroacetyl-α-amino acid methyl esters separated on *N*-lauroyl-L-valyl-t-butylamide capillary column	142
N-Trifluoroacetyl-α-amino acid isopropyl esters separated on *N*-trifluoroacetyl-L-phenylalanyl-L-leucine cyclohexyl ester capillary column	143
N-Trifluoroacetyl-α-amino acid isopropyl esters separated on *N*-trifluoroacetyl-L-valyl-L-valine cyclohexyl este. capillary column	144
Use of several perfluoroacyl reagents for derivatization of amino acids separated as esters on *N*-trifluoroacetyl-L-phenylalanyl-L-leucine cyclohexyl ester capillary column	145
Separation of amino acids on peptide stationary phases — structural correlations; dipeptide and higher polymeric phases on capillary columns	146
N-Trifluoroacetyl-L-valyl-L-leucine cyclohexyl ester as capillary column stationary phase for separations of 11 amino acids as *N*-trifluoroacetyl isopropyl esters	147
N-Caproyl-L-valine-*n*-hexyl amide as stationary phase for GC separation of *N*-trifluoroacetyl amino acid isopropyl esters on a capillary column	148
N-Trifluoroacetyl-L-α-amino-*n*-butyryl-α-amino-*n*-butyric acid cyclohexyl ester stationary phase for separation of amino acid isopropyl esters (*N*-trifluoroacetyl) on a capillary column	149
Separation of *N*-trifluoroacetyl amino acid isopropyl esters on a capillary column coated with *N*-trifluoroacetyl-L-norvalyl-L-norvaline cyclohexyl ester	150
Resolution of *N*-trifluoroacetyl-(±)-tert. leucine isopropyl ester on *N*-trifluoroacetyl-L-norvalyl-L-norvaline cyclohexyl ester and *N*-trifluoroacetyl-L-alanyl-L-alanine cyclohexyl ester capillary columns	151
Comparison of separation characteristics of several dipeptide stationary phases for derivatized amino acid enantiomers, using capillary columns	152
Comparison of separation characteristics of three dipeptide phases for resolution of naturally occurring amino acids, using capillary columns	153
Use of packed GC columns coated with chiral diamide stationary phases to separate *N*-trifluoroacetyl methyl and isopropyl amino acid esters (two phases with relatively high — 190°C — maximum operating temperatures)	154
Methionine dipeptide stationary phases for separation of amino acid enantiomers on a capillary column	155
Resolution of amino acid esters (as *N*-trifluoroacetyl derivatives) on *N*-lauroyl-L-valyl-L-valine lauryl ester capillary column	156
Separation of racemic amino acids on a *N*-trifluoroacetyl-L-phenylalanyl-L-asparagine acid-bis-cyclohexyl ester capillary column	157
Separation of amino acid enantiomers on optically active dipeptide carborane stationary phases on capillary columns — reasonably high (150°C) maximum operating temperature	158
N-Trifluoroacetyl-L-prolyl-L-proline cyclohexyl ester as a stationary phase for separation of amino acid derivatives on a capillary column	159
Diamides derived from L-valine as stationary phases for separation of (±)-α-amino acid derivatives on capillary columns	160
Use of polysiloxanes with chiral groups as stationary phases for short-time resolution of many amino acid derivatives on capillary columns	161

p-nitrobenzyl esters. More recently Souter[176] separated α-methylbenzylamines, amphetamines, and related amines as the (+)-10-camphorsulfonates in several solvent systems.

The ratios of enantiomers of citronellic and related acids were determined by Valentine et al. by HPLC as well as by NMR.[177] Koreeda et al.[178] used HPLC for prepar-

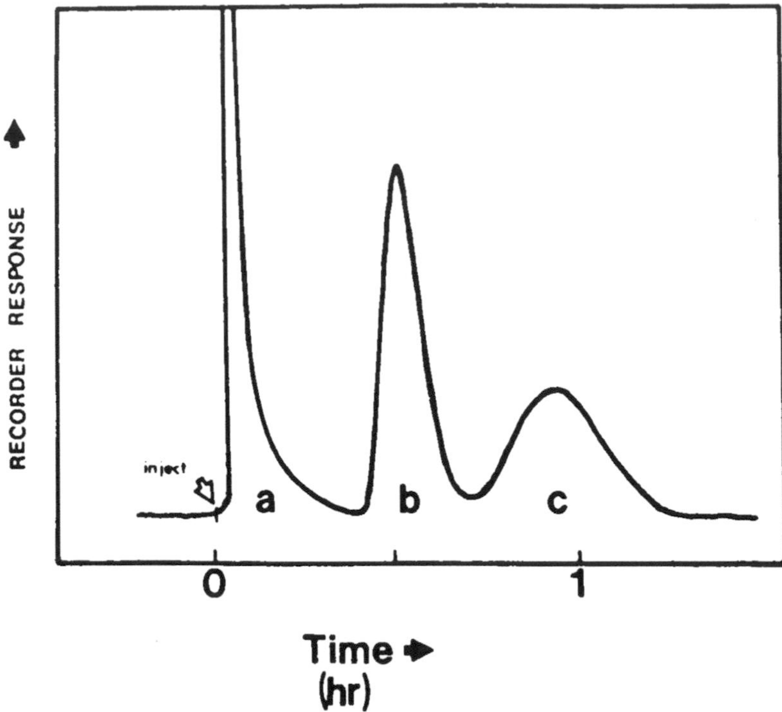

FIGURE 22. Anomalous peak shapes observed for the separation of N-PFP (DL) α-methylbenzylamine on a 5.0% carbonyl-bis-(L-valine isopropyl ester) column at 96.9°C (on the first smectic phase). a, solvent; b, peak for the (S) or D (+) solute; c, peak for the (R) or L(−) solute. (From Lochmuller, C. H., and Souter, R. W., *J. Chromatogr.*, 88, 41, 1974. With permission.)

ative separation of some *cis*-diol enantiomers in an effort to establish the absolute configuration of natural (+)-abscisic acid. In the latter work, (+)-α-methoxy-α-trifluoro-methylphenylacetyl chloride was used to form diastereomeric esters. The same resolving agent was used by Nakanishi et al. for separation of some diastereomers in connection with studies of insect juvenile hormone.[179]

Some excellent separations of acyclic isoprenoid acids were reported via the R(+)- or S(−)-α-methyl-*p*-nitrobenzylamine derivatives.[180] Several advantages of using HPLC rather than NMR determination for these compound types have been discussed.[177] Some of the features of molecular structure which tend to contribute to α values for such separations are also discussed.

Amino acid separation work was extended by Furukawa et al.[181] to other amino acids, still using D-10-camphorsulfonyl *p*-nitrobenzoates. To shorten analysis times, a variety of solvents and gradient systems were tested. Retention times of L-amino acid derivatives were found to be consistently shorter than those of corresponding D-amino acid derivatives. This may indicate potential use for configurational assignments as well as optical purity assessment by HPLC.

Separation of diastereomeric *N*-(−)-α-methoxy-α-methyl-1-naphthaleneacetyl amino acid methyl esters was accomplished by Goto et al.[182] Very large relative retention (α) values were reported using a μ-Porasil® column with cyclohexane/ethyl acetate solvent systems for 17 amino acids. The optical purity obtained for the chiral derivatizing reagent used [(−)-α-methoxy-α-methyl-1-naphthaleneacetic acid] was better than 99.5%.

Carboxylic acid enantiomers have been resolved by HPLC of diastereomeric amides.[174,183,184] Separations were achieved on silica gel columns in hexane/ethyl acetate solvent systems. It was pointed out that protic solvents had detrimental effects, and hydrocarbon mixtures with polar aprotic solvents yielded best results.

Of current interest is the presence of benzo[a]pyrene as an environmental pollutant; its carcinogenic effects in animals have caused much concern.[185] It requires metabolism by microsomal enzymes to exert its toxicity, mutagenicity, and carcinogenicity.[185] HPLC methodology was recently reported for facile resolution of optical isomers of the metabolites, allowing elucidation of the detailed benzo[a]pyrene activation pathways.[186] The di-(−)-methoxyacetate diastereomers were prepared and injected onto a DuPont "Zorbax" SIL® (silica adsorbent) column for separation.

Ten racemic helicenes and two double helicenes have been resolved by HPLC recently.[187] R(−) and S(+)-2-(2,4,5,7-tetranitro-9-fluorenylideneaminoxy) propionic acid and three R(−)-homologues derived from butyric, isovaleric, and hexanoic acids were used as chiral charge-transfer complex-forming stationary phases, *in situ* coated on silica microparticles. Recycle-HPLC was shown to be effective for resolution of difficultly separable [5]-helicene enantiomers. Molecular structure and selector-selectand interaction were discussed.

3. Liquid Column Chromatography

The resolution of racemic amino acids was first reported by Kotake et al.[188] Dalgliesh subsequently studied the structural features needed for resolution to occur in aromatic amino acids.[189] Resolution of the pharmacologically important D,L-3,4-dihydroxyphenylalanine (DOPA) was reported by Baczuk et al.[190] using an asymmetric adsorbent synthesized by linking a commercial polydextran medium to L-arginine. Many attempts followed at synthesis of polymer resins wherein an optically active group was introduced into the polymer backbone for service as adsorbents for resolution of racemates. For an extensive listing of separations of optical isomers by adsorption chromatography, the review by Buss and Vermeulen should be consulted.[191] Other more recent related work includes that by Blaschke et al. wherein optically active polyacrylic esters and amides of ephedrine derivatives were investigated,[192] optically active suspension polymers were used,[193] and polyacryl- and polymethacrylamides from 1-phenylethylamine were created[194] for various racemate separations.

Davankov et al. reported recently a technique which they term "ligand-exchange chromatography".[195] The ligand-exchange method differs from adsorption or ion-exchange in that the sorbate-stationary phase interaction is due to formation of a coordination bond (or bonds) inside the coordination sphere radius of a complexing metal ion. The complexing metal ion may be bound to mobile ligands. In this case, separation is based on the chromatography of several ligand-metal complexes. If the metal ion is bound to nonmobile ligands of a polymeric stationary phase, then separation arises from chromatography of mobile phase-borne ligands. Resolutions of a number of racemates have been attempted with this technique.[196-198]

4. Other Chromatographic Techniques

Thin-layer chromatography (TLC), paper chromatography (PC), and electrophoresis have all been applied to resolution of stereoisomers, especially optical isomers. Buss and Vermeulen[191] published a large table of representative resolutions of optical isomers by paper chromatography; most applications are for amino acid racemates. Thin-layer chromatography has been extensively applied to diastereomers and geometrical isomers; for typical examples, the reader is referred to a recent publication by Palamereva et al.[199] and to a volume by Niederwieser and Pataki.[200] Most TLC work involves diastereomers of varied compound types: 3-methyl-tryptophans,[201] di- and tripeptides,[202] some aliphatic substances,[203] and thymidine "hydrates".[204]

Only one report of paper electrophoretic resolution of diastereomers has been made: the recent work by Jakovljevic et al.[205] A variety of compound types was studied.

C. Spectroscopy

1. Nuclear Magnetic Resonance Spectroscopy (NMR)

Diastereotopic nuclei, or those which cannot be interchanged by a symmetry operation,[206] exist in diastereotopic environments in either chiral or achiral solvents. In principle, they are anisochronous and ought to have different chemical shifts and coupling constants. In contrast, enantiomeric nuclei are equally screened in achiral solvents and are isochronous. However, in chiral solvents or in achiral solvents containing a chiral complexing reagent, these nuclei become diastereotopic. These symmetry relationships have been used in NMR methods to determine optical purity by measuring diastereomer ratios derived from the reaction of a mixture of enantiomers with an optically pure reagent, and also by measuring enantiomer ratios in chiral solvents or in achiral solvents with a chiral complexing reagent.

To determine enantiomeric purity by integration of diastereomeric nuclei requires that no racemization occurs during derivatization of the compound under study. Any optically pure resolving agent which meets this requirement is applicable in principle, but it must produce in practice diastereomers with a large enough chemical shift difference between the diastereotopic nuclei to permit accurate integration. The simpler the spectral patterns of the diastereotopic nuclei (i.e., doublets or singlets), the more accurate the integration becomes. Diastereotopic nuclei in the derivative which have different chemical shifts may arise from part of the structure of the original compound under study or from part of the derivatizing agent. The acid chloride of (R)-O-methylmandelic acid has been used to determine optical purities of amines and alcohols.[206-208]

Dale et al. introduced the reagent α-methoxy-α-trifluoromethylphenylacetic acid for derivatizations[209] which circumvented problems of racemization encountered with some applications of O-methylmandelic acid. The reagent of Mosher generally exhibited excellent separation of diastereotopic fluorine and proton resonances, marked stability to racemization of derivatives, and the presence of a CF_3- group, permitting use of fluorine NMR. This reagent has been used for the determination of optical purity of a series of secondary amines and alcohols,[209] phenylethylene glycol,[210] and phenyltrimethylsilylcarbinol.[211]

Besides chiral derivatizing reagents, optically active solvents have also been used for NMR determination of enantiomeric purity. Solution of a racemic solute in a racemic solvent causes formation of transient diastereotopic solvates. On the NMR time scale, solute-solvent interactions are averaged, and the diastereomeric interactions in the two enantiomeric solvate sets produce only one set of solute signals. In an optically active solvent, only one set of diastereotopic interactions exists, and the enantiotopic nuclei of the solute then are unequally screened leading, in principle at least, to different chemical shifts and a direct measure of enantiomeric purity.

Optically active solvents which cause the largest chemical shift differences for enantiotopic solute nuclei are those which contain a group of high diamagnetic anisotropy near the solvent asymmetric center and which interact with or strongly solvate the solute. The solvent should be optically pure since as optical purity decreases, the chemical shift difference of enantiotopic solute nuclei decreases. Solvents which have been widely used are α-(1-naphthyl)ethylamine[212] and 2,2,2-trifluoro-1-phenylethanol.[213] The latter is more useful since less solvent resonances appear in the proton NMR spectra. It has been used for optical purity determinations for many types of compounds including amines,[214] sulfoxides,[215] other sulfur-containing species,[216] and α-amino esters.[217] In the latter case, the optical purities calculated by NMR agreed well with those determined by other methods.

Pirkle and Hoekstra[218] recently reported on carbon-13 nonequivalence of enantiomers in chiral solvents. Several chiral benzylic amines were used to promote ^{13}C nonequivalence of enantiomeric α-substituted benzylic alcohols.

Lanthanide shift reagents complex strongly with a variety of functional groups and cause remarkable changes in chemical shifts.[219] Optically active lanthanide shift reagents effect different chemical shift changes for enantiomeric substrates. Regis Chemical Company[220] makes available a number of Kiralshift® reagents for optical purity measurements by NMR. Table 8 is a tabulation of a few applications of optically active shift reagents for enantiomeric purity assessment.[221-225]

2. Mass Spectrometry and IR Spectroscopy

Enantiomers give identical mass spectra because all bond energies and all fragments which can be formed are identical. One may possibly find small differences on comparison of mass spectra of diastereomers if two assymetric centers are present in the molecule. Differences between such diastereomers are more significant in alicyclic compounds for several reasons.[226] Mass spectra of diastereomers have been discussed with several examples.[226] Very recently McMahon and Sullivan[227] reported simultaneous measurement of plasma levels of D-propoxyphene and L-propoxyphene using stable isotope labels and mass fragmentography.

Additive absorbance was used to quantitate geometrical isomers by IR in the pharmaceutical compound clomiphene citrate.[228] Although the spectra of the two isomers (Z and E, structures below)

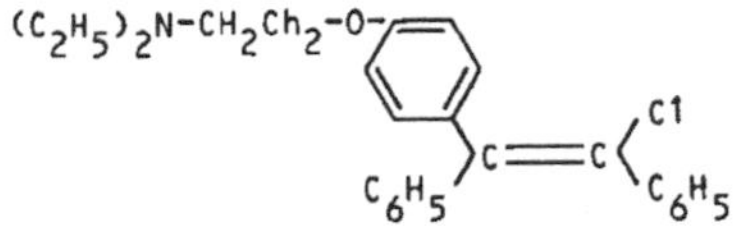

$(C_2H_5)_2N-CH_2Ch_2-O-$ ... C_6H_5, Cl, C_6H_5 — Z-Clomiphene

$(C_2H_5)_2N-CH_2CH_2-O-$ ... C_6H_5, Cl, C_6H_5 — E-Clomiphene

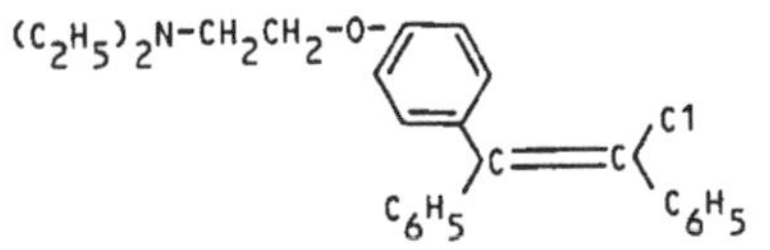

are very similar giving small isomer shifts, careful choice of the analytical measurement frequencies allowed determination of the isomer ratio to about ±1% accuracy. The results were in good agreement with those of an HPLC method, and the IR work required less sample preparation.

The stereoisomers of clomiphene citrate may also be quantitatively determined by differential IR absorbance. Absorbance spectra of the pure Z and E isomers (as shown above) are subtracted from the absorbance spectrum of the mixture at the same concentration. The ratio of the maximum value of one difference curve to the minimum value of the other difference curve is equal to the ratio of the mole factors of isomers present. An IR Fourier transform spectrometer coupled to a computer data system was used for this work for two reasons: a highly repeatable frequency is necessary for small band shifts, and spectra for pure isomers were stored for long periods of time since they were not measured for each determination.

D. Miscellaneous Techniques

The application of thermal analysis, particularly differential thermal analysis (DTA) to isomeric purity analysis of pharmaceuticals has been demonstrated.[230] In one example, the presence of mixed *cis-* and *trans-* isomers of an amino acid compound was detected by TGA easily at the 2% level.[230]

Table 8
DETERMINATION OF ENANTIOMERIC PURITY BY NMR BY THE USE OF CHIRAL SHIFT REAGENTS

Application	Ref.
Use of the europium complex of 3-hepta-fluorobutanoyl-(±)-camphor to separate NMR signals of racemic alcohols, sulfoxides, an epoxide, and an aldehyde	221
Application of several chiral shift reagents for various compounds' optical isomer assessment	222
Chiral chelating reagents for determination of enantiomeric purity of various organic substances	223
Determination of enantiomeric purity of some L-phenylalkanols using a chiral NMR shift reagent	224
Enantiomeric composition of bicyclic ketal insect pheromone components by use of a chiral shift reagent	225

Resolving agents and resolution of organic compounds has been reviewed by Wilen[231] who discussed such techniques as resolution via molecular complexes and inclusion compounds. In addition, a large section is devoted to resolving agents for acids, bases, alcohols, carbonyls, amino acids, and other organics.

The resolution of racemic modifications is discussed in terms of biochemical separation, asymmetric transformation, chromatography, and channel complex formation (among others) by Finar.[232]

Recently enantiomeric differentiation by chiral macrocyclic polyethers derived from D-mannitol and binaphthol has been accomplished.[233]

III.D. EXAMPLES OF STEREOISOMER SEPARATIONS OF PHARMACEUTICAL INTEREST

The challenge of measuring stereoisomeric purity for pharmaceutical compounds can be met with a variety of analytical approaches. While thin-layer chromatography (TLC), electrophoresis, and packed column chromatography are certainly useful, the most powerful techniques to date have been GC and HPLC. Table 9 presents selected applications of stereoisomer separations of pharmaceutical interest. A few applications will be discussed as examples in the remainder of this chapter.

Benoxaprofen®, an anti-inflammatory drug, was the subject of two interesting optical purity investigations. Gas chromatography procedures were used for determination of isomeric impurities in *p*-chlorobenzoyl chloride and two intermediates from a synthesis of Benoxaprofen®.[307] Benoxaprofen® and its four possible isomeric impurities shown in Figure 23 may be separated as the methyl esters on a liquid crystalline GC stationary phase.[308]

HPLC has been used to resolve the carbinol diastereomer intermediates from the synthesis of propoxyphene.[299] Resolution and quantitative analysis of the α- and β-carbinols was done with good precision in crude carbinol sample lots, and camphorsulfonic acid salts were also assayed. Figures 24a and 24b indicate that the carbinol isomers are baseline-resolved at high or low levels of β in the desired α product, so a sensitive assessment of isomeric purity was easily made.

In a study of the pharmacokinetics of the antibiotic moxalactam, an HPLC method was developed to determine its diastereomer levels in serum and urine.[292] Quantitation was done by use of an internal standard technique, and both recovery and stability studies were reported. A typical chromatogram is shown in Figure 25.

Table 9
SELECTED STEREOISOMER SEPARATIONS OF PHARMACEUTICAL INTEREST

No.	Sample description	Remarks	Ref.
Acids			
1.	Urinary 2-hydroxycarboxylic acid enantiomers	Separated as diastereomeric derivatives; biomedical applications (GC)	234
2.	Bile acids (26 hydroxylated 5β-cholanic acids)	Resolved as methyl ester-TMS ether derivatives on 3 columns of varying polarity (GC)	235
3.	Enantiomers of α-vinyl-α-aminobutyric acid	Resolved as N-TFA-O-methyl esters; determination in human body fluids using commercially available Chirasil-Val™ capillary column (GC)	236
Alcohols and Phenols			
1.	Chiral secondary alcohols (from metabolized ketones)	Separated as diasteromeric urethanes; simple, rapid procedure for samples from biological media (GC)	237
2.	C-24 epimeric sterols (9 pairs)	Resolved as TMS ethers (GC)	238
3.	α- and β-Naltrexol and α- and β-naloxal from reduction of naltrexone	Semiquantitative, electron-capture sensitive method (GC)	239
4.	Labetalol (antihypertensive) diastereomers	Separated as n-butyl boronates; method to assess human urinary excretion (GC)	240
5.	Pharmaceuticals with adrenergic, and particularly β-blocking effects (O-acetyl alcohol enantiomers)	Resolution on two chiral phases as N, O-*bis*-heptafluorobutyryl derivatives (GC)	241
6.	Epimeric (25*R*)- and (25*S*)-26-hydroxycholesterol	Diastereomers resolved directly as the 3β,26-diacetates (HPLC)	242
7.	Epemeric alcohols formed upon reduction of pregnenolone and progesterone	Preparative isolation (HPLC)	243
Amines			
1.	Amphetamine and related primary amines	Polar vs. nonpolar columns compared for derivatized samples using new amino acid chlorides (GC)	244
2.	Amphetamines and related primary amines, directly as enantiomers, or as diastereomers	New cyclic resolving agents tested; new chiral stationary phase also tested (GC)	245
3	Ephedrines and pseudoephedrines, primarily	Separated as diastereomers formed with N-TFA L-prolyl chloride; suitability for biological samples (GC)	246
4.	Fenfluramine, norfenfluramine	Enantiomer resolution after N-PFP L-prolyl chloride derivatization; electron-capture detection and use for biological samples (GC)	247
5.	Amphetamine enantiomers	Resolved as diastereomeric derivatives (GC)	248
6.	Amphetamine enantiomers	Urine extracts derivatized to form diastereomer products (GC)	249
7.	Amphetamine enantiomers	Quantitation from raw materials form or tablets (GC)	250
8.	Amphetamine and related primary amines	Chiral reagent used to form diastereomers; determination in rat urine (GC)	251
9.	Primary aliphatic amines and N-hydroxylamine metabolite enantiomers	Separation after conversion to diastereomers; determination at levels near 1 mcg/ml (GC)	252

Table 9 (continued)
SELECTED STEREOISOMER SEPARATIONS OF PHARMACEUTICAL INTEREST

No.	Sample description	Remarks	Ref.
10.	Nortilidine and bisnortilidine enantiomers	Nitrogen-sensitive detection of N-TFA L-leucyl diastereomers formed by reaction with human plasma extracts; nanogram levels (GC)	253
11.	Ephedrines	Separation by conversion to diastereomeric amides (GC/MS)	254
12.	Propanolol enantiomers	Diastereomeric derivatives were formed by reaction with chiral 1-phenylethyl isocyanate (GC)	255
13.	β-Adrenoceptor agonist enantiomers (amino alcohols)	Reacted with N-perfluoroacyl prolyl chlorides to form diastereomers (GC)	256
14.	Amine drugs like amphetamine, substituted tryptamines, ephedrines	Enantiomers resolved as the diastereomeric N-TFA L-prolyl derivatives (GC/MS)	257
15.	Tocainide (antiarrythmic) enantiomers	Direct resolution as N-HFB derivatives using Chirasil-Val™ column; human plasma and urine (GC)	258
16.	Tocainide enantiomers in blood plasma	Derivatized to diastereomers using (*S*)-α-methoxy-α-trifluoromethylphenylacetyl chloride; quantitation using electron capture detection (GC)	259
17.	Tricyclic antidepressants and antihistamines with propylidene side chains: *cis - trans* isomers	Benzothiepine-series compounds including dosulepine, medosulspine, dithiadene were resolved in about 10 min (HPLC)	260
18.	(*E*) and (*Z*) isomers of doxepin hydrochloride	Precision, accuracy, and selectivity were examined (HPLC)	261
19.	(*E*) and (*Z*) isomers of clopenthixol and a N-dealkyl metabolite	Determination in serum; quantitation by internal standard technique; sensitivity, precision and specificity were evaluated (HPLC)	262
20.	Fenfluramine	Preparative resolution of diastereomers formed with (−)-menthyl chloroformate (HPLC)	263
21.	2,5-Dimethoxy-4-methylamphetamine in plasma	Resolved as diastereomeric amide derivatives (HPLC)	264
22.	Amphetamines and related amines	Resolved as (+)-10-camphorsulfonyl diastereomeric amides; comparison on several eluant systems (HPLC)	265
23.	Amphetamines	Resolved as diastereomeric reaction products of several reagents (HPLC)	266
24.	Diltiazem Hydrochloride	Resolution of enantiomers as the diastereomeric reaction products of chiral 2-(2-naphthyl)-propionyl chloride (HPLC)	267
Amino Acids			
1.	3-(3,4-Dihydroxyphenyl)alanine (DOPA) and α-methyl DOPA	Resolved as diastereomers after reaction with diazomethane and a chiral reagent (GC)	268
2.	Penicillamine enantiomers and other applications of biological interest	Direct determination of the two enantiomer percentages by use of the commercial Chirasil-Val™ column (GC)	269
3.	Thyroxines in human serum	Resolved using L-proline as a chiral additive to the eluant; amperometric detection (HPLC)	270

Table 9 (continued)
SELECTED STEREOISOMER SEPARATIONS OF PHARMACEUTICAL INTEREST

No.	Sample description	Remarks	Ref.
4.	L-Isomer in D-penicillamine	Quantitative analysis after derivatization (HPLC)	271
5.	Labeled DOPA enantiomers	Resolved as dipeptides by TLC	272
Amino Alcohols			
1.	Norephedrine	Analytical and preparative work with Pirkle covalent phenyglycine column; enantiomers resolved as 2-oxazolidone derivatives (HPLC)	273
2.	Epinephrine	Pirkle type 1-A chiral column was used to resolve enantiomers as cyclic oxazolidine derivatives (HPLC)	274
3.	Alprenolol, oxprenolol, metoprolol, propranolol, hyosyamine, and the amine terodilin	(+)-10-Camphorsulfonic acid and analogues were added to the mobile phase to resolve enantiomers (HPLC)	275
4.	Analgesic, 1-[2-(3-hydroxyphenyl)- 1-phenylethyl]-4-(3-methyl-2-butenyl) piperazine	β-Cyclodextrin was added to the mobile phase to resolve the enantiomers (HPLC)	276
5.	Propranolol in human plasma	Enantiomers were resolved as diastereomeric N-TFA L-prolylamides; fluorometric detection (HPLC)	277
6.	Propranolol and other β-adrenergic antagonists	Enantiomers were resolved following derivatization to diastereomers (HPLC)	278
7.	Propranolol enantiomers	Resolved as diastereomeric derivatives; determination in human plasma at about 1 ng/ml (HPLC)	279
Carboxylic and Sulfonic Acids			
1.	α-Methylarylacetic acids (including ibuprofen, naproxen, fenoprofen, and benoxaprofen) as 1-naphthalene-methylamides and other chiral amides	Enantiomers were resolved on Pirkle column (HPLC)	280
2.	Benoxaprofen and other α-methyl-aryl-acetic acids	Samples were resolved as the diastereomeric (−)-α-methybenzylamides (HPLC)	281
3.	Naproxen	Enantiomers were derivatized to diastereomers with (+)-2-octanol and determined quantitatively (HPLC)	282
4.	Clinofibrate (hypolipidemic agent)	Enantiomers were separated as the diastereomeric β-(+)-α-methyl-benzylamides (HPLC)	283
5.	Carprofen (anti-inflammatory agent)	Analysis as diastereomeric (*S*)-(−)-α-methylbenzylamides from human biological samples (HPLC)	284
Miscellaneous Compounds			
1.	Steroid epimers	Separated on nematic liquid crystal phase (GC)	285
2.	Calcium gluceptate epimers	Determined in aqueous solutions (GC)	286
3.	Ergot peptide and eburnane alkaloids	Enantiomers resolved using *d*-camphorsulfonic acid as an additive to the mobile phase (HPLC)	287, 288
4.	Nine indole alkaloids, some of which have therapeutic value	Enantiomers resolved on a chiral β-cyclodextrin polymer gel bed with mildly acidic citrate buffer (HPLC)	289

Table 9 (continued)
SELECTED STEREOISOMER SEPARATIONS OF PHARMACEUTICAL INTEREST

No.	Sample description	Remarks	Ref.
5.	DOPA, methydopa, carbidopa, and tryptophan	Enantiomers resolved using L-phenylalanine in the eluant (HPLC)	290
6.	Naproxen in serum	Resolved the enantiomers as diastereomeric derivatives; detection by fluorescence (HPLC)	291
7.	Moxalactam antibiotic in serum and urine	Diastereomers were resolved and determined quantitatively (HPLC)	292
8.	Penicillin and cephalosporin diastereomers	Samples include ampicillin, cephalexin, and phenoxyethylpenicillin (HPLC)	293
9.	Four isomeric cocaines	Diastereomers quantitated and identified using flow programming (HPLC)	294
10.	*R*- and *S*-epimers of the β-lactam antibiotic, moxalactam	Application to pharmaceuticals and to human urine samples; quantitation (HPLC)	295
11.	α- and β-Anomers of streptozocin (antineoplastic agent)	Quantitation of drug diasteromers in sterile powder and of the bulk drug substance (HPLC)	296
12.	Propoxyphene α- and β-isomers	Diastereomers resolved in pharmaceutical preparations (HPLC)	297
13.	Diastereomers of tricyclic neuroleptic drugs, including flupenthixol, clopenthixol, doxepine, and promazine	Comparison of batches of raw materials and of dosage forms; quantitation (HPLC)	298
14.	Propoxyphene carbinol diastereomers	Analytical separation; sensitivity, precision, and linearity were examined (HPLC)	299
15.	Cyclothiazide diastereomers	Method for screening urine samples for thiazide diuretics; evaluation of recovery and specificity (HPLC)	300
16.	Diastereomeric 7-ureidoacetamido Cephalosporins	Large variety of structures of compounds (HPLC)	301
17.	Pharmacologically active sulfoxides	Enantiomers resolved on immobilized bovine serum albumin stationary phase; some very large resolution factors (HPLC)	302
18.	Propoxyphene diastereomers	Comparison of conventional TLC and HPTLC	303
19.	Enantiomers of antitumor agent cyclophosphamide	Resolved after diastereomer formation with chiral 1-phenethyl alcohol (TLC)	304
20.	Chloroquine (antimalarial)	Enantiomers resolved on chiral polyamide packed column	305
21.	Thalidomide enantiomers	Resolved on packed column of chiral polyamide	306

Souter and Jensen[303] showed high performance thin-layer chromatography (HPTLC) to be an effective means of resolving the α- and β-diastereomers of propoxyphene hydrochloride. This was significant because it not only demonstrated the application of the HPTLC technique for stereoisomer separations, but it was also potentially useful for the fast α-propoxyphene identification in forensic[297,309] and drug chemistry laboratories.[310] This TLC system was also an important improvement over earlier work[309] wherein spots appeared to tail significantly. Figure 26 demonstrates the spot tailing on plates developed in acetone[309] and the lack of tailing in the system from Souter and Jensen.[303]

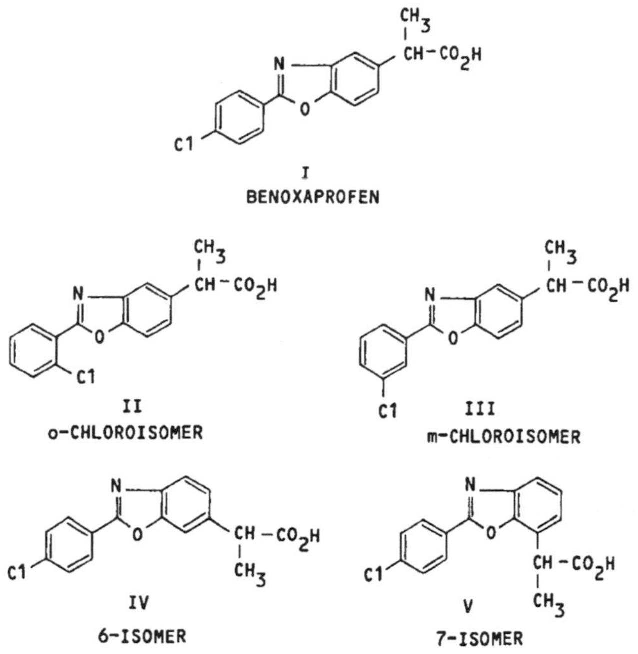
CH3
CH-CO2H
N
O
C1
I
BENOXAPROFEN
II
o-CHLOROISOMER
III
m-CHLOROISOMER
IV
6-ISOMER
V
7-ISOMER

FIGURE 23. Stuctures of Benoxaprofen® and its four possible synthetic intermediate impurities. (From Hall and Mallen, *J. Chromatogr.*, 118, 268, 1976. With permission.)

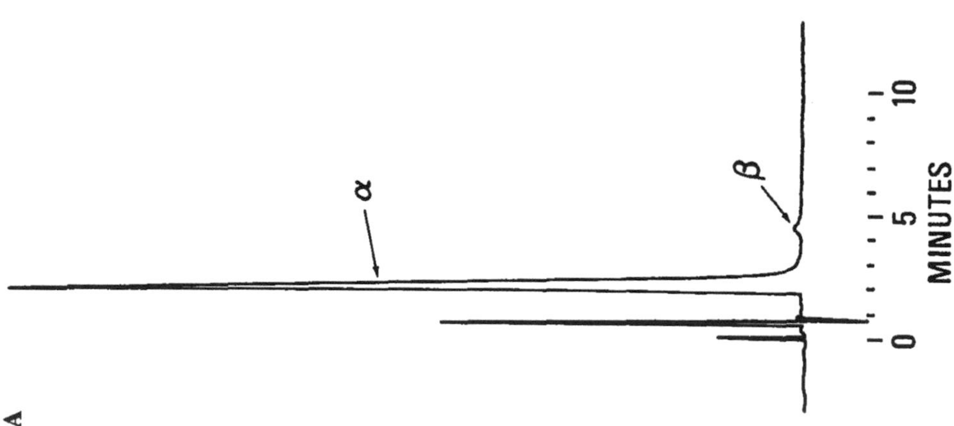

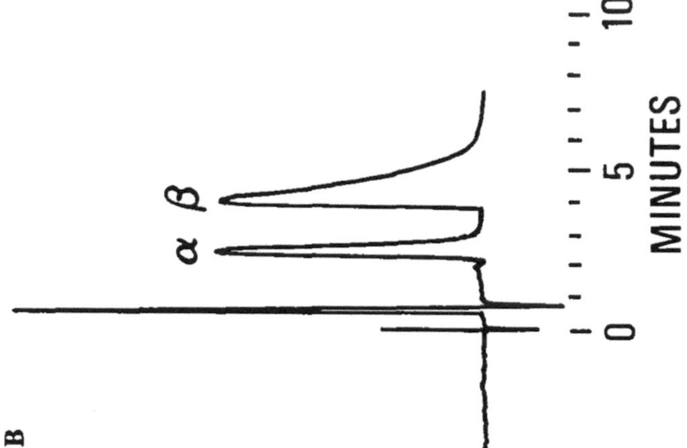

FIGURE 24. (A) High performance liquid chromatogram for 1% β carbinol in α carbinol. (B) High performance liquid chromatogram for 70% β carbinol in α carbinol. (From Souter, R. W., *J. Chromatogr.*, 134, 187, 1977. With permission.)

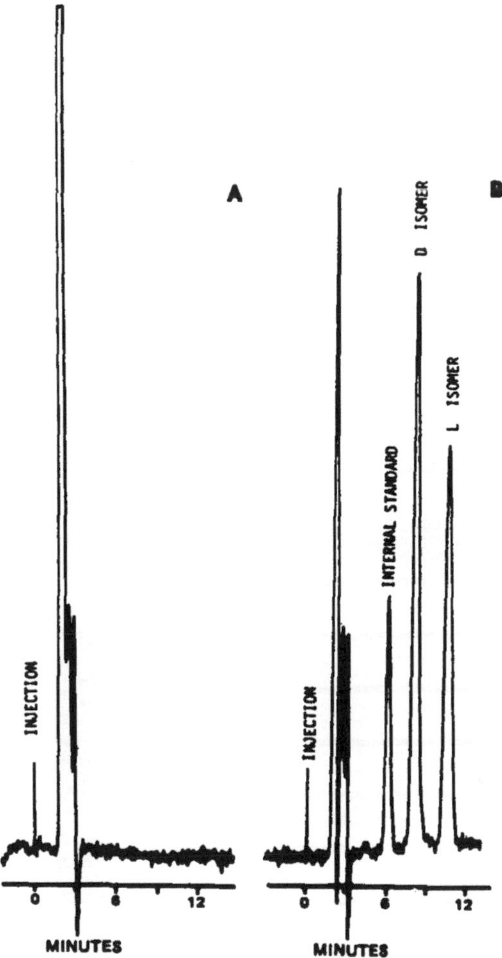

FIGURE 25. Chromatograms obtained after extracting 0.5 ml of a blank patient plasma (A) and a moxalactam plasma (B). The peak heights of each isomer correspond to a plasma concentration of 10 μg/mℓ. (From Ziemniak, J. A., Miner, D. J., and Schentag, J. J., *J. Pharm. Sci.*, 71, 399, 1982. Reproduced with permission of the copyright owner.)

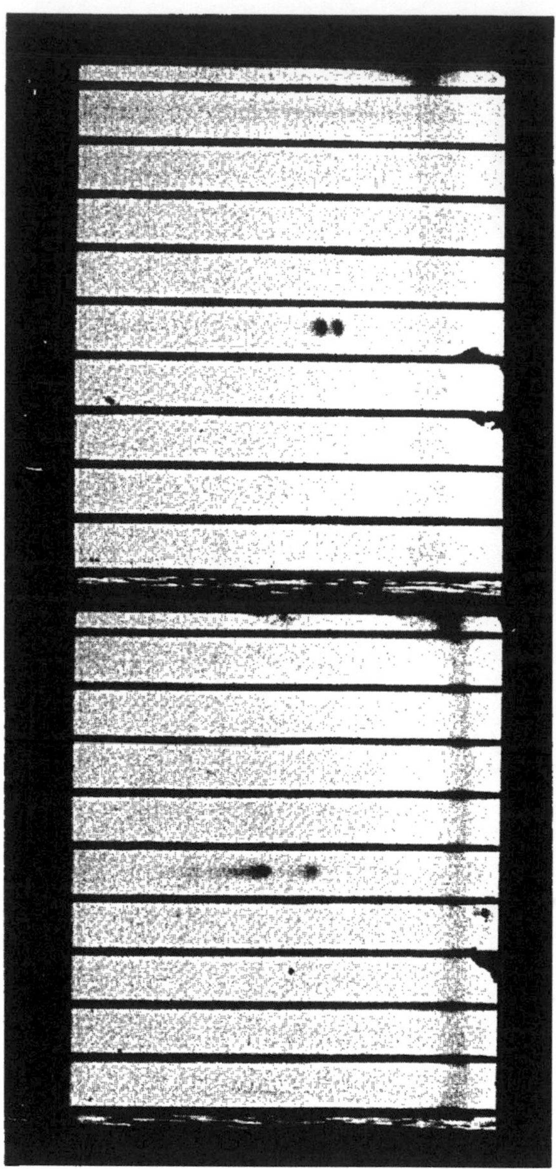

FIGURE 26. HPTLC separations of α- and β-propoxyphene hydrochloride in (left) the solvent system from this work and in (right) acetone described in Reference 309; spot size 50 μg/isomer. (From Souter, R. W. and Jensen, E. C., *J. Chromatogr.*. 281, 386, 1983. With permission.)

REFERENCES

1. **Foye, W. O., Ed.,** *Principles of Medicinal Chemistry,* Lea & Febiger, Philadelphia, 1974, 81.
2. **Mislow, K.,** *Introduction to Stereochemistry,* W. A. Benjamin, New York, 1966.
3. **Modell, W., Schild, H., and Wilson, A.,** *Applied Pharmacology,* W. B. Saunders, Philadelphia, 1976, 392.
4. **Melmon, K. and Morrelli, H., Eds.,** *Clinical Pharmacology, Basic Principles in Therapeutics,* Macmillan, New York, 1972, 492.
5. Medical Literature Dept., Merck, Sharp, and Dohme Laboratories, *Codeine and Certain Other Analgesic and Antitussive Agents: A Review,* Merck and Co., Rahway, N.J., 1970, 2.
6. **Bickerman, H. A.,** *Med. Clin. North Am.,* 45, 805, 1961.
7. *U.S. Patent,* 2,728,779, 1955.
8. **Wilson, C. O., Griswold, O., and Doerge, R. F., Eds.,** *Textbook of Organic Medicinal and Pharmaceutical Chemistry,* 6th ed., Lippincott, Philadelphia, 1971, 32.
9. *Chromatography/Lipids,* Vol. 9, No. 3, Supelco, Bellefonte, Pa., July 1975, 1.
10. *Chromatography/Lipids,* Vol. 9, No. 3, Supelco, Bellefonte, Pa., July 1975, 4.
11. **DiCorcia, A. and Samperi, R.,** *Anal. Chem.,* 46, 140, 1974.
12. Catalog No. 11, Supelco, Bellefonte, Pa., 1977.
13. Catalog 20, Applied Science Laboratories, State College, Pa., 1977, 24.
14. **Littlewood, A. B.,** *Gas Chromatography. Principles, Techniques, and Applications,* Academic Press, New York, 1970, 420.
15. *Chromatography/Lipids,* Bulletin 740B, Supelco, Bellefonte, Pa., 1974.
16. **Gray, G. W.,** *Molecular Structure and the Properties of Liquid Crystals,* Academic Press, New York, 1962.
17. **Kelker, H. and von Schivizhoffen, E.,** *Advances in Chromatography,* Vol. 6, Giddings, J. C. and Keller, R. A., Eds., Marcel Dekker, New York, 1968, 247.
18. **Dewar, M. and Schroeder, J.,** *J. Am. Chem. Soc.,* 86, 5235, 1964.
19. **Dewar, M. and Schroeder, J.,** *J. Org. Chem.,* 30, 3485, 1965.
20. **Schroeder, J., Schroeder, D., and Katsikas, M.,** *Liquid Crystals and Ordered Fluids,* Johnson, J. F. and Porter, R. S., Eds., Plenum Press, New York, 1970, 169.
21. **Andrews, M., Schroeder, D., and Schroeder, J.,** *J. Chromatogr.,* 71, 233, 1972.
22. **Richmond, A. B.,** *J. Chromatogr. Sci.,* 9, 571, 1971.
23. **Cook, L. and Spangelo, R.,** *Anal. Chem.,* 46, 122, 1974.
24. **Grushka, E. and Solsky, J.,** *J. Chromatogr.,* 112, 145, 1975.
25. **Mattsson, P. and Nygren, S.,** *J. Chromatogr.,* 124, 265, 1976.
26. **Krÿgsman, W. and VanDeKamp, C.,** *J. Chromatogr.,* 131, 412, 1977.
27. *Chromatography/Lipids,* Bulletins 738 and 742A, Supelco, Supelco Park, Bellefonte, Pa., 1974.
28. **Souter, R. and Bishara, R.,** *J. Chromatogr.,* 140, 245, 1977.
29. **Mostecky, J., Popl, M., and Křiž, J.,** *Anal. Chem.,* 42, 1132, 1970.
30. **Fell, V. and Lee, C.,** *J. Chromatogr.,* 121, 41, 1976.
31. **Apon, J. and Nicolaides, N.,** *J. Chromatogr. Sci.,* 13, 467, 1975.
32. **Buser, H.,** *J. Chromatogr.,* 114, 95, 1975.
33. **Schwetz, B. A., Norris, J. M., Sparschu, G. L., Rowe, V. K., Gehring, P. J., Emerson, J. L., and Gerbig, C. G.,** *Environ. Health Perspect.,* 5, 15, 1973.
34. **Hoshika, Y. and Takata, Y.,** *J. Chromatogr.,* 120, 379, 1976.
35. **Petrowitz, H. J.,** *Progress in Thin-Layer Chromatography and Related Methods,* Vol. 3, Niederwieser, A. and Pataki, G., Eds., Ann Arbor Science Publishers, Ann Arbor, Mich., 1972, 1.
36. **Jakovljevic, I., Bishara, R., and Kress, T.,** *J. Chromatogr.,* 134, 238, 1977.
37. **O'Doherty, G. and Fuhr, K.,** *Belg. Pat.,* 764, 591, 1971.
38. **Bowden, R. and Seaton, T.,** U.S. Patent 3,725,414, 1973.
39. **Stringham, R. and Torba, F.,** U.S. Patent 3,557,124, 1971.
40. **Bishara, R., Born, G., and Christian, J.,** *J. Chromatogr.,* 64, 135, 1972.
41. **Melick, W. F., Escue, H. M., Maryka, J. J., Mezera, R. A., and Wheeler, E. P.,** *J. Urol.,* 74, 760, 1955.
42. **Gorrod, J. W. and Carey, M. J.,** *Biochem. J.,* 119, 52P, 1970.
43. **Jakovljevic, I., Zynger, J., and Bishara, R.,** *Anal. Chem.,* 47, 2045, 1975.
44. **LeRosen, A. L., Carlton, J. K., and Moseley, P. B.,** *Anal. Chem.,* 25, 666, 1953.
45. **Fishbein, L.,** *J. Chromatogr.,* 27, 368, 1967.
46. **Gillio-Tos, M., Previtera, S., and Vimercati, U.,** *J. Chromatogr.,* 13, 571, 1964.
47. **Petrowitz, H. J.,** *Chimia,* 18, 137, 1964.
48. **Petrowitz, H. J.,** *Chem. Ztg.,* 89, 7, 1965.

49. **Zweig, G. and Sherma, J., Eds.,** *Handbook of Chromatography,* Vol. 1, CRC Press, Boca Raton, Fla., 1972, 435.
50. *Programmed Multiple Development Apparatus,* Regis Chemical Co., Morton Grove, Ill., 1974.
51. **Perry, J. A., Haag, K. W., and Glunz, L. J.,** *J. Chromatogr. Sci.,* 11, 447, 1973.
52. Regis PMD News, No. 13, Regis Chemical Co., Morton Grove, Ill., November 1975.
53. **Snyder, L. R.,** *Principles of Adsorption Chromatography,* Marcel Dekker, New York, 1968, 167.
54. *Paired-Ion Chromatography,* Waters Assoc. Bulletin D61, Waters Assoc., Milford, Mass., May 1976.
55. **Sved, S., McGilveray, I., and Beaudoin, N.,** *J. Pharm. Sci.,* 65, 1356, 1976.
56. **Casy, A. F. and Armstrong, N. A.,** *J. Med. Chem.,* 8, 57, 1965.
57. **Casy, A. F., Beckett, A. H., and Iorio, M. A.,** *Tetrahedron,* 23, 1405, 1967.
58. **Casy, A. F.,** *PMR Spectroscopy in Medicinal and Biological Chemistry,* Academic Press, London, 1971, 34.
59. **Cerfontain, H., Koeberg-Telder, A., Kruk, C., and Ris, C.,** *Anal. Chem.,* 46, 72, 1974.
60. **Levy, G. C. and Nelson, G. L.,** *Carbon-13 Nuclear Magnetic Resonance for Organic Chemists,* Wiley-Interscience, New York, 1972.
61. **Levy, G. C., Nelson, G. L., and Cargioli, J. D.,** *Chem. Commun.,* 506, 1971.
62. **Levy, G. C. and Nelson, G. L.,** *Carbon-13 Nuclear Magnetic Resonance for Organic Chemists,* Wiley-Interscience, New York, 1972, 84.
63. **Retcofsky, H. L. and Friedel, R. A.,** *J. Phys. Chem.,* 72; 290, 2619; 1968.
64. **Retcofsky, H. L. and Friedel, R. A.,** *J. Phys. Chem.,* 71, 2513, 1967.
65. **Streitweiser, A., Jr. and Fahey, R. C.,** *J. Org. Chem.,* 27, 2352, 1962.
66. **Bauman, R. P.,** *Absorption Spectroscopy,* John Wiley & Sons, New York, 1962, 403.
67. **Cerfontain, H., Duin, H., and Vollbracht, L.,** *Anal. Chem.,* 35, 1005, 1963.
68. **Arends, J. M., Cerfontain, H., Herschberg, I. S., Prinsen, A. J., and Wanders, A. C. M.,** *Anal. Chem.,* 36, 1802, 1964.
69. **Mantel, M. and Stiller, M.,** *Anal. Chem.,* 48, 712, 1976.
70. **Gifford, A. P., Rock, S. M., and Comaford, D. J.,** *Anal. Chem.,* 21, 1062, 1949.
71. **Rylander, P. N., Meyerson, S., and Grubb, H.,** *J. Am. Chem. Soc.,* 79, 842, 1957.
72. **Biemann, K.,** *Mass Spectrometry,* McGraw-Hill, New York, 1962, 151.
73. **Biemann, K.,** *Mass Spectrometry,* McGraw-Hill, New York, 1962, 297.
74. **Sullivan, H. R., Marshall, F. J., McMahon, R. E., Änggård, E., Gunne, L., and Holmstrand, J.,** *Biomed. Mass Spectr.,* 2, 179, 1975.
75. **Sullivan, H. R., Wood, P. G., and McMahon, R. E.,** *Biomed. Mass Spectr.,* 3, 212, 1976.
76. **Heller, W. and Curmè, H.,** *Physical Methods of Chemistry,* Part IIIC, Weissberger, A. and Rossiter, B., Eds., Wiley-Interscience, New York, 1972, 169.
77. **Horeau, A.,** *Tetrahedron Lett.,* 3121, 1969.
78. **Lochmüller, C. H. and Souter, R. W.,** *J. Chromatogr., Chromatogr. Rev.,* 113, 283, 1975.
79. **Gil-Av, E. and Nurok, D.,** *Advances in Chromatography,* Vol. 10, Giddings, J. C. and Keller, R. A., Eds., Marcel Dekker, New York, 1974, 99.
80. **Buss, D. R. and Vermeulen, T.,** *Ind. Eng. Chem.,* 60, 12, 1968.
81. **Dichiaro, J., Bate, R., and Keller, R.,** *Separation Techniques in Chemistry and Biochemistry,* Keller, R. A., Ed., Marcel Dekker, New York, 1967, 301.
82. **Petrowitz, H. J.,** *Progress in Thin-Layer Chromatography and Related Methods,* Vol. 3, Niederwieser, A. and Pataki, G., Eds., Ann Arbor Science Publishers, Ann Arbor, Mich., 1972, 1.
83. **Wilen, S. H.,** *Topics in Stereochemistry,* Vol. 6, Allinger, N. L. and Eliel, E. L., Eds., Wiley-Interscience, New York, 1971, 110.
84. Gas-Chrom Newsletter, Vol. 15, No. 1, Applied Science Laboratories, State College, Pa., 1974.
85. Gas-Chrom Newsletter, Vol. 15, No. 6, Applied Science Laboratories, State College, Pa., 1974.
86. *Chromatography/Lipids,* Vol. 8, No. 5, Supelco, Bellefonte, Pa., 1977.
87. **Magidman, P., Barford, R., Saunders, D., and Rothbart, H.,** *Anal. Chem.,* 48, 44, 1976.
88. **Gil-Av, E. and Nurok, D.,** *Proc. Chem. Soc. (London),* 146, 1962.
89. **Gil-Av, E., Charles-Sigler, R., and Fisher, G.,** *J. Chromatogr.,* 17, 408, 1965.
90. **Charles-Sigler, R. and Gil-Av, E.,** *Tetrahedron Lett.,* 35, 4231, 1966.
91. **Weygand, F., Prox, A., and Schmidhammer, L.,** *Angew. Chem. Int. Ed. Engl.,* 2, 183, 1963.
92. **Halpern, B. and Westley, J.,** *Biochem. Biophys. Res. Commun.,* 19, 361, 1965.
93. **Halpern, B. and Westley, J.,** *Tetrahedron Lett.,* 21, 2283, 1966.
94. **Dabrowiak, J. and Cooke, D.,** *Anal. Chem.,* 43, 791, 1971.
95. **Ayers, G. S.,** *Diss. Abstr.,* 33, 1936-B, 1972.
96. **Bonner, W. A.,** *J. Chromatogr. Sci.,* 10, 159, 1972.
97. **Iwase, H. and Murai, A.,** *Chem. Pharm. Bull.,* 22, 8, 1974.
98. **Iwase, H. and Murai, A.,** *Chem. Pharm. Bull.,* 22, 1455, 1974.

99. Iwase H., *Chem. Pharm. Bull.*, 22, 1663, 1974.
100. Iwase, H., *Chem. Pharm. Bull.*, 22, 2075, 1974.
101. Nambara, T., Goto, J., Taguchi, K., and Iwata, T., *J. Chromatogr.*, 100, 180, 1974.
102. Bonner, W. A., VanDort, M. A., and Flores, J. J., *Anal. Chem.*, 46, 2104, 1974.
103. Whelan, J. K., *J. Chromatogr.*, 111, 337, 1975.
104. Iwase, H., *Chem. Pharm. Bull.*, 23, 1604, 1975.
105. Flores, J. J., Bonner, W. A., and VanDort, M. A., *J. Chromatogr.*, 132, 152, 1977.
106. König, W. A., Rahn, W., and Eyem, J., *J. Chromatogr.*, 133, 141, 1977.
107. VanDort, M. A. and Bonner, W. A., *J. Chromatogr.*, 133, 210, 1977.
108. Gordis, E., *Biochem. Pharmacol.*, 15, 2124, 1966.
109. Gunne, L., *Biochem. Pharmacol.*, 16, 863, 1967.
110. Wells, C. E., *J. Assoc. Off. Anal., Chem.*, 53, 113, 1970.
111. Souter, R. W., *J. Chromatogr.*, 108, 265, 1975.
112. Souter, R. W., *J. Chromatogr.*, 114, 307, 1975.
113. Gal, J., *J. Pharm. Sci.*, 66, 169, 1977.
114. Halpern, B. and Westley, J., *Chem. Commun.*, 34, 1966.
115. Karger, B. L., Stern, R. L., Keane, W., Halpern, B., and Westley, J. W., *Anal. Chem.*, 39, 228, 1967.
116. Pereira, W., Jr. and Halpern, B., *Aust. J. Chem.*, 25, 667, 1972.
117. Murano, A. and Fujiwara, S., *Agric. Biol. Chem.* 37, 1977, 1973.
118. Cross, J., Putney, B., and Bernstein, J., *J. Chromatogr. Sci.*, 8, 679, 1970.
119. Pereira, W., Bacon, V. Patton, W., Halpern, B., and Pollock, G., *Anal. Lett.*, 3, 23, 1970.
120. Anders, M. W. and Cooper, M. J., *Anal. Chem.*, 43, 1093, 1971.
121. Westley, J. and Halpern, B., *J. Org. Chem.*, 33, 3978, 1968.
122. Hamberg, M., *Chem. Phys. Lipids*, 6, 152, 1971.
123. Juliá-Arechaga, S., Irurre-Pérez, J., and Sanz-Burata, M., *Afinidad*, 28, 833, 1971.
124. Juliá, S., Pons, R., and Sanz, M., *Afinidad*, 32, 971, 1975.
125. Maestas, P. D. and Morrow, C. J., *Tetrahedron Lett.*, 14, 1047, 1976.
126. Murano, A., *Agric. Biol. Chem.*, 36, 2203, 1972.
127. Rickett, F. E., *Analyst (London)*, 98, 687, 1973.
128. Horiba, M., Kobayashi, A., and Murano, A., *Agric. Biol. Chem.*, 41, 581, 1977.
129. Rose, H. C., Stern, R. L., and Karger, B. L., *Anal. Chem.*, 38, 469, 1966.
130. Westley, J., Halpern, B., and Karger, B., *Anal. Chem.*, 40, 2046, 1968.
131. Pollock, G. E. and Jermany, D. A., *J. Gas Chromatogr.*, 6, 412, 1968.
132. Sanz-Burata, M., Irurre-Pérez, J., and Juliá-Arechaga, S., *Afinidad*, 27, 698, 1970.
133. Karger, B., Herliczek, S., and Stern, R., *Chem. Commun.*, 625, 1969.
134. Abo, I., Wasa, T., and Musha, S., *Japan Analyst*, 23, 1409, 1974.
135. Lafosse, M. and Durand, M., *Analysis*, 3(7), 403, 1975.
136. Kates, M., Hancock, A., and Ackman, R., *J. Chromatogr. Sci.*, 15, 177, 1977.
137. Pirkle, W. and Hauske, J., *J. Org. Chem.*, 42, 1839, 1977.
138. Gil-Av, E., Feibush, B., and Charles-Sigler, R., *Gas Chromatography*, 1966, Littlewood, A. B., Ed., Institute of Petroleum, London, 1967, 227.
139. Gil-Av, E. and Feibush, B., *Tetrahedron Lett.*, 35, 3345, 1967.
140. Feibush, B. and Gil-Av, E., *Tetrahedron*, 26, 1361, 1970.
141. Parr, W., Yang, C., Bayer, E., and Gil-Av, E., *J. Chromatogr. Sci.*, 8, 591, 1970.
142. Feibush, B., *Chem. Commun.*, 544, 1971.
143. Koenig, W. A., Parr, W., Lichtenstein, H. A., Bayer, E., and Oró, J., *J. Chromatogr. Sci.*, 8, 183, 1970.
144. Nakaparksin, S., Birrell, P., Gil-Av, E., and Oró, J. *J. Chromatogr. Sci.*, 8, 177, 1970.
145. Pan, W., Yang, C., Pleterski, J., and Bayer, E., *J. Chromatogr. Sci.*, 50, 510, 1970.
146. Corbin, J., Rhoad, J., and Rogers, L., *Anal. Chem.*, 43, 237, 1971.
147. Parr, W. and Howard, P., *Chromatographia*, 4, 162, 1971.
148. Grohmann, K. and Parr, W., *Chromatographia*, 5, 18, 1972.
149. Parr, W. and Howard, P. Y., *J. Chromatogr.*, 71, 193, 1972.
150. Parr, W. and Howard, P. Y., *J. Chromatogr.*, 67, 227, 1972.
151. Parr, W. and Howard, P. Y., *J. Chromatogr.*, 66, 141, 1972.
152. Parr, W. and Howard, P. Y., *Anal. Chem.*, 45, 711, 1973.
153. Howard, P. Y. and Parr, W., *Chromatographia*, 7, 283, 1974.
154. Charles, R., Beitler, U., Feibush, B. and Gil-Av, E., *J.Chromatogr.*, 112, 121, 1975.
155. Andrews, F., Brazell, R., Parr, W., and Zlatkis, A., *J. Chromatogr.*, 112, 197, 1975.
156. Iwase, H., *Chem. Pharm. Bull.*, 23, 1608, 1975.

157. **König, W. A.,** *Chromatographia,* 9, 72, 1976.
158. **Brazell, R., Parr, W., Andrawes, F., and Zlatkis, A.,** *Chromatographia,* 9, 57, 1976.
159. **Stolting, K. and Konig, W.,** *Chromatographia,* 9, 331, 1976.
160. **Beitler, U. and Feibush, B.,** *J. Chromatogr.,* 123, 149, 1976.
161. **Frank, H., Nicholson, G., and Bayer, E.,** *J. Chromatogr. Sci.,* 15, 174, 1977.
162. **Feibush, B. and Gil-Av, E.,** *J. Gas Chromatogr.,* 5, 257, 1967.
163. **Corbin, J. A. and Rogers, L. B.,** *Anal. Chem.,* 42, 974, 1970.
164. **Feibush, B., Gil-Av, E., and Tamari, T.,** *J. Chem. Soc. Perkin Trans.,* 2, 1197, 1972.
165. **Rubenstein, H., Feibush, B., and Gil-Av, E.,** *J. Chem. Soc. Perkin Trans.,* 2, 2094, 1973.
166. **Lochmüller, C. H., Harris, J. M., and Souter, R. W.,** *J. Chromatogr.,* 71, 405, 1972.
167. **Lochmüller, C. H. and Souter, R. W.,** *J. Phys. Chem.,* 77, 3016, 1973.
168. **Lochmüller, C. H. and Souter, R. W.,** *Am. Lab.,* 11, 25, 1973.
169. **Lochmüller, C. H. and Souter, R. W.,** *J. Chromatogr.,* 87, 243, 1973.
170. **Lochmüller, C. H. and Souter, R. W.,** *J. Chromatogr.,* 88, 41, 1974.
171. **Lochmüller, C. H. and Souter, R. W.,** *J. Chromatogr.,* 87, 243, 1973.
172. **Gil-Av, E.,** *J. Mol. Evol.,* 6, 131, 1975.
173. **Weinstein, S., Feibush, B., and Gil-Av, E.,** *J. Chromatogr.,* 126, 97, 1976.
174. **Helmchen, G. and Struburt, W.,** *Chromatographia,* 7, 713, 1974.
175. **Furukawa, H., Sakakibara, E., Kamei, A., and Ito, K.,** *Chem. Pharm. Bull.,* 23, 1625, 1975.
176. **Souter, R. W.,** *Chromatographia,* 9, 635, 1976.
177. **Valentine, D., Jr., Chan, K., Scott, C., Johnson, K., Toth, K., and Saucy, G.,** *J. Org. Chem.,* 41, 62, 1976.
178. **Koreeda, M., Weiss, G., and Nakanishi, K.,** *J. Am. Chem. Soc.,* 95, 239, 1973.
179. **Nakanishi, K., Schooley, D., Koreeda, M., and Dillon, J.,** *Chem. Commun.,* 1235, 1971.
180. **Scott, C., Petrin, M., and McCorkle, T.,** *J. Chromatogr.,* 125, 157, 1976.
181. **Furukawa, H., Mori, Y., Takeuchi, Y., and Ito, K.,** *J. Chromatogr.,* 136, 428, 1977.
182. **Goto, J., Hasegawa, M., Nakamura, S., Shimada, K., and Nambara, T.,** *Chem. Pharm. Bull.,* 25, 847, 1977.
183. **Helmchen, G., Ott, R., and Sauber, K.,** *Tetrahedron Lett.,* 3873, 1972.
184. **Helmchen, G., Völter, H., and Schühle, W.,** *Tetrahedron Lett.,* 1417, 1977.
185. **Gelboin, H., Kinoshita, N., and Wiebel, F.,** *Fed. Proc.,* 31, 1298, 1972.
186. **Yang, S., Gelboin, H., Weber, J., Sankaran, V., Fischer, D., and Engel, J.,** *Anal. Biochem.,* 78, 520, 1977.
187. **Mikěs, F., Boshart, G., and Gil-Av, E.,** *J. Chromatogr.,* 122, 205, 1976.
188. **Kotake, M., Sakan, T., Nakamura, N., and Senoh, S.,** *J. Am. Chem. Soc.,* 73, 2973, 1951.
189. **Dalgliesh, C.,** *J. Chem. Soc.,* 3940, 1952.
190. **Baczuk, R., Landram, G., DuBois, R., and Dehm, H.,** *J. Chromatogr.,* 60, 351, 1971.
191. **Buss, D. and Vermeulen, T.,** *Ind. Eng. Chem.,* 60, 12, 1968.
192. **Blaschke, G.,** *Chem. Ber.,* 107, 237, 1974.
193. **Blaschke, G. and Donow, F.,** *Chem. Ber.,* 108, 1188, 1975.
194. **Blaschke, G. and Donow, F.,** *Chem. Ber.,* 108, 2792, 1975.
195. **Davankov, V., Rogozhin, S., Semechkin, A., Baranov, V., and Sannikova, G.,** *J. Chromatogr.,* 93, 363, 1974.
196. **Bernauer, K., Jeanneret, M., and Vonderschmitt, D.,** *Helv. Chim. Acta,* 54, 297, 1971.
197. **Rogozhin, S. and Davankov, V.,** *Chem. Commun.,* 490, 1971.
198. **Snyder, R., Angelici, R., and Meck, R.,** *J. Am. Chem. Soc.,* 94, 2660, 1972.
199. **Palamareva, M., Kurtev, B., and Haimova, M.,** *J. Chromatogr.,* 132, 73, 1977.
200. **Niederwieser, A. and Pataki, G., Eds.,** *Progress in Thin-Layer Chromatography and Related Methods,* Vol. 3, Ann Arbor Science Publishers, Ann Arbor, Mich., 1972.
201. **Frahn, J. and Mills, J.,** *Aust. J. Chem.,* 12, 65, 1959.
202. **Hubert, P. and Dellacherie, E.,** *J. Chromatogr.,* 80, 144, 1973.
203. **Palamareva, M., Haimova, M., Stefanovsky, J., Viteva, L., and Kurtev, B.,** *J. Chromatogr.,* 54, 383, 1971.
204. **Cadet, J. and Téoule, R.,** *J. Chromatogr.,* 115, 191, 1975.
205. **Jakovljevic, I., Bishara, R., and Souter, R.,** *Chromatographia,* 2, 23, 1978.
206. **Raban, M. and Mislow, K.,** *Topics in Stereochemistry,* Vol. 2, Allinger, N. and Eliel, E., Eds., Interscience, New York, 1967, 199.
207. **Raban, M. and Mislow, K.,** *Tetrahedron Lett.,* 3961, 1966.
208. **Jacobus, J., Raban, M., and Mislow, K.,** *J. Org. Chem.,* 33, 1142, 1968.
209. **Dale, J., Dull, D., and Mosher, H.,** *J. Org. Chem.,* 34, 2543, 1968.
210. **Dale, J. and Mosher, H.,** *J. Org. Chem.,* 35, 4002, 1970.

211. **Biernbaum, M. and Mosher, H.,** *J. Org. Chem.*, 36, 3168, 1971.
212. **Burlingame, T. and Pirkle, W.,** *J. Am. Chem. Soc.*, 88, 4294, 1966.
213. **Pirkle, W.,** *J. Am. Chem. Soc.*, 88, 1837, 1966.
214. **Pirkle, W., Burlingame, T., and Beare, S.,** *Tetrahedron Lett.*, 5849, 1968.
215. **Pirkle, W. and Beare, S.,** *J. Am. Chem. Soc.*, 90, 6250, 1968.
216. **Pirkle, W., Beare, S., and Muntz, R.,** *J. Am. Chem. Soc.*, 91, 4575, 1969.
217. **Pirkle, W. and Bear, S.,** *J. Am. Chem. Soc.*, 91, 5150, 1969.
218. **Pirkle, W. and Hookstra, M.,** *J. Magn. Reson.*, 18, 396, 1975.
219. **Campbell, J.,** *Aldrichimica Acta*, 4, 55, 1971.
220. Regis Biochemical Catalog, Regis Chemical Co., Morton Grove, Ill., 1978.
221. **Fraser, R., Petit, M., and Saunders, J.,** *Chem. Commun.*, 1450, 1971.
222. Regis Lab Notes, No. 13, Regis Chemical Co., Morton Grove, Ill., January 1973.
223. **McCreary, M., Lewis, D., Wernick, D., and Whitesides, G.,** *J. Am. Chem. Soc.*, 96, 1038, 1974.
224. **Yamamoto, K., Hayashi, T., and Kumada, M.,** *Bull. Chem. Soc. Jpn.*, 47, 1555, 1974.
225. **Stewart, T., Plummer, E., McCandless, L., West, J., and Silverstein, R.,** *J. Chem. Ecol.*, 3, 27, 1977.
226. **Biemann, K.,** *Mass Spectrometry. Organic Chemical Applications*, McGraw-Hill, New York, 1962, 144.
227. **McMahon, R. and Sullivan, H.,** *Res. Commun. Chem. Pathol. Pharmacol.*, 14, 631, 1976.
228. **Gendreau, R., Griffiths, P., Ellis, L., and Anfinsen, J.,** *Anal. Chem.*, 48, 1907, 1976.
229. **Gendreau, R. and Griffiths, P.,** *Anal. Chem.*, 48, 1910, 1976.
230. **Schwenker, R. and Garn, P., Eds.,** *Thermal Analysis*, Academic Press, New York, 1969.
231. **Wilen, S.,** *Topics in Stereochemistry*, Vol. 6, Allinger, N. and Eliel, E., Eds., Wiley-Interscience, New York, 1971, 107.
232. **Finar, I.,** *Organic Chemistry, Vol. 2, Stereochemistry and the Chemistry of Natural Products*, 5th ed., Wm. Clowes and Sons, Ltd., London, 1975, 110.
233. **Curtis, W., King, R., Stoddart, J., and Jones, G.,** *J. Chem. Soc. Chem. Commun.*, 284, 1976.
234. **Kammerling, J. P., Duran, M., Gerwig, G. J., Ketting, D., Briunvis, L., Vliengenthart, J. F. G., and Waldman, S. K.,** *J. Chromatogr.*, 222, 276, 1981.
235. **Iida, T., Chang, F. C., Matsumoto, T., and Tamura, T.,** *J. Lipid Res.*, 24, 211, 1983.
236. **Haegele, K. D., Schoun, J., Alken, R. G., and Huebert, N. D.,** *J. Chromatogr.*, 274, 103, 1983.
237. **Gal, J., DeVito, D., and Harper, T. W.,** *Drug Metab. Dispos.*, 9, 557, 1981.
238. **Thompson, R. H., Jr., Patterson, G., Thompson, M. J., and Slover, H. T.,** *Lipids*, 16, 694, 1981.
239. **Malspeis, L., Bathala, M. S., Ludden, T. M., Bhat, H. B., Frank, S. G., Sokoloski, T. D., Morrison, B. E., and Reuning, R. H.,** *Res. Commun. Chem. Pathol. Pharmacol.*, 12, 43, 1975.
240. **Goromaru, T., Matsuki, Y., Matsuura, H., and Baba, S.,** *J. Pharm. Soc. Jpn. Yakugaku Zasshi*, 103, 974, 1983.
241. **Oi, N., Takai, R., and Kitahara, H.,** *J. Chromatogr.*, 256, 154, 1983.
242. **Redel, J. and Capillon, J.,** *J. Chromatogr.*, 151, 418, 1978.
243. **Allenmark, S. and Boren, H.,** *J. Liquid Chromatogr.*, 4, 1797, 1981.
244. **Souter, R. W.,** *J. Chromatogr.*, 108, 265, 1975.
245. **Souter, R. W.,** *J.Chromatogr.*, 14, 307, 1975.
246. **Beckett, A. H. and Testa, B.,** *J. Pharm. Pharmacol.*, 25, 382, 1973.
247. **Caccia, S. and Jori, A.,** *J. Chromatogr.*, 144, 127, 1977.
248. **Gordis, E.,** *Biochem. Pharmacol.*, 15, 2124, 1966.
249. **Gunne, L.-M,** *Biochem. Pharmacol.*, 16, 863, 1967.
250. **Wells, C. E.,** *J. Assoc. Off. Anal. Chem.*, 53, 113, 1970.
251. **Gal, J.,** *J. Pharm. Sci.*, 66, 169, 1977.
252. **Sternson, L. A. and Hincal, F.,** *J. Chromatogr.*, 153, 167, 1978.
253. **Hengy, H., Vollmer, K.-O., and Gladigan, V.,** *Clin. Chem.*, 14, 692, 1978.
254. **Gilbert, M. T. and Brooks, C. J. W.,** *Biomed. Mass. Spectrom.*, 4, 226, 1977.
255. **Thompson, J. A., Holtzman, J. L., Tsuru, M., Lerman, C. L., and Holtzman, J. L.,** *J. Chromatogr.*, 238, 470, 1982.
256. **Caccia, S., Chiabrando, C., DePonte, P., and Fanelli, R.,** *J. Chromatogr. Sci.*, 16, 543, 1978.
257. **Liu, R. H., Ku, W. W., and Fitzgerald, M. P.,** *J. Assoc. Off. Anal. Chem.*, 66, 143, 1983.
258. **McErlane, K. M. and Pillai, G. K.,** *J. Chromatogr.*, 274, 129, 1983.
259. **Sedman, A. J. and Gal, J.,** *J. Chromatogr.*, 306, 155, 1984.
260. **Slais, K. and Subert, J.,** *J. Chromatogr.*, 191, 137, 1980.
261. **Whall, T. J. and Dokladalova, J.,** *J. Pharm. Sci.*, 68, 1454, 1979.
262. **Aaes-Jorgensen, T.,** *J. Chromatogr.*, 183, 239, 1980.
263. **Arnould, M.-L., Serkiz, B., and Volland, J.-P.,** *Analusis*, 8, 76, 1980.
264. **Goto, J., Goto, N., Hikichi, A., and Nambara, T.,** *J. Liquid Chromatogr.*, 2, 1179, 1979.
265. **Souter, R. W.,** *Chromatographia*, 9, 635, 1976.

266. M.-Hatch, K. J., Ames, M. M., and Gal, J., *Abstr. Pap. Am. Chem. Soc.*, 185, 69, 1983.
267. Shimizu, R., Ishu, K., Tsumagari, N., Tanigawa, M., Matsumoto, M., and Harrison, I. T., *J. Chromatogr.*, 253, 101, 1982.
268. Gal, J. and Ames, M. M., *Anal. Biochem.*, 83, 266, 1977.
269. Frank, H., Nicholson, G. J., and Bayer, E., *J. Chromatogr.*, 146, 197, 1978.
270. Hay, I. D., Annesley, T. M., Jiang, N. S., and Gorman, C. A., *J. Chromatogr.*, 226, 383, 1981.
271. Nachtmann, F., *Int. J. Pharm.*, 4, 337, 1980.
272. Barooshian, A. V., Lautenschleger, M. J., and Harris, W. G., *Anal. Biochem.*, 49, 569, 1972.
273. Wainer, I. W., Doyle, T. D., Hamizadeh, Z., and Aldridge, M., *J. Chromatogr.*, 268, 107, 1983.
274. Wainer, I. W., Doyle, T. D., Hamizadeh, Z., and Aldridge, M., *J. Chromatogr.*, 261, 123, 1983.
275. Pettersson, C. and Schill, G., *J. Chromatogr.*, 204, 179, 1981.
276. Nobuhara, Y., Hirano, S., and Nakanishi, Y., *J.Chromatogr.*, 258, 276, 1983.
277. Hermansson, J. and Von Bahr, C., *J. Chromatogr.*, 221, 109, 1980.
278. Sedman, A. J. and Gal, J., *J. Chromatogr.*, 278, 199, 1983.
279. Hermansson, J., *Acta Pharm. Suec.*, 19, 11, 1982.
280. Wainer, I. W. and Doyle, T. D., *J. Chromatogr.*, 284, 117, 1984.
281. McKay, I. W., Mallen, D. N. B., Shrubsall, P. R., Swann, B. P., and Williamson, W. R. N., *J. Chromatogr.*, 170, 482, 1979.
282. Johnson, D. M., Reuter, A., Collins, J. M., and Thompson, G. F., *J. Pharm. Sci.*, 68, 112, 1979.
283. Nakazawa, H., Kanamaru, Y., and Murano, A., *Chem. Pharm. Bull. (Jpn.)*, 27, 1694, 1979.
284. Stoltenborg, J. K., Puglisi, C. V., Rubio, F., and Vane, F. M., *J. Pharm. Sci.*, 70, 1207, 1981.
285. Zielinski, W. L., Jr., Johnston, K., and Muschik, G. M., *Anal. Chem.*, 48, 907, 1976.
286. Suryanarayanan, R. and Mitchell, A. G., *J. Pharm. Sci.*, 73, 78, 1984.
287. Szepesi, G., Gasdag, M., and Ivanscics, R., *J. Chromatogr.* 241, 153, 1982.
288. Szepesi, G., Gasdag, M., and Ivanscics, R., *J. Chromatogr.* 244, 33, 1982.
289. Zsadon, B., Szilasi, M., Tudos, F., and Szejtli, J., *J. Chromatogr.*, 208, 109, 1981.
290. Gelber, L. R. and Neumeyer, J. L., *J. Chromatogr.*, 257, 317, 1983.
291. Goto, J., Goto, N., and Nambara, T., *J. Chromatogr.*, 239, 559, 1982.
292. Ziemniak, J. A., Chiarnonte, D. A., Miner, D. J., and Schentag, J. J., *J. Pharm. Sci.*, 71, 399, 1982.
293. Salto, F., *J. Chromatogr.*, 161, 379, 1978.
294. Lewin, A. H., Parker, S. R., and Carroll, F. I., *J. Chromatogr.*, 193, 371, 1980.
295. Konaka, R., Kuruma, K., Nishimura, R., Kimura, Y., and Yoshida, T., *J. Chromatogr.*, 225, 169, 1981.
296. Oles, P. J., *J. Pharm. Sci.*, 67, 1300, 1978.
297. Soni, S. K. and VanGelder, D., *J. Assoc. Off. Anal. Chem.* 64, 875, 1981.
298. Po, A. L. W. and Irwin, W. J., *J. Pharm. Pharmol.*, 31, 512, 1979.
299. Souter, R. W., *J. Chromatogr.*, 134, 187, 1977.
300. Tisdall, P. A., Moyer, T. P., and Anhalt, J. P., *Clin. Chem.*, 26, 702, 1980.
301. Young, M. G., *J. Chromatogr.*, 150, 221, 1978.
302. Allenmark, S., Bomgren, B., Boren, H., and Lagerstrom, P.-O., *Anal. Biochem.*, 136, 293, 1984.
303. Souter, R. W. and Jensen, G. C., *J. Chromatogr.*, 281, 386, 1983.
304. Jarman, M. and Stec, W. J., *J. Chromatogr.*, 176, 440, 1979.
305. Blaschke, G., Kraft, H.-P., and Schwanghart, A.-D., *Chem. Ber.*, 111, 2732, 1978.
306. Blaschke, G., Kraft, H.-P., Fickenstscher, K., and Kohler, F., *Arzneim. Forsch. Drug Res.*, 29, 1610, 1979.
307. Cockerill, A., Hall, M., Mallen, D., Osborne, D., and Prime, D., *J. Chromatogr.*, 129, 339, 1976.
308. Hall, M. and Mallen, D., *J. Chromatogr.*, 118, 268, 1976.
309. Newby, N. R. and Hughes, R. B., *J. Forensic Sci.*, 25, 646, 1980.
310. Barkan, S. and Wainer, I. W., *J. Chromatogr.*, 240, 547, 1982.

INDEX

A

D

E

F

G

H

I

K

L

M

N

O

P

Q

R

S

T

U

V

W

X

Y

Z